TECHNOLOGY AND SOVIET ENERGY AVAILABILITY

Also of Interest from Westview Press

Oil Strategy and Politics, 1941–1981, Walter J. Levy, edited by Melvin A. Conant

Technology Transfer to the USSR, 1928–1937 and 1966–1975: The Role of Western Technology in Soviet Economic Development, George D. Holliday

Soviet Economic Planning, 1965–1980, Fyodor I. Kushnirsky

Technology, International Economics, and Public Policy, Hugh H. Miller and Rolf P. Peikarz

OPEC and Natural Gas Trade Prospects and Problems, Bijan Mossavar-Rahmani and Sharmin B. Mossavar-Rahmanı

OPEC: Twenty Years and Beyond, Ragaei El Mallakh

† *The Soviet Economy: Continuity and Change,* edited by Morris Bornstein

† Available in hardcover and paperback.

TECHNOLOGY AND SOVIET ENERGY AVAILABILITY

Office of Technology Assessment

Routledge
Taylor & Francis Group

LONDON AND NEW YORK

First published 1982 by Westview Press, Inc.

Published 2019 by Routledge
52 Vanderbilt Avenue, New York, NY 10017
2 Park Square, Milton Park, Abingdon, Oxon OX14 4RN

Routledge is an imprint of the Taylor & Francis Group, an informa business

Copyright © 1982 Taylor & Francis

Library of Congress Catalog Card Number: 82-50892

ISBN 13: 978-0-367-28967-6 (hbk)
ISBN 13: 978-0-367-30513-0 (pbk)

Foreword

This assessment was undertaken in response to requests from the House Committee on Foreign Affairs; the Senate Committee on Banking, Housing, and Urban Affairs; and the House Committee on Science and Technology to examine the contribution of American and other Western technology and equipment to Soviet energy availability in the present decade.

In November 1979, OTA published an assessment of *Technology and East-West Trade.* Among the conclusions of that work was the suggestion that the United States enjoys significant technological advantages over the U.S.S.R. in petroleum equipment and computers, two areas that have been accorded high priority in Soviet economic planning. It therefore appeared that these technologies might offer the United States opportunities to use export controls for purposes of exercising political leverage over the Soviet Union.

The present study addresses in detail the significance of American petroleum equipment and technology to the U.S.S.R. and the resulting options for U.S. policy. However, the scope of this assessment is far broader. It examines the problems and opportunities that confront the U.S.S.R. in its five primary energy industries—oil, gas, coal, nuclear, and electric power. It discusses plausible prospects for these industries in the next 10 years; identifies the equipment and technology most important to the U.S.S.R. in these areas; evaluates the extent to which the United States is the sole or preferred supplier of such items; and analyzes the implications for both the entire Soviet bloc and the Western alliance of either providing or withholding Western equipment and technology.

OTA is grateful for the assistance of its project advisory panel chaired by Sen. Clifford Case, as well as for the advice of numerous reviewers in agencies of the U.S. Government, academia, and industry. However, it should be understood that OTA assumes full responsibility for its report, which does not necessarily represent the views of individual members of the advisory panel.

JOHN H. GIBBONS
Director

Technology and Soviet Energy Availability

Advisory Panel

Clifford Case, *Chairman*
Curtis, Mallet-Prevost, Colt & Mosle

E. C. Broun, Jr.
Hughes Tool Co.

Robert Campbell
Indiana University

Leslie Diencs
University of Kansas

John Garrett
Gulf Oil Exploration and Production

Marshall Goldman
Wellesley College and Harvard University

Gregory Grossman
University of California at Berkeley

Robert Jackson
Dresser Industries

Stanley Lewand
Chase Manhattan Bank

Richard Nehring
The Rand Corporation

Richard Pipes (until January 1981)
Harvard University

Dankwart Rustow
City University of New York

Henry Sweatt
Honeywell

Allen S. Whiting
University of Michigan

Project Staff

Lionel S. Johns, *Assistant Director*, OTA
Energy, Materials, and International Security Division

Peter Sharfman, *Program Manager*
International Security and Commerce Program

Ronnie Goldberg, *Project Director*
Martha Caldwell   Mildred Turnbull*

Administrative Staff
Helena Hassell   Dorothy Richroath   Jackie Robinson

Project Contractors
Battelle Columbus Laboratories
L. W. Borgman & Co.
EG&G Corp.
William McHenry and Seymour Goodman,
  University of Virginia
Thane Gustafson
Edward Hewett

Eric Anthony Jones
Angela Stent
Stephen Sternheimer
Williams Brothers Engineering
Thomas A. Wolf,
  Center for Science and
  Technology Policy, New York University

OTA Publishing Staff

John C. Holmes, *Publishing Officer*

John Bergling   Kathie S. Boss   Debra M. Datcher   Joe Henson

---

*OTA contractor.

Acknowledgments

This report was prepared by the staff of the International Security and Commerce Program of the Office of Technology Assessment. The staff wishes to acknowledge the contribution of OTA's contractors in the collection, analysis, and preparation of material for the report; and to thank the following individuals and Government agencies for their generous assistance:

Central Intelligence Agency
Department of Commerce
Centrally Planned Economies Project,
  Wharton Econometric Forecasting
  Associates, Inc.
Michael Hammett
John Hardt
M. K. Horn,
  Cities Service Co.
William Kelly

Yuri Ksander and Charles Cooke,
  Engineering Societies Commission
  on Energy
Paul Langer
Theodore Shabad
Jonathan Stern
Myron Uretsky
Rafail Vitebsky
Joseph Yager

Conversions (except where otherwise noted)

1 million tons standard fuel $\times$ 4.582 = million barrels of oil
1 million metric tons hard coal $\times$ 5.051 = million barrels of oil
1 million metric tons of hard coal $\times$ 0.6859 = million metric tons of oil
1 million tons of standard fuel = 0.9091 million metric tons hard coal
1 million metric tons of oil $\times$ 7.33 = million barrels oil equivalent
1 million barrels oil equivalent per day $\times$ 365 $\div$ 7.33 = million tons oil equivalent
1 billion cubic meters natural gas $\times$ 0.8123 = million tons oil equivalent
1 billion cubic meters natural gas $\times$ 5.982 = million barrels oil equivalent
1 billion cubic feet $\times$ 23.01 = million metric tons oil equivalent
1 cubic meter = 264.172 American gallons
1 cubic foot = 0.0283168 cubic meters
1 American barrel = 42 American gallons

Abbreviations

million metric tons — mmt
billion cubic meters — bcm
million barrels per day — mbd
million barrels of oil equivalent — mboe
million barrels of oil equivalent per day — mbdoe
British thermal unit — Btu
kilowatt-hour — kWh
Megawatts — MW

Contents

CHAPTER 1

Summary: Issues and Findings

CONTENTS

TABLE

FIGURE

Summary: Issues and Findings

INTRODUCTION

The Soviet Union occupies the largest span of territory in the world and is abundantly endowed with energy resources. It is the world's largest oil producer and a major exporter of both oil and gas. Despite this enviable position, however, controversy has arisen in the past few years over whether the U.S.S.R. itself, or the Soviet bloc as a whole, may face an energy shortage during the present decade.

This possibility has provoked a debate among U.S. policymakers over whether it is in the best interest of the United States to assist the Soviet Union in its energy development. Those who favor such a policy believe it is justified to bolster American exports; to increase the world's total available supply of energy; to obviate extensive Communist pressure on world energy markets; and/or to reduce the likelihood that the U.S.S.R. would intervene in the Middle East to acquire oil it could no longer produce in sufficient quantities at home.

Adherents of the opposing view contend that to assist in the development of Soviet energy resources would be to help strengthen the economy of an adversary and/or that such assistance may convey direct or indirect military benefits, either: 1) because it might lead to the transfer of dual-use technologies that have military application; 2) because oil itself is a strategic commodity; or 3) because it could enhance the U.S.S.R.'s ability to exert "energy leverage" over West European nations if it placed the Soviet Union in a position to threaten to withhold energy exports.

Both of these perspectives entail certain unstated assumptions. Primary among these is the assumption that it is in the power of the United States to significantly affect the outcome of Soviet energy development in the near or midterm. Thus, focusing on the issue of whether or not the United States *should* assist the U.S.S.R. in its energy development tends to lead to the neglect of more basic questions. Among these is the issue of what course Soviet energy production will take if present policies—in both the West and the U.S.S.R. itself—remain unchanged. This is a controversial question in the West, and perhaps within the U.S.S.R. as well. Moreover, there are the central issues of *whether, how,* and *to what extent* the United States, either itself or in concert with its Western allies, *could* affect the energy future of the Soviet Union.

This study, undertaken at the request of the House Committee on Foreign Affairs; the Senate Committee on Banking, Housing, and Urban Affairs; and the House Committee on Science and Technology, was designed to investigate the latter set of technical issues so that the policy debate might be placed on a firmer footing. Specifically, it addresses the following questions:

First, what opportunities and problems confront the U.S.S.R. in its five primary energy industries—oil, gas, coal, nuclear, and electric power—and what are plausible prospects for these industries in the present decade?

Second, what equipment and technology are most needed by the U.S.S.R. in these areas; of this, how much has been or is likely to be purchased from the West; and to what extent is the United States the sole or preferred supplier of such items?

Third, given the evidence regarding the previous two questions, how much difference could the West as a whole and/or the United States alone make to Soviet energy availability by 1990; and what are the implications of either providing or withholding such assistance for both the entire Soviet bloc and for the West?

As will become clear, the U.S.S.R. faces both problems and opportunities in the years ahead. On the one hand, the Soviet Union is the world's largest oil producer and it has the world's most extensive gas reserves. It has the advantage of long experience with a large and complex petroleum industry, and it also has vast yet-to-be- explored territories that may contain energy resources. On the other hand, the development of these resources is constrained by two important and interrelated problems.

The first is the cost of energy exploitation. A diminishing proportion of Soviet oil and coal reserves are located in readily accessible areas. As older deposits are depleted, the U.S.S.R. must look to increasingly remote and difficult areas for proven reserves or promising sites for new discoveries. While proven gas reserves are more than ample, production is constrained by the pipeline capacity available to transport the gas. Construction of new pipelines—in this case across Siberia—is time-consuming and expensive since most of the pipe and other equipment must be purchased from the West. In short, the development and installation of technology and infrastructure to exploit the Soviet Union's remaining energy will take some time.

The time required for this exploitation and the severity of the problem itself will be affected by the second constraint on Soviet energy development. This arises from the nature of the Soviet economy—the rigidities introduced by the system of central planning and the problems caused by price and incentive structures that inhibit efficiency and productivity in both the energy sector and its supporting industries. While nonmarket economies such as that of the U.S.S.R. do have the advantage of allowing maximum marshaling of resources in priority sectors, **there is an important sense in which the major inhibitor of Soviet energy development is the Soviet economic system, which not only** produces conditions under which domestic solutions to energy industry problems become more difficult, but which also limits the extent to which the U.S.S.R. is willing or able to turn to the West for assistance.

PRINCIPAL FINDINGS

ENERGY AND THE STATE OF THE SOVIET ECONOMY

The rate of Soviet economic growth over the past quarter century has generally declined, and **Western experts are virtually unanimous in predicting a continued slowing in the near term.** To the extent that this economic slowdown signifies stagnation or decline in the rate of growth of per capita consumption in the U.S.S.R.—i.e., in the improvement of living standards for the Soviet populace—it may create political difficulties for the Soviet leadership.

Easily accessible and abundant energy played an important role in generating high growth rates in the past. The U.S.S.R. now faces the possibility of a plateau or even decline in oil output. The latter would certainly cause Soviet economic growth to slow even more, although the magnitude of such a slowdown is difficult to calculate. The impact of falling energy supplies will depend on a system of complex interrelationships in the economy and on Soviet policy regarding the composition of future energy balances and foreign trade patterns. Every policy option carries with it some costs and benefits to the Soviet economy.

If the U.S.S.R. is able to maintain levels of energy production close to the "best" cases posited here (stable or slightly increased oil production, and large increases in gas output), the Soviet economy could continue to

grow at the modest rate of the past 5 years; to supply Eastern Europe with energy at 1980 levels; and to increase the amount of oil and/or gas available for export to the West for hard currency. Under "worst" case assumptions, Soviet economic growth would slow considerably, and the ability of the U.S.S.R. to increase its real nonenergy imports from the West would be seriously impaired. This would negatively affect the overall growth prospects for East-West trade and would place further strains on the Soviet economy. **Actual conditions will probably fall between these extremes.**

In constructing best and worst case scenarios for 1990, OTA assumed that Western assistance in the development of Soviet energy resources would have its greatest quantitative impact on production after 1985. **With extensive Western assistance in energy (particularly gas) development, Soviet hard currency earnings could rise substantially by the end of the decade.** In the worst case scenario, with little or no Western assistance, Soviet exports of energy for hard currency would disappear by 1990.

If these cases are indeed close to the range of plausible outcomes, it appears that the simultaneous maintenance of a politically feasible rate of economic growth in the U.S.S.R., **the further expansion of real energy exports to Eastern Europe after 1985, and a reasonably high rate of growth of East-West trade may hinge importantly on whether or not the West plays a significant part in developing Soviet gas resources in the 1980's.**

SOVIET ENERGY POLICY

Despite the centralized nature of the Soviet system, policymaking takes place in a political context in which individuals and groups compete for resources and influence. **There is ample evidence of debates over the relative priority that should be accorded different energy sectors.** While the decisions made have naturally reflected the choices of the Communist party and its ruling executive committee (Politburo), a number of state

Planning and administrative organizations, and ministerial, regional, and scientific groups also play identifiable roles in the formulation of energy policy, and are critical to the implementation of policy once formulated.

At one time, Soviet leaders placed some stress on the importance of the coal industry, and there were indications that it would receive priority in investment. This may at least partly have been due to the influence of the late Premier Kosygin, an advocate of coal development. The current Five Year Plan (FYP) indicates that this is no longer the case. **Emphasis has now been placed on gas production and, to a lesser extent, on development of the nuclear industry.** However, the energy debates of the past few years suggest that competition for resources among energy sectors may well reappear, particularly when the impending change in the aging Soviet leadership takes place.

THE ENERGY SITUATION OF EASTERN EUROPE

While Eastern Europe is much less dependent on imported energy than is Western Europe, these nations are constrained by geologic conditions that offer only limited prospects for increased domestic energy production, relatively energy-intensive economies, and limited ability to increase hard currency exports to pay for energy on world markets. **In the past, heavily subsidized exports of Soviet oil have been crucial to East European economic development. If this subsidy were abruptly removed, the impact on Eastern Europe as a whole would be disastrous.** The U.S.S.R. does appear to be beginning a transition, however; it has already announced that its oil exports to Eastern Europe will remain at 1980 levels, and it seems to be increasing the level of exports of gas priced at world market rates.

If the countries of Eastern Europe succeed in their plans for increased domestic production of coal and nuclear power (plans that may well engender growing environmental concerns) and for energy conservation and substitution measures, *and* if the

U.S.S.R. continues its oil exports at 1980 levels, Eastern Europe could make it to the end of the decade without a major energy-driven crisis. In the more likely case that these programs are only moderately successful, there will be pressure on the U.S.S.R. to increase its energy exports to Eastern Europe. In the absence of such assistance from the Soviet Union, pressure for economic reform within Eastern Europe could be expected to grow.

However, it is a mistake to think of Eastern Europe as a monolith. The situations of the six countries examined in this study vary significantly, and range from that of Romania, which appears to be facing the most difficult economic prospects even assuming a number of "optimistic" developments with respect to its energy situation, to Hungary, which would seem to be best able to withstand even a number of "worst case" conditions. The case of Poland is also noteworthy. Polish coal production has allowed it to be Eastern Europe's sole net energy exporter. To the extent that the present political and economic difficulties in Poland constrain coal output, the adverse repercussions on the energy situation of the region as a whole could be significant.

SOVIET ENERGY PRODUCTION IN THE 1980'S

Figure 1 summarizes Soviet primary energy production over the past 30 years and Soviet plans for 1985. The pattern exhibited here shows that for many years the bulk of energy output was in coal. The rate of growth in coal production began to decline in the 1960's, when oil overtook it as the predominant fuel. Coal production is now virtually stagnant, and the rate of growth in oil has markedly declined from that of the previous two decades. The fuel of the future is clearly gas, production of which, according to the U.S.S.R.'s own projections, will nearly equal that of oil in energy value by 1985. The following sections examine in more detail the current state of and prospects for Soviet energy industries in the present decade.

Oil

Projections of Soviet oil production in 1985 span an enormous range (see table 1). The Central Intelligence Agency's (CIA) most recent forecast maintains that output could **decline** by nearly 17 percent, while **increases** of roughly the same magnitude have been foreseen by the British Economist Intelligence Unit. The U.S.S.R. itself in its current FYP envisages slightly increase production, and the U.S. Defense Intelligence Agency endorses the feasibility of the Soviet target. The disparities among forecasts for 1990 are even more striking. CIA believes production will decline more than 40 percent from 1980 levels, while others contend that the Soviet oil industry could actually produce 25 percent more oil in 1990 than it did in 1980.

These predictions are all based on different interpretations of fragmentary Soviet information, different subjective evaluations of Soviet oil industry practice, and different judgments about the future of the Soviet economy and its capabilities. **OTA does not believe that it is a useful exercise to attempt to determine which, if any, of these predictions is "correct."** Indeed, given the poor record of forecasters even of U.S. production, it seems foolish to attempt to assert with any degree of assurance the outcome of complex processes in the U.S.S.R. 10 years hence.

OTA has instead attempted to identify *plausible* best and worst cases for Soviet oil production. These are not predictions; they are intended solely to provide a context within which the range of possible outcomes for Soviet energy availability in this decade can be discussed. **OTA finds the upper range of the U.S.S.R.'s own target—which sets a goal of modest growth by 1985—to be a not-unreasonable best case.** On the other hand, given that many things can simultaneously go wrong in the Soviet oil industry, it is reasonable to base discussion of worst case outcomes on the upper end of the CIA range. For 1990, even using best and worst case projections as a basis for analysis is a highly tenuous exercise. OTA has chosen as a best

Figure 1.—Soviet Primary Energy Production (1950-80)

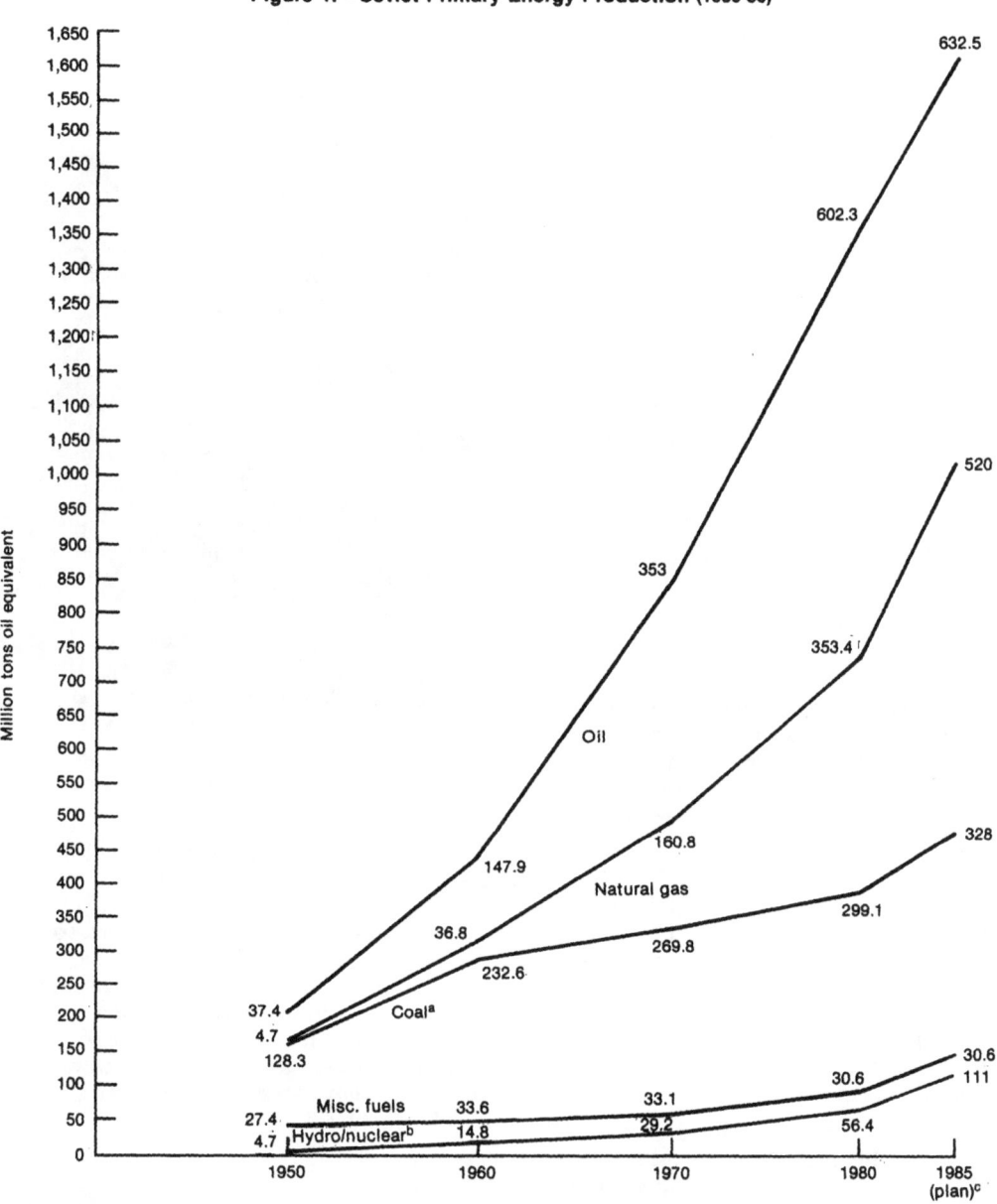

aCoal for 1980 = 716 MT × .67 = Standard Fuel × .9091 = MTHC × .6859 = MTOE.

bHydroelectric and nuclear electricity are converted at fuel rates for central thermal stations.

cMidpoint of plan range is plotted.

SOURCES: L. Dienes and T. Shabad, *The Soviet Energy System* (Washington, D.C.: V. H. Winston & Sons), 1979, p. 32; and the Office of Technology Assessment.

Table 1.—1985 Soviet Oil Production Forecasts[a]

Million tons	Million barrels per day	Date of forecast
1. 500-550	10-11	April 1981
2. 560-610	11.2-12.2	June 1981
3. 600	12	1979
4. 605-655	12.1-13.1	1979
5. 612-713	12.3-14.3	1978
6. 620-645	12.5-12.9	1980
7. 620-645	12.5-12.9	August 1981
8. 650-670	13-13.5	1979
9. 700	14	1980

[a]Soviet oil production in 1980 was 603 million tons.

SOURCES:
1. CIA, as reported in Joseph A. Licari, "Linkages Between Soviet Energy and Growth Prospects for the 1980's," paper presented at the 1981 NATO Economics Directorate Colloquium, Apr. 8-10, 1981. These numbers replace the 1977 estimates of 400 to 500 mmt.
2. OECD, Committee for Energy Policy, "Energy Prospects of the U.S.S.R. and Eastern Europe," June 26, 1981.
3. Robert Ebel, "Energy Demand in the Soviet Bloc and the PRC," June 1979.
4. Leslie Dienes and Theodore Shabad, The Soviet Energy System (Washington, D.C.: V. H. Winston, 1979), table 53, p. 252.
5. Herbert L. Sawyer, "The Soviet Energy Sector: Problems and Prospects," Harvard University, January 1978, quoted in George W. Hoffman, "Energy Projections—Oil, Natural Gas and Coal in the U.S.S.R. and Eastern Europe, Energy Policy, pp. 232-241.
6. Soviet Eleventh FYP target.
7. U.S. Defense Intelligence Agency, "Allocation of Resources in the Soviet Union and China—1981." Statement of Maj. Gen. Richard X. Larkin before the Joint Economic Committee, Subcommittee on International Trade, Finance, and Security Economics, Sept. 3 1981.
8. Jeremy Russell, Shell Oil.
9. David Wilson, Soviet Oil and Gas to 1980, Economist Intelligence Unit Special Report No. 90. This report was published just after the Soviet plan target was released. In a foreword, the author reasserts his belief that oil production of 700 mmt is achievable and attributes the lower Soviet plan to an apparent decision to divert resources from oil to gas production.

Photo credit: TASS from © SOVFOTO

Coal-loaded trains leave Karaganda

case hypothesis oil production remaining stable at the Soviets' own 1985 production target, and as a worst case, production declining, but remaining at a level above the 40 percent drop forecast by CIA.

Gas

Given the problems in the oil industry, and the fact that it is possible—indeed likely—that oil production will not rise greatly, **gas is the key to the Soviet energy future in this decade.** This is the energy sector with by far the best performance record in the past 5 years, and it appears to have been given priority in investment in the present FYP.

Proven Soviet gas reserves are tremendous; they may be likened to the oil reserves of Saudi Arabia. Gas, therefore, has a good potential for replacing oil both in Soviet domestic consumption and as a hard currency earning export. **Gains in gas output could more than compensate for the apparent slowing of growth in oil production.** The ex-tent to which the U.S.S.R. can capitalize on its gas potential will depend on its ability to substitute gas for oil. This in turn rests on two factors: its ability to convert to gas in boiler and industrial applications, and its ability to add to the gas pipeline network. **The rate of construction of new pipelines, both for domestic use and for export, will be the most important parameter in determining the extent to which Soviet gas can be utilized.**

Coal

High-quality, easily accessible Soviet coal reserves have become depleted, and the volume of coal output in the last several years has actually declined. **Even the relatively modest coal production targets in the**

Eleventh FYP seem excessively optimistic, and gains in overall coal production will be offset to some degree by the fact that the quality of much of the new coal being mined is low, In fact, the quantity of coal mined could increase at the same time as the total energy derived from it (its standard fuel equivalent) actually declined.

The difficulties facing the Soviet coal industry are compounded by the fact that a number of problems must be addressed simultaneously if production is to increase meaningfully. These problems include low labor productivity, lagging additions to mine capacity, insufficient quality and quantity of mining equipment, insufficient coal transport capacity, and inability to use the low-quality Siberian coals that are making up an increasing share of production. Massive investment in the coal industry would be required to achieve gains in most or all of these areas, and even then success in terms of dramatic production increases could not be assured. This point underlines the relative cost effectiveness of relying on gas, rather than coal, as a substitute fuel.

Nuclear

Soviet nuclear power production has increased greatly in the past 5 years, and the U.S.S.R. is committed to an ambitious nuclear program. It foresees nuclear energy contributing as much as 14 percent of electricity production by 1985, and perhaps 33 percent by 1990. But Soviet targets for installed nuclear capacity, while attainable in principle, are probably overly optimistic.

More than in any other energy industry, progress here will depend on the efficiency and production capacity of equipment manufacturers. The growth of Soviet nuclear power will not be constrained by lack of know-how, nor is it likely to be inhibited by the kind of safety and environmental concerns so prevalent in the West. Very little is known about available Soviet uranium supplies, but it is probably safe to assume that these are adequate to support the nuclear growth that the Soviets themselves envis-

age, even given the competing claims of the military sector. Thus, the critical variable for the success of the Soviet nuclear power program will be the ability of support industries to construct nuclear power stations and of reactor and other equipment manufacturers to deliver on time and in sufficient quantities.

Electricity

To a great extent, the performance of the electric power industry in the present decade will be tied to the success of the nuclear program. Should planned additions to installed nuclear capacity fall seriously behind schedule—not an unlikely eventuality—fossil-fired generation will be called upon to cover the shortfalls. This prospect raises potential difficulties. Although, on the face of it, the industry appears to be sufficiently flexible, the extent to which fossil-fuel capacity can serve as a buffer for nuclear capacity is limited, to an unknown extent, by the degree to which low-quality Siberian coal can be utilized and absorbed by the electric power system, and by Soviet ability to complete the Unified Power System, which eventually will link all of the nation's regional electricity grids. The fate of the grid will be tied to the future of long-distance electricity transmission. The U.S.S.R. has amassed great experience in power transmission, including long-distance, high-voltage (250 to 1,000 kV) direct current transmission. However, Soviet power engineering is moving into a relatively new field—ultrahigh voltage transmission (over 1,000 kV)—which, at least initially, will entail high investment and operating costs.

Despite these problems, the evidence suggests that the U.S.S.R. will have sufficient reserve capacity in its power generation system in the 1980's and will be able to compensate for some shortfalls in nuclear capacity—provided that the necessary fuel supplies are available. Given the problems of the coal industry, however, meeting the latter condition cannot be taken for granted—unless gas can be used much more extensively.

THE CONTRIBUTION OF WESTERN EQUIPMENT AND TECHNOLOGY TO SOVIET ENERGY INDUSTRIES

Oil and Gas

There is no question that Soviet oil production has been assisted by American and other Western technology and equipment, although the impact of this assistance is impossible to quantify. In 1979, the Soviet Union devoted approximately 22 percent of its trade with its major Western trading partners (some $3.4 billion) to energy-related technology and equipment. The vast majority of these purchases—about $2.7 billion—was destined for the Soviet oil and gas sector (and most of this was for pipe and pipeline equipment). Western exports in the past have helped to compensate for shortfalls in the production of Soviet domestically produced equipment, and for the fact that the quality of equipment is usually inferior to that which can be obtained in the West.

It is also true, however, that the impact of Western assistance has been lessened by at least two important factors. First, whether for lack of hard currency, a lack of perceived need, or a fear of dependence on the West, the U.S.S.R. has never imported *massive* amounts of oilfield equipment. Second, imported equipment and technology is usually less productive in the U.S.S.R. than it would be in Western nations. This may be due to a combination of factors. The Soviet Union has not often allowed hands-on training by Western suppliers to be carried out in the field, and suppliers themselves may be unwilling to meet Soviet conditions for the supply of spare parts, maintenance, or training services. In addition, Soviet maintenance and production practices are often not conducive to prolonging the useful life and promoting the efficiency of imported equipment.

The one area in which Soviet petroleum equipment and technology purchases might be described as "massive" is large diameter pipe and other equipment (compressor stations and pipelaying equipment) for the construction and operation of gas pipelines. There is no evidence that reliance on the West in this area will lessen in the present decade. Indeed, given the crucial importance of increased gas production and gas exports to the short- and medium-term Soviet energy future, there is reason to believe that such dependence will increase.

There is no doubt that the Soviet petroleum industry will benefit from continued—and increased—infusion of Western equipment and technology imports. But the expenditures of hard currency and the extent of the hands-on Western involvement necessary to make such imports maximally effective would be unprecedented Soviet behavior. With the possible exception of gas pipeline construction, there is little evidence that the U.S.S.R. is ready to make such changes.

Other Energy Sectors

In the past, the U.S.S.R. has been virtually self-sufficient in coal, nuclear, and electric power technology and equipment. Purchases in these areas have been small and spotty and appear to have been intended to compensate for specific deficiencies in the quantity or quality of domestic equipment, rather than to acquire new technological know-how.

Should the U.S.S.R. decide to reverse past practice and begin purchasing significant amounts of equipment in these areas, these purchases might consist of surface mining equipment, complete nuclear reactors and/or plants, and computers. Although every energy sector could profit from more extensive computerization and from the availability of Western (especially American) computer hardware and software, this would be of particular benefit to the electric power industry for management and control of the Unified Power System.

THE PROSPECTS FOR ENERGY CONSERVATION AND SUBSTITUTION IN THE U.S.S.R.

Conservation should be an extremely promising policy for the U.S.S.R. It could be accomplished both through a centralized "high-investment" strategy—i.e., through industrial modernization—and through a "low-investment" or housekeeping strategy—i.e., improving the efficiency of operation of equipment already in place.

Despite evidence of interest in conservation, emphasis to date has been on producing, not saving, energy. To the extent that the latter has been accorded official attention, stress has been on exhortation to industry and individual consumers to conserve, a strategy which is unlikely to produce major results quickly because of weaknesses in the price structure, the prevailing incentive system, the enforcement mechanism, and the ways in which consumption is monitored and measured. **In short, the U.S.S.R. has accomplished major energy savings, and opportunities for more still exist. But rigidities in the political and economic structure have prevented Soviet policymakers from taking full advantage of them.** The situation may be ameliorated somewhat by an increase in energy prices scheduled for 1982, but the extent of the impact of such a reform on energy consumption cannot at this stage be predicted.

In addition to policies that result in overall energy savings, the U.S.S.R. is interested in lowering domestic oil consumption in order to free oil for highly lucrative export to world markets. This can be accomplished by substituting other fuels for oil in domestic use. **Opportunities for substitution with coal are severely constrained** both by the difficulties besetting coal production and by environmental concerns. **Gas is the most promising alternative fuel but, as noted, the extent to which it can be utilized domestically is limited by the internal gas pipeline network,** expansion of which will encounter the same difficulties as expansion of pipelines for gas exports.

THE FOREIGN AVAILABILITY OF WESTERN TECHNOLOGY AND EQUIPMENT

The Concept of Foreign Availability

The Export Administration Act of 1979 contains the provision that decisions regarding the control of U.S. equipment and technology should be affected by the availability of similar or equivalent items in other countries. It is left to the Department of Commerce to establish the capacity to gather and assess the information upon which determination of "foreign availability" will be made. **There remain serious conceptual and practical problems that must be resolved before "foreign availability" can become a viable criterion for export licensing decisions.** As of this writing, these problems have yet to be taken up in a comprehensive or systematic manner by either Congress or the Department of Commerce.

There is no commonly accepted definition of "foreign availability," nor is there a central repository of information, or system for gathering such information, in place. OTA's own judgments of the foreign availability of items of energy-related technology and equipment must, therefore, be understood to be highly generalized. The term as it is used here denotes the existence in Western Europe and/or Japan of items with similar technical parameters and capabilities as those available from firms in the United States.

Oil and Gas Equipment and Technology

The United States is not the predominant supplier of most petroleum-related items imported by the U.S.S.R. OTA has identified numerous foreign firms which supply oil and gas equipment to the Soviet Union, reinforcing the theme of the international nature of the petroleum industry. **Technology devel-**

oped in the United States is quickly diffused throughout the world through an extensive network of subsidiaries, affiliates, and licensees.

There are a few items of oil and gas equipment and technology that are solely available from the United States, and a few others for which the United States is generally considered a preferred supplier. With the exception of advanced computers, however, the U.S.S.R. is either not purchasing these items, is on its way to acquiring the capacity to produce them itself, or has demonstrated that they are not essential to its petroleum industry. **The United States continues to represent the ultimate in quality for some equipment, but the extent of that lead is diminishing, and the U.S.S.R. can and does obtain most of what it needs for continued development of its oil and gas resources from outside the United States.** This is particularly true of what appears to be the most crucial import for this decade—large diameter pipe and pipeline equipment. The Soviet Union procures these items from Japan, West Germany, Italy, and France. Indeed, **the United States does not produce the large diameter pipe that constitutes the U.S.S.R.'s single most important energy-related import.**

Other Equipment and Technology

As noted above, if the U.S.S.R. reversed its present policy and began to import more extensively in its other energy sectors, one area that might receive particular attention is coal surface mining. The United States is a world leader in this field, and should the Soviet Union seek large amounts of the best and largest capacity surface mining equipment, it would be likely to turn to the American firms.

THE POLICIES OF AMERICA'S ALLIES TOWARD SOVIET ENERGY DEVELOPMENT

OTA examined the energy relations (including trade in both energy-related equipment and technology and in fuel itself) between the U.S.S.R. and five of America's principal allies—Japan, West Germany, France, Italy, and Great Britain. **With the exception of Britain, trade with the Soviet Union has generally been more important for all of these nations than it has for the United States.** While all cooperate in controlling the export to the Soviet Union of equipment and technology with direct military relevance, it is nevertheless true that **these nations are far more inclined than is the United States to consider trade with the U.S.S.R. a desirable element in their foreign policy and commerce, and to eschew the use of export controls for political purposes.**

Although trade with the U.S.S.R. makes up only a small portion of the overall trade of these countries, in 1979 energy-related exports constituted nearly one-half of Japanese, one-third of Italian, approximately one-quarter of West German and French, and about 10 percent of British exports to the U.S.S.R. The comparable figure for the United States was 7 percent. In absolute amounts, this translates into more than $1 billion worth of energy-related exports in 1979 for Japan, nearly that amount for West Germany, and almost one-half billion each for France and Italy. (This rank order has changed markedly in the past 5 years, with Japan overtaking West Germany as the U.S.S.R.'s major Western trading partner.) Most of these exports were destined for the Soviet petroleum industry, and an important part of this trade was in the large diameter steel pipe used in the U.S.S.R. for gas pipelines. Indeed, West German steel corporations are among the most vociferous in promoting such trade with the U.S.S.R. There is evidence that employment in several of West Germany's largest steel firms might be seriously affected by the loss of the Soviet market.

These nations also import energy from the Soviet Union. The most important Soviet energy commodity for Western Europe now and for Japan in the future is gas. **In 1979, about 24 percent of Italy's and about 16 percent of West Germany's total gas require-**

ments came from the U.S.S.R. These figures may be interpreted in several different ways, however. Italy, which had the highest reliance on Soviet energy, purchased about 10 percent of all the primary energy it used from the U.S.S.R. The comparable figures for West Germany, France, the United Kingdom, and Japan respectively were 6, 5, 1, and 2 percent. In no country examined in this study did Soviet energy constitute more than 9 percent of 1979 total energy imports. But although overall "dependence" on the U.S.S.R. is low, some countries might face significant disruptions if Soviet gas became unavailable.

The most important and controversial example of West European (and possibly Japanese) energy cooperation with the U.S.S.R. is the proposed new pipeline, which will carry gas from West Siberia to as many as 10 West European nations. (This pipeline was originally planned to carry gas from the Yamburg field. The U.S.S.R. has now decided to delay development of that field and concentrate instead on further development of the Urengoy field, both for incremental domestic needs and export commitments.) **The scale of this project guarantees that it will raise the level of East-West energy interdependence in qualitative as well as quantitative terms.**

Barring unexpected political or economic developments—probably even in the face of active diplomacy on the part of the United States—the gas pipeline project is likely to proceed. **West Germany, France, and Italy all look to Siberia as a way to increase and diversify energy supplies while at the same time increasing energy equipment and tech-**nology exports. The latter consideration may also be important for Japan. Moreover, the pipeline project has political implications for each of the participants, and these too are important motives for proceeding. West Germany, for instance, has a vital interest in providing the U.S.S.R. with incentives to moderate its behavior in Europe and to help to foster improved relations with East Germany. Japan looks to its trade and energy relations with the U.S.S.R. as an important counterweight to its growing relationship with the Peoples Republic of China.

If the West Siberian Pipeline is developed as currently envisaged, West Germany, France, and Italy will certainly become more dependent on Soviet gas, **although this gas will to some extent replace the Soviet oil they presently import.** In any case, dependence on the U.S.S.R. would still be significantly smaller than dependence on OPEC. A cutoff of Soviet gas would impact each country differently (each, mindful of the risks entailed in the deal, has made different contingency arrangements), but **none would be immune from hardship, particularly in the context of a tightened world oil market or other energy crisis.** Each of the three would benefit from the development of more effective contingency plans, allowing for substitution of alternative energy supplies in the event of Soviet shortfalls, and thereby diminishing the opportunities for the U.S.S.R. to make use of any sort of "gas weapon" to exert political pressure on its gas customers. The most effective contingency planning would be that undertaken by West European nations as a bloc—but as yet there are no serious prospects that this will occur in any formal sense.

FOUR ALTERNATIVE U.S. POLICY PERSPECTIVES

Suggestions for U.S. policy regarding Soviet energy development can be categorized around four alternative strategies. This section briefly sets out the basic tenets espoused by adherents of these strategies and indicates OTA's findings with respect to each.

1. *The embargo perspective* seeks to severely curtail or eliminate the ability of

U.S. firms to sell energy-related (especially petroleum) equipment and technology to the U.S.S.R., either because these items may have direct military relevance or because oil and gas are considered to be strategic commodities. In this connection, it is often asserted that helping the U.S.S.R. to develop these resources is helping bolster the economy of an adversary nation.

OTA found that very few items of oil and gas technology and equipment could be diverted to direct military applications. Computers are the most important exception here, and these are already subject to both U.S. and multilateral export controls. **Exercise of this policy option, justified by the inherent importance of petroleum, would be tantamount to pursuing a policy of economic warfare against the U.S.S.R.** The United States attempted this after World War II. It formally abandoned the effort in 1969, in recognition of the facts that the United States was sole supplier of few of the items sought by the U.S.S.R. from the West, and that United States allies were not willing to participate in such restrictive policies. The unenthusiastic response of Western Europe and, to a lesser extent, Japan to President Carter's post-Afghanistan technology embargo against the U.S.S.R. indicates that this attitude has probably not changed. **It is possible to posit circumstances under which the United States could persuade its allies to reverse their own policies, but this would likely take a dramatic change in the political climate as well as a major policy initiative on the part of the United States.** The latter might have to include concrete suggestions for energy supply alternatives to Soviet gas.

2. *The linkage perspective* most closely describes present U.S. policy toward trade with the U.S.S.R. Linkage is a policy that seeks to use the prospect of expansion or curtailment of trade as a "carrot" or "stick" to exact policy concessions from the trading partner. This perspective accommodates a number of different opinions as to how and under what circumstances linkage should be attempted, but in one form or another it has

influenced U.S. trading policy with the U.S.S.R. since at least the Nixon era.

There is no unambiguous evidence regarding the effects, if any, that linkage vis-a-vis the Soviet Union has ever had on Soviet domestic or foreign policies; thus, no final determination of its success or failure can be made. In the case of petroleum equipment and technology, **the effectiveness of a linkage policy would be limited by the fact that the United States is the sole supplier of very few items crucial to the Soviet oil and gas industry, and in those cases in which it is a preferred supplier (e.g., pipelaying equipment), the U.S.S.R. has available alternatives that, albeit second-best choices, could produce the desired results. The limitation of linkage are well illustrated by the fact that the import most crucial to Soviet energy development in the present decade—large diameter pipe—is not produced in the United States.**

3. *The energy cooperation perspective* assumes both that American technology and equipment could make a significant positive contribution toward increasing Soviet energy availability in the present decade and that such a development would be in the interests of the United States in that it would help to reduce Western dependence on OPEC and relieve pressure on world energy markets.

OTA's findings suggest that although American technology and equipment have assisted the Soviet petroleum industry, the United States is not the only—indeed, perhaps not even the most important—Western nation to provide such assistance. **For United States exports to make more of a difference, not only would the United States have to be willing to sell massive amounts of equipment and technology to the U.S.S.R. on attractive terms—probably involving export credits—but the U.S.S.R. would itself have to be willing to purchase in large amounts, to utilize the imported items in a more efficient manner, and to allow the United States and other Western firms to**

Photo credit: TASS from ©SOVFOTO

U.S. pipelaying equipment used in the construction
of the Northern Lights gas pipeline

provide greater hands-on training and to participate more fully in Soviet energy projects.

4. *The commercial perspective* rests either on the belief that trade and politics should remain separate and/or on the judgment that regardless of the export control policy it adopts, the United States is unlikely to be able to significantly affect the U.S.S.R.'s energy situation on this decade. Those who espouse these views therefore believe that U.S. firms should be permitted to reap whatever economic benefits can be gained from selling nonmilitarily relevant items to the U.S.S.R.

Such a policy might allow significant sales for individual firms, but unless it were accompanied by the extension of official export credits, it is highly unlikely that it would result in enough trade to have any direct impact on the overall foreign trade or competitive position of the United States. On the other hand, **the lack of pronounced economic gains resulting from such a policy could be at least partially outweighed by potential political benefits derived from removing the issue of Soviet energy from the arena of conflict between the United States and its allies.**

The Soviet Oil and Gas Industry

CONTENTS

LIST OF TABLES

LIST OF FIGURES

The Soviet Oil and Gas Industry

In 1980, the Soviet Union was both the world's largest producer of oil and its largest gas exporter. It is ironic, therefore, that much of the discussion of Soviet energy that has taken place in the West centered until recently on a debate over the continued viability of Soviet energy independence, at least in the present decade. This debate was occasioned by the 1977 Central Intelligence Agency (CIA) projection that Soviet oil production would peak and begin to decline sharply by the early 1980's. It is by now clear that this outcome is unlikely. The rate of growth of Soviet oil production has slowed markedly, but output does not appear to have peaked. Indeed, the CIA has revised its forecast, pushing back the anticipated decline until after 1985. The focus of interest has now shifted from oil to gas—and from the potential consequences of a Soviet oil shortage to the implications, both for the U.S.S.R. and the West, of an abundance of Soviet gas.

This chapter attempts to elucidate the grounds of past controversies and illuminate present uncertainties. It examines the present condition of, and potential for, the Soviet oil and gas industries, with special emphasis on the impact of the West on oil and gas production. After a brief historical introduction it surveys the U.S.S.R.'s oil- and gas-producing regions. It then describes the state of each industry sector—exploration, drilling, production, transportation of oil and gas, refining, and offshore activities—including the past and potential contributions of Western equipment and technology. Finally, it summarizes the controversy over the future of Soviet oil production and posits plausible best and worst case estimates of oil and gas output for 1985 and 1990.

INTRODUCTION

The Russian oil industry is one of the world's oldest.[1] When Baku, the historical center of the industry, was ceded to Russia by Persia in 1813, oil was already being produced from shallow hand-dug pits. Development of these sites near the Caspian Sea languished until after 1862, when production began to rise, aided by the introduction of drilling (1869), the end of the state monopoly on production (1872), and an influx of foreign entrepreneurs, such as Robert and Ludwig Nobel (1873). Russian oil production peaked in 1901 at nearly 12 million metric tons per year (mmt/yr) (240,000 barrels per day (bd) or 0.24 million barrels per day (mbd) (see table 2) and then dropped rapidly. This drop was due to a number of factors, including labor unrest surrounding the 1905 revolution.

Although drilling technology was introduced at about the same time in the United States and in Russia, the Russians soon fell behind. In 1901, the Russians relied on wooden drilling tools that could achieve well depths up to 300 ft. In contrast, European concerns using metal drilling technology could drill wells of over 1,800 ft; in 1909, rotary drills in the United States were reaching depths of 2,400 ft.

---

[1] Historical material is taken from Mashall I. Goldman, *The Enigma of Soviet Petroleum* (London: George Allen & Unwin, 1980); Robert W. Tolf, *The Russian Rockefellers: The Saga of the Nobel Family and the Russian Oil Industry* (Stanford, Calif.: Hoover Institution Press, 1976); William J. Kelly and Tsuneo Kano, "Crude Oil Production in the Russian Empire: 1818-1919," *The Journal of European Economic History*, vol. VI, No. 2, fall 1977, pp. 307-338; and William J. Kelly, "Crisis Management in the Russian Oil Industry: The 1905 Revolution," *The Journal of European Economic History*, forthcoming.

**Table 2.—Russian and Soviet Oil and Gas
Production, Selected Years**
(million tons, billion cubic meters)

	Oil	Gas
1860	0.004	—
1870	0.033	—
1880	0.382	—
1890	3.9	—
1901	12	—
1910	11.3	—
1920	3.9	—
1930	18.5	—
1940	31.1	3.2
1950	37.9	5.8
1960	147.9	45.3
1970	353.0	198
1975	491.0	289
1980	603.0	435

SOURCE: Goldman, op. cit., pp. 14-15, 22, 23; *Soviet Geography*, April 1981, pp. 273, 276; Dienes and Shabad, op. cit, p. 70.

Oil production fell precipitously after the Bolshevik Revolution and the resulting confiscation and nationalization of the oilfields. Soon, however, the new Soviet Government began to open these fields to foreign technology and investment. By 1927, with prodigious Soviet effort and the help of American, French, Japanese, German, and British firms, production surpassed the 1901 level. By 1932, with production at 22.4 mmt (0.45 mbd), petroleum exports accounted for 18 percent of Soviet hard currency receipts. It was at this point that foreign involvement was curtailed.

The year 1940 marked the beginning of another temporary drop in Soviet oil production. During the course of World War II, many oilfields were destroyed, and postwar recovery was slow. Indeed, this recovery was only accomplished with a second infusion of imported equipment and technology, technology that helped to create the present modern, nationwide industry.

The first important oil discoveries outside of Baku had been made around 1932 in the Volga-Urals region. The war, as well as a shortage of drilling equipment, delayed further exploration and development in this area, but as the oilfields around Baku peaked and began to decline in the early 1950's, the Soviet industry shifted its emphasis northeastward to the Volga-Urals. By 1955, drilling activities were concentrated there, and the rate of oil production again began to climb sharply.

The pattern has repeated itself. As the Volga-Urals fields peaked in the 1970's, the Soviets were able to offset declining production by bringing new discoveries online, this time in West Siberia. The first discovery in West Siberia was made in 1960. As table 3 shows, by 1970 its fields accounted for almost 10 percent; 6 years later, over one-third; and now more than one-half of total Soviet oil production. It is appropriate, therefore, to begin this chapter's survey of major oil-producing regions with West Siberia.

**Table 3.—The Growing Importance of West Siberian
Oil Production** (million tons)

Year	U.S.S.R.	West Siberia	West Siberia as a share of U.S.S.R. (%)
1965	242.9	1.0	0.4
1970	353.0	31.4	9.8
1975	490.8	148.0	30.2
1976	519.7	181.7	35.0
1977	545.8	218.0	39.9
1978	571.4	254.0	44.5
1979	586	284	48.5
1980[a]	640	315	49.2
1980[b]	603	312	51.7
1985 (plan) ...	620-645	385-395	61-62

[a]Original FYP target.
[b]Actual production.

SOURCE: Wilson, op. cit., pp. 5-6; *Soviet Geography*, April 1981, p. 273.

LOCATION OF MAJOR PETROLEUM REGIONS[2]

WEST SIBERIA

The West Siberian lowlands lie between

[2]Throughout this report, "petroleum" is taken to mean both oil and gas.

the Ural mountains on the west and the Central Siberian Platform on the east (see fig. 2). The petroleum-producing regions are found in the Tyumen province (*oblast*) and part of the adjacent Tomsk oblast. This area pre-

Figure 2.—Major Petroleum Basins, Oilfields, and Gasfields

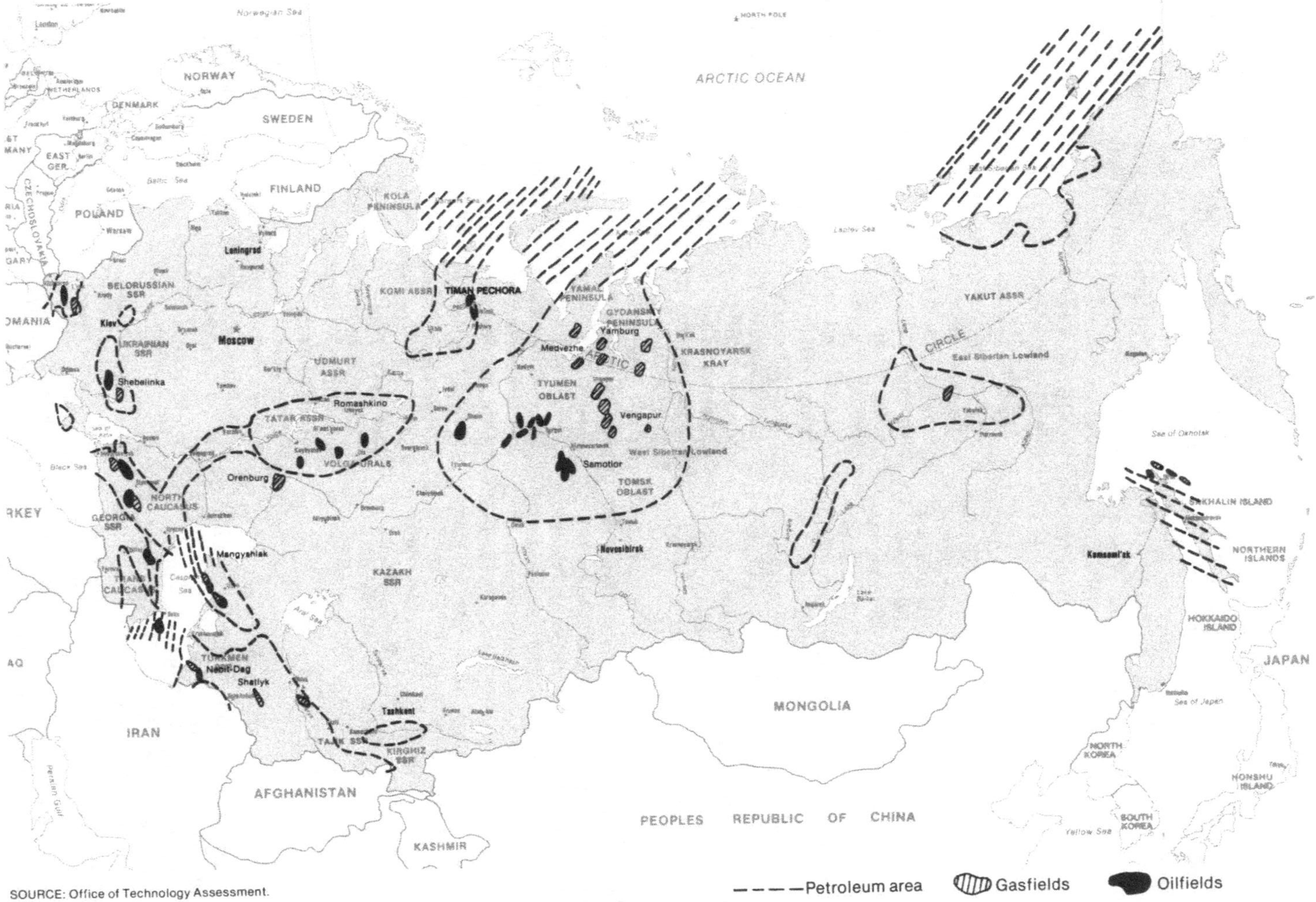

SOURCE: Office of Technology Assessment.

sents formidable natural obstacles to oil and gas exploration and production. The terrain is extremely flat, and most of the area is a vast swamp, interspersed with sluggish streams and occasional dry ground. The Vasyugan swamp on the left bank of the Middle Ob River alone covers 100,000 square miles (mi^2), equivalent in area to New York, New Jersey, and Pennsylvania combined. Other swamps and bogs abound.

Western Siberia typically has 6 to 8 months each year of below-freezing temperatures. At high water in the Ob system there are 19,000 miles of navigable rivers, but the Ob is blocked for 190 to 210 days per year with ice 30 to 60 inches thick. The shipping season thus lasts only about 5 months. In addition, much of the area is underlain by permanently frozen subsoil (permafrost) that impedes drainage, stunts plant roots, and makes forestry, farming, stock raising, mining, oil drilling, excavation, pipelaying, and most construction activities difficult and expensive.

The map in figure 2 shows few towns of significance in this area; names on Soviet maps often represent mere riverside clearings with a few houses. Large areas are unsettled, virtually inaccessible in summer because of swamps and mosquitoes. In these inhospitable surroundings lies one of the world's largest oil- and gas-producing regions. Here the Soviet Union produced more than one-half of its oil in 1980—312 mmt, about 6.24 mbd—and here too lie the enormous gasfields on which the future of the Soviet gas industry rests.

Oil

Oil was first discovered in West Siberia in 1960 near Shaim on the Konda river, a tributary of the Ob. By 1969, after an intensive exploration effort, 59 fields had been identified in the Middle Ob River region. These included nine large fields (defined as having recoverable reserves of 50 to 100 mmt or 366 million to 733 million barrels (bbl) each); nine giant fields (with recoverable reserves of 100 to 500 mmt or 733 to 3,665

million bbl); and the supergiant Samotlor. "Supergiant" is a designation for fields having recoverable reserves of more than 500 million tons. In this case, the Soviets broke their own self-imposed silence regarding the size of reserves, reporting that Samotlor, the largest oilfield in the U.S.S.R., contains "about 2 billion recoverable tons" (14.7 billion bbl) of oil.[3]

Soviet planners have concentrated on West Siberian development in an effort to maximize production, and Samotlor has dominated Soviet oil production in the past 5 years. In 1980, Samotlor alone yielded 150 million tons (3 mbd), approximately one-quarter of total oil production for the year. Another way of describing the contribution of this field is through its contribution to incremental output. During the Ninth Five Year Plan (FYP) (1971-75), Samotlor provided over 65 percent[4] of the production increase of West Siberia, and over half of the growth in oil output for the entire U.S.S.R. Much of the controversy over the future of Soviet oil production rests on the question of how long output at Samotlor can be maintained, and whether there is a sufficient number of small deposits to replace it once it does decline. The first billion tons of oil had been recovered from Samotlor by July 1981. At the current rate of production, the field is expected to give out by the late 1980's.

The Soviets are anticipating the inevitable peaking and decline of Samotlor, whenever that may be, by developing a number of smaller West Siberian fields. There is no question that there are many deposits, both in the Ob Valley and in the more remote area of the West Siberian plain, which will be brought into production. One problem with developing such fields, however, is the provision of infrastructure, both to accommodate oilfield workers and to transport the oil itself.

[3]*Sovetskaya Rossiya*, Feb. 28, 1970, quoted in Leslie Dienes and Theodore Shabad, *The Soviet Energy System: Resource Use and Policies* (Washington, D.C.: V. H. Winston & Sons, 1979), p. 58.

[4]Dienes and Shabad give a figure of 66 percent; David Wilson, in *Soviet Oil and Gas to 1990* (London: The Economist Intelligence Unit Ltd, November 1980), p. 9 estimates 73 percent.

Photo credit: Oil and Gas Journal

Oil rig near Surgut in West Siberia's Middle Ob region

The pervasive permafrost conditions in Siberia require special, difficult, and expensive construction techniques. For this reason, railroad and road construction has been kept to a minimum. The railroad reached Surgut, for instance, only in 1975, 10 years after the start of oil production there. Hard surfaced roads are largely confined to the area around Surgut and Nizhnevartovsk. It has been estimated that 1 mile of surfaced road in this region costs between 500,000 and 1 million rubles compared to 100,000 to 150,000 rubles in the European part of the U.S.S.R.

The Soviets employ the "work-shift" method in West Siberia; crews are shuttled by helicopter from base cities like Surgut or Nizhnevartovsk to isolated drilling sites where they live in dormitories for the period of their shift.[5] This is not very different from industry practice in the West in difficult areas such as the North Slope, but the obvious disadvantages of this life have required substantial bonuses and incentive schemes to attract workers. Nevertheless, there are still labor shortages and very high rates of labor turnover. As the deposits being worked move farther east, new base cities, or at least permanent settlements, will be required.

Gas

The gasfields of West Siberia extend for over 1,000 miles from the Yamal Peninsula in the north deep into the Tyumen oblast. Production in the area did not begin in any substantial amount until after 1970, by which time pipelines had been constructed to transport the gas. The location of the Tyumen fields is shown in figure 2 while table 4 summarizes production from deposits that are already online. The Tyumen fields are immensely important to the Soviet gas industry. The increase in production of natural gas at Tyumen between 1975 and 1980 (about 120 billion cubic meters (bcm)) amounted to more than 82 percent of the increase for the entire country (146 bcm), and

[5]Ibid., pp. 59-60.

Table 4.—Production of Gas in Tyumen Oblast
(billion cubic meters)

	1975	1979	1980	1981 (plan)	1985 (plan)
Medvezhe	29.9	71	70	70	NA
Vyngapur	NA	13	15	15	NA
Urengoy	NA	26	53	88	250
Others	3.6	2	2	NA	NA
Casinghead gas	2.2	11	14	17	NA
Total gas production	35.7	123	156	190	330-370

NA = not available.
NOTE: Totals may not add due to rounding.
SOURCE: *Soviet Geography*, April 1981, p. 276.

1985 plans call for production here to nearly *double*.

Gas extraction in West Siberia began in 1963 with a group of small gas deposits on the left bank of the lower reaches of small gas deposits on the left bank of the lower reaches of the Ob River. Three years later, the world's largest gasfield was discovered at Urengoy. Urengoy is one of several supergiant fields (i.e., fields containing reserves larger than 1 trillion cubic meters) in this region, and it is the focus of development for the present FYP. Other large fields include Medvezhye, which began producing in 1972, and Yamburg, for which the controversial planned pipeline to Western Europe was originally named. The gas for this pipeline, at least initially, will now come from Urengov.

The West Siberian gas industry has been beset with problems that have caused Urengoy to underfill its plan targets. These have nothing to do with the reserves, which make the field the largest single concentration of gas in the world, but rather are the result of inadequate infrastructure. This rests in part on the difficult conditions described in the previous section, exacerbated by the northern location of most of the gasfields, and also in part on the fact that the development of gas has generally lagged behind that of West Siberian oil because alternative sources of supply were already available from Central Asia.[6] But there is also evidence of planning failures in the gas industry.[7]

[6]See Jonathan P. Stern, *Soviet Natural Gas Development to 1990* (Lexington, Mass.: Lexington Books, 1980), pp. 35-6.
[7]Ibid., pp. 13-14.

Photo credit: Oil and Gas Journal

Gas treatment facility in the Medvezhye field

One large problem, for instance, has been delays in construction of gas-processing plants. These are necessary to treat the gas before it can be transported. Transportation itself can only be accomplished after the installation of pipelines and the compressor stations that move the gas through the pipe. This sequence has not always been well-planned. Even with the entire infrastructure in place, there are reports of compressor stations remaining idle because gas is being sent in quantities insufficient to justify their operation. In addition, there are large short-falls in the construction of housing, medical facilities, places of entertainment, etc., for gasfield workers; and poor organization at drilling sites that has led to difficulties in moving rigs from one location to another and to accidents and blowouts. These problems will have to be solved before West Siberia's tremendous gas potential is fully realized.

VOLGA-URALS

The Volga-Urals region lies in the far more accessible and temperate European portion of the U.S.S.R., between the Volga River and the western edge of the Ural mountains. Its major petroleum producing areas are found in the Tatar and Bashkir Republics, and Kuybyshev, Perm, and Orenburg oblasts.

These provinces together formed the U.S.S.R.'s most important oil-producing region until the late 1970's. Volga-Urals production peaked in 1975 and was exceeded by West Siberia in 1977. It is now in decline (see table 5). But even today, the Volga-Urals accounts for about one-third of Soviet oil production and is the third largest oil-producing province of the world.

The climate of the Volga-Urals is similar to that of the Canadian plains, and the extreme conditions that hamper the extraction and transportation of petroleum in West Siberia do not pertain there. In fact, except for some areas of the Perm oblast, the Volga-Urals deposits are readily accessible by road, rail, the Volga and Kama rivers, and close to major petroleum-using industrial centers. In addition, a large part of the Soviet refining industry is located in the region, although since the decline of the Volga-Urals fields these refineries use oil brought in by pipeline from West Siberia.

Oil

As was noted above, concentration on Volga-Urals oil did not begin until after World War II when development was fostered by the movement of heavy industry into the region. The giant Romashkino field in the Tatar Republic was discovered in 1948. Other large fields were found in Bashkir Republic and in Kuybyshev oblast. Between 1956 and 1958, these three prov-

Table 5.—Oil Production in the Volga-Urals Region
(million tons)

	1975	1976	1977	1978	1979	1980 annual plan
Tatar Republic......	103.7	100.0	97.8	97.2	85.8	83
Bashkir Republic....	40.3	40.2	40.1	39.6	39.7	38
Kuybyshev.........	34.8	33	31	27	25	25
Perm..............	22.3	23.5	24	24	24	22
Orenburg	13.9	12.6	12.8	12	12	10
Udmurt Republic....	3.7	4.4	5.5	6.5	7.4	8
Volgograd	NA	NA	NA	NA	NA	NA
Saratov...........	8	8	8	7	7	7
Total	226.7	221.7	219.2	213.0	201.0	193

NA = not available.

SOURCE: Wilson, op. cit., p. 15.

inces became the first, second, and third largest producers in the Soviet Union, a situation that persisted into the late 1970's. Even now, the Volga-Urals is important to Soviet oil industry planners, who hope to slow the region's decline by allocating resources to open new smaller and deeper deposits and by applying tertiary recovery methods in existing deposits.

Gas

Gas production in the Volga-Urals is centered at Orenburg in the Orenburg oblast.[8] Discovered in 1966, Orenburg is of special importance to the Soviet gas industry, for it is the latest supergiant deposit to be located in the more temperate European part of the country. Its gas is therefore more easily accessible to industrial users. But only part of Orenburg's gas goes to domestic consumers. The rest is now being transported to Eastern Europe through the 1,700 mile (2,750 km) Orenburg or "Soyuz" pipeline. This pipeline stretches from Orenburg to the Czechoslovakian border at Uzhgorod where it connects with the existing Brotherhood or "Bratstvo" pipeline system. Orenburg was built as a joint Council for Mutual Economic Assistance (CMEA) project, with East European countries supplying labor and materials in return for eventual repayment in gas deliveries. When it reaches full capacity, the Orenurg-Uzhgorod line is scheduled to carry 28 bcm/yr of gas, nearly all of it to be exported; 15.5 bcm of this will be divided between Czechoslovakia, East Germany, Hungary, and Poland, and the rest sold in Western Europe.[9] Bulgaria and Romania's shares are being delivered in another pipeline from the Ukraine.

The advantages of Orenburg's favorable location have to some extent been offset by the technical obstacles posed by the fact that its gas contains both condensate and corrosive sulfur. These must be removed in gas-processing plants before the gas can be transported. There are three processing complexes at Orenburg, two for treatment of gas used domestically and one that processes gas for the pipeline. Production is obviously linked to the capacity of these plants, as well as to the capacities of the pipelines that carry the gas to both domestic and foreign consumers. Construction delays occurred in both of these areas. In addition, housing, transport, and equipment shortages hampered exploration and drilling activities. Meanwhile, Orenburg is providing valuable experience in dealing with sulfurous gas, and reserves explored to date are sufficient to maintain 1979 production rates (48 bcm/yr) until the year 2000.

An additional major source of gas, also sulfurous, has now been discovered southwest of Orenburg at Karachaganak. This field is expected to be developed to replace any decline in Orenburg production.

THE CASPIAN BASIN AND NORTH CAUCASUS

As noted above, the first Russian oil was produced in the Baku district near the Caspian Sea. These sites are now part of an oil-producing region that spans the North Caucasus, Georgia, Azerbaidzhan Republic, Kazakhstan, and part of Turkmen Republic. At Baku, oil is being produced offshore in the Caspian Sea. Together these areas form an oil province of over 1 million km^3, containing hundreds of oilfields.

Oil

The importance of the Caspian basin oilfields has been steadily diminishing. Indeed, many of the older fields, producing since the turn of the century, are now virtually depleted. The Soviets have attempted to stem the decline through offshore development, deeper drilling, and use of water injection and enhanced recovery techniques, but nevertheless, production has continued to fall.[10] Table 6 chronicles this decline during

---

[8]See Wilson, op. cit., pp. 21-22; Stern, op cit., p. 31; and Dienes and Shabad, op. cit., pp. 77-79.
[9]Each East European country will receive 2.8 bcm, except Romania which will receive 1.5 bcm.

[10]Wilson, op. cit., p. 17.

Table 6.—Oil Production in the Caspian Region
(million tons)

	1975	1979	1980[a] (annual plan)	1980	1985 (plan)
Azerbaidzhan	17.2	14	19.7	14	NA
Kazakhstan	23.9	18	26.9	18.4	23
Turkmen Republic	15.6	9.5	18.6	8	6
Total	56.7	41.5	65.2	40.4	NA

[a]Estimate from Wilson, op. cit., p. 17.
NA = not available.
SOURCE: *Soviet Geography*, April 1981, p. 273.

the last FYP period. The shortfall of 25 mmt (500,000 bd) between the 1980 plan and actual production figures is a significant portion of the shortfall of 37 mmt (743,000 bd) from the original national 1980 plan (640 mmt or 12.8 mbd planned; 603 mmt or 12.1 mbd actual production.) The entire Caspian region produced less than 7 percent of the country's oil in 1980; it is not expected to resume a major producing role.

Gas

The gasfields of the North Caucasus have declined in much the same fashion as the area's oilfields. Two important groups of deposits at Stavropol and Krasnodar began producing in the late 1950's. These fields peaked in 1968, and the rate of depletion since then has been very high. Between 1970 and 1975, output fell by 5 bcm/yr at Stavropol and 16 bcm/yr at Krasnodar. The two deposits combined now produce less than 20 bcm of gas and appear to be declining at an ever-increasing rate.[11]

UKRAINE

Oil

Ukrainian crude oil is of high quality (i.e., it is low in tar, paraffin, and other pollutants and has a high yield of light distillates) and is located close to consumers in the European U.S.S.R. But the Ukraine's oil industry is beset with depleting reserves, new oil being found primarily at great depths and under difficult geologic conditions.[12] Ukrainian oil

[11]Stern, op. cit., pp. 30-31.
[12]Wilson, op. cit., p. 23; Dienes and Shabad, op. cit., p. 53.

production peaked in 1972 at 14.5 mmt (0.291 mbd), and while the Tenth FYP called for a decline to 8.6 mmt (0.173 mbd) by 1980, production was 8.3 mmt (0.167 mbd) in 1979 and 7.7 mmt (0.154 mbd) in 1980. The Ukraine has never contributed more than 4 percent of Soviet oil production, and its importance is not expected to increase.

Gas

In contrast, between 1960 and 1975, the Ukraine was the major Soviet gas-producing region, its output largely sustained by the giant Shebelinka field, which came onstream in 1956 and was supplying 68 percent of Ukrainian gas by 1965. The Ukraine also has a number of smaller deposits. However, these were not able to stave off decline once Shebelinka peaked in 1972. Now, like the North Caucasus, the Ukraine's gas production is in absolute decline, and output declined from 68.7 bcm in 1975 to 51 bcm in 1980. There are indications that the Soviets will continue to invest in the Ukrainian fields in order to maintain production in the accessible west of the country for as long as possible, Western analysts disagree, however, over the potential of the area. On one hand, new deposits have been announced at Shebelinka, in the Black Sea, and in the Dnepropetrovsk oblast.[13] On the other hand, a downturn in economic indicators over the last plan period—e.g., production costs doubled and labor productivity fell—together with the continued declines in production, have led others to conclude that the region will become increasingly less important in the future as rapid declines persist. Indeed, the 1981 annual plan foresees a production decline to 47 bcm.[14]

CENTRAL ASIA

Gas

The Central Asian desert is a gas-, rather than oil-, producing region, with vast fields

[13]Wilson, op. cit., p. 23.
[14]Stern, op. cit., p. 30; *Soviet Geography* (April 1981), p. 276.

lying near the Iran and Afghanistan borders in the Uzbek and Turkmen Republics. Central Asian gas has been important in the U.S.S.R. since the mid-1960's when the giant Gazli deposit in Uzbekistan was brought online. In 1965, Gazli alone produced 12 percent of Soviet gas. Since Gazli peaked in 1971, the region has declined in relative importance, but as table 7 demonstrates, production there has remained stable, largely through the development of sulfurous gas reserves. The sulfur is being recovered at the Mubarek gas-processing complex and the gas then transmitted into a pipeline system that extends from Central Asia to Central Russia. Before the beginning of development of these sulfurous reserves, it was thought that the level of output might not be maintained much longer.[15]

Gas production in Turkmenistan rose very rapidly in the early 1970's, but the rate of increase now appears to have leveled off. This republic includes the giant Shatlyk deposit, one of the 10 largest in the world, which alone accounts for nearly one-half of the area's production. Shatlyk is now producing at full capacity, and as table 7 indicates, output for the republic as a whole is increasing slowly. It is expected to rise to over 80 bcm as a result of the development of the newly discovered Dauletabad (Sovetabad) field.[16] Thus, Central Asian gas may continue to replenish the southern supplies depleted by the exhaustion of the North Caucasus fields.[17]

---

[15]Stern, op. cit., p. 34; *Soviet Geography* (April 1981), p. 276.
[16]Stern, op. cit., p. 35; *Soviet Geography* (April 1981), p. 279.
[17]Dienes and Shabad, op. cit., p. 84.

Table 7.—Gas Production in Central Asia
(billion cubic meters)

	1975	1979	1980	1981 (annual plan)	1985 (plan)
Uzbek Republic	37.2	37	39	37	—
Turkmen Republic	51.8	70	70.5	69	81-83
Total Central Asia	89.0	107	109.5	106	—

SOURCE: *Soviet Geography*, April 1981, p. 276.

KOMI

The Komi Republic lies in the Timan-Pechora region, north of the Volga-Urals, on the edge of the Barents Sea. It is an area of taiga forest and tundra, technically part of the European U.S.S.R., but having a climate similar to that of West Siberia. Nevertheless, this area of 250,000 km² lies 1,000 km west of Tyumen and is thus significantly closer than Siberia to centers of energy consumption. Komi is one of the Soviet Union's older oil and gas regions; its first commercial oil was produced in 1930. It is also virtually the only such older region that is not now in decline.

Oil

Komi's first commercial oil was produced near Ukhta in the southwest part of the region, although yields were negligible until the development of three large fields (West Tebuk, Dzhyer, and Pashnya) between 1962 and 1970. This development caused Komi oil production to rise sevenfold between 1960 and 1970, from 0.806 to 5.6 mmt (from 0.016 to 0.113 mbd).[18]

Exploration efforts then shifted northward. There development of two large fields, Usinsk and Vozey, began in 1973, and production continued to rise. Table 8, which shows Komi oil production over the Tenth FYP period, reflects this growth in output. However, Komi failed to meet its 1980 FYP target. The shortfall appears to have been due to such difficulties as early loss of reservoir pressure and infrastructure problems. The latter included construction of a railway branch line, for as in West Siberia, the development of the region is hampered by lack of roads.

While output is continuing to increase in Komi, the long-term prospects for the region seem to rest on exploration activities currently centered further north at the mouth

---

[18]Komi liquid hydrocarbons include a substantial amount of gas condensate. Oil output data for the region therefore often include condensate. The 1970 figure for oil plus condensate is 7.6 million tons. Ibid., p. 55.
[19]Ibid., pp. 55-6; Wilson, op. cit., p. 22.

Photo credit: TASS from © SOVFOTO

Oilfield equipment installation in the Komi taiga

Table 8.—Oil Production in Komi
(million tons, oil and gas condensate)

1975	11
1979	19
1980	21
1985 (plan)	26[a]

[a]Estimate.

SOURCE: *Soviet Geography*, April 1981, p. 273.

of the Pechora river, and offshore in the Barents Sea. This activity is currently being supported by a settlement of some 20,000 people, but the high expectations of the Soviets may perhaps be evidenced by reported plans for building a new town for 60,000 people in an area presently occupied mainly by tundra-dwelling reindeer herders.[20]

[20]Dienes and Shabad, op. cit., p. 56.

Gas

Komi became an important producer of gas after the giant Vuktyl gas and gas condensate deposit came online in 1968. Vuktyl, which lies 120 miles east of Ukhta on the right bank of the Pechora River, accounts for most of the region's natural gas production. As table 9 shows, this was scheduled to amount to some 22 bcm in 1980, but actual

Table 9.—Natural Gas Production in Komi
(billion cubic meters)

1975	18.5
1976	18.9
1977	18.9
1978	19
1979 (plan)	19
1980 (plan)	22

SOURCE: Wilson, op. cit., p. 22.

output reached only 18 bcm. The outlook is for a gradual decline in the 1980's.

The development of Vuktyl led to the construction of another major gas pipeline, the Northern Lights, which ultimately became a system of pipelines carrying vast flows of gas from West Siberia. The Northern Lights system was carrying 70 bcm in 1981 and is being expanded to a capacity of 90 bcm. It stretches westward across the European U.S.S.R., intersecting with the Moscow-Leningrad line at Torzhok and going on through Minsk to the Czechoslovakian border at Uzhgorod. The 1980 Northern Lights traffic included 18 bcm/yr of gas from Komi (most of it from Vuktyl) and over 50 bcm from West Siberia.[21]

The Komi region is rich in other gas deposits, but aside from the further development of Vuktyl, its future is uncertain. This is due both to the fact that much of Komi's gas lies in very deep reserves, and to the lack of infrastructure in this harsh, hitherto untouched, territory. Soviet long-term plans called for production to rise to 40 bcm by 1990. This suggests that newly discovered fields are expected to be brought onstream, but also that Komi will be depended on for only a fraction of the amount of gas that is slated to come from West Siberia.[22]

SAKHALIN

Sakhalin is a Far Eastern island situated between the Sea of Okhotsk and the Sea of Japan. It lies close both to the Pacific coast of Siberia and to the Japanese northern island of Hokkaido. Commercial oil production in Sakhalin began in 1921 when the island was under Japanese military occupation. Onshore production amounted to only about 3 mmt (0.06 mbd) in 1978, and offshore exploration has yet to be completed. The major importance of Sakhalin lies in its potential. Through Soviet-Japanese cooperation, it is hoped that Sakhalin may produce enough oil both to export to Japan and to supply some of the needs of the Soviet Far East, presently calculated at about 15 mmt/yr (0.301 mbd).[23] The Sakhalin project is discussed in detail in chapter 11.

SUMMARY

The center of Soviet oil production has moved progressively eastward over the last century. Once in the European portion of the U.S.S.R.—Baku on the Caspian Sea, and then the Volga-Urals region—the focus of this production now lies in West Siberia. This shift has meant that, increasingly, oil must be extracted far from major population and industrial centers and transported long distances to consumers. Moreover, conditions in West Siberia are harsh, the costs of extracting the oil higher, and erecting the infrastructure necessary to find, produce, and transport it concomitantly more difficult and expensive than in older producing regions. These factors affect the rapidity with which Siberian oil can be exploited. For these reasons, the Soviets continue to devote significant resources to slowing the decline and prolonging the productive life of the more westerly fields, particularly in the Volga-Urals region. A small contribution to this effort to maximize the production of relatively more accessible oil is made by Komi, which is the only producing area in the European part of the U.S.S.R. not yet in decline.

Similarly, the future of Soviet gas production lies in the less hospitable eastern regions. In the case of gas, however, significant contributions to production increases may be expected from the Volga-Urals, i.e., from Orenburg. Some production can still be maintained in older deposits in the Ukraine and North Caucasus, although this is becoming increasingly expensive, and there is disagreement over how long it can continue. In addition, Central Asian gasfields can contribute substantially to the gas available in the southern part of the country.

[21]Ibid., p. 86.
[22]Ibid., p. 87.

[23]Wilson, op. cit., p. 24.

EXPLORATION

Having briefly surveyed the major oil- and gas-producing areas of the U.S.S.R., this chapter now examines the manner in which oil and gas are discovered, produced, and transported in the Soviet Union. For the purposes of this analysis, the oil and gas industry has been divided into six segments or phases—exploration, drilling, production and enhanced recovery, transportation of oil and gas, refining, and offshore activities. Each of these segments will be discussed in terms of current Soviet practice, technological requirements, and the degree to which the U.S.S.R. has in the past or could beneficially in the future utilize Western technology.

INTRODUCTION

In order for oil and gas production to be sustained over the long term, additions must be made to reserves that compensate for the petroleum taken out of the ground. To replace the reserves produced during the Tenth FYP period (1976-80) the U.S.S.R. would have had to add to reserves, both from new finds and additions to existing fields, an additional 2.9 billion tons of oil (21.1 billion bbl). This estimate exceeds estimated gross discoveries during 1971-75 by about 50 percent.[24] Yet official emphasis in the U.S.S.R. in the last 15 years appears to have been largely on production from known deposits, rather than on exploration for and preparation of new areas. It has, therefore, been common in the Western literature to find the U.S.S.R. criticized for neglect of exploration efforts.[25] It could be argued that, until relatively recently, any lag in Soviet exploration activities was caused by the fact that discoveries such as Samotlor made extensive exploration efforts unnecessary, at least from a short-term perspective. For the past decade, however, it is more likely that the progress of Soviet oil and gas exploration

has been impeded by a general lack of availability of appropriate equipment.

Exploration for oil and gas in the Soviet Union seems to be handicapped by a lag, not in knowledge, but in its application. The U.S.S.R. has relatively few personnel skilled in advanced exploration techniques, and inadequate stocks of technologically advanced equipment. Some of these problems could certainly be remedied in the short run through purchases of foreign equipment and technology. However, given the leadtimes necessary to develop new fields, it may be too late for such purchases—which might enable the U.S.S.R. to explore at greater depths and in more difficult terrain—to much affect production prospects for the 1980's.

This section describes the methods by which exploration takes place; evaluates, to the degree that this is possible, the Soviet state of the art in these methods; and discusses the past and potential contribution of Western technology in this area. More attention is paid here to exploration for oil than for gas. This is a reflection of the fact that gas reserves in the U.S.S.R. are commonly acknowledged to be more than sufficient to sustain planned increases in production. This is not the case with oil reserves, the present extent and future prospects of which are matters of some controversy.

THE EXPLORATION PROCESS

Exploration for both oil and gas generally takes place in three phases: regional surveys that identify promising geological conditions for the presence of hydrocarbons; detailed geophysical surveys that evaluate specific areas in the regions identified in phase one; and exploratory drilling to test the findings of the first two phases.

Regional Surveys

Regional surveys are conducted in an effort to outline areas that might contain thick

[24]CIA, "Prospects for Soviet Oil Production: A Supplemental Analysis," July 1977, p. 23.

[25]Goldman, op. cit., p. 121; Robert W. Campbell, *Trends in the Soviet Oil and Gas Industry* (Baltimore: Johns Hopkins University Press, 1976), pp. 9-10; CIA, op. cit., pp. 1, 5.

sediments of hydrocarbons in structural traps. This is done with instruments or sensors, usually mounted in aircraft, which measure from the air changes in the magnetic fields and variations in the Earth's gravity. Sometimes, overflights of prospective areas are supplemented by ground-level measurements. Recently, both the United States and the U.S.S.R. have experimented with satellite surveys.

Detailed Surveys

The principle method of conducting a detailed analysis of an area is by seismic survey. Either an explosive or a device that vibrates the Earth is used to generate sound waves, which are reflected and refracted by the underground geological formations. The echoes are detected by seismographs or geophones, and recorded on magnetic tape. The result is a two-dimensional view of the subsurface structures. Seismic surveys produce large quantities of information that must be processed on large computers in order to generate these maps of underground geology, but minicomputers are now used to preprocess the data before it is passed on to a data processing center.

Drilling

Once a promising prospective area is located, the next step is exploratory drilling. Indeed, despite the sophistication of much geophysical seismic work, drilling remains the only means of positively verifying the presence or absence of hydrocarbons in structures. Exploratory drilling utilizes the same technology and equipment as does production drilling, although decisions as to the number, location, and depth of the wells will naturally differ depending on whether or not they are being drilled for exploratory purposes. Drilling technology itself is discussed in a later section of this chapter.

EVALUATION OF EXPLORATION EFFORTS

The success of exploratory activities is determined by additions to reserves—the amount of oil and gas found. In the case of the U.S.S.R., where oil reserves are a state secret, it is obviously difficult to evaluate the adequacy of exploration technology and equipment. The best that can be presented here are some qualitative impressions.

Exploration Technology

The U.S.S.R. is believed to possess adequate domestic capabilities for regional surveys, but its detailed seismic work may be inhibited by equipment that Western observers describe as bulky, difficult to transport, and of comparatively low quality. In general, the U.S.S.R. produces detailed survey equipment inferior in accuracy and capability to Western models. (For example, American experts who have examined Soviet geophone cables have found that they introduce extraneous "noise" into the data, a problem that makes results more difficult to interpret.[26]) These models appear to have been adequate in the past, but as the U.S.S.R. is driven to explore for oil at increasing depths, the need for greater quantities of higher quality seismic equipment will grow. The capabilities of the Soviet seismic equipment manufacturers to meet such a need is uncertain.

In the United States, a shift to the collection of seismic information by digital means began in 1962 and was essentially complete 10 years later (half of the crews had switched by 1968; 80 percent by 1972). In the U.S.S.R., roughly 40 percent of collection work is still done by traditional analog methods. Analog methods provide high-quality results, but these cannot easily be subjected to further processing, and they are far less efficient than digital methods. Consequently, the amount of work that can be done is smaller and the results are much less sophisticated.

As important to the success of detailed survey efforts is the quality and availability of computer facilities. Geophysical exploration is extremely computer intensive, and Soviet planners seem to be aware of the im-

---

[26]Private communication to OTA.

portance of making available both minicomputers to produce a rough picture of the geology of the area and large computers to further refine it. In each of these areas, the U.S.S.R. remains at least several years behind the West. However, it must be noted that most of the large oil discoveries in the world were made with seismic technology available by 1960. There is, therefore, no necessary correlation between state-of-the-art equipment and the size of potential finds.

Two philosophies for the processing of seismic information have emerged in the United States. In the first, a substantial amount of processing is done in the field at locally based minicomputer-equipped computer centers. The second uses centralized computer centers with large "number-crunching" high-speed (and often state-of-the-art) computers. The former philosophy has been mostly pursued by the independent geophysical contractors in the United States, while a combination of both has been employed by the major oil companies.

The Soviets have followed both paths. Thus, a small number of field systems utilizing minicomputers began to be introduced in the 1970's. By 1978, the Minister of Oil noted that second- and third-generation processing systems were being introduced,[27] and 22 systems were to be added by 1980. Some of these systems must be operated by highly skilled crews, which are in short supply. In addition, problems have arisen in coordinating the production and supply of spare parts and services for the minicomputers.

The Soviets clearly wish to increase minicomputer use. A program has been initiated to this end by encouraging the Ministries of Oil and Geology, Minpribor (Ministry of Instrument Making, Automation Equipment, and Control Systems), and the U.S.S.R. Academy of Sciences to work together on

minicomputer standardization and production.[28]

The Soviet capability to process digital seismic information using "number-crunchers" also lags considerably behind that of the United States. Until the mid-1970's the development of advanced geophysical techniques in the U.S.S.R. was made more difficult by an undeveloped computer base in general: only one high-speed computer model was known to be serially produced.

Although this machine was almost the equivalent of Western computers in speed when it first appeared in 1964, limited peripherals, a small core size, and very limited software degraded its performance significantly.[29] The other machines that were available were not well-suited for geophysical processing. For example, programs that take 45 minutes to run on a second-generation Soviet computer would take less than 2 minutes on a modestly high-speed U.S. model.[30] Furthermore, hundreds of the Soviet machines are needed just to process the data from one oilfield.[31] There is evidence that a special processor for geophysical data has been developed, but it is doubtful that it has appeared in any quantities.[32]

Although third-generation Soviet computers started appearing in 1972, the fastest, most powerful models were delayed until the late 1970's. In 1977, there were 18 processing centers for geophysical data in the Ministry of Oil.[33] Financing for computer-

[27]N. A. Maltsev, "From Well-site to Ministry," *Ekonomicheskaya gazeta*, No. 32, 1978, p. 15; V. Knayzev, "Gas Under Barkhany," *Trud*, June 3, 1980, p. 1.

[28]*Pribory i sistemy upravleniya*, No. 3, 1979, pp. 44-45; O. A. Potapov, "The Problem of Processing Large Masses of Geological and Geophysical Data and Ways to Solve It," in S. B. Gurevich, ed., *Holography and Optimal Processing of Information in Geology and Geophysics*, Order of Lenin Physical-Technical Institute imeni A.F. Joffee, Moscow, 1979.
[29]George Rudins, "Soviet Computers: A Historical Survey," *Soviet Cybernetics Review*, January 1970, pp. 6-44.
[30]*Sotsialisticheskaya industriya*, Mar. 7, 1978, p. 2.
[31]Potapov, op. cit.
[32]*Razvedochnaya geofyzyka*, No. 77, 1977, pp. 27-33.
[33]O. M. Rynskiy, "Automation of Economic Planning Calculations—The Most Important Directions in the ASU—Neft!" *Ekonomika neftyanoi promyshlennosty*, 1977.

related expenditures in the Ministry of Geology was substantially increased for 1975-80, but by the late 1970's, work was just beginning on using large computers for geophysical processing.[34] Thus, it is possible that between the two Ministries, the Soviets really only began to do a sizable amount of digital processing using large computers in the past few years.

Activity Levels

Soviet economic planning places heavy emphasis on attaining output targets. The practical consequence of this for oil and gas exploration efforts has been that those ministries charged with exploration—the Ministry of Geology and the Ministries of Oil and Gas—tend to focus on fulfilling their FYP targets, even when such relatively short-term considerations may be at odds with the maximization of oil production over the longest period of time. Those drilling teams and equipment devoted to exploration are unavailable for the drilling of producing wells—wells that yield petroleum that counts toward the fulfillment of output targets. Therefore, it may be more attractive to drill appraisal wells close to already producing regions than exploratory wells in remote areas. Moreover, the fact that drilling targets are expressed in terms of meters drilled, rather than oil or gas found, creates disincentives for deep drilling that is slower and more difficult than drilling a greater number of shallower wells.

At least partly as a consequence of systemic factors such as these, the number of meters drilled in exploratory wells actually declined between 1967 and 1975,[35] from 5.8 million to 5.4 million meters. Whereas in 1964 and 1965 Soviet oil output increased by 13.6 mmt for every million meters drilled, in 1976 this figure fell to 5.6 mmt.[36] This is a

reflection of the fact that no giant oil discoveries in the U.S.S.R. have been reported since the early 1970's.

Moreover, drilling targets have been consistently underfulfilled. In West Siberia, for instance, only about 80 percent of the planned volume of exploratory drilling was carried out in 1974, and drillers failed to fulfill their plans in each of the 5 years from 1971 to 1975.[37] During these years, large finds compensated for the level of exploratory effort and further encouraged the devotion of larger shares of drilling efforts to development rather than exploration. Rigs engaged in development drilling are about four times more productive than those used for exploration. This is because depths are shallower; the infrastructure is better; and less time is needed to move between locations.[38]

The Tenth FYP (1976-80) obviously recognized the need to step up exploratory activities. It called for efforts to find additional reserves in Siberia, Central Asia, and Kazakhstan, as well as offshore and in traditional producing areas. Total drilling targets were also raised significantly. In West Siberia alone a more than threefold increase was called for.[39] Given past performance, the likelihood of meeting such a target is at least questionable, but even a 10 percent increase in exploratory drilling would represent a significant investment in exploration activities that have hitherto been stagnant.[40] Moreover, there is evidence that exploration teams have been moving further and further away from the established centers of the industry into more remote areas of Siberia and the Arctic Circle.[41]

---

[34]Potapov, op. cit.

[35]Campbell, *Trends . . .*, op. cit., pp. 10-11; see also Goldman, op. cit., p. 122.

[36]Goldman, op. cit., 122.

[37]Wilson, op. cit., p. 45.

[38]CIA, op. cit., p. 23.

[39]Ibid.

[40]Robert W. Campell, "Implications for the Soviet Economy of Soviet Energy Prospects," *ACES Bulletin,* 20(1), spring, 1978, p. 40.

[41]See Wilson, op. cit., pp. 44-50.

THE CONTRIBUTION OF WESTERN EQUIPMENT AND TECHNOLOGY TO EXPLORATION

The U.S.S.R. has been virtually self-sufficient in regional survey equipment. In the area of detailed survey work, it has purchased geophones from the West, but not in very large numbers. Nor has it ordered the replacement parts for these geophones that U.S. industry experts assert must certainly be required. This leads to the inference that at least this Western equipment may now be inoperable or unreliable.

By far the largest contribution of the West to Soviet exploration activities, however, has been in the area of computers and related software and equipment. In general, the indirect reliance of the Soviets on U.S. computer developments has been large.[42] Oil and gas exploration has benefited both directly and indirectly from this dependence.

The advantage of Western computing equipment is that it can be purchased in complete ready-to-use sets. The Soviets have made major purchases of collection- and processing-related geophysical equipment, from firms, mostly in the United States. The American firm Geosource has completed a $5 million to $6 million deal that included outfitting three complete digital crews with 24 off-the-road exploration vehicles, a portable field recording unit, eight remote processing minicomputer centers, and processors used in conjunction with the minicomputer system.[43] An option for six more crews, including 49 additional vehicles, was exercised by the Soviets as part of a $13 million sale for 1978 delivery.[44] The post-Afghanistan technology embargo has now put further

such sales in limbo, and automated display equipment and geophysical equipment, including five more minicomputer systems, sold in 1979 for $9 million[45] have not yet been delivered. Even without this sale, a significant number of the estimated 300 digital collection crews in the U.S.S.R. have been outfitted with equipment supplied, not only by the United States but also by West Germany and France. However, the latter are almost all based on American equipment; Hungarian and East German systems are also available to the U.S.S.R., but these tend to be inferior to, and more costly than, those produced in the West. In addition, the U.S.S.R. has purchased a fully equipped French exploration ship, and has had other ships outfitted in the West.

Through these purchases, the Soviets have acquired advanced Western techniques. According to industry sources, there are as many as 30 to 35 U.S. minicomputer systems in the U.S.S.R. that have been specially designed for geophysical work. These include simple 16-bit dedicated array processors.[46] At least some of the U.S. minicomputer systems were shipped with software packages and Geosource trained 80 Soviet operators in the United States. This firm also installed the systems in the U.S.S.R. and gave extensive field training there.

As a result of these sales the Soviets are in some places using seismic techniques that were current in the United States about 1975. Present practice in the United States has moved far ahead of this level. In the United States array processors are now outfitted with programmable microprocessors that increase throughput by a factor of five. Main and secondary storage sizes have been increased substantially. The Soviets had not as of 1980 mastered multichannel techniques, which enhance the exploration of those deep structures that lack shallower ex-

[42]See Seymour E. Goodman, "Soviet Computing and Technology Transfer: An Overview," *World Politics*, vol. XXXI, No. 4, July, 1979, pp. 539-570; and N. C. Davis and S. E. Goodman, "The Soviet Bloc's Unified System of Computers," *Computing Surveys*, vol. 10, No. 2, June 1978, pp. 93-122.
[43]*Soviet Business and Trade*, May 25, 1977, pp. 1, 3, and Mar. 2, 1977, p. 10.
[44]Ibid., Aug. 3, 1977, p. 1.

[45]Geosource Inc., *1979 Annual Report*, p. 5.
[46]An array processor is a computer designed to handle very computer-intensive calculations by simultaneously manipulating large matrices of numbers, and is very useful for seismic applications.

pressions.[47] They only began serial production of their own array processors in 1980.[48] Although geophysical array processor designs may have been available, it is unlikely that a nonmilitary sector could have acquired them.

Again, however, any correlation between state-of-the-art seismic equipment and significant oil production increases has yet to be demonstrated. Thus, although the Western minicomputer systems that have been sold do not represent the state-of-the-art, it is not clear that the latest equipment is vital to the U.S.S.R. On the other hand, the magnitude of these sales implies that U.S. computer technology has played a significant role in aiding the Soviets to collect and process digital seismic data efficiently. This has proven true in the area of large "number-crunchers" as well.

An indirect dependence on U.S. large computer technology is evident in the third-generation Soviet- and East European-made Ryad computers that gradually became available in the 1970's. These are essentially functional duplicates of IBM models. The Soviets pursued this course in order to minimize risk and design decisions, to acquire the ability to tap the great body of software available in the West, and to use Western secondary storage and peripheral devices.[49] But it took the U.S.S.R. almost as long to duplicate these models as it took IBM to develop them. The first, small models did not begin to appear until 1972. Thus, the gradual improvement in Soviet seismic data processing capability between 1972 and 1980 can be equated to that of some of the major U.S. oil companies between 1965 and 1973—with the exception of array processors.

The delivery of large U.S. computers for geophysical processing began in the early 1970's, soon after the 1969 Export Administration Act lifted more stringent export control guidelines.[50] Several large computers were purchased from Xerox in 1973, and between 1975 and 1980 the Soviets purchased at least six major U.S. computers from CDC and IBM. CDC supplied a computer for the Ministry of Geology's All-Union Research Institute; another was to be used for the processing of offshore drilling information on Sakhalin; and two were for processing centers in Irkutsk and Tyumen.[51] Two IBM computers have been sold, one for use in the construction of offshore drilling rigs and the processing of information for offshore exploration, and one to the Ministry of Oil.[52] An export license for the latter sale was held back until IBM agreed to scale down the array processors that were to be included.[53] This also happened with other purchases described above.[54] These computer sales have all included software, supplied by both French and American companies.[55]

The U.S.S.R. has never sent appreciable amounts of seismic data to the West for processing, although it has sent small batches, apparently to test-check its own software. Its reluctance to send data seems to stem from a combination of secrecy, pride, and a reluctance to use hard currency for services.

Although the Soviets have obtained U.S. computers and Western software, and may therefore be able to do some sophisticated processing, it seems that they have so far been unable to implement some of the most advanced algorithms. This is a reflection of a much larger Soviet difficulty in software development, an area in which the U.S.S.R. has been notoriously weak. The reasons for this weakness are largely systemic. The Soviet economy is simply not structured to facilitate—indeed, even to allow—the close

[47]Ted Agres, "U.S. Builds Soviet War Machine," *Industrial Research and Development*, July 1980 p. 3.

[48]*Soviet Business and Trade*, Aug. 1, 1979, p. 5.

[49]Davis and Goodman, op. cit.

[50]See Office of Technology Assessment, *Technology and East-West Trade* (Washington, D.C.: U.S. Government Printing Office, 1979).

[51]*Soviet Business and Trade*, July 21, 1975, p. 2; Mar. 15, 1978, p. 1; June 6, 1979, p. 2.

[52]*Soviet Business and Trade*, Dec. 7, 1977, pp. 1-2.

[53]*Soviet Business and Trade*, Aug. 11, 1979, p. 7.

[54]*Soviet Business and Trade*, June 6, 1979, p. 2.

[55]*Soviet Business and Trade*, Nov. 10, 1976, p. 1; Nov. 8, 1978, p. 3.

and constant interaction between users and suppliers which is necessary to the implementation of appropriate software. Scientists, designers, and theoreticians are unable to communicate directly with systems users. Moreover, these users have little or no incentive to risk even temporary productivity or output declines in order to assimilate innovations.

The software problem is pervasive and has been felt in the seismic exploration area. One Soviet author, for instance, has asserted that "a number of important and necessary algorithms for the processing of geological and geophysical data are often not realized in practice."[56] Many of these applications are based on the use of sophisticated multiple-function array processors and very large capacity disks (in the range of 300 MBytes or more), which have not been made available to the Soviets primarily because of their importance in military applications.

The need for large capacity disks stems from the large size of data sets that are now being collected at high sampling rates. Very thin structures, usually found in small fields, may be missed at lower sampling rates, but it is difficult or impossible to split up the data sets taken at higher rates (such as 0.5 milliseconds) onto separate disks for processing. The Soviets have so far only been able to master the production of small quantities of 100 MByte disks (with oxides from West Germany), but these have been of poor quality. An emerging technology in the United States is acoustic holography (three-dimensional wave analysis), which allows the geophysicist to "see" structures three dimensionally. Large capacity disks are indispensable for this application.

Array processors are used in conjunction with large computers as well as with minicomputers. They are critical for offshore exploration, which yields roughly 50 times the data of onshore operations. Permafrost also presents massive complications and requirements for processing power. Since analog methods are still used extensively in the U.S.S.R., the overall throughput for seismic exploration is much slower than in the United States, and the computing power in use in the United States is still far greater. For example, a single oil company in the United States uses 10 large dedicated mainframes, 30 array processors, and over twenty-five 300 MByte disks. As Soviet hydrocarbons become harder to find, the more advanced computer-related technologies will become more important.

Given the fact that the Soviets have been very slow to introduce digital seismic equipment, the paucity of suitable computers until very recently, and the volume of Western sales, it is clear that Western equipment, especially U.S. computers and associated hardware, has filled an important gap in Soviet ability to process geophysical information. Large U.S. computers are located in all the major oil-producing areas—the Far East, Eastern and Western Siberia, in Baltic and Caspian Sea offshore drilling—as well as in Moscow.

SUMMARY AND CONCLUSIONS

The Soviet Union has expressed its intention to reverse past neglect of exploration activities. In this effort, it will face difficulties associated with the fact that it has insufficient quantities of seismic equipment, that this equipment is not up to Western standards, and that prevailing incentive systems tend to work against the allocation of resources to exploration. These problems will inhibit the U.S.S.R. in its attempts to survey more territory, in harsh terrain, and to prospect for oil at increasing depths. But it is not clear that improved seismic equipment alone would necessarily lead in the end to higher oil production. The number of giant oil discoveries remaining and the environments in which they most likely exist (both subjects of controversy among Western geologists) are at least as important in determining the success of exploration activities as the availability of the technology and equipment to identify them.

---

[56]Gurevich, op. cit.

The Soviet Union has relied on the West, and particularly on the United States, for assistance in developing the computers and computer-related equipment necessary to sophisticated seismic work. Although the Western equipment in the U.S.S.R. does not represent the state-of-the-art, such equipment has not been necessary in the past to locate major oil deposits. The U.S.S.R. is still seeking Western aid in this area, but it is unlikely that it will feel pressured to turn to the West in the 1980's to the same degree that it did in the previous decade. In the near term, Soviet ability to explore for new reserves is likely to hinge at least as much on the number of field crews it can deploy, the availability of highly skilled personnel, and its ability to assemble integrated sets of equipment for data collection in the field.

In the United States, experience has shown that computer techniques have allowed production declines to slow. In the U.S.S.R. such techniques might improve the efficiency of exploratory activities (i.e., the success rate), and thus, as the decade proceeds, advanced computer systems and software could similarly help to sustain production. Although it appears that the U.S.S.R. is moving ahead with the development of its own systems, systemic problems may delay their development and introduction. If this is the case, there will be significant pressure to acquire such systems from the West, probably toward the end of the decade. The Soviets may seek high-density, fast-transfer secondary memory devices, programmable array processors, integrated sets of equipment for data collection, and information display devices. The prime motivating factor for hardware purchases may be to get working software. If such items are unavailable, and if past practice continues, the U.S.S.R. will likely do without or use what is available, albeit in a suboptimal, more expensive manner, after significant delays.

DRILLING

INTRODUCTION

It is common in the West for energy experts to be critical of drilling practices in the U.S.S.R. It has been asserted that the inferior quality of Soviet-made drilling equipment will hinder progress in drilling unless "quantum improvements" are made; and that weaknesses exist in all elements of Soviet drilling technology and in the organization and supply of drilling operations.[57] In part, these evaluations rest on the fact that the U.S.S.R. has chosen a different—and demonstrably less efficient—technological path in its drilling operations from that pursued in the West, and on the unevenness of Soviet industrial standards and production which creates obstacles for drilling teams.

This section describes the methods by which oil and gas wells are drilled, evaluates the Soviet state-of-the-art in these methods, and discusses the past and potential contribution of Western technology to Soviet drilling. It must be noted that this discussion rests on incomplete and sometimes inconsistent data. Recent information on the annual number of meters drilled or drill bits produced is difficult to obtain from Soviet sources, but it is also surprisingly difficult to acquire similar figures for the United States. The data provided here have been verified to the degree that this was possible, but should not be considered conclusive beyond an indication of orders of magnitude.

THE DRILLING PROCESS

Oil and gas wells are drilled with a bit, i.e., a tool that bites into the earth and progressively deepens the bore hole. In the earliest days of the petroleum industry, hand digging was replaced by a system utilizing a chisel-shaped bit that traveled up and down

---

[57]CIA, op. cit., p. 21; Campbell, *Trends . . . ;* op. cit., p. 19.

on the end of a rope and simply pounded the well deeper. Today, technology has progressed to the point where sophisticated metals and alloys are used to create a wide array of precision tools, designed specifically for different types of rock and drilling conditions.

Equally important developments have taken place in the other technologies and equipment necessary for drilling. The ropes by which drill bits were raised and lowered in the well have evolved into drill pipe (still sometimes referred to as drill "string") made of high-quality steel; the muddy water that was pumped into the well to cool the bit and help to flush up debris has been replaced by "drilling mud" that is a chemically designed mixture of water and/or finely divided material such as special clays, barites, and chemicals. Drilling rigs—the hoists and derricks from which the drill pipe and bit are suspended and which support, raise and lower them—have developed into large, heavy-duty structures capable of bearing and hoisting weights of several hundred tons. Wooden stakes, inserted into the well to prevent it collapsing, have disappeared in favor of tubular steel casings that are cemented in with special oil-well cement to prevent corrosion and leakage and to reinforce the structure; a safety device called a blowout preventer may be attached to this casing at the surface of the well to prevent sudden explosive escapes of gas or liquid caused by high pressures. Finally, sophisticated electronic "well logging" instruments are now available. These are lowered into the well on cables and measure the density and permeability of the geological structure surrounding the well, allowing geologists to estimate the quantity and recoverability of potential reserves.

In the West, the most commonly used drilling technique employs a rotary drill. This is a system in which both the hollow drill pipe and the bit are rotated at the surface of the well by a rotary table. Drilling mud is pumped down the pipe and out through fluid courses in the bit, and this fluid conveys the rock cuttings to the surface where they can be examined for early traces of oil. With this method, the bit usually remains in the hole until it becomes too dull to be effective. At that point, the drill pipe and bit are drawn up together and the bit replaced. During the drilling process additional lengths of pipe may be added while the bit remains in the ground. A variety of drill bit designs make rotary drilling effective in both soft and hard rock formations.

SOVIET DRILLING EQUIPMENT AND PRACTICE

Turbodrilling

Although the Soviet Union originally employed rotary drilling techniques, it proved unable to produce sufficient quantities of the high-quality pipe necessary to withstand the torque applied in rotary methods.[58] Use of low-quality pipe leads to pipe breakage and the consequent loss of drilling time in retrieving the remaining pipe and bit from the borehole. To overcome these problems, the Soviet Union developed the turbodrill.

The turbodrill is a series of multistage turbine sections through which the drilling mud or fluid is passed. The turbine is placed at the end of the drill pipe, just above the bit, and the power required to rotate the turbine is provided by the fluid. The need to turn the entire drill string is thus eliminated, and far less stress or torque is applied to the pipe, which is either not rotated at all or is rotated only very slowly.

The turbodrill is a major Soviet engineering feat, one that was achieved at a time (the late 1940's) that Western engineers were trying and failing to resolve the problems associated with the turbine concept. This system has enabled Soviet drilling teams to dig farther and deeper than would otherwise have been possible given the stress on the drill pipe entailed in the rotary method. The turbodrill works particularly well in soft rock

---

[58]Campbell, op. cit., p. 21.

formations and is well-suited to directional drilling, a technique which allows the bit to be oriented at the bottom of the borehole in a predetermined direction. Directional drilling has been particularly useful to the Soviets in offshore Caspian drilling.[59] At one time, about 86 percent of Soviet drilling was done by turbodrill; this may now have fallen to about 80 percent.[60]

But the turbodrill is not without drawbacks, and it is responsible for much of the criticism leveled at the performance of the Soviet oil and gas industry. The efficient use of turbodrills requires high capacity rig mud pumps, the best of which are produced in the United States. Soviet mud pumps are greatly inferior to these. More importantly, turbodrills operate at three to four times the speed of rotary drills (120 to 600 rpm v. 30 to 150 rpm), a fact that promotes more rapid wearing of the drill bit, especially in hard rock. Replacing the bit is a time-consuming process, as the bit and drill string must be withdrawn from the ground. As well depths increase, time loss becomes even more of a problem. Thus, drilling in the U.S.S.R. takes longer than elsewhere in the world. Soviet drilling teams are said to devote an average of only about 15 percent of their time to actually drilling; the remainder is spent withdrawing and reinserting the drill string and replacing the drill bit.[61]

Other problems associated with the turbodrill are the fact that it cannot be used under high-stress conditions; that it requires more frequent maintenance when operated in high temperature formations; and that its efficiency deteriorates when it is used with certain drilling muds.[62]

Given these problems, several options are open to the Soviets. A turbodrill could be designed that would operate at lower speeds and withstand higher bit weight; the quality—and hence the longevity—of the drill bits could be improved; or rotary drilling could be substituted, at least for deeper wells. Each of these would entail basic improvements in Soviet drilling equipment and technology, a subject that is discussed in more detail below. However, it must be noted that there is no evidence to suggest that the U.S.S.R. plans a wholesale replacement of turbo with rotary drilling.

Although rotary drilling has been introduced in those areas where local conditions provide a particularly strong rationale (e.g., in areas with deep wells and high temperatures), this has been the exception rather than the rule. Not only do the problems that initially led to the development of the turbodrill persist, but much of the existing stock of ground equipment would have to be replaced to be compatible with rotary drills. Moreover, the U.S.S.R. has a significant amount of pride invested in turbodrilling. In short, it seems more likely that the Soviets will push for incremental improvements in their existing equipment rather than replace it with essentially Western technology.

Drill Bits

Contradictory reports have appeared in the West over both the number and the quality of Soviet domestically produced drill bits. In 1977, the CIA estimated that the U.S.S.R. was producing 1 million rock drill bits annually, compared with only about 400,000 in the entire rest of the world.[63] This may be contrasted with a more recent report that cites Soviet production figures of 421,000 in 1970; 352,000 in 1975; and approximately 400,000 in 1980.[64] Part of the discrepancy here lies in the fact that the CIA figure is for rock bits "of all types," i.e., those used for other purposes than oil and gas drilling. Similarly, the latter figures are for one type of drill bit only—rolling cutter rock bits. These, it is true, are used in the

[59]V. I. Mishchevich, "Drilling Operations in the U.S.S.R. for 60 Years," *Neftyanoye khozyaystvo,* No. 10, 1977, pp. 24-30.

[60]Goldman, op. cit., p. 42.

[61]Ibid., p. 41.

[62]R. W. Campbell, *The Economics of Soviet Oil and Gas* (Baltimore: Johns Hopkins University Press, 1968), pp. 106-114.

[63]CIA, op. cit., p. 26.

[64]Wilson, op. cit., p. 60.

vast majority of Soviet oil and gas drilling—96 percent of development and 88 percent of exploratory drilling; nevertheless the output figure is somewhat understated.

It has been claimed that these output figures may be misleading, not only because the quality of the bits produced is so poor (a matter discussed below), but also because the U.S.S.R. may not produce a sufficient variety of bits to allow the sophisticated matching of drilling equipment to drilling conditions that is standard practice in the West. In this connection, it has been claimed that "a typical Soviet factory produces 255,000 bits a year, but only two models. In the United States, a typical factory produces only 70,000. In part this is because production is frequently interrupted to allow the firm to tool up for the 600 models it offers."[65] If this is meant to imply that only a few types of drilling bits are available in the U.S.S.R., it is clearly misleading. There is evidence that between 1971 and 1975, 35 new types of bit were produced in the Soviet Union, and that during the last FYP period, more than 30 new models were developed, including 20 models of a bit made from ultrahard alloys, designed to drill to depths of 4,000 to 5,000 m and to operate at a faster rate of penetration than conventional rock bits.[66] Whether this variety is sufficient to maximize efficient bit use is another matter, however.

The number of bits or of models available may be less important than the quality of the bits produced. Quality is clearly an issue that has troubled Soviet planners, and may have been one of the chief motives for the import of a facility for the production of tungsten-carbide drill bits from Dresser Industries in the United States (see below). Whereas the majority of drill bits produced and used in the West are "journal bearing," the U.S.S.R. continues to employ an older technology, i.e., most of the bits used are "roller-bearing" models. The chief difference between these two designs is that roller bearings contain a number of small rotating elements, whereas the more technologically advanced journal bearing appears simpler, consisting mainly of two close-fitting parts. Journal bearing bits offer a larger surface area and they tend to be longer lived than roller bearing bits. The fact that the majority of the bits to be produced in the Dresser plant are roller bearing suggests that the U.S.S.R. may find these more suitable for use with the turbodrill.

Soviet efforts to improve the quality of domestically designed drill bits have centered in at least three areas: development of natural and synthetic diamond bit technologies; improvement of roller bearing designs; and work with hard alloy based and coated bits.

Natural diamond bits have been in use in the U.S.S.R. for some 25 years, but their utility has been limited by both economic and technical considerations. Natural diamond bits are costly to produce. In addition, they are prone to failure under the high vibrations that are a byproduct of turbodrilling. On the other hand, diamond bits reportedly last longer than other models, and may therefore be better suited to deep drilling. The U.S.S.R. is now testing synthetic diamond bits that could be produced more cheaply than natural ones.[67]

The Soviet literature also records efforts to improve roller bearing technology. In 1979, the Minister of the Oil Industry reported that Soviet research institutes had developed 122 varieties of new bits suitable for both low- and high-speed drilling, and that 75 of these were being put into production.[68] There are already indications of improved results with such new bits, including

---

[65]Goldman, op. cit., p. 41.
[66]Wilson, op. cit., p. 60.

[67]Campbell, *Trends . . . ,* op. cit.; D. Dadshev "An Important Reserve for Deep Drilling," *Vyshka,* Oct. 20, 1978, p. 2; V. V. Petrov, "The Drillers of Checheno-Ingushetiya in the Struggle for Technical Progress," *Neftyanoye khozyaystvo,* No. 10, 1977, pp. 67-70.
[68]N. A. Maltsev, "New Frontiers of the U.S.S.R. Oil Industry," *Neftyanoye khozyaystvo,* No. 9, 1979, pp. 5-10.

a report of 20 percent improvement in average meters drilled per bit.[69]

In addition to new designs, the U.S.S.R. is also developing more durable materials for its bits. Since 1977, about 30 percent of all bits produced have had hard alloy teeth in their rock-crushing elements, and as of 1979, factories were reportedly beginning to produce bit parts from steels that had undergone electroslag and vacuum arc remelting.[70] These are processes that remove impurities from the molten metal.

The most promising results have come from bits incorporating new superhard alloys that have a high resistance to wear. Claims for one bit utilizing such an alloy include the assertion that it can replace 40 to 70 conventional or two to three diamond bits, and that individual models have lasted over 1,100 m in production and 500 m in exploratory drilling. These bits are expensive to produce and have been found to be most cost effective at depths of 2,000 to 5,500 m. When used with turbodrills, it is claimed that they can reduce operating costs between 27 and 51 percent, largely because of reductions in downtime. Such bits are reprocessed to recover the alloy.[71]

In the final analysis, however, the real test of improved bit quality is the number of meters drilled each year in both development (producing) and exploratory wells. Soviet statistics show a marked improvement in drill bit productivity over the past 10 years which, as table 10 indicates, more than doubled between 1970 and 1978. Whether the U.S.S.R. can achieve the extremely ambitious target of raising this productivity a further 2.6 times by 1985 seems more problematic, given past performance. In general, bit productivity has been higher in Western Siberia than in the country as a whole because wells there are drilled into formations

Table 10.—Average Bit Runs (meters per bit)

	1970	1975	1978	1980[a]	1985[a]
Development drilling	33.7	54.2	76.1	77.2	198.5[b]
Exploratory drilling	19.8	26.6	28.1	33.6	—

[a]Plan.
[b]Pledge in *Trud*, June 6, 1980.
SOURCE: Soviet data reported in Wilson, op. cit., p. 59.

that are softer than those encountered elsewhere. However, as depths increase, hard rock is encountered even in these deposits. It is significant, therefore, that bit productivity here has grown in spite of increasing average well depths. In other regions of the U.S.S.R., such productivity declined as wells got deeper.[72] Whether these trends will continue in the face of the probability that new finds in Western Siberia are likely to come at everincreasing depths remains to be seen.

Drilling Mud

Both rotary and turbodrill equipment require lubrication with drilling fluid or mud. It is important to use muds that are chemically appropriate, i.e., which will not react with the underground rock formations in such a way that the formations are damaged or oil and gas zones overlooked. Scientifically designed muds are known in the U.S.S.R., and their production was slated to increase during the Tenth FYP as a number of new compounds became available. However, Soviet practice with respect to the use of these fluids is uneven. Western experts have observed cases of drilling crews simply using water mixed with local clay when proper chemical muds were in short supply. Whether or not this practice is widespread or confined mainly to remote exploration sites is unknown, but it is certain that it can cause damage to subsurface strata. Moreover, the careless reuse of poor mud—apparently a practice more common in the U.S.S.R. than elsewhere in the world—can inhibit the effectiveness of well logging equipment. Used mud, unless properly treated and processed,

[69]M. Abramson and V. Pozdnyakov, "The Series 1AN Bits Are Also Effective for Tubine Drilling," *Neftyanik*, No. 9, 1977, pp. 10-11.
[70]V. I. Pavlov, K. A. Kuznetsov, and N. P. Umanchik, "Oil Industry Machine Building," *Khimicheskoye i neftyanoye mashionostroyeniye* No. 11, 1977, pp. 18-22; Maltsev, op. cit.
[71]Dadashev, op. cit.; *Pravda Ukrainy*, Jan. 13, 1980, p. 2.

[72]Wilson, op. cit. p. 59.

contains oil, gas, or rock from previously drilled sites, thus distorting the data gathered from the present site.

Drill Pipe

Soviet difficulties in producing adequate quantities of high-quality drill pipe are well-documented in both Western and Soviet literature. At its most general level, this problem is part of a set of difficulties common to the entire Soviet civilian economy: an incentive structure that emphasizes quantity over quality, combined with a complex array of infrastructural problems that leads to shortages of the materials, workers, and equipment necessary to fulfill production targets. The result is often pipe with defective threads and joints that cannot withstand extreme temperatures and fail to protect pipes from corrosion and paraffin buildup.[73] Poor quality pipe can cause the drill string to break, dropping the bit and other parts into the well and requiring time consuming "fishing expeditions" to recover them. Wells then remain idle while replacement parts—which are not always available—are sought.

Soviet drill pipe seems to be adequate for wells down to about 2,500 m, but the weight and stress on the string at greater depths lead to frequent pipe failures. This has obvious implications for the average well depths achievable in the U.S.S.R.—an issue made all the more important by the fact that new finds are likely to be made at deeper levels. There is ample evidence that the Soviets are able to drill very deep wells—8,700 m and greater[74]—and the average depth of wells[75] has been increasing. For development wells the average grew from 1,772 m in 1970 to 1,994 in 1978; and for exploration wells from 1,928 m in 1960 to 2,775 in 1975. The average depth of exploration wells has now stabilized (it was 2,797 in

1978), but it is impossible to determine whether this is the result of an inability to drill deeper or a decision that deeper wells are not necessary. In any case, Soviet ability to provide enough quality equipment (including both bits and pipe) to quickly and efficiently drill a large number of deep wells has been questioned. In 1977, CIA estimated that on average it took Soviet drillers more than a year to drill 3,000 m,[76] while in the West this could be accomplished in one-half to one-quarter of that time.

It is difficult to evaluate these figures. While deep drilling claims a great deal of attention because of its cost and complexity, in 1981 only about 1 percent (some 800) of the wells drilled in the United States will be deeper than 4,840 m. In 1979, the average depth of a United States exploration well was 1,811 m and of a development well 1,361 m. The key question is not the depth at which technology allows one to drill, but rather the depth at which resources will be found. Moreover, most "deep" drilling is for gas—economic oil finds are generally made at shallower depths. Generalizations about the relation of Soviet deep drilling capabilities to oil production prospects should, therefore, be made with extreme care.

Rigs and Hoisting Equipment[77]

The U.S.S.R. has produced about 500 oil drilling rigs per year over the past 30 years, but output has been declining since 1975, from 544 rigs in that year to 505 in 1978. Similarly, the size of the Soviet "rig park" has declined. In 1970, 2,083 rigs were operating in the U.S.S.R., 1,124 of them belonging to the Ministry of Oil; in 1978, the total had declined to 1,915, of which 1,013 belonged to the Ministry of Oil. One important determinant of the size of the rig park is the number of rigs retired each year. The average life of a rig in the United States, where older equipment is repaired, is 15 to 20 years. In the U.S.S.R., where scrapping ap-

---

[73]N. Safyullin, "Where Metal 'Flies'," *Pravda,* Feb. 28, *1978, p.* 2.
[74]See "Transfer of Technology and the Dresser Industries Export Licensing Actions," hearing before the Permanent Subcommittee on Investigations of the Committee on Governmental Affairs, U.S. Senate, Oct. 3, 1978, pp. 173, 178.
[75]Wilson, op. cit., p. 55.

[76]CIA, op. cit., p. 26.
[77]Wilson, op. cit., pp. 55-59. Unless otherwise noted, Soviet data in this section is from this source.

pears to be far more common, the average life of a rig is about 6 years. This does have an advantage. If the entire rig park is replaced nearly twice in every decade, its quality can be rapidly upgraded.

The declining size of the rig park is apparently of some concern to Soviet planners, who have included increases in rig production in the Eleventh FYP. On the other hand, the Soviet Union has consistently exported about a quarter of its domestically produced drilling rigs each year. Presumably, if concern about the number of rigs available for exploratory and development oil and gas drilling were intense, a portion of those designated for export could be diverted. There is no evidence that this is occurring.

Moreover, despite a smaller rig park, the total number of meters drilled has increased over the past 10 years. Table 11 shows selected data which demonstrate this increase in drilling rig productivity. If these figures are accurate, they show a phenomenal rise in such productivity, particularly in contrast to what has been achieved in the United States.

The U.S.S.R. appears to be counting on improving the average size and technical characteristics of the new rigs that are slated to replace those being scrapped. Plans include producing new models for deep drilling adjacent to the Caspian (up to 15,000 m); modifying existing rigs for rotary drilling; and designing rigs especially for "cluster drilling" in West Siberia. Cluster drilling is a

process particularly suited to the soft rock conditions of West Siberia. It entails building artificial islands from which clusters of up to 20 inclined wells are drilled in the clay or sand. However, it must be noted that at least some of the "improvements" to Soviet rigs consist of additions to rig design—such as higher horsepower motors for hoisting, improved brakes, and increased diesel engine performance—which are incremental changes to basic rig designs of the 1950's. A more revolutionary change—shifting from traditional block and tackle hoists to a hydraulic hoist system—appears to be stymied by bureaucratic problems.[78]

Although both the quality and quantity of rigs are increasing, there is evidence too that the U.S.S.R. has not always achieved an optimum mix of equipment. In 1976, for instance, one drilling association complained that it was oversupplied with rigs designed to drill wells below 5,000 m, but did not have enough lighter rigs for the shallower depths normally required.[79]

One important barrier to increasing drill rig productivity is the time entailed in setting up and tearing down rigs as they move from one exploratory site to another. Thus, the availability of portable or "unitized" rigs is important. The U.S.S.R. has attempted to improve its situation with respect to unitized rigs both by importing them from the West (see below), and by creating its own unitized rigs with new cranes and transport equipment. These rigs can reportedly be assembled by a single crew (as opposed to several different crews of carpenters, earth movers, etc.), are 60 tons lighter than other Soviet rigs, and can drill to 3,300 m at speeds 33 percent faster than were hitherto achievable. But while experimental modular rigs have been successfully tested, serial pro-

Table 11.—Drilling for Oil and Gas (thousand meters)

	Exploration	Development	Total
1950	2,127	2,156	4,283
1960	4,023	3,692	7,715
1970	5,146	6,744	11,890
1975	5,419	9,751	15,170
Drilling by the Ministry of Oil only			
1975	2,733	8,927	11,659
1976	2,500	9,600	12,070
1977	2,400	10,400	12,800
1978	2,400	11,700	14,100
1979	2,500	13,000	15,500
1980 plan	2,500	17,000	19,500

SOURCE: Soviet data in Wilson, op. cit., p. 56.

[78]A. Sherstnev, "Extra-Plan Innovation," *Sovetskaya Rossiya*, Dec. 11, 1977, p. 2. Courtesy of Battelle Columbus Labs.

[79]A. P. Gorkov and G. F. Lisovskaya, "Analysis of Estimated Costs of Drilling and Ways of Reducing Them in the 'Ukrneft' Association," *Ekonomika neftyanoye promyshlennosty*, No. 5, 1976, pp. 7-11. Courtesy of Battelle Columbus Labs.

duction is only just beginning, and it is not clear how long it will take to produce such rigs in significant numbers.

Well Logging Equipment

The poor quality of Soviet well logging has been attributed to two basic problems: the extensive use of the turbodrill, and the lack of quality field instrumentation. The action of the turbodrill is such that it occasionally produces erratic and irregular walls in the borehole. Uneven walls cause the drilling mud to be forced into the resulting cracks and fissures. When probing this type of borehole for hydrocarbon content or permeability, it is difficult to separate the contributions from the mud from the actual geological structure. Soviet domestic well logging instruments in the field are essentially copies of American equipment acquired as part of lend-lease after World War II. Although Soviet research institutes have developed instruments comparable to the Western state-of-the-art, these do not appear to have been tested or put into operation. The result is that the accuracy of available Soviet well logging instruments is generally inferior to that of Western models.

Blowout Preventers

In the United States, blowout preventers are considered basic safety devices, and their use is required by law. In the Soviet Union, they are employed usually only in initial drilling in new regions or where underground conditions (mainly very high pressures) or corrosion are expected to cause problems. Once these initial wells are drilled, the use of blowout preventers is infrequent. Although this equipment may be necessary to cap runaway wells, it does not boost production.

Computers

In the United States, a comprehensive set of computer-based aids is usually used to optimize drilling operations. This consists of both onsite and remote monitoring, including online systems connected to a large central data base for advice on drill bit and mud selection and other parameters, and faster than real-time analysis. Computer-based drilling systems to select correct muds and bits and optimize equipment maintenance schedules can speed up the drilling process and help to eliminate drilling deficiencies—provided that crews have the incentive to do more drilling and have the appropriate range of muds and bits from which to choose. It is not clear that these conditions always pertain in the U.S.S.R.

The available evidence indicates that the degree of onsite Soviet drilling optimization is not very great. Applications are primarily related to data processing and acquisition, which involve calculations of geological formations and conditions, well-angling, and the selection of muds and drill bits.[80] It has been claimed that remote monitoring of drilling paramters is taking place on a "wider and wider" scale in the U.S.S.R.,[81] but the availability of sensing devices is limited, and indeed this practice is relatively new in the United States. The Minister of the Oil Industry in 1978 pointed out that although designs exist, output of "the necessary apparatus has not been organized."[82] There is available a system to monitor and control drilling, and another that predicts drill bit wear on the basis of drill stem torque measurements,[83] but it is impossible to say whether these are in widespread use.

Similarly, there is little indication of the use of computerized well-logging devices. The reservoir modeling routines that exist are unable to handle complicated structures. There is evidence only of a few instances of computers being used for the overall planning of drilling strategies.

---

[80]Yu. V. Vadetskiy, *The Drilling of Oil and Gas Wells: 4th Edition With Additions and Corrections* (Moscow: Izd. "Nedra" 1978), p. 289.

[81]Ibid., p. 302.

[82]Ibid., p. 303; Maltsev, "From Well-Site...," op. cit., p. 15.

[83]Vadetskiy, op. cit.; and P. M. Chegolin, A. G. Yarusov, and E .N. Yefimov, *Upravlyayuschiye sistemy i mashiny*, No. 4, July/August 1977.

THE CONTRIBUTION OF WESTERN EQUIPMENT AND TECHNOLOGY TO DRILLING

Drill Bits

Although the U.S.S.R. has not purchased a significant number of drill bits from the West, Soviet concern about drill bit quality is obviously reflected in the purchase of a U.S. drill bit manufacturing facility from Dresser Industries. Once fully operational, this plant will produce 100,000 bits each year, 86,000 of which are to be tungsten carbide insert bits (10,000 journal bearing, 74,000 sealed roller bearing, and 2,000 non-sealed roller bearing). In all, Dresser furnished designs for 37 separate bits to be produced in the plant. According to the company, all designs incorporated technology as it existed in Dresser plants at the time the contract was signed (1978). In addition to manufacturing equipment, the sale included product drawings, bills of materials, material specifications, and inplant process and heat treatment specifications for the 37 specific designs.

Soviet motives for acquiring this facility are open to differing interpretations. The decision might indicate that the planners are reasonably satisfied with domestic capabilities to produce more conventional milled tooth bits, but lack the manufacturing capacity for the tungsten carbide designs. On the other hand, once the need for more and better tungsten carbide bits was recognized, a new plant might have been seen as simply the most expeditious way of acquiring additional capabilities.

The bits to be produced in the Dresser plant should operate for long periods at the high rotation speeds of Soviet turbodrills. In fact, it has been estimated that each of these bits will substitute for at least two, and perhaps as many as four, Soviet-made bits. It is a highly speculative exercise to translate this into estimated production increases, but a rough idea of the potential contribution of this technology transfer may be gleaned by assuming that, once the plant is producing at full capacity, and without additional rigs, the new bits allow an increase in meterage drilled of 10 to 20 percent. Assuming that this equates to 10 to 20 percent more new wells with a 30 percent success rate, oil production increases of 3 to 6 percent as a result of this plant are *possible*. (This assumes constant productivity.)

But such increases are by no means certain. Improvements in drill bit quality cannot be translated directly into production increases in isolation from such factors as the incentives provided to drilling teams and the availability and quality of rigs and other equipment. In addition, even if the Dresser plant opens as originally scheduled in 1982, it is not clear that it will achieve the same volume and quality of bits as would be the case in the United States. This problem may be exacerbated by the fact that, as part of the post-Afghanistan technology embargo, the U.S. Government prevented Dresser from providing onsite training of Soviet personnel. (This occurred despite the fact that all licensed equipment had been delivered.) The most that can be said with confidence is that, all other things being equal, the new plant should eventually have measurable impact on Soviet ability to drill more efficiently. Whether this additional drilling capacity could translate into significantly higher production would, however, depend on trends in well productivity.

Drill Pipe

Recent purchases of drill pipe from Japan, Germany, and France have allowed the Soviets to drill deeper than is generally possible with domestically produced pipe. Western exports of drill pipe are included in more comprehensive categories in trade statistics (see ch. 6), and it is therefore difficult to estimate the magnitude of Soviet drill pipe imports. It is probably safe to assume that this pipe is reserved for deeper wells, which presently account for about 5 to 10 percent of Soviet oil production. The importance of Western pipe will increase if this proportion changes in the future.

Rigs

Aside from rigs used in offshore operations, a subject discussed separately below, the U.S.S.R. is believed to have purchased a sizable number of drilling rigs from Canada during the 1970's, and at least 15 portable rigs from Finland. Soviet emphasis on drilling faster, deeper, and in more locations in the present decade will require additional changes in the composition of the rig park (i.e., the variety and quality of available rigs) as well as increased rig production. It is likely that these demands will lead to continued imports.

Well Logging Equipment

Soviet logging instruments, as Soviet seismic hardware, lag Western equipment both in accuracy and efficiency, i.e., the number of sensors downhole at a given time. The U.S.S.R. has purchased items of this equipment from both U.S. and French firms, but in amounts that do not seem to significantly alter its overall capabilities. Logging operations in those wells supplied with Western equipment may be completed 3 to 10 times faster and with greater accuracy than otherwise, but this does not necessarily contribute to production. In order to significantly increase its overall logging time, the U.S.S.R. would have to purchase enough hardware to equip at least 100 crews and also allow Western technicians in for training and to operate the equipment. Even this number of crews would have difficulty logging the more than 20,000 new wells drilled annually.

Blowout Preventers

The U.S.S.R. has purchased small quantities of American blowout preventers, but probably imports most of this equipment from Romania. U.S. industrial representatives who have examined both Romanian and Soviet-made blowout preventers have found them inferior to those produced in the United States, a situation that may change in the future, as a U.S. firm has sold to Romania a new design for blowout preventers which may improve their quality.

SUMMARY AND CONCLUSIONS

Although Soviet commitment to the turbo—as opposed to rotary—drill makes good sense given the U.S.S.R.'s present manufacturing capabilities, the speed and efficiency of Soviet drilling have been inhibited by this commitment, as well as by the low quality of drill bits and drill pipe, and the size and composition of the rig park. There is no reason to believe that the U.S.S.R. will attempt a wholesale switch to rotary drilling, but it has placed increased emphasis on improving the quality of bits and pipe. In the former case, this has consisted of both stressing domestic design and production of new types of higher quality bits, and more importantly of importing an American-designed facility for the production of large numbers of high-quality bits. In the case of drill pipe, the U.S.S.R. has relied almost entirely on imports from Europe and Japan to compensate for domestic production. Some imports have augmented the Soviet rig park.

In none of these cases does the problem appear to be a lack of scientific or technical knowledge on the part of the Soviet Union. Rather, drilling equipment deficiencies seem to stem from the same systemic problems which pervade all Soviet industries, among them the continued emphasis on quantity over quality of output. The Soviet Union produces drill bits in very large quantities, but there are indications of insufficient variety, and evidence that many bits are so poorly made that they may last one-tenth to one-half as long as Western bits.

Despite these problems, the Soviet Union has managed increases in meterage drilled, drill rig productivity, and in its accomplishments in deep drilling. However, plans for the 1980's call for enormous improvements in each of these areas, improvements at least on the scale of those achieved in the 1970's. Clearly, increased investment is being devoted to the oil and gas sector, but it is impossible to determine how much of this will go into manufacturing drilling equipment. There is no sign of impending basic changes in the incentive system. Thus, given past performance, both in the oil and gas industries and in those industries that manufacture equipment for oil and gas drilling, it is difficult to see how dramatic improvements in production will be accomplished without stepped-up imports of Western drilling equipment.

PRODUCTION

INTRODUCTION

Perhaps even more controversial than Soviet drilling practices and capacities are Soviet oil production techniques. These can differ considerably from those common in the West and have occasioned the charge that in the interest of obtaining maximum shortrun output to achieve plan targets, the Soviets have consistently employed methods that damage their fields and ultimately lead to less oil being recovered. Much of the debate over the future of Soviet oil production, in fact, centers on the practice of waterflooding. Equally, much of the claim for the importance of Western oilfield equipment concerns the provision of Western pumps and other technology for use in fields where substantial waterflooding has taken place.

This section briefly explains the petroleum production process and the role of waterflooding in this process; describes Soviet methods for developing fields, including the level of Soviet domestic production technology; and discusses the past and potential role of the West in this area.

THE PRODUCTION PROCESS[84]

Oil and gas production are affected by the porosity and permeability of the reservoir in which they are found, by the water and gas content of the reservoir, and by the viscosity (i.e., thickness) of the oil. The way in which petroleum deposits must be developed and the extraction techniques applied vary importantly according to these factors. A petroleum reservoir consists of a stratum of porous rock, usually sandstone, limestone, or dolomite, capped by a layer of impervious rock. Oil and gas are stored in the small spaces or pores in the porous layer and contained by the cap rock. Fractures or fissures add to the storage capacity of the reservoir. In order for oil to enter or leave porous rock, there must be free connection between the pores. The ability of the rock to allow the passage of fluids through its interstices depends on the size of the channels which connect the pores, i.e., on permeability. The rate at which petroleum can be extracted from a reservoir depends largely on its permeability, but both porosity and permeability may vary over relatively small areas. Thus, wells located in different parts of the same reservoir may have different producing rates.

The first stage in the process of developing oilfields and gasfields is to drill appraisal wells. These are used to determine the permeability of the rock, the amount of water in the reservoir, the properties of the petroleum, etc. Such information helps to dictate the size of the surface production facilities brought to the field. Actual development of a field may begin before the appraisal process is complete. Development wells are drilled in patterns that reflect the contours of the reservoir: they may be in grid formations, straight lines, or rings. In general, the location of the wells is such as to enhance the producing life of the field. This might mean, for instance, that wells be initially drilled close to water zones so that as oil or gas in the area is depleted they can be turned into water injection wells. Numerous items of

[84]See the British Petroleum Co., Ltd., *Our Industry Petroleum* (London: The British Petroleum Co., Ltd., 1977), pp. 124-136.

equipment are required at the wellhead to "complete" the well. These include a variety of valves, casings, and tubings designed to control the well and the petroleum it is producing.

Oil

Oil that collects in structural traps usually occurs in association with both water and gas. The pores in the reservoir rock were originally occupied by water, which was partially displaced when petroleum migrated into the upper part of the rock. The percentage of remaining water is obviously an important factor in determining the volume of oil in the reservoir. Sometimes the water underlays the entire oil zone. When a considerable body of water underlays the oil in the same sedimentary bed, it is referred to as the "aquifer."

Oil under pressure contains dissolved gas in amounts governed by reservoir pressure and temperature. The oil is "saturated" if it cannot dissolve any more gas at a particular temperature and pressure; it is "undersaturated" if it could dissolve more gas under the same conditions. In those cases where there is more gas in the reservoir than the oil is capable of holding in solution, the extra gas, which is lighter than the oil, rises and forms a "gas cap" above the oil accumulation. Moreover, if for any reason the pressure in a saturated oil reservoir is reduced, gas will come out of solution and change the production conditions.

The viscosity of oil can depend on the quantity of gas that it holds in solution. Crude oil in a reservoir can range from very viscous (if it contains little or no dissolved gas) to extremely light and thin (containing large amounts of gas under high pressure). The thinner the oil, the more readily it will flow through the pores and interstices of the rock into the bottom of the well.

In order for this movement of the oil to take place, the pressure under which the oil exists in the reservoir must be greater than the pressure at the bottom of the well. So long as this difference in pressure can be maintained, the oil and its associated dissolved gas will continue to flow into the well hole. The rate at which oil or gas moves towards the borehole depends on the reservoir permeability and, in the case of oil, viscosity.

As the well begins producing, reservoir pressure decreases, and the rate of production will decline unless the pressure can somehow be sustained. There are three natural ways in which reservoir pressure is maintained: hydrodynamics, dissolved gas associated with the oil, and the free gas in the gas cap. These production mechanisms are referred to as "water drive," "solution gas drive," and "gas cap drive." The natural drainage of the oil through the reservoir rock under its own gravity provides a further mechanism, and a combination of any or all of these may operate in the same reservoir. The oil obtained as a result of these natural production mechanisms is known as "primary recovery," and a field is said to be in the primary phase of recovery so long as there is sufficient pressure left in the reservoir to bring the oil to the bottom of the producing well without outside interference.

At some time in the life of a producing well, primary recovery mechanisms will become insufficient and the reservoir pressure will fall to the point where it can no longer force the oil from the rock into the well. This stage can be reached long before the reservoir is depleted, but it once meant the abandonment of the well. There are now artificial means, known as secondary and tertiary recovery, of maintaining reservoir pressure and forcing more oil out of the pore spaces of the reservoir rock. As much as 50 to 90 percent of the oil in a reservoir may be left in place after the end of the primary recovery phase. Secondary and tertiary recovery techniques now make it technically possible to recover 30 to 90 percent of the oil in a deposit.

Secondary recovery involves the direct displacement of oil with a fluid which is cheaper and easier to obtain than the oil itself. The obvious substances are those that imitate the primary production mechanisms—water and gas. When water is used, the secondary recovery process is known as "waterflooding;" when gas is used the process is called "gas drive" or "gas injection."[85]

Waterflooding is the most successful and extensively used secondary recovery technique—so much so, in fact, that it is now considered an integral part of the development of most fields. Water is introduced under pressure into the reservoir via injection wells. These wells may be located adjacent to producing wells to penetrate the reservoir below the oil/water level in the periphery of the oil zone, or they may be drilled in a line across the reservoir or in a grid pattern. The method chosen usually depends on the type of reservoir and rock and fluid characteristics. In some reservoirs, there is considerable variation in the permeability of the rock, and in these instances the rate of injection must be carefully controlled to avoid trapping and leaving behind large quantities of oil. Similarly, it is important to ensure that the injection water is compatible with the natural reservoir water and that it is free from impurities that might block the pores in the reservoir rock. Filters and forms of chemical treatment may be employed to achieve maximum efficiency in this respect.

Where waterflooding has been employed, it is likely that the reservoir pressure will be so low that mechanical assistance will be required to bring the oil and water to the surface. This is usually accomplished with pumps, the simplest and most common being sucker-rod pumps, which work like plungers. They are run into the well at the bottom of a length of tubing, and powered by a pumping jack at the surface. Far more efficient, especially for deep wells, are electric submersible pumps which, together with their motors and electric cables, are lowered into the well on the tubing through which the oil is to be produced. Electric pumps have a much greater capacity than those of the plunger type and are used when high pumping rates are desired. An alternative to pumps is gas-lift equipment, which injects gas into the oil column in the well bore. This method is preferred where the crude oil contains considerable amounts of sand or suspended solids that could damage mechanical pumps.

More complex and sophisticated variations of these secondary recovery techniques may be applied to achieve an even greater degree of recovery of the oil in the reservoir. Known as "tertiary" recovery, these usually involve the treatment of reservoir rock with chemicals or heat. Research and field testing are being conducted in these techniques, but tertiary recovery is relatively new and is still seeing rather limited commercial applications in the West.

Gas

It is possible to have a free gas accumulation with no underlying oil zone, especially in deeper portions of basins. Sometimes gas is contained in a closed reservoir where there is no water closely associated with it, and is driven out of the pores of the reservoir rock by its own expansion. As gas escapes, the reservoir presure declines. It is possible to maintain reservoir pressure through water injection, but unlike oil, gas recovery as such is not improved sufficiently by water displacement to justify the expense of this operation. It is preferable, therefore, for the gas to be contained in a reservoir where it is in direct contact with an aquifer possessing a sufficient natural water drive mechanism. The process here is the same as in oil wells— the gas is driven out by the expansion of the aquifer water into the vacated pores, and there is no marked decrease in the reservoir pressure or in the producing capacity of the well. Care must be taken that production rates are not so high as to cause damage from infiltrating water, a consideration which applies equally to oil wells.

---

[85]Some experts would hold that secondary recovery is limited to waterflooding and that the use of gas is a "tertiary" recovery technique. OTA here follows the industry usage as expressed in British Petroleum, op. cit.

SOVIET EQUIPMENT AND PRACTICE

Secondary Recovery: Waterflooding

Gas injection is not an important form of secondary recovery in the Soviet Union,[86] but in 1980 over 85 percent of the oil produced in the U.S.S.R. (v. about 50 percent of the oil in the United States) was extracted with the aid of waterflooding. In West Siberia, 99 percent of all oil is obtained with waterflooding.[87] Soviet oilfields are often injected with water at high pressures from the beginning of their development. The dual effect has been both to raise initial recovery rates and to reduce the number of producing wells required per unit of land. The latter allows the U.S.S.R. to conserve capital and reduce the amount of drilling per ton of oil produced.

There is little doubt that waterflooding produces more oil in the short run, but controversy exists over the degree to which this practice contributes to maximizing the ultimate recovery possible from a given field. Western observers have argued that, depending on the rate of injection and field pressure, water can prematurely break through the oil-bearing formations into the producing wells. The net result is that total output over the life of the field is reduced.[88] Soviet experts, however, contend that waterflooding allows the U.S.S.R. to ultimately recover a much higher percentage of oil in place than has been possible in the West. Soviet ultimate recovery rates of 50 to 60 percent have been claimed, with the average reportedly as high as 40 to 50 percent.[89] This may be compared to a U.S. average in 1977 of 32 to 33 percent.[90]

It may well be that these figures are not directly comparable, and that the U.S.S.R.

calculates its ultimate recovery by using a base other than that employed in the West.[91] Moreover, there is evidence in the Soviet technical literature of numerous and increasing problems associated with waterflooding, and indications that at least some ultimate recovery targets have been scaled down.[92] On the other hand, the Soviet Union shows no sign of abandoning its long-held waterflood practices. Indeed, current plans are to raise recovery rates still more, and it has been announced that this will be achieved through more intensive waterflooding, albeit with "improved" methods.[93] With Soviet oil production in mid-1981 still not having peaked, it would appear that this is one area of disagreement in which no final verdict is yet possible.

A similar debate concerns the level of the "water-cut" in the U.S.S.R., i.e., the percentage of water in the oil-water mixture that comes out of producing wells. Table 12 shows Soviet water-cut figures for the past 15 years.

It is clear from table 12 that the average water-cut has been increasing, and that at least between 1965 and 1976 the increase in Western Siberia particularly was precipitous. These figures show that for the U.S.S.R. as a whole in 1980, 57 percent of the total output of producing wells was expected to consist of water. If the 1980 water-cut target of 57 percent were reached, this would mean that in order to produce 603 million tons of oil, over 1,400 million tons of oil and water had to be extracted. Moreover, the water-cut of individual fields can far exceed national or regional averages. There are fields in the Ukraine, for example, in which by 1975 water represented from 60 to nearly 80 percent of total output.[94]

Such figures are difficult to interpret or extrapolate, however. In 1977, CIA attempted a simple extrapolation based on assumed increases in water-cut of, alternatively, 3 and

[86]W. Kelly, H. L. Shaffer, and J. K. Thompson, *Energy Research and Development in the U.S.S.R.* (Battelle Columbus Labs, 1980), pp. IV 14-IV 32., unpublished manuscript.
[87]Wilson, op. cit., p. 75.
[88]CIA, op. cit., p. 13.
[89]Kelly, et al., op. cit., p. IV-30; Wilson, op. cit., p. 78.
[90]CIA, op. cit.

[91]Kelly, et al., op. cit.
[92]Ibid., pp. 13-32; CIA, op. cit., p. 14.
[93]Wilson, op. cit., pp. 78-81.
[94]Kelly, et al., op. cit., p. IV-31.

Table 12.—Waterflooding

	1965	1970	1975	1976	1977	1978	1980[a]
Average water-cut	41.2	43.9	48.2	49.9	50.8	51.8	57.0
Average water-cut, West Siberia	1.1	5.0	14.5	15.8	NA	NA	NA
Average water-cut, U.S.S.R. excluding							
West Siberia............................	41.2	46.2	57.0	59.0	NA	NA	NA

NA = not available.
[a]Annual plan target.
SOURCE: Wilson, op. cit., p. 76.

6 percent per year. This exercise yielded projections of average water-cuts of 65 and 80 percent respectively for the U.S.S.R. in 1980. In fact, the 1978 average national water-cut was 51.8 percent and the actual 1980 figure probably somewhere between that and the target of 57 percent.[95] Indeed, projections of this sort are complicated by the fact that the water-cut does not rise regularly for the country as a whole, or even for individual deposits or wells. Large annual increases may be recorded at certain recovery rates, but these may fall to between 1 and 2 percent per year at certain points in the life of a deposit. Nor do the Soviets appear to operate on the basis of a simple or single cutoff point beyond which the water-cut makes further production uneconomic. Eventually, the water-cut may rise to 97 or 98 percent, in which case the cost of pumping fluid will exceed the value of the oil obtained, and the well will be shut down. When the water-cut in an entire field reaches this point, the field may have to be redrilled. (Romashkino has been redrilled four times for this reason). But Soviet experts contend—and U.S. practice has verified—that some deposits can operate for many years with water-cuts of 80 or 90 percent, depending on the value of the oil produced and the costs associated with the waterflooding.

Fluid Life and Pumping Requirements

Regardless of the unresolved issue of the wisdom and propriety of waterflooding, there is general agreement that present Soviet practice entails enormous fluid-lift requirements, and that the higher the water-cut in the future, the greater this problem will become. Both sucker rod and electric submersible pumps are produced and used in the Soviet Union, but the latter have a far greater capacity and are much preferred, especially in Siberia where wells have high water-cuts and therefore high fluid-lift requirements. But electric pumps are in relatively short supply. In 1975, for instance, there were 68,000 producing oil wells in the Soviet Union. Some 54,000 of these were being pumped—9,100 with electric pumps and the rest with sucker rod pumps. In all, the U.S.S.R. had an inventory of about 60,000 of the latter, about 75 percent of which were operational.[96]

The U.S.S.R. has also encountered difficulties with the quality of its domestically produced electric pumps. One problem is pump capacity, which hitherto has been insufficient for use in West Siberia. Many of the wells at Samotlor, for instance, yield 2,000 m³ or more of fluid (petroleum plus water) a day. Soviet pumps have capacities of only 700 m³/day, although a 1,000 m³ model has very recently been tested. Another problem is the frequency of equipment breakdowns, particularly in areas with high salt deposition where the average interrepair period lasts 60 days. Improvements in both quantity and quality of electric submersible pumps are apparently being sought in the present FYP period.

Gaslift has not been employed extensively in the U.S.S.R., but there are indications that plans now call for an acceleration in its use, particularly in those areas where sub-

---

[95]CIA, op. cit., pp. 16-17; Wilson, op. cit., p. 77.

[96]Wilson, op. cit., p. 68.
[97]Ibid., p. 66.

mersible pump capacity has been inadequate. There has been some urgency, for instance, in attempting to transfer gaslift to Samotlor where, as described above, both high salinity and fluid lift requirements make Soviet submersible pumps ineffective. Gaslift has been used in West Siberia for at least 10 years, but it has been introduced slowly and accounts for only a small volume of the fluid raised. Part of the reason may be the high capital cost, particularly of installing the necessary compressor units. If this, rather than inadequacies in the design and quality of Soviet-built equipment, is the primary difficulty, there is a chance that the increases in the price of crude oil scheduled for 1982 may make gaslift more practicable. Meanwhile, the U.S.S.R. has ordered gas-lift equipment from France and is also attempting to improve the quality of its domestically built equipment.[98]

Tertiary Recovery[99]

Tertiary recovery techniques are very expensive and many are still relatively experimental, even in the West. There is evidence of Soviet experimentation with a variety of methods, including steam injection and polymer flooding, and indications that attempts will be made to apply tertiary recovery in older producing regions during the 1980's. The Soviets themselves have reported that, on the basis of experiments in tertiary recovery techniques, a 10 to 15 percent increase in recovery rates can be forecast and "billions of tons" of oil can be reclassified as active reserves. It seems highly unlikely, however, given the Soviet system, that the experiments carried out during the Tenth FYP could affect production significantly for several years at least.

In the long run, there appears to be no serious technical barrier to an expansion in the role of tertiary recovery. Rather, there is an economic barrier in the high cost of tertiary methods. Tertiary recovery must be accorded adequate investment, and economic considerations will influence which methods and geographic sites are developed. Perhaps more important, tertiary recovery results may be affected by the same systemic problems, particularly those stemming from the incentive system, that crop up continually in explanations for Soviet inefficiency. The variety of conditions encountered between and within petroleum fields means that tertiary recovery methods must be chosen and used sparingly and with care. Lack of skilled personnel together with incentives for maximum output over the short term, however, have tended in the past to cause Soviet petroleum industry workers to apply crude and unsuitable techniques in efforts to achieve "quick fixes."

Computers

Computers can be used in oilfield and gasfield operations to monitor wells by signaling when flow rates change, to control the action of pumps, and to optimize enhanced recovery techniques. The increase in production made possible by oilfield automation can be substantial. For example, one oil company in Texas was able to get within 3 bbl of the maximum allowable rate per day using computers, an increase of about 450,000 tons per year. Soviet systems seem to be limited to monitoring flow rates. The largest saving here is in personnel. The Soviet news agency TASS has claimed that by 1980, 85 percent of Soviet oil output would be automated,[100] presumably with this system. However, the head of the Ministry of Oil has noted that the lack of terminals is hindering its operation,[101] and in any case it falls well short of Western standards.

THE CONTRIBUTION OF WESTERN TECHNOLOGY AND EQUIPMENT TO PRODUCTION

Submersible Pumps

The Soviet Union obviously has a substantial domestic capacity for producing

---

[98]Ibid., pp. 66-68.
[99]See Kelly, et al., op. cit., pp. IV-41-42.

[100]*Soviet Business and Trade,* No. 12, Nov. 9, 1977, p. 8.
[101]Maltsev, op. cit.

electric submersible pumps, but since these have lower capacities and require more maintenance than their Western counterparts, it has in the past purchased quantities of pumps from the United States—the only other country in the world where they are manufactured. U.S. pumps lift up to 1,000 tons of fluid per day, and can last up to five times longer than Soviet pumps.

By 1978, the U.S.S.R. had purchased about 1,500 American pumps, with deliveries staggered over several years. Only about 2,000 pumps are produced each year in the United States, and back orders and limited manufacturing capacity had restricted deliveries to the U.S.S.R. to about 30 pumps per month. The CIA has estimated that these American pumps may have enhanced Soviet oil production by as much as 1 mbd.

In retrospect, this figure seems misleading. The U.S.S.R. has purchased *no* submersible pumps since 1978, yet its oil production has not only failed to decline by 1 mbd; it has risen. The American pumps seem to have a lifetime of 3 to 6 months in Soviet service before major overhaul is required. But although the Soviets asked their American suppliers for training in pump repair, the U.S. firms refused. This suggests that, unless the U.S.S.R. has developed unprecedented capabilities in learning to repair foreign equipment with no information from the supplier, the U.S. pumps were probably used until they failed and then put aside. If this is the case, it is likely that no American pumps were still in service by 1979, and the U.S.S.R. must have been able to substitute its own equipment.

This is not to suggest that U.S. pumps were not important to the Soviets or that they might not be so in the future. If pumps could be purchased at previous rates or increased by a factor of two or three *and* if these pumps were replaced or repaired when they failed, the impact on Soviet oil production could be substantial, although difficult to quantify. Estimates by U.S. industry ex-perts show an enormous range—from 8 to 20 percent increases in production.

Gaslift Equipment

Soviet gaslift efforts have been importantly enhanced by the sole gaslift sale reported in the West—a $200 million deal made with two French firms in 1978 for equipment for approximately 2,400 wells, including gas compressors, high-pressure manifolds, and control valves. (Although the French firms were the general contractors for this project, American equipment, built in Ireland, formed part of the package. U.S. computers have also been used for operating the gaslift equipment at the surface.) This equipment has been employed at the high-priority Samotlor where the downhole life of submersible pumps is particularly short. However, this quantity of equipment will equip only about 20 percent of Samotlor's wells, and probably has accounted for some 1 to 3 percent of current production.

Computers

The Soviets have purchased sophisticated automation systems for oilfields and gasfields from the United States and France. Much of this has gone to Western Siberian oilfields (Samotlor and Fedorovsk), including a multimillion dollar multilevel process control system that regulates and optimizes gas-lift operations on several thousand oil wells. Another gas monitoring system was purchased for the Orenburg gasfield. The equipment used in these systems only recently went into production in the Soviet bloc, and considering the leadtimes needed to develop such systems, it is clear that these purchases allowed capabilities far in advance of those that would otherwise have been possible. The Samotlor and Fedorovsk sales are particularly important because those fields are entering critical secondary recovery stages.

These systems may not be necessary if the fields are pumped continuously at maximum possible rates. But if the Soviets do or intend to plan oilfields so as to optimize overall recovery rates, automation equipment will be needed. Automated systems for water and gas injection operations will also be in greater demand as more and more fields enter these stages of production and as labor becomes scarce.

The Soviets have recently started producing the computers needed for sophisticated reservoir analysis, i.e., computers with large, fast main memories. Damaged (over-flooded) fields and overestimates of reserves have partially been due to poor reservoir modeling. Future needs will be increased by the number of fields entering secondary and tertiary recovery phases, and by the switch in the late 1980's to fields with more difficult and complex geologies.

Although the U.S.S.R. has been producing all the equipment needed for such multilevel oilfield and gasfield control systems, it does not have much experience in building them, and it may be several years before the Soviets can organize such systems themselves. They may, therefore, continue to turn to the West for such systems over the short term. Over the long term, the U.S.S.R. is unlikely to purchase large computers for reservoir modeling, since it has recently mastered their production. However, purchasing the software may save considerable time and lend new insights into reservoir models, especially since combining very complex geophysical know-how and software is one of the most difficult tasks in the geophysical field. The United States remains the world leader in this technology.

SUMMARY AND CONCLUSIONS

Oilfield development in the Soviet Union employs different techniques than are common in the West. Most importantly, the U.S.S.R. initiates secondary recovery, and particularly waterflooding, at an earlier stage in the producing life of its fields. Although this practice is widely used in the West, the extent to which it is employed in the U.S.S.R. together with documented cases of its misuse have led Western experts to label it potentially damaging to overall extraction prospects. On the other hand, many Soviet petroleum experts continue to believe that extensive waterflooding actually enhances ultimate recovery rates. So long as Soviet oil production continues to rise in fields like Samotlor, which as a high water-cut, this debate is unlikely to be resolved.

While it is misleading to generalize about the water-cut rate for the U.S.S.R. as a whole because of important variations between regions and fields, there is no doubt that poor management has led to damage in some fields; or that the fluid-lift requirements occasioned by waterflooding are burdensome. This problem is intensified in the U.S.S.R. because of its poor domestic capability for producing large numbers of high-quality electric submersible pumps for removing the oil and water mixture from wells. Pumps imported from the United States have been important in alleviating this problem, but the U.S.S.R. has demonstrated that it is not entirely dependent on such imports.

TRANSPORTATION OF OIL AND GAS

INTRODUCTION

Once oil and gas are brought to the surface, they must be conveyed to processing facilities and then to refineries, storage, or to ports. In the U.S.S.R., as in the West, this is accomplished by rail, road, water, and pipeline, although the latter has proved to be the most efficient and cost-effective mode of transport. To a large extent, Soviet plans for increased gas production and gas exports rest on the further extension of the gas pipeline system. The length and capacity of Soviet pipeline networks has expanded significantly, an achievement that has been accomplished with extensive imports of Western equipment and technology, particularly large diameter pipe. This section describes the way in which petroleum pipelines function, details Soviet progress in constructing and operating them, and discusses the role of the West in oil and gas transportation.

TRANSPORTATION OF OIL AND GAS BY PIPELINE

Pipelines are generally the most cost effective way of conveying large volumes of petroleum over long distances by land. Although they require a high initial capital investment, in the long run operating and maintenance costs are low, and the cost of pipeline transport drops rapidly with increases in the diameter of the pipe, and therefore the quantity of petroleum that can be transported. A difficulty with pipelines is their inflexibility. Once laid, it is impossible to change their routes, although provision

can be made for additions to pipeline capacity.

Pipelines carry oil from the well to field processing centers, and from there to refineries and onward. Sometimes these pipes are lined to protect them from corrosive materials in the petroleum; sometimes they may require insulation or the installation of heating facilities along their route to prevent oil from congealing.

Separate gas pipelines transport gas, which may be independently produced, found in association with crude oil, or produced during the refining process. In the past, associated gas produced in oilfields was often flared off or allowed to escape into the atmosphere. With higher gas prices, this natural gas is now transported to markets.

Important characteristics of the pipe from which oil and gaslines are constructed are

Photo credit: TASS from ©SOVFOTO

Soviet domestically produced steel pipe

the strength of the material, the technique of manufacture, and the diameter. One improvement in pipe technology has been to achieve strong pipe with thin walls, thus allowing a decrease in production costs. A second has been the introduction of seamless pipe, which avoids the weaknesses introduced by welded seams. Seamless pipe of wide diameter (40 inches and above) is difficult to manufacture, but advances in metallurgical technology have led to the ability to produce long lengths of wide diameter, thin-walled rolled pipe that can withstand high pressures.

Both oil and gas are moved along the pipeline with the aid of mechanical devices—pumps and compressor stations. Oil pipelines may be equipped with any of a variety of pumps—centrifugal, steam turbines, diesel engines, etc.—the appropriateness of which are determined by the volume to be transported, the viscosity of the oil, the pressure required, and the availability of fuel. Pumping stations situated along the route of the pipeline can be maintained manually by mechanical controls, but as distances increase, remote automation becomes more efficient.

Natural gas is pushed through the pipeline by the pressure obtained from compressing the gas. Gas from the pipeline itself is normally used to fuel the compressor engines. Valves are installed every 10 to 30 miles along the pipeline to make it possible to isolate sections for maintenance and repair and to close automatically in response to rapid large drops in pressure. Whereas in a level oil line of constant diameter, pressure will decrease uniformly with distance, in a gas line it decreases according to parabolic law. Pressures are highest at the outlet of a compressor station and drop between them. The efficiency of gas compressors depends on where along the line they are situated. The result is that gas pipeline capacity, unlike that of oil lines, is related to route length as well as pipe diameter. The capacity of a gas pipeline can be increased by adding compressor stations at close (50 to 100 mile) intervals along the route.

SOVIET OIL PIPELINES

West Siberian oil has constituted an increasing share of total Soviet production in the past decade. This has necessitated the construction of major oil pipelines to bring the crude hundreds of miles to refineries and consumers.[102] Nevertheless, rail has remained an important means of transporting oil products in the U.S.S.R. The usual Soviet practice is to use pipelines to carry crude oil from field to refinery and the railroad thereafter. (Truck transport, which is used extensively in the United States, is confined to the distribution of products over relatively short distances to consumers.) This heavy use of rail transport is expensive and inefficient, particularly for the usually short hauls. Product pipelines (i.e., pipelines designed to carry refined products from the refinery to local distribution points) would rapidly pay for themselves, but delays in pipeline construction have meant that the volume of oil and products carried by railroad has continued to grow. In 1979, 35 percent of all oil freight was transported in this manner[103] (see table 13).

Present Soviet policy appears to place priority on phasing out the use of rail for crude oil transport before extending product pipeline capacity. In any event, establishing an extensive products system will require intricate planning, as such a system must serve a variety of distribution points and cope with a number of different products.

[102]This is because the only railway running from the oil-producing region of West Siberia is working at full capacity. Wilson, op. cit., p. 32.

[103]Ibid., p. 25.

Table 13.—Transport of Oil and Oil Products
(million tons)

	1975	1976	1977	1978	1979
Volume of oil and products transported by:					
Pipeline	499	532	559	589	609a
Rail	389	394	406	412	NA
River	39	38	37	40	NA
Sea	91	101	104	109	NA
Total	1,018	1,065	1,107	1,150	NA

NOTE: Totals may not add due to rounding.
NA = not available.
aCited in *Izvestiya*, Jan. 26, 1980.
SOURCE: Wilson, op. cit., p. 25.

Despite the fact that 1980 plan targets for oil pipeline construction were not met,[104] the length and capacity of the Soviet oil pipeline system have grown extensively over the past 20 years (see table 14). By the end of 1979, there were 67,400 km (about 41,900 miles) of crude oil and product pipeline in the U.S.S.R.

The rate at which additions to oil pipeline capacity can be made depends importantly on the terrain to be covered and the diameter of the pipe being laid. Pipelines are expensive and slow to build in the difficult conditions of West Siberia, where high winds, sand erosion, and swamps inhibit construction. The larger the diameter of the pipe, the higher the capacity of the pipeline, and the U.S.S.R. has placed emphasis on increasing its use of wide diameter pipe. This growth, as well as the corresponding increase in pipeline capacity, is shown in table 15.

The location of major Soviet oil pipelines and their capacities is shown in figure 3. As would be expected, the evolution of this network has been dictated by the movement of the center of Soviet oil production from the Caucasus to the Volga-Urals, and then to West Siberia; and by the location of refineries and export markets. Lines from the Volga-Urals fields, for instance, followed three basic directions before the development of West Siberia:[105] eastward into Siberian refineries at Omsk and Angarsk; westward to Eastern Europe (this is the Friendship pipeline that carries crude oil to refineries in Poland, East Germany, Czechoslovakia, and Hungary; branch lines also serve domestic refineries); and northwestward to refineries in the central and northwest portions of the country.

The onrush of West Siberian oil required large-scale network expansion using increasingly large pipe diameters. West Siberian lines also now follow three basic directions: southeastward into eastern Siberia;[106] southward to Kazakhstan and Central Asia; and southwestard to the European U.S.S.R. In addition, a new pipeline corridor runs due west across the Urals. Construction on a 48 inch, 3,300 km (over 2,000 mile) segment of this line began in 1977. It would appear that this line is intended to handle expected increases in West Siberian production. The first stage (from Surgut to Perm) was completed in 1979 and, despite the difficulties of construction, the entire pipeline was finished

---

[104]Ibid., p. 26.

Table 14.—Length of Oil and Oil Product Pipelines
(thousand kilometers)

	1960	1965	1970	1975	1978	1979	1980 (plan)
Crude........	13	22	31	46	NA	NA	NA
Products.....	4	7	7	10	NA	NA	NA
Total.......	17	29	37	56	64	67	75

NOTE: Totals may not add due to rounding.
NA = not available.
SOURCE: Dienes and Shabad, op. cit., p. 62; Wilson, op. cit., p. 26.

---

[105]Ibid., p. 66.

[106]This line appears to have been planned as the initial link in a system carrying West Siberian oil to the Pacific coast for export to Japan. This plan has been dropped and such oil flows, if and when they exist, will be handled by the Baykal-Amur Mainline (BAM) railway. Dienes and Shabad, op. cit.

Table 15.—Diameters and Capacity of Oil Pipelines

Diameter (inches)	Annual capacity (million tons)	Length of pipelines (thousand kilometers)						
		1940	1950	1955	1960	1965	1970	1975
Below 20........	—	4.1	5.4	7.5	9.2	9.5	10.8	12.1
20	8	—	—	3.0	6.3	9.9	10.3	15.9
28	17	—	—	—	1.7	6.1	9.5	11.0
32	25	—	—	—	0.05	1.8	2.9	5.9
40	45	—	—	—	—	1.3	3.9	6.4
48	75	—	—	—	—	—	—	4.9
Total length of pipelines		4.1	5.4	10.5	17.3	28.6	37.4	56.2

SOURCE: Dienes and Shabad, op. cit., p. 63.

Figure 3.—Major Oil Pipelines

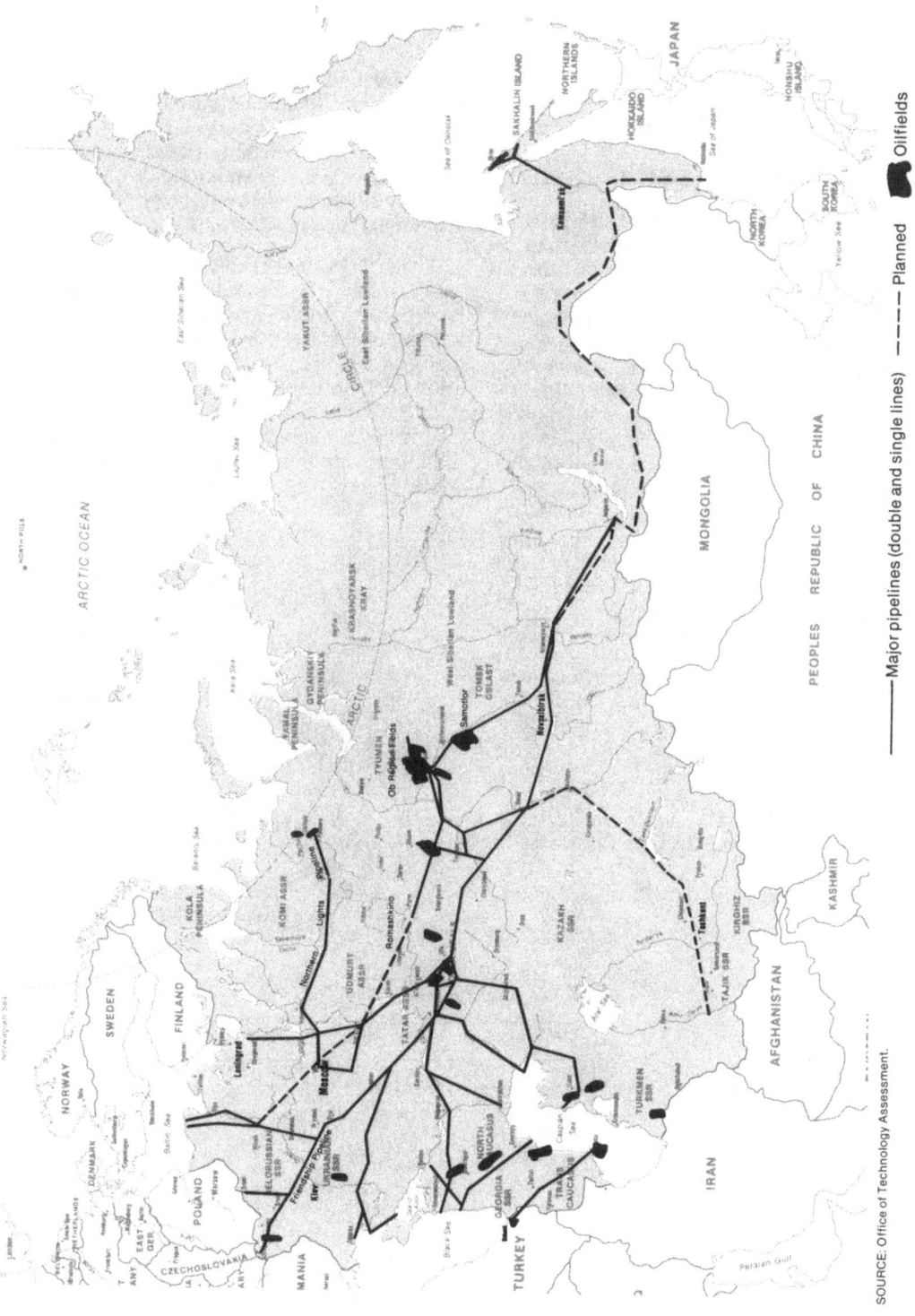

——— Major pipelines (double and single lines)   ----- Planned   ⬛ Oilfields

SOURCE: Office of Technology Assessment.

in 1981.[107] The completed line will both serve refineries and provide oil for export through the Baltic Sea terminal at Ventspils.

GAS PIPELINES

In 1980, the U.S.S.R. produced 435 bcm of gas and the Eleventh FYP has called for a 50-percent increase in gas production, mostly from West Siberia, in part to support greatly expanded gas exports. The success of these plans will largely rest on the capacities of the pipelines that are required to transport this gas. The existing gas pipeline network grew from about 99,000 km (61,000 miles) in 1975 to about 130,000 km (80,000 miles) in 1980, (see table 16) and has increasingly employed large diameter pipe. Nevertheless, the low capacities of gas pipelines present particular problems. A 48-inch oil pipeline, for instance, can carry 75 mmt of oil each year; a 56-inch gas pipeline can carry only 23 mmt of oil equivalent (23 bcm of gas). Present plans to raise West Siberian gas production, therefore, require enormous increases in the length of the pipeline network. The Eleventh FYP calls for about 40,000 km of new pipeline, an increase of 30 percent, including 25,000 km of 56-inch pipe carrying gas from the Urengoy field.[108]

The Soviet gas network is shown in figure 4. While important lines in this system serve

[107]Theodore Shabad, "News Notes," *Soviet Geography,* April 1981, p. 275.
[108]Ibid.

the fields of Central Asia and the giant Orenburg deposit in the Volga-Urals (from which gas is carried to Eastern Europe), the most important part of the network connects the enormous West Siberian fields to consumers. These fields will be supplying most of the increment to production in the next decade.

Three Siberian trunk systems presently carry most of this burden. They are the Northern Lights system, one branch of which runs to the Czech border from whence gas is exported; the Tyumen-Moscow line; and the Tyumen-South Urals system, which may be able to feed Siberian gas to the Caucasus to make up for the cessation of Iranian gas imports. (A pipeline project that would have supplied Iranian gas to the U.S.S.R. has now been abandoned.) Branch lines from these major systems serve a number of major towns along their route. All of these lines employ 56-inch pipe. An important projected pipeline is the one which will bring additional gas to Western Europe. This is discussed in detail in chapter 12.

Given the fact that pipe with diameters exceeding 56 inches has not been mass-produced anywhere in the world, there are three ways in which the U.S.S.R. might improve its pipeline capacity: cooling the gas to increase its density, raising the pressure of pipelines from the present 75 atmospheres to 100 or 125 atmospheres, and reducing the distance between (and therefore increasing the number of) compressor stations. While research into the technology for the first two

Table 16.—Length and Diameters of Gas Trunk Pipelines

Diameter (inches)	Optimal annual capacity (bcm)	Length of pipelines (thousand of kilometers at year-end)											
		1940	1950	1955	1960	1965	1970	1975	1976	1977	1978	1979	1980 (plan)
< 28		0.33	2.3	4.9	11.8	20.2	29.9	36.2	—	—	—	—	—
28	4.0	—	—	—	6.2	10.6	12.9	16.1	—	—	—	—	—
32	6.2	—	—	—	2.3	4.0	5.0	7.0	—	—	—	—	—
40	8.7	—	—	—	0.7	7.5	15.9	20.6	—	—	—	—	—
48	13.3	—	—	—	—	—	3.8	15.7	—	—	—	—	—
56	20.2	—	—	—	—	—	—	3.6	—	23	—	—	—
Total length of pipelines		0.33	2.3	4.9	21.0	42.3	67.5	99.2	103.5	111.7	117.6	124.4	134.2

SOURCE: Dienes and Shabad, op. cit., p. 83; Wilson op. cit., p. 26; Leslie Dienes, private communication.

Figure 4.—Major Gas Pipelines

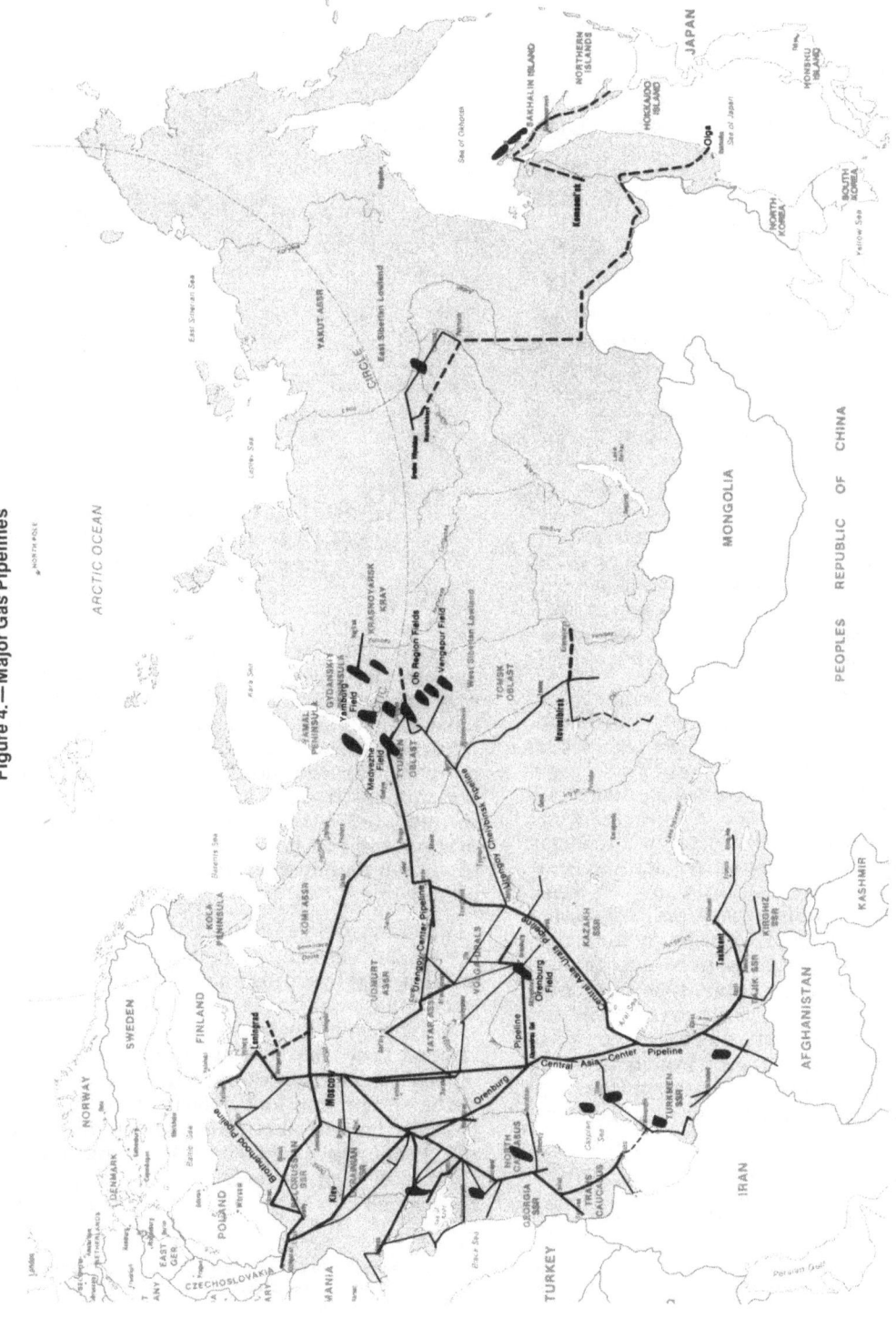

SOURCE: Office of Technology Assessment.

Major lines (dual or more pipelines) ——— Branch Lines (single pipelines) ——— Planned lines - - - - Gasfields ●

of these options is ongoing, and cooling technology in gas transportation appears particularly promising, their widespread application does not appear to be imminent. Increasing the number of compressor stations will therefore play the most crucial role in increasing gas pipeline capacity.

Unfortunately, the design and quality of Soviet compressors are poor and construction of compressor stations has chronically lagged behind plan. Moreover, there is apparently an ongoing debate within the U.S.S.R. as to whether or not the distance between compressor stations should indeed be shortened. The argument against such practice is that construction of these stations is highly labor intensive. Reducing their number will shorten construction periods and accelerate the delivery of gas.[109] Whether or not this view will prevail remains to be seen. Should a decision to increase the number of stations be reached, however, it is likely that this is another area in which Western equipment might play an important role.

Other factors inhibiting the construction of West Siberian gas pipelines are labor shortages, the inadequacy of electricity supplies, and shortages of excavating and pipelaying equipment. Delays in the construction of permanent settlements for the large number of workers required to lay gas pipelines have contributed to constant labor turnover. Meanwhile, the North Tyumen region of West Siberia, location of the most important giant gasfields, does not have a permanent electricity supply. Not only does this mean that each compressor station must have its own mobile power unit and the personnel to maintain it, but frequent power failures cause expensive interruptions in compressor operation and can damage compressor units. Finally, pipeline construction is seriously affected by the failure of enterprises producing engineering, excavating, and construction equipment to fulfill their obligations.

It has been asserted that despite these difficulties, Soviet pipelaying work is "fast and efficient."[110] It is difficult to evaluate this statement. Certainly Siberian conditions impose constraints that should affect the success criteria by which any such enormous enterprise is measured. However, the ultimate test will be the extent to which the U.S.S.R. is able to meet its own goals—with or without massive purchases from the West.

THE ROLE OF WESTERN PIPELINE EQUIPMENT AND TECHNOLOGY

Pipe[111]

The Soviets have been heavily dependent on imports of Western pipe of 40 inch and greater diameter, most of which seems to be used in main-line high-pressure transport of natural gas, and on Arctic quality pipe. Comparisons of Soviet domestic pipe production and import figures suggest that, at least through 1975, this dependence exceeded 50 percent and was growing. In 1979, the value of steel pipe imported by the U.S.S.R. rose by 29 percent over 1978.[112]

The Soviet steel industry is capable of producing, 40-, 48-, and 56-inch pipe, and indeed a substantial part of the domestic gas distribution system uses domestic pipe, albeit mostly of small diameter. It would therefore appear that the massive imports are designed to avoid bottlenecks arising from insufficient production capacity and to compensate for the fact that Soviet domestic pipe is of lower quality in yield strength, wall thickness, and general workmanship that that which can be purchased abroad. Soviet

[109]See Robert W. Campbell, "Soviet Technology Imports: The Gas Pipeline Case," The California Seminar on International Security and Foreign Policy, February 1981, pp. 14-20; I. Ia. Furman, "Determining the Comparative Effectiveness of Variants in Designing Gas Transmission Lines," *Gazovaya promyshlennost, Seriya Ekonomika, organizatsiya i upravlenie v. gazovoy promyshlennosty*, No. 4, 1981, pp. 20-21.

[110]Wilson, op. cit., pp. 29-31.
[111]This section is based on Campbell, "Soviet Technology Imports . . .," op. cit., pp. 10-12.
[112]Wilson, op. cit., p. 27.

welding practices apparently add to the problem of quality. The U.S.S.R. is attempting to upgrade its pipe manufacturing capabilities. To this end, it has purchased a seamless pipe manufacturing plant from French and German firms. The plant has an annual capacity of 170,000 metric tons (in 1976, Soviet domestic output of 40 inch and larger pipe was 2.6 million tons[113]). In addition, a new pipe-rolling plant destined to produce 56-inch pipe for use at 100 atmospheres is using imported steel plate.

Pumps and Compressors

The U.S.S.R. has purchased pipeline boosters, pumping stations, and gas compressors from the West, but the area in which Western technology appears to have been most important is in compressors for gas pipelines. These do not seem to be produced in adequate quantity and quality in the U.S.S.R. itself. In 1976, the average size of a Soviet turbine-powered centrifugal compressor unit (produced under license from the U.S. Dresser Industries and making up 71 percent of installed gas compression capacity) was slightly over 4 MW. The U.S.S.R. now widely produces units of 5-, 6-, and 10-MW capacity, but ones of 16 and 25 MW, mass-produced routinely in the West, were only scheduled to begin serial production in 1981.[114] The large units needed to increase installed capacity must, therefore, still come almost entirely from the West.

The U.S.S.R. has been importing substantial amounts of gas-turbine-powered compressor equpment since 1973; by 1976, 3,000 MW of such units (about one-third of installed capacity of all types of compressor equipment) had been imported from Austria, Great Britain, Japan, Norway, and the United States.[115] While it is obviously impossible to quantify the net benefits generated by these purchases, one Western expert has estimated that pipe and compressor imports together may well have paid for themselves within 2 years, given the acceleration of gas transport capacity they allowed.[116]

Other Equipment

The U.S.S.R. has purchased American pipeline inspection equipment, but in general has not relied extensively on the West for other material for pipeline cleaning. More important are imports of pipelaying and excavation equipment, particularly that designed for cold climates. Pipelayers have been purchased both from the West and from Eastern Europe, but there are signs that the U.S.S.R. may wish to accelerate such imports for construction of the proposed new pipeline to Western Europe (see ch. 12).

Computers

The U.S.S.R. has the ability to build reasonably advanced computer-based control systems for oil and gas pipelines, but the majority of pipelines probably still employ older, less efficient and more labor-intensive technologies. For instance, even the best Soviet systems do not use microprocessors and minicomputers as they are used in Western state-of-the-art systems. These are connected in a multilevel hierarchy with a central mainframe, and produce greater reliability, quicker response times, and lower labor requirements.

The U.S.S.R. has purchased in excess of $10 million of pipeline-related computer equipment from the West, mostly for pipelines such as Orenburg, for which large amounts of other Western equipment has also been purchased. Announced plans call for the introduction of at least 10 domestically produced oil pipeline automation systems. Such systems could reduce operational mistakes and allow better pipeline maintenance, speedier leak detection, and better overall management of oil and gas flows. If these systems are indeed about to be intro-

---

[113]Campbell, "Soviet Technology Imports . . . ," op. cit., p. 12.
[114]Wilson, op. cit., p. 29.
[115]Campbell, "Soviet Technology Imports . . . ," op. cit., pp 15, 21.

---

[116]Ibid., pp. 22-24.

duced, and if delays in the provision of domestic equipment for them occur, the U.S.S.R. may turn to the West for assistance. However, it is not clear if this is a priority area for the acquisition of Western computers or the expenditure of hard currency.

SUMMARY AND CONCLUSIONS

The Soviet oil and gas pipeline system has grown extensively in the past decade. But if the large increases in West Siberian natural gas production so important to Soviet energy plans as a whole are to be realized, further expansion of the gas pipeline network is crucial.

Oil and gas pipelines have benefited extensively from Western pipe and compression equipment. Although the U.S.S.R. does manufacture some large diameter pipe and is seeking to upgrade its capabilities in this area, no end to its dependence on Western imports of such pipe is in sight. Similarly, although there are Soviet-made compressors, including some new models with capacities that were hitherto available only in the West, Soviet purchases of such equipment from the West still appear to be cost effective. Indeed, for some time to come, such purchases will probably also be necessary as domestic manufactures remain inadequate in both quality and quantity.

REFINING

INTRODUCTION

Crude oil is not a homogenous substance. Depending on the nature of the deposit from which it comes, it can consist of a variety of compounds and exhibit a wide range of properties. These properties determine a number of both liquid and gas products which can be produced from the crude in primary distillation. The refineries at which these products are made can employ both primary and secondary processing techniques so that ideally each refinery can produce the optimum product mix for the crude oil which comes to it. Common products are jet, diesel, and residual fuel oil; gasoline; kerosine or paraffin; lubricating oils; and bitumen. During the oil refining process, considerable volumes of gas may also be released. These gases—methane, ethane, propane, and butane—can be used as fuel for the refining process or marketed separately.

Natural gas too is processed both to obtain marketable products and to purify it. Some gases have a high content of natural gas liquids, which can be separated out. In addition many natural gases contain hydrogen sulfide, sometimes in amounts ranging to more than 75 percent. "Sour" gases, i.e.,

those with high hydrogen sulfide content, are toxic and corrosive, and must be treated before they are used.

Relatively little is known about the Soviet refining industry. The U.S.S.R. issues no statistics on oil refining, and information on throughput—in terms of both product quality, variety, and quantity—must be gathered from indirect sources and based on estimates. This section will briefly summarize major characteristics of this industry.

THE SOVIET OIL REFINING INDUSTRY

Major difficulties in the Soviet refining industry have resulted from the geographic distribution and capacity of existing refineries, and from the quality and mix of the refined products produced. There are indications that the U.S.S.R. recognizes and is attempting to remedy these problems, but that much improvement in the refining sector is still necessary.

Figure 5 shows the location of major Soviet refineries. There are some 44 operating refineries in the U.S.S.R.[117] These are

---
[117]Wilson, op. cit., p. 37.

Figure 5.—Soviet Oil Refinery Sites

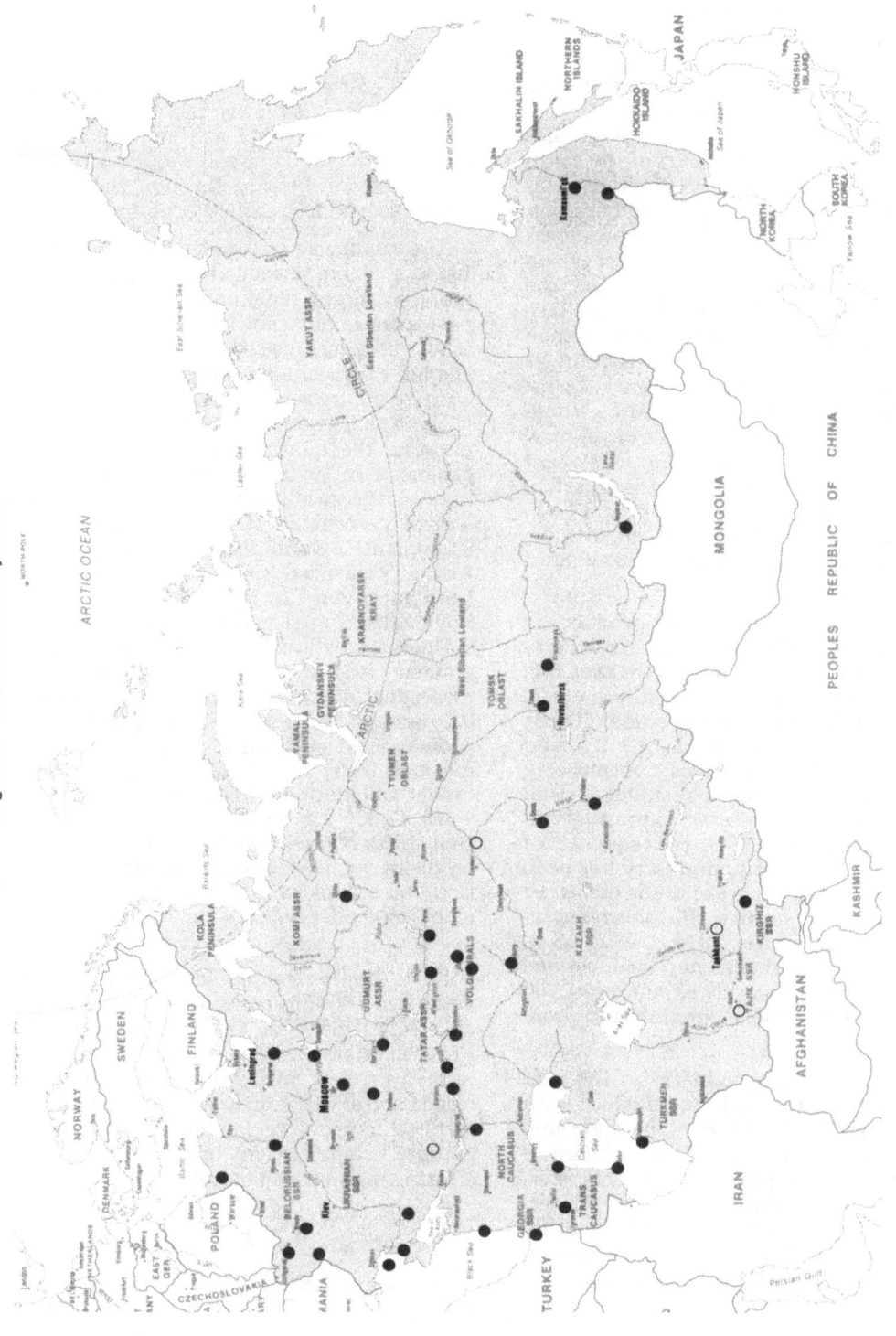

● Operating   ○ Under construction

SOURCE: Office of Technology Assessment.

well distributed, with some bias toward the older producing regions—Central Russia, Volga-Urals, North Caucasus, and Azerbaidzhan regions in the western part of the country. One very large refinery is situated in West Siberia. The geographic distribution puts a heavy strain on the overburdened Soviet rail system which, as the previous section indicated, is relied upon for the transport of refined oil products. This fact may have contributed to an apparent change in refinery building policy. While the Soviet predilection has been for expanding refinery capacity by adding to existing sites, there are now plans for the construction of new refineries. Four of these came onstream during the Tenth FYP period in the Ukraine, Belorussia, northeast Kazakhstan, and Lithuania,[118] but plans for others in Central Asia, Eastern Siberia, and southern Kazakhstan have experienced delays.

The quality of Soviet refined products is notoriously poor, the result of inadequate planning and investment in a sector that was ill-prepared to deal with the rapidly increasing volume and variety of the crude oil that began to be produced in the 1950's and 1960's. A large share of refinery output continues to be in the form of residual fuel oil (called "mazut") that is used extensively in electricity generation. One consequence of inadequacy in the refining industry has been an emphasis on the export of crude oil rather than products. Even with the enormous increase in oil prices after 1974, the U.S.S.R. could earn more hard currency from refined product exports. However, export potential is constrained by product quality and product mix.

Nor has the refining industry in the past coped in an altogether optimum fashion with domestic needs. A chronic and pressing problem, for instance, is the large share of heavy fuel oil in the product mix. This is accounted for by the lack of appropriate secondary refining capacity to produce other products. One result is that fuel oil is burned in cases where natural gas would be a more rational and economic fuel. The task facing the U.S.S.R. is to lower the overall share of fuel oil in the product mix and raise that of other products. This will require rationalizing selected refineries through the installation of secondary refining equipment.

The quality as well as the mix of products must also be raised. One example can be found in motor fuels and lubricants. In the past, poor quality automotive products have tended to decrease the efficiency and life span of the machinery and to increase the requirements for repairs and maintenance. The U.S.S.R. has thus made a concerted effort to improve the quality and quantity of such products. In 1970, for instance, the share of high-octane gasoline in total gasoline output was 50 percent; in 1979, this share reached 94 percent.[119] Similarly, the quality of diesel fuel, as measured by its sulfur content, has been improving. In 1965, only 40 percent of Soviet diesel fuel had a sulphur content of 0.5 percent or less. The rate is now over 95 percent, and about 47 percent of Soviet diesel fuel has a sulfur content of less than 0.2 percent. Some sense of the practical consequences of such an improvement may be gleaned from the fact that, according to Soviet calculations, an engine that will run some 57,000 km on diesel fuel with a 1-percent sulfur content will last nearly 89,000 km on diesel with 0.2 percent sulfur.[120] In addition, reductions have been made in the losses of oil and oil products during the refining process.

Such improvements have required advances in refining technology and considerable investment in the refining industry. Further alteration and improvement of the refined product mix will require additional large capital expenditures and additional efforts to improve technology. The past and potential contributions of the West in these efforts are discussed below.

---

[118]Ibid.

[119]Campbell, *Trends . . .*, op. cit., p. 47; Wilson, op. cit., p. 39.
[120]Ibid.

THE SOVIET GAS-PROCESSING INDUSTRY

The associated gas produced with oil, "casinghead gas," can be used in unprocessed form to fuel power stations or it can be processed to separate the liquid petroleum gases. In the past, the U.S.S.R. typically vented or flared this gas. Now, however, efforts are being made to collect and process it. Between 1975 and 1979, production of casinghead gas rose from 29 to 36 bcm, an overall utilization rate of some 69 percent. The Tenth FYP specifically addressed the problem of utilizing the large amounts of casinghead gas being produced (and flared) in West Siberia by planning construction of gas-processing facilities in that region. This work has fallen behind schedule, but the present FYP calls for additional refineries and envisages that by 1985 West Siberian casinghead gas will be fully utilized.

WESTERN TECHNOLOGY IN THE SOVIET REFINING INDUSTRY

The U.S.S.R. has purchased large amounts of oil refining equipment and technology from East Germany and Czechoslovakia, as well as from Japan, France, Italy, and Britain. Western purchases have tended to consist of entire refineries rather than of component parts, and the primary U.S. contribution has been in provision of design and engineering services to Italian and Japanese construction firms (see ch. 6).

To implement planned improvements in refining, the U.S.S.R. will probably require additional Western assistance, although with the exception of computing that might boost efficiency, the technologies involved are not advanced. These include secondary refining techniques such as hydrocracking and catalytic cracking. In the West, microprocessors and minicomputers are used extensively in refinery operations. Although the Soviets use computers in all of their refineries, microprocessors are found only in the largest and most important, and are, con-

Photo credit: TASS from ©SOVFOTO

French technicians at a French hydrocracking plant in the Bashkir ASSR

servatively speaking, several years behind those used in the West. The need for microprocessor- and minicomputer-based refinery control systems will continue to grow in the U.S.S.R., particularly as the product mix is restructured. So far, the purchase of such systems does not seem to have been accorded high priority, although some minicomputer- and microprocessor-based systems have been included in sales of larger units destined for refineries. The U.S.S.R. has the hardware base to design and implement such systems itself, but it lacks experience in building the software and associated control devices.

It is likely that systemic constraints and incentives will impede the development and introduction of computerized refinery process control systems much more than will problems with the technology. Western purchases may accelerate improvements in efficiency, but from the present signals, it appears that the U.S.S.R. will continue to rely predominantly on domestic developments.

SUMMARY AND CONCLUSIONS

Discussion of the Soviet oil refining industry is inhibited by lack of data, but a few generalizations are possible. Soviet priorities in this sector lie in building more new refineries in West Siberia and the East, near sources of supply and markets. In addition, emphasis is now being placed on improving both the refinery product mix and quality. If past patterns persist, these changes in the refining industry will take place with the aid of infusions of Western technology and equipment, particularly of complete refineries. On the other hand, if oil production does not continue to increase, expansion in refining capacity will not be necessary.

OFFSHORE

INTRODUCTION

The extensive development of offshore oil and gas deposits, i.e., deposits in lakes, seas, and oceans, is a relatively new phenomenon in both the U.S.S.R. and the West. Although it is generally believed that the U.S.S.R. has a promising offshore potential, production from offshore deposits has not yet made a very noticeable impact on overall output.

Most of the equipment and technology employed in the Soviet offshore sector has been either directly purchased from the West or reproduced from Western designs. As new offshore deposits are identified and developed, the need for additional sophisticated offshore equipment will grow. The degree of priority to be accorded offshore development in the Eleventh and Twelfth FYP's is not clear.

This section briefly reviews the techniques and equipment necessary to find, produce, and transport offshore petroleum, surveys the state of Soviet practice in these areas, and describes the past and potential contribution of Western technology to offshore development.

OFFSHORE EXPLORATION AND PRODUCTION

Offshore and onshore exploration of oil involve essentialy the same processes. An energy source, usually compressed air, is used to generate an impulse capable of penetrating the Earth's crust. As the energy is reflected and refracted by the underlying geological structures, it returns to the surface in the form of echoes that are detected by sensors (hydrophones). Arrays of hydrophones towed behind a seismic survey ship are used to detect the returning echoes. The information is then processed in roughly the same manner as onshore data are processed.

Modern seismic survey ships employ sophisticated computer systems to control the precise timing of the bursts of compressed air and to preprocess the data collected from the hydrophones. Computers, linked with satellite navigation systems, provide a precise fix on the position of the ship. The positioning of the hydrophone arrays is also controlled by the computer.

Offshore exploratory drilling is accomplished using basically the same equipment as onshore, the major difference being the platform on which the drilling equipment is placed. Three major types of movable platforms are presently available—jackups, semisubmersibles and drilling ships.

Jackup rigs generally employ three or four hydraulically operated "legs." The rig is towed to the drilling site where the legs are extended downward to the ocean floor. The entire rig is then raised clear of the water line to permit drilling operations. Modern jackups are currently limited to working in depths no greater than 300 ft.

The semisubmersible rig is a refinement of the jackup. Semisubmersibles are constructed on two or more pontoons on which the rig floats, and either self-powered or towed to the drilling site. Once over the site, the pontoons are partially flooded to provide a stable platform from which to drill. These rigs are either anchored into position or employ a dynamic positioning system. They operate in water depths up to 1,500 ft and can drill to 25,000 ft.

Drill ships represent the state-of-the-art in offshore drilling. The ship is usually of standard design with a drilling rig mounted in the middle of the deck. When the ship is over the drill site, it is either moored with anchors or a dynamic positioning system is employed. A dynamic positioning system uses a series of computer-controlled thrusters in conjunction with a set of sonar beacons placed on the ocean floor to maintain the ship's position over the drill site. A modern dynamically positioned drill ship is capable of exploratory drilling in water depths up to 6,000 ft.

After the exploratory drilling phase is complete, the same well logging process as in onshore wells is used to determine the size of the reservoirs and the possible flow rates. If production is warranted, the reservoir can be developed from an artificial island or a fixed platform production rig, the latter being most common. Once the well is dug and the casing cemented, special subsea wellhead completion equipment is fastened to the casing. This unit is remotely controlled and establishes the rate of production from the well.

Fixed platform rigs differ from exploration rigs in their degree of mobility. These rigs are erected on platforms that have been firmly embedded in the ocean floor. Once the platform is in place, up to 65 development wells may be drilled from the platform using offset directional drilling techniques. Where ice floes make the use of a fixed platform im-

practical, however, artificial islands may be employed. Artificial islands and fixed platforms are limited to use in water depths of no more than 1,500 ft. Production drilling to greater depths will require new designs such as tension leg platforms or compliant guyed towers. Finally, the petroleum is brought to shore either via pipeline or tanker.

THE SOVIET OFFSHORE INDUSTRY

The offshore regions of the U.S.S.R. offer enormous potential for oil and gas production. It has been estimated that of the 8 million km^2 of total Soviet shelf area, 2.5 million km^2 are promising for the discovery of petroleum. The most attractive regions are the Arctic, Baltic, Black, Caspian, and Okhotsk Seas.[121] Some drilling has taken place in the Sea of Okhotsk off Sakhalin Island (using Japanese equipment; see ch. 11), and in the Black and Baltic Seas, and limited exploration of Arctic waters in the Barents and Kara Seas has begun, but Soviet offshore experience so far has been largely confined to production from shallow waters in the inland Caspian Sea.

Development of the Caspian dates to the 1920's, when earthen causeways were built into the Sea. These were later replaced by fixed pile-supported and trestled platforms, and drilling proceeded from these and from small natural islands. The full potential of the Caspian is only now beginning to be realized, however, as the U.S.S.R. develops the capacity to explore and drill in waters greater than 200 m (600 ft). A major discovery in 1979, the "28th April" deposit, has spurred interest in continuing Caspian exploration and development.[122]

Exploitation of the Caspian, and even more importantly, of the other promising offshore regions is constrained by the status of Soviet deep-water technology. Although the

[121]Campbell, *Trends . . .*, op. cit., pp. 23-44.
[122]Wilson, op. cit., pp. 49-50.

Tenth FYP called for the modernization and augmentation of offshore drilling capacity, this upgrading has not proceeded as rapidly as planned. The Soviet stock of domestic offshore equipment currently consists of seven jackups (three of which are obsolete) and two semisubmersible rigs.[123] The inadequacy of this equipment base is attested to by the extent of Soviet dependence on Western offshore equipment.

THE CONTRIBUTION OF WESTERN OFFSHORE TECHNOLOGY

The U.S.S.R. has purchased or contracted for offshore exploration, drilling, and production equipment from a wide variety of countries, including the United States, France, Holland, Finland, and Japan (see ch. 6). Its own recently acquired ability to build jackups and semisubmersibles is the product of Western technology imports, and it has contracted with a Finnish firm for three dynamically positioned drill ships for use in Arctic waters. These ships are based on a Dutch design, and are being fitted with the latest Western drilling and subsea completion equipment. It is believed that these ships will provide the Soviets with their first

deep drilling and subsea completion capabilities.[124]

In addition to purchases of equipment and technology, the U.S.S.R. has entered into joint offshore development projects with other nations. These include the Sakhalin project with Japan, and the Petrobaltic consortium, which at present involves cooperation between the U.S.S.R., Poland, and East Germany in exploring in the Baltic Sea. The consortium is now using a jackup rig built by a Dutch firm and furnished with drilling equipment of U.S. origin.

SUMMARY AND CONCLUSIONS

The U.S.S.R. obviously wishes to expand its offshore activities and capitalize on its great potential. However, its own offshore capabilities are still in their infancy, and purchases of Western equipment and technology will probably continue to be crucial to offshore development in the foreseeable future. Given the fact that exploration in most offshore regions has not even begun, it is difficult to imagine significant offshore production occurring before the end of the present decade.

---

[123]Ibid., p. 96.

[124]*Directory of Marine Drilling Rigs*, 1980–81.

THE PROSPECTS FOR SOVIET OIL PRODUCTION IN 1985

The prospects for Soviet oil production in the next 5 years have been the subject of controversy ever since CIA's 1977 prediction that Soviet oil output would peak at 550 to 600 mmt (11 to 12 mbd) and then drop sharply to 500 mmt (10 mbd) or less by 1985. The CIA has since revised its estimates, but the fact remains that its original work has largely set the terms for the entire Soviet energy debate, with experts ranged on different sides of what have become *the* central Soviet energy questions: will oil output peak in the 1980's, and if so, when; can a production plateau be maintained, and if so, how

long; if oil output begins to drop, how sharp a decline can be expected?

Leaving aside for the present the issue of whether oil production does indeed constitute the key to the future of the Soviet energy balance (this point is treated below), this section will discuss the prospects for Soviet oil production in 1985 and 1990, identifying for the purposes of this analysis reasonable best and worst case estimates.

Table 17 summarizes recent projections of Soviet oil production. As the table shows, the lower limit is represented by the most re-

Table 17.—Soviet Oil Production Forecasts, 1985
(million metric tons)

Million tons	Date of forecast
1. 500-550	April 1981
2. 560-610	June 1981
3. 600	1979
4. 605-655	1979
5. 612-713	1978
6. 620-645	1980
7. 620-645	August 1981
8. 650-670	1979
9. 700	1980

SOURCES:
1. CIA, as reported in Joseph A. Licari, "Linkages Between Soviet Energy and Growth Prospects for the 1980's," paper presented at the 1981 NATO Economics Directorate Colloquium, Apr. 8-10, 1981. These numbers replace the 1977 estimates of 400 to 500 mmt.
2. OECD, Committee for Energy Policy, "Energy Prospects of the U.S.S.R. and Eastern Europe," June 26, 1981.
3. Robert Ebel, "Energy Demand in the Soviet Bloc and the PRC," June 1979.
4. Leslie Dienes and Theodore Shabad, *The Soviet Energy System* (Washington, D.C.: V. H. Winston, 1979), table 53, p. 252.
5. Herbert L. Sawyer, "The Soviet Energy Sector: Problems and Prospects," Harvard University, January 1978, quoted in George W. Hoffman, "Energy Projections—Oil, Natural Gas and Coal in the U.S.S.R. and Eastern Europe, *Energy Policy*, pp. 232-241.
6. Soviet Eleventh FYP target.
7. U.S. Defense Intelligence Agency, "Allocation of Resources in the Soviet Union and China—1981." Statement of Maj. Gen. Richard X. Larkin before the Joint Economic Committee, Subcommittee on International Trade, Finance, and Security Economics, Sept. 3 1981.
8. Jeremy Russell, Shell Oil.
9. David Wilson, *Soviet Oil and Gas to 1980*, Economist Intelligence Unit Special Report No. 90. This report was published just after the Soviet plan target was released. In a foreword, the author reasserts his belief that oil production of 700 mmt is achievable and attributes the lower Soviet plan to an apparent decision to divert resources from oil to gas production. EIU's 1990 target of 750 mmt, which is discussed below, is not disclaimed.

cent forecast of the CIA and the upper by a publication from the British Economist Intelligence Unit (EIU). The U.S.S.R.'s own Eleventh FYP figures fall between. Using this universe, OTA has posited best and worst case scenarios for Soviet oil production in 1985 of 645 and 550 mmt (12.9 and 11 mbd) respectively. **These are not predictions.** Rather, OTA has selected plausible high and low output figures that can be used to illuminate the energy problems and opportunities facing the U.S.S.R. The basis for these scenarios can best be understood through an examination of the assumptions and reasoning behind the differing forecasts shown here. The following sections, therefore, compare the arguments bolstering the table's high and low estimates—700 mmt (14 mbd) and 500 to 550 mmt (10 to 11 mbd).

Differing estimates of future Soviet oil production have been based on two separate but related sets of arguments. One concerns Soviet oil reserves; the other the state of Soviet oil production practice and technology. Seemingly irreconcilable differences between diverging forecasts can be traced to different assumptions and expectations with respect to each of these.

SOVIET OIL RESERVES

Introduction

Oil production over any given period of time obviously depends in part upon the quality, quantity, and accessibility of the resources in the ground. Reserves are the portion of this resource base that has been identified. They are important in determining both cumulative production and annual rates of output. Unless the rate of additions to reserves keeps pace with or exceeds the rate of production, output cannot remain stable indefinitely or rise over long periods.

Discussion of oil reserves anywhere in the world can be confusing, first because the standards of classification and definition employed by analysts are not always identical; and second, because the concept of a reserve is meaningful only within the context of some standard of economic feasibility. This means that as economic conditions or available technology change, amounts of oil ascribed to different reserve categories will change. This complex situation is compounded in discussions of the U.S.S.R. both because the Soviet system of reserve classification and nomenclature is very different from that employed in the West, and because Soviet oil reserve information is an official state secret. Western analysts must, therefore, calculate their estimates of Soviet reserves from intermittent and sometimes inconsistent bits of information. It is hardly surprising that analysts working from different data bases should arrive at differing conclusions.

In sum, two major points must be stressed. First, estimates of reserves are not static. As additional research, exploration, and development drilling proceed, reserves in any category may be redesignated to higher or lower classifications. Second, given the variety and subtlety of reserve classification systems, extreme care must be taken in interpreting differing reserve estimates. It is important to ascertain the definitions and other assumptions upon which such estimates are based.

The U.S. and Soviet Systems of Classification

Briefly, two different nomenclatures are employed in the United States. According to the American Petroleum Institute, reserves may be either measured, inferred, or indicated. These are referred to as proved, probable, or possible by the U.S. Geologic Survey. In both cases, categorization is by two sets of criteria, the degree of geological assurance that the oil exists, and the economic feasibility of producing it.

Soviet categories—A, B, C_1, C_2, D_1, and D_2— are also broken out according to the degree of exploration and appraisal drilling that has been carried out and some inexplicit criteria of economic recoverability. Only in the most general sense can Soviet reserve categories be made to correspond to those used in the West. Indeed, Western analysts have disagreed over the relations between the two classification systems.

Low and High Soviet Reserve Estimates

The CIA and EIU figures for Soviet oil reserves for the most part form the lower and upper limits of the estimates that have appeared in the West. While the CIA has estimated approximately 4.1 bmt (30 billion bbl) of what it calls "proven" reserves, the EIU study gives a range of 14 to 15 bmt (102 to 110 billion bbl) of what *it* calls "proven and probable" reserves. Only one reserve figure has slightly exceeded EIU's—16 bmt (120 billion bbl), published by the Swedish group Petrostudies.[125] On closer examina-

tion, these figures appear to be based on different nomenclatures, different standards of comparability between Western and Soviet definitions, and different evaluations of the abilities of the Soviet oil industry.

The CIA produced a figure for proved reserves that it believes represents a realistic estimate of economically recoverable oil given present technology. It must be noted that this definition yields a conservative estimate. Nor is CIA sanguine about Soviet prospects for additionals to reserves. To replace the oil produced between 1976 and 1980, the U.S.S.R. would have had to have found 2.9 bmt (21 billion bbl). According to CIA, this amount exceeded gross discoveries during 1971–75 by roughly 50 percent.[126] Reversing the decline in discovery rates would require increasing exploratory drilling to what CIA believes to be an unlikely extent. The best hope for reserve additions in the 1980's in this analysis, therefore, becomes luck—new finds of giant or supergiant fields near enough to existing infrastructure to allow their quick development.

At the opposite end of the spectrum, EIU does not believe that the present Soviet oil reserve situation is a constraint on near- and medium-term production. The EIU study adopts the same vocabulary as the CIA—"proved," and "probable"—but it offers no explanation of the meanings assigned to these words or how they are correlated to Soviet categories.

Conclusions

The basis on which these high and low reserve estimates rest is in large part that of subjective judgments about the level of technology, the production costs, and the amount of time necessary to exploit deposits of oil in different stages of development and exploration. CIA, given its own evaluation of Soviet petroleum technology, has applied very strict criteria to its estimate. EIU appears to be far more sanguine about the ability of the Soviets to recover more oil from their existing fields and to develop new fields

---

[125]Petrostudies has received widespread publicity for forecasts of Soviet oil reserve and production potential far in excess of those found elsewhere. Most Soviet energy experts regard Petrostudies' claims as so sweeping as to be highly unreliable.

[126]CIA, "Prospects . . . ," p. 23.

within the time frame of current planning periods. This difference in outlook and interpretation is also reflected in the two assessments of other aspects of Soviet oil industry practice.

SOVIET OIL PRODUCTION CAPABILITIES

In addition to its low estimation of Soviet ability to make sufficient additions to reserves to support increased production in the absence of new giant discoveries, CIA's analysis of the prospects for Soviet oil output rely heavily on its evaluation of Soviet production practices and its expectations that the U.S.S.R. would be unable to significantly improve these by 1985 or 1990. The EIU study on the other hand is optimistic about Soviet ability to meet plans to improve its oil industry performance and technology. Among the most important areas of disagreement between the two analyses are waterflooding and other enhanced recovery techniques, drilling, and equipment manufacture.

CIA appears to have departed from some aspects of its original 1977 treatment of waterflooding which, as noted above, presented an overly simplistic—and therefore misleading—picture of the magnitude, trends, and consequences of this practice. But the overall judgment would seem to remain: The U.S.S.R. has engaged in a basically short-sighted policy of overexploiting its largest deposits through a method that can lead to sharp production declines once a field has been exhausted. Should this occur at Samotlor, the consequences for the entire Soviet oil industry would be immense—particularly since new finds do not appear to have been keeping pace with the rate of production. Moreover, emphasis on maximum production has led to concentration on development drilling over exploratory drilling. CIA doubts the Soviet's ability to achieve the increases in both of these activities necessary to improve the discovery rate and to continue to increase production. This judgment rests in turn on doubts over

the availability of Soviet drilling crews and the quality and quantity of Soviet drilling equipment—including bits, pipe, and rigs—which can be produced during the present plan period. Nor does CIA believe that Western equipment and technology can be imported in sufficient quantities and utilized efficiently enough to make up for these lacks.

The EIU analysis takes precisely the opposite tack. Although it notes past cases of underfulfillment of oil industry targets, it emphasizes those areas in which the U.S.S.R. has made progress (in number of meters drilled, for instance) in the past 5 years. More significantly, it reports Soviet Eleventh FYP targets and development plans—for number of drilling crews, number of new exploratory and development wells to be drilled, drill crew productivity, average new well yield, production of more and/or better bits, drill pipe and rigs, utilization of sophisticated tertiary recovery techniques—and forms its assessment of the future of the industry on the assumption that these will be fulfilled.

These "pessimistic" and "optimistic" assessments of Soviet oil production prospects therefore rest in large part on judgments of Soviet ability to greatly improve on a wide variety of oil industry parameters in the next 5 years. Past patterns of plan underfulfillment, poor performance in a number of areas (equipment quality, drill rig and crew productivity, for example), and above all the inefficiencies and obstacles introduced by the nature of the Soviet economy and incentive system, all tend to advise caution in counting too heavily on fundamental changes in likely Soviet achievements. On the other hand, improvements have been made in a number of areas, and there is enough evidence of increased investment in the oil industry to suppose that such improvements will continue.

The Soviets' own Eleventh FYP reflects diminished expectations for oil production. Its upper limit of 645 mmt (12.9 mbd) is, after all, only slightly higher than the original underfulfilled 1980 target of 640 mmt. Moreover, the average annual rates of growth envisaged in the plan are substantially lower than the rate of 1.9 percent achieved between 1979 and 1980. Whether the new target is more the reflection of a deliberate investment decision to grant lower priority to oil in favor of gas (and possibly nuclear power development) or more a response to Soviet recognition that greater oil production would be unachievable in the absence of giant new finds, it is impossible to tell. In any case, those who before the plan targets were revealed believed that gains to as high as 700 mmt by 1985 were technically achievable are unlikely to abandon that view.

OTA has chosen the Soviet Union's own target of 645 mmt for its best case scenario.

Such an outcome seems achievable *if* a number of things go well for the U.S.S.R., including new finds, improvements in domestic manufacturing capacities, improvements in oil industry productivity, and continued—if not greater—imports of Western equipment to compensate for domestic shortfalls and inadequacies. The CIA projection is a plausible worst case, given the opposite set of expectations about Soviet domestic accomplishments and the continuation of present relatively modest levels of Western imports. However, the likeliest outcome is probably somewhere between 550 and 645 mmt— i.e., for the next 5 years, the U.S.S.R. through increased effort and investment in its oil industry may well be able to hold production fairly stable at 1980 rates.

THE PROSPECTS FOR SOVIET OIL PRODUCTION IN 1990

The variables affecting possible Soviet oil production in the latter part of the decade are numerous and complex, and the exercise even of selecting plausible best and worst cases for analytic purposes is highly speculative. Here too, the universe of responsible estimates that have appeared in the West are bounded by CIA and EIU, CIA projecting a fall to about 350 to 450 mmt (7 to 9 mbd) and EIU an increase to 750 mmt (15 mbd). The central question once again is whether production will fall, rise, or remain stable. Assuming no new giant finds in the immediate future, no massive infusions of Western equipment, and no hands-on assistance in offshore development, OTA believes that projections that see Soviet oil production rising significantly between 1985 and 1990 are excessively optimistic. A more realistic best case assumes that output could

be held stable at 1985 levels or slightly below. A plausible worst case would see an absolute decline in production, although perhaps not as serious as that envisaged by CIA. However, it must be emphasized that the difficulties that the U.S.S.R. will probably still be experiencing in 1990 may not be permanent. As the U.S.S.R. develops capabilities to explore and exploit its offshore and East Siberian resources, oil production eventually could begin to rise once more. This at least is the expectation of both the U.S. Defense Intelligence Agency and the U.N.'s Economic Commission for Europe.[127]

---

[127]U.S. Defense Intelligence Agency, "Allocation of Resources in the Soviet Union and China—1981." Statement of Maj. Gen. Richard X. Larkin before the Joint Economic Committee, Subcommittee on International Trade, Finance, and Security Economics, Sept. 3, 1981.

THE PROSPECTS FOR SOVIET GAS PRODUCTION, 1985 AND 1990

The attention surrounding the Soviet oil production controversy has until recently obscured the significance of gas in the Soviet energy future. That significance should not be underestimated. Gas has been called the "ace in Soviet energy plans . . . a critical cushion for the uncertainties faced by the planners with respect to other sources of (energy) supply."[128] Indeed, the key question for Soviet energy availability in the present decade may not be whether oil production is about to decline, but rather whether the U.S.S.R. can exploit its tremendous gas reserves quickly enough for gas to become the critical fuel in the CMEA energy balance. While it is important that gas not be regarded as the U.S.S.R.'s easy energy panacea—as this chapter has pointed out, gas development faces numerous obstacles— discussion of the prospects for gas production to 1990 is essential to understanding the opportunities confronting the Soviet Union during this period. This section discusses the range of estimates of Soviet gas production in 1985 and 1990 and, as with oil, posits OTA's best and worst case scenarios.

Soviet proven natural gas reserves are enormous. This fact is uncontroversial. In 1980, these were variously estimated in the West at 25 to 33 trillion cubic meters (tcm), some 40 percent of the world's proven reserves.[129] The U.S.S.R.'s own estimate is some 39 tcm. This is the thermal equivalent of 31.6 billion tons or 231.6 billion bbl of oil. These orders of magnitude have not been disputed, and the size of the proven reserve base alone is enough to support many years of substantially increased production. Indeed, Soviet gas reserves are equivalent in magnitude to Saudi Arabian oil reserves.

Constraints on gas output, therefore, rest not on resources, but on the ability of the gas industry to deliver its product to consumers in the U.S.S.R. and in Eastern and Western Europe. Specifically, these constraints concern the ability of the U.S.S.R. to substantially increase its gas pipeline network—and this in turn entails obtaining quantities of large diameter pipe and compressor stations—and to provide adequate infrastructure for the industry and its workers. These problems are exacerbated by the fact that the bulk of the U.S.S.R.'s established reserve base lies in West Siberia.

The Eleventh FYP calls for increasing gas output from its 1980 level of 435 bcm to 600 to 640 bcm by 1985, and Soviet projections supplied to the Secretariat of the United Nations Economic Commission for Europe for 1990 are 710 to 820 bcm. These and the range of production estimates that have appeared in the West are summarized in table 18. As this table shows, there is very little disagreement over these figures. For 1985, it seems reasonable to adopt the range in the Soviet plan as worst and best cases. Actual production will probably fall in between. The

Table 18.—Soviet Natural Gas Production Estimates, 1985-90 (bcm/mbdoe)

(bcm)	1985	(mbdoe)	(bcm)	1990	(mbdoe)
1. 605		9.9	750		12.3
2. 600		9.8	700		11.5
3. 660		10.8	—		—
4. 600		9.8	750		12.3
5. 598-647		9.8-10.6	—		—
6. 600		9.8	765-785		12.5-12.9
7. 600-640		9.8-10.5	710-820		11.6-13.4

SOURCES:
1. Leslie Dienes and Theodore Shabad, *The Soviet Energy System* (Washington, D.C.: V. H. Winston, 1979), table 53, p. 252.
2. CIA, as reported in Joseph A. Licari, "Linkages Between Soviet Energy and Growth Prospects for the 1980's," paper presented at the 1981 NATO Economics Directorate Colloquium, Apr. 8-10, 1981.
3. Herbert L. Sawyer, "The Soviet Energy Sector: Problems and Prospects," Harvard University, January 1978, quoted in George W. Hoffman, "Energy Projections—Oil, Natural Gas and Coal in the U.S.S.R. and Eastern Europe, *Energy Policy*, pp. 232-241.
4. David Wilson, *Soviet Oil and Gas to 1980*, Economist Intelligence Unit Special Report No. 90.
5. "Situation et Perspectives du Bilan Energetique des Pays de L'Est," *Le Courier des Pays de L'Est*, No. 216, March 1978, median and low hypotheses only.
6. Jonathan P. Stern, *Soviet Natural Gas Development to 1990* (Lexington, Mass.: Lexington Books, 1980), table 15.1, p. 178.
7. Soviet Eleventh FYP and projections submitted to the Secretariat of the U.N. Economic Commission for Europe.

---

[128] Dienes and Shabad, op. cit., p. 287.
[129] Stern, op. cit., pp. 22-3; Wilson, op. cit., p. 44.

situation for 1990 is more complex. Gas production in the 1980's will be determined, above all, by the level of Western exports of pipe and compressors available to the U.S.S.R., and a worst case should posit little or no Western assistance. OTA has, therefore, chosen as its worst case assumption production of about 665 bcm. This level of output is approximately halfway between the high case for 1985 and the low end of the Soviets' own 1990 projection. OTA's best case for 1990 is 765 bcm, the midpoint of the Soviet range. The midpoint, rather than the upper end of the target was chosen in the face of the extraordinary magnitude of the Soviet range.

SUMMARY AND CONCLUSION

The heart of the Soviet oil and gas industry is now firmly established in West Siberia. Despite the harsh climate, difficult terrain, and the remoteness of this region, the U.S.S.R. has over the past 10 years built an extensive petroleum industry there. In 1980 West Siberia accounted for over half of oil and nearly 36 percent of gas output. While there is general agreement that the prospects for gas are bright, the continued viability of the oil sector, as well as the ability of the U.S.S.R. to remain the world's foremost oil producer and a net oil exporter, has recently been the subject of controversy in the West.

This controversy was initiated by a 1977 CIA report that contended that Soviet oil production would drop precipitously to as low as 400 mmt (8 mbd) by 1985, occasioning the possibility that the CMEA as a bloc or even the U.S.S.R. itself would have to import oil. CIA has now raised its estimates and has clarified its position on CMEA oil imports, i.e., it does not foresee the U.S.S.R. itself buying oil on world markets although it still contends that the CMEA as a bloc would be in a net deficit position. In fact, CIA's basic argument has changed little: it believes that in the absence of new major finds, Soviet oil output will peak and decline sharply before the end of the decade.

CIA production estimates are the lowest to appear in the West. Others have projected substantial increases in Soviet oil production in the present decade. These judgments appear to be based on a different interpretation of the available Soviet reserve data, different evaluations of Soviet oil industry practice, and a higher estimation than CIA's of the ability of the U.S.S.R. to overcome its own institutional problems in improving a variety of industry parameters.

OTA's own survey of the state of Soviet oil equipment and technology has shown that, while the U.S.S.R. has selectively utilized Western imports in a number of areas, it produces large quantities of most of the items it needs. For the most part the Soviet difficulty is not that it lacks the know-how to provide for itself, but that the structure of its economy and its incentive system are such that it has difficulty producing sufficient quantities of high-quality equipment. Items purchased from the West are usually of higher standard than those that can be produced at home, and this reinforces their ability to compensate for domestic shortfalls. In addition, there are three areas in which the West probably can make contributions to the Soviet state-of-the art. These are computers, associated software, and integrated equipment sets for oil exploration; offshore technology, still in its infancy in the U.S.S.R.; and high capacity submersible pumps.

Soviet oil industry performance as measured in equipment productivity and number and depth of new wells is improving, and Western equipment must clearly have contributed to these improvements, although such contributions are impossible to quantify. There is every sign that the U.S.S.R. would like to continue to benefit from Western help, but it does not appear to be ready to depart from past practice and import massive volumes of equipment or request hands-on assistance.

OTA believes that CIA's estimates may be taken as a worst case outcome for Soviet oil production, but that it is also possible that the Soviets could achieve their own target for 1985 of 645 mmt (12.9 mbd), which would represent a modest increase over 1980 production of 603 mmt. Likelier than either of these extremes, however, is that production will remain about stable. This assumes no major changes in Western export or Soviet import policy, and no major new discoveries of oil.

Gas production is far less controversial. Here the proven reserve base is immense and production is constrained mainly by the lack of pipelines to carry the gas to consumers.

Western large diameter pipe and pipeline compressors have made dramatic contributions in the past, and there is every indication that continuation of such imports will be crucial throughout the decade. The Soviets' own plan targets of 600 to 640 bcm gas production in 1985 (equivalent to 9.8 to 10.5 mbd of oil, up from 435 bcm in 1980) seem to bound plausible worst and best outcomes in this sector. By 1990, production could exceed 750 bcm (12.3 mbd of oil equivalent), contingent on continued infusions of Western pipe and pipeline equipment.

While these large increases in Soviet gas production will not signify the end of the U.S.S.R.'s near- and medium-term energy problems, gas can certainly help to compensate—both in domestic consumption and in export—for oil production that has leveled off or even slightly declined.

CHAPTER 3

The Soviet Coal Industry

CONTENTS

LIST OF TABLES

LIST OF FIGURES

The Soviet Coal Industry

Prior to World War II, coal was the dominant source of fuel in the Soviet Union, as it was elsewhere in the world. In 1940, it supplied 75 percent of Soviet energy needs. Since then, oil and natural gas have become increasingly important and by the late 1970's, coal's share of total Soviet energy consumed had fallen to approximately 29 percent. There are incentives now to reverse this trend. Oil exports earn the Soviet Union the hard currency it needs to finance imports of Western grain and technology, and it is not surprising that Soviet energy planners have shown a strong interest in substituting other fuels for oil, particularly in electric power generation and in boiler applications.[1] Coal is such a substitute.

Unfortunately, the Soviet Union's reserves of easily obtainable high-quality coal are now seriously depleted, and the Soviet coal industry has experienced serious difficulties in simply maintaining production. The expansion of the industry which would be required for coal to be widely substituted for oil now seems extremely unlikely. The purpose of this chapter is to describe the current state and potential of the Soviet coal industry, including: 1) the characteristics of major coal deposits; 2) the technological and infrastucture problems facing the coal industry; 3) the degree of reliance of the Soviet coal industry on Western technology; and 4) the prospects for the industry in the next decade.

[1]See A. Troitskiy, "Electric Power: Problems and Perspectives," *Planovoye khozyaystvo*, No. 2, February 1979, p. 20.

INTRODUCTION

Soviet coal production increased steadily between 1970 and 1975, growing approximately 16 million metric tons (mmt) per year. The Tenth Five Year Plan (FYP) (1976-80) proclaimed that coal would replace oil wherever possible, and additional yearly increases averaging about 20 mmt were targeted. But the coal industry has encountered problems. Production peaked at 723.6 mmt in 1978, far short of the original goal, and has been declining since. Production in 1980 (716 mmt) was 89 mmt short of the original plan target and 29 mmt below the revised 1980 annual plan (see table 22, below.) The coal industry has consistently had difficulties meeting its output goals, and these difficulties cannot be expected to disappear in the foreseeable future.

In terms of sheer magnitude, the U.S.S.R. has substantial coal reserves. Table 19 shows the World Energy Conference Survey of Energy Resources estimates of world coal reserves. According to this survey the Soviet Union has over half of the world's resources of coal that could be successfully exploited and used within the foreseeable future, and approximately one-quarter of world explored reserves recoverable under present local economic conditions and available technology.[2] The difficulties faced by the Soviet coal industry lie not in the size of the resource base, but rather in the location of coal reserves and in the quality of the coal now being or expected to be mined in the foreseeable future.

The first coalfields to be exploited in the U.S.S.R. were located near the major population centers of the Western European part of

[2]For the purposes of this chapter, Soviet "explored" coal reserves, i.e., those relevant for present planning purposes, are roughly equivalent to proved, probable, and possible categories in Western nomenclature.

Table 19.—Estimated Coal Reserves of the World
(billion tons)

	Recoverable reserves[a]	Percent world total	Total reserves[b]	Percent world total	Total resources[c]	Percent world total
U.S.S.R.	150.6	23.1%	301.2	19.2%	6,298.2	53.1%
United States[d]	200.4	30.7	400.8	25.6	3,223.7	27.2
Canada	6.1	0.9	10.0	0.6	119.9	1.0
People's Republic of China....	88.2	13.5	330.7	21.1	1,102.3	9.3
India........................	12.8	2.0	25.5	1.6	91.5	0.8
Rest of Asia	6.6	1.0	19.1	1.2	27.6	0.2
Federal Republic of Germany .	43.6	6.7	109.7	7.0	315.4	2.7
United Kingdom	4.3	0.7	109.0	7.0	179.5	1.5
Poland	25.0	3.8	42.9	2.7	66.8	0.6
Rest of Europe	66.9	10.3	91.0	5.8	108.0	0.9
South Africa..................	11.7	1.8	26.7	1.7	48.9	0.4
Rest of Africa	5.6	0.9	6.7	0.4	16.0	0.1
Australia[e]	26.8	4.1	81.9	5.2	218.9	1.8
Rest of Oceania	0.2	—	0.4	—	1.2	—
Latin America	3.1	0.5	10.1	0.7	36.3	0.3
World Total	651.7	100.0%	1,565.6	100.0%	11,854.1	100.0%

[a]Amount of reserves in place that can be recovered under present local economic conditions and available technology.
[b]The portions of total resources that have been carefully measured and assessed as exploitable under local economic conditions and available technology.
[c]Total amount available in the Earth which can be successfully exploited and used within the foreseeable future.
[d]Estimates of U.S. coal reserves here may not agree with other domestic data.
[e]Does not include additional resources for Queensland.
SOURCE: National Coal Association, *Coal Facts*.

the country. Some of these have been mined for so long that the thick, easily accessible coal seams are rapidly nearing depletion. The remaining seams are not readily susceptible to existing methods of mechanization. They are one-third thinner than the national average and lie considerably deeper in the ground. As mine depths increase, so do the costs of extraction and the risks from gas and explosions. Further problems arise because the equipment installed in these mines has become increasingly seam-specific. As seams are worked out, the equipment cannot be transferred.

As many mines have become difficult or expensive to operate, new ones have been opened. Those in the eastern part of the country, like the eastern oilfields and gasfields, are located in sparsely populated, inhospitable regions from which the cost of transporting coal to consumers is much higher. The Soviets now look increasingly to surface mining as a source of growth in coal production because surface-mined coal is cheaper to extract than underground coal— it can be mined with higher productivity equipment requiring less labor. But the relative share of surface mining in the U.S.S.R. is still low—about 37 percent in 1980 (as opposed to about 52 percent in the United States in 1978.)[3] For the present, this low level is adversely affecting overall labor productivity growth and output in the industry. For the future, the Kansk-Achinsk, Ekibastuz, Kuznetsk, and South Yakutia basins are the favored sites for expanded surface mining. But these basins are all

[3]*Ugol'*, No. 2, 1980, p. 4; U.S. Department of the Interior, Bureau of Mines Information Circular, "Coal Mine Equipment Forecast to 1985," 1976, p. 5.

located at considerable distances from the consuming centers in the European U.S.S.R. and high transportation costs would at least partially offset the lower costs of extraction.

Moreover, the quality of the coal in some of these new basins is very poor. Coal, which is formed as the result of millions of years of physical and chemical changes to moist vegetable matter, is a complex heterogeneous material.[4] It varies by type (depending on the kind of original plant materials from which it was formed), rank (based on the carbon and oxygen content, degree of moisture, volatile matter, etc.), and the type and amount of impurities that it contains. In general, anthracite and various grades of bituminous coal are preferable to lignite or brown coal because they have a higher heat content per unit. Large portions of Soviet coal reserves are comprised of the less desirable deposits, and the calorific value of an average ton of Soviet coal has been declining (see table 20). Between 1970 and

1978, the calorific value of Soviet coal declined from 4,853 kilocalories per kilogram (kcal/kg to 4,711 kcal/kg, a drop of 3 percent. The decline could be even greater in the future. This is due, in part, to the depletion of higher grade coals and the increasing role of lignite, primarily from the Kansk-Achinsk basin. In consequence, part of any future growth in coal production will be offset by declines in the calorific content and thus in the heat value of the coal shipped to consumers.

Indeed, Soviet coal production figures must be treated warily, for they are given in terms of "run-of-mine," i.e., coal which has not yet been cleaned. This may cause output figures to be overstated by as much as 20 to 40 percent.[5] The most common impurities found in coal are sulfur, stones, and ash. Sulfur forms oxides which cause pollution; stones and ash (noncombustible material that remains after the coal has been burned) provide no heat and add to transportation costs. Coal, particularly lignite, may also contain considerable moisture which inflates its true weight.

In sum, the success of the Soviet coal industry seems to rest on the expansion of surface mining. Although the Tenth FYP sought to raise underground output, this actually fell by 23 mmt during the plan period, while surface mining production rose some 36 mmt and came closer to meeting its target. Unfortunately, however, Soviet surface-mined coal is often of poor quality. The prospects for the industry, therefore, strongly depend on the degree to which surface mining can be expanded and the success with which the coal thus mined can be treated, transported, and used. The survey of the major Soviet coal basins which follows provides the context for evaluating these two issues.

---

[4]See Charles Simeons, *Coal: Its Role in Tomorrow's Technology* (Oxford: Pergamon Press, 1978); and Bernard Cooper, "Research Challenge: Clean Energy From Coal," *Physics Today*, January 1978.

Table 20.—Coal Production in Natural Units and in Standard Fuel
(million metric tons, except calorific value)

Year	Coal in natural units (1)	Coal in standard fuel[a] (2)	Calorific value, kcal/kg[b]
1940	165.9	140.5	5,928
1945	149.3	115.0	5,392
1950	261.1	205.7	5,515
1955	389.9	310.8	5,580
1960	509.6	373.1	5,125
1965	577.7	412.5	4,998
1970	624.1	432.7	4,853
1975	701.3	471.8	4,709
1976	711.5	479.0	4,713
1977	722.1	486.0	4,711
1978	723.6	487.0	4,711
1979	718.7	483.9	4,713
1980	716.0	479.7	4,690

[a]One metric ton of standard fuel equals 27.8 million Btu or 7 gigacalories.
[b]Column (2) divided by column (1) and multiplied by 7×10^3 kcal/kg of standard fuel.

SOURCES: U.S.S.R. Central Statistical Administration, *Narodnoye khozyaystvo SSSR v 1978 g.* (Moscow: Izd. "Statistika," 1979), p. 144; *Ibid* (1975), p. 219 and (1980), pp. 170-171.

---

[5]See Robert W. Campbell, *Soviet Energy Technologies: Planning, Policy, Research, and Development* (Bloomington, Ind.: Indiana University Press, 1980); and V. V. Strishkov, George Markon, and Zane E. Murphy, "Soviet Coal Productivity: Clarifying the Facts and Figures," *Society of Mining Engineers Journal*, May 1973.

MAJOR SOVIET COAL-PRODUCING REGIONS

Figure 6 shows the location of the Soviet Union's major coal-producing areas. The geographic distribution of Soviet coal is unfortunate. The heavily populated and industrialized European part of the U.S.S.R. contains only 6 percent of the nation's coal reserves. The rest are located in the Arctic, Siberia, or Kazakhstan where climatic conditions make coal extraction and transportation difficult and expensive. Tables 21 and 22 summarize the extent of explored reserves and recent coal production by basin. The following survey briefly describes the chief characteristics of each of these basins.

BITUMINOUS BASINS

Donets (Donbass) (No. 2 on map)

The Donets basin covers some 60,000 square kilometers (km²) mainly in the Ukraine, and has explored reserves of over 40 billion metric tons (bmt) (see table 21). It

Table 21.—Geographical Distribution of Soviet Recoverable Coal Reserves, 1967 (billion metric tons)

Basin/field	Proved, probable, possible		Potential	
U.S.S.R. total	255	(180)	170	(99.8)
Kansk-Achinsk	72.6	(71.1)	43.0	(35.4)
Kuznetsk	59.5	(33.0)	60.8	(25.1)
Donets	40.4	(7.7)	17.2	(1.2)
Pechora	7.9	(4.1)	6.9	(1.9)
Ekibastuz	7.4	(—)	—	(—)
Karaganda	7.6	(3.5)	1.8	(0.1)
Irkutsk	7.1	(7.1)	13.3	(13.2)
Turgay	6.3	(5.6)	0.4	(0.4)
Moscow.............	4.8	(4.8)	2.3	(2.3)
Minusinsk	2.8	(2.8)	43.0	(35.4)
Dnepr	2.6	(2.6)	—	(—)
South Yakutia.......	2.6	(2.5)	3.2	(3.0)
Tungus	1.9	(1.7)	3.0	(2.9)
Maykyuben...........	1.8	(0.9)	—	(—)
South Urals	1.3	(—)	—	(—)
Lena	1.1	(0.7)	1.4	(1.4)

NOTE: Column figures in parantheses ( ) denote coal down to 300 meters.

(—) denotes not available.

SOURCES: V.A. Shelest, *Regionalnyye energoekonomicheskiye problemy SSSR* (Moscow: Izd. "Nauka," 1975), pp. 113-116; and *Sovetskaya geologiya* (April 1970), p. 57.

Table 22.—Soviet Coal Production
(million metric tons)

	1970	1971	1972	1973	1974	1975	1976	1977	1978	1979	1980	1981 (plan)
U.S.S.R. total[a]	624.1	640.9	655.2	667.6	684.5	701.3	711.5	722.1	723.6 (748)	719 (752)	716 (790-810)	
Minugleprom SSSR, of which	—	634.3	648.9	661.4	678.1	694.6	704.7	715.7	—	—	—	
Donets	218.0	217.5	217.4	219.4	219.5	221.5	223.7	222.0	—	208	203	(213)
Kuznetsk	113.3	115.5	119.2	123.3	128.3	134.0	138.9	141.9	(153.7)	154.9)	(162.4)	(149)
Karaganda	38.4	39.8	41.7	43.3	45.3	46.3	47.4	48.2	—	(47)	(48.6)	(49)
Pechora..........	21.5	22.0	22.5	23.0	23.4	24.2	25.8	26.7	—	28.9	28	(29)
Ekibastuz	22.6	—	—	—	—	45.8	(49.4)	(50-53.5)	(57.0)	59.2	66.8	(72)
Kansk-Achinsk	—	—	—	—	—	27.9	29.1	31.6	—	33	34.5	(46)
Moscow...........	36.2	36.7	36.7	36.1	35.1	34.1	30.9	29.5	—	27	25	(23)
Degree of plan fulfillment percent	—	103.5	103.5	103.2	102.5	102.5	102.4	101.7	96.7	95.6	88.2-90.5[b]	—

NOTE: Column figures in parentheses ( ) denote plan targets.

[a]Total includes coal produced outside Minugleprom, the Soviet Coal Ministry. Figures taken from *Narodnoye khozyaystvo* for various years.

[b]Based on original FYP targets.

SOURCES: Most production figures are from the April issues of *Ugol'* for given years. Other data are from the following:

1979 plan figures: *Ekonomicheskaya gazeta*, No. 5 (February 1979), p. 1.

1979 total production: *Pravda Ukraina*, (Jan. 26, 1980), p. 2.

1980 plan total: *Narodnoye khozyastvo Kazakhstana*, (October 1978), pp. 38-45.

1977 Ekibastuz plan: *Partinyaya zhizn Kazakhstana*, (January 1978), pp. 34-35. [JPRS 71, 127, (May 17, 1978), p. 9.]

1976 and 1977 Ekibastuz plan targets: *Ugol'*, (January 1978), pp. 16-20.

1970 production: V.A. Shelest, *Regionalnyye energoekonomicheskiya problemy SSSR*, (Moscow: Izd "Nauka," 1975), p. 26.

1975 Kansk-Achinsk and Ekibastuz production and 1980 basin plan targets: A.M. Nekrasov and M.G. Pervukhin, *Energetika SSSR v. 1976-1980 g.*, (Moscow: Izd, "Statistika," 1977), p. 146.

1981 plan: *Ekonomicheskaya gazeta*, No. 15 (1981), p. 2.

See also *Soviet Geography*, April issues.

Figure 6.—Soviet Coal Basins

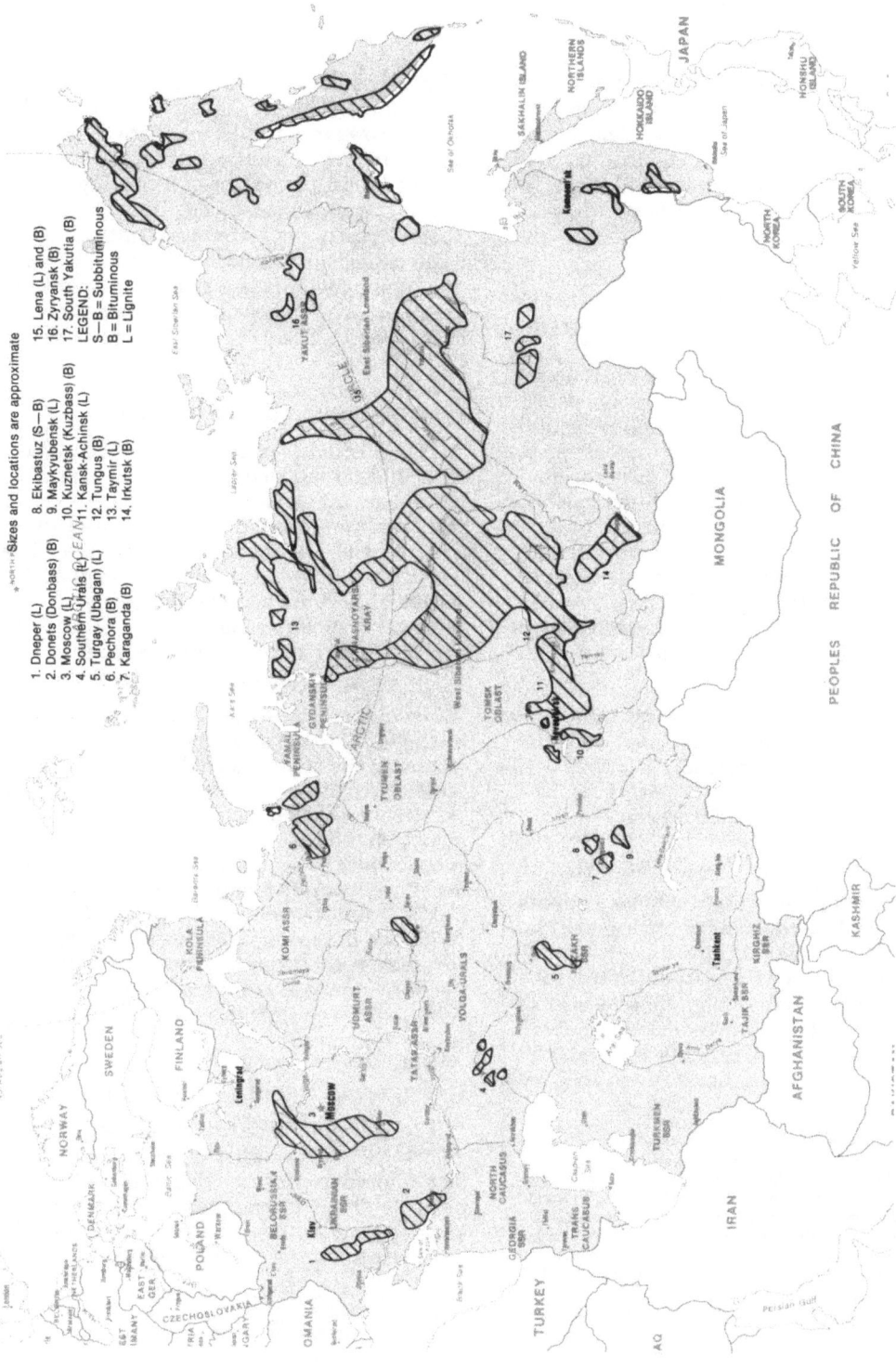

*Sizes and locations are approximate

1. Dneper (L)
2. Donets (Donbass) (B)
3. Moscow (L)
4. Southern Urals (B)
5. Turgay (Ubagan) (L)
6. Pechora (B)
7. Karaganda (B)
8. Ekibastuz (S—B)
9. Maykyubensk (L)
10. Kuznetsk (Kuzbass) (B)
11. Kansk-Achinsk (L)
12. Tungus (B)
13. Taymir (L)
14. Irkutsk (B)
15. Lena (L) and (B)
16. Zyryansk (B)
17. South Yakutia (B)

LEGEND:
S—B = Subbituminous
B = Bituminous
L = Lignite

SOURCE: Office of Technology Assessment.

is one of the oldest sites of underground mining in the U.S.S.R., and, as table 22 shows, the country's leading producer. The basin contains high-grade coals, including coking coal and anthracite, and is located close to consuming industries. It is, therefore, particularly important to the Soviet coal industry.

Past mining in the Donets concentrated on thicker coal seams close to the surface. Much of this coal is now depleted, and miners must work thin seams at ever-increasing depths. In fact, the average depth of working faces in 1978 was well over 500 meters (m), and this depth was increasing at 15 m per year, nearly twice the national average. In addition over 80 percent of Donets coal lies in seams less than 1-m thick (coal which would not even be counted in U.S. reserves), and many of these are steeply pitched which makes them difficult to work. Deteriorating mining conditions have also led to increasing ash and stone contents in Donets coal.[6]

It is not surprising, therefore, that production in the Donets has declined over the past decade, from 216 mmt in 1970, to 203 in 1980. The 1981 production target is 213 mmt, lower than actual 1970 output.[7]

Kuznetsk (Kuzbass) (No. 10 on map)

The Kuznetsk basin is the Soviet Union's second largest hard coal-producing region, covering some 26,000 km² in southwest Siberia. This basin is especially important because of its reserves of high-quality coking coal, much of which can be surface mined. Explored reserves at Kuznetsk are some 60 bmt (see table 21). Production here rose by

25 percent between 1970 and 1977, reaching 141.9 mmt. The 1980 plans called for 162.4 mmt. However, the latter were almost certainly underfulfilled (see table 22), and lately Soviet literature has been reporting production problems in the basin. These seem to be at least partly due to failure to introduce new mine capacity. Indeed, in the past 18 years, only one new mine has gone into operation.[8] There are also indications of labor shortages in the basin. The 1981 plan target was only 149 mmt.[9]

Pechora Basin (No. 6 on map)

The Pechora basin covers 120,000 to 130,000 km² in the extreme northeast of the European U.S.S.R., north of the Arctic Circle. Much of the coal here is located in permafrost areas, and has been only superficially studied. The basin contains explored reserves of 7.9 bmt, much of it coking coal (see table 21). A large percentage of this lies below 300 m, but in general, the coal is closer to the surface than in the Donets basin.

Development of the Pechora basin began in the early 1940's, using forced labor. But although the basin was able to supply coal to northern Russia during World War II, the extremely cold climate has made mine construction difficult. Pechora mines are also susceptible to gas explosions. Production here rose from 21.5 mmt in 1970 to 26.7 mmt in 1977, and was slated to reach 29.8 mmt by 1979, but in recent years the basin has failed to meet plan targets. About two-thirds of Pechora's output is high ash content coking coal, which requires cleaning before use; the rest is steam coal.

Karaganda Basin (No. 7 on map)

The Karaganda basin, covering 3,000 km² in northwest Kazakhstan, contains 7.6 bmt of explored reserves. Over half of this lies below 300 m. Karaganda has both coking and steam coal. Here, the steam coal is high in ash content and difficult to enrich.

[6]A. V. Sidorenko, *Mining Science and the Rational Utilization of Raw Mineral Resources* (Moscow: Izd. "Nauka," 1978), p. 47; Joseph K. Wilkinson (ed.), "Soviet Coal Strives for Expansion," *Coal Age,* April 1978, p. 86; A. V. Tyzhnov, "Geological Reserves of Coal in the U.S.S.R.," *Sovetskaya geologiya,* No. 4, April 1970, p. 64; and Leslie Dienes, "Regional Dimensions of Soviet Energy Policy," paper prepared for the American Association of Geographers, 1979, p. 38.

[7]*Soviet Geography,* April 1981, p. 276; *Ekonomicheskaya gazeta,* No. 15, 1981, p. 2.

[8]G. Shumkin, "Let's Look For and Find the Reserves," *Trud,* Sept. 15, 1978, p. 1.

[9]*Ekonomicheskaya gazeta,* No. 15, 1981, p. 2.

Large-scale production in Karaganda began in the 1930's, when the area was first reached by railway and coal could be shipped to the iron and steel industries in the Urals. Now local iron and steel plants are major consumers. Karaganda's output grew steadily between 1970 and 1977, reaching 48.2 mmt. No production figures after 1977 have been published, but it is highly probable that growth in output has been slowing since the mid-1970's, a fact reflected in the 1981 production target of 49 mmt.[10] In March 1979, it was reported that the "Karagandaugol" mining association was producing below plan goals in January and February of that year, and that the association had also been under plan for 1978.[11] A Kazakh Party official reported in October 1978, that the "50th Anniversary of the U.S.S.R." mine, one of the basin's best, was below plan for nearly all indicators, including output, and was even producing below the 1977 level for the same months.[12] There are indications of equipment problems, shortages of labor, and inadequate new mine construction.[13] Past planning mistakes also haunt the Karaganda basin. The city of Karaganda is located over valuable reserves—11 beds with 1 bmt of coal. Consequently, mined-out seams here have been packed with rubble to prevent subsidence of the city, an operation which diverts needed labor away from production.[14]

South Yakutia (No. 17 on map)

The South Yakutian basin lies in a remote area of the Soviet Far East. In 1967 explored reserves here were set at a relatively low level, 2.6 bmt, but more recent work may have significantly expanded these estimates.

In any case, the contribution of South Yakutia lies in the future. Although it is now producing only very small amounts of coal, it is the site of a major Soviet-Japanese energy cooperation project that is expected to yield about 85 million tons of medium-quality coking coal for export to Japan by the year 2000. (For details of this project, see ch. 11.)

SUBBITUMINOUS AND LIGNITE BASINS

Ekibastuz (No. 8 on map)

Ekibastuz is a small—160 km²—area in northeast Kazakhstan containing 7.4 bmt of explored reserves. Production here rose from 22.7 mmt in 1970 to 59 mmt in 1979 (6 mmt short of the plan target). Ekibastuz has abundant coal suitable for surface mining, and in 1978, it alone accounted for 22 percent of Soviet surface-mined coal.[15] Labor productivity in Ekibastuz is high, but the ease and consequent low cost of extraction is somewhat offset by the poor quality of the coal, which has a high ash content (averaging 40 percent, but reaching as high as 48 to 56 percent in some cases) and thus a low calorific value per unit. Some Ekibastuz coal is used locally as steam coal, but in 1979, over 60 percent of the basin's output was shipped outside Kazakhstan.[16]

Kansk-Achinsk (No. 11 on map)

The Kansk-Achinsk basin is located to the east of the Kuznetsk basin. Its explored reserves, the largest in the U.S.S.R., have been set at 72.6 bmt. This coal can be surface mined at low cost. Unfortunately, however, it is mostly lignite, which is characteristically low in heat value, and high in moisture content. Kansk-Achinsk coal also tends to self-ignite when dried. For these reasons, its transportation is difficult, and demand for it is low. In 1975, about 90 percent of the basin's output was used locally. This coal is difficult to use even locally, however, and

[10]Ibid.; Leslie Dienes and Theodore Shabad, *The Soviet Energy System: Resource Use and Policies* (Washington, D.C.: V. H. Winston & Sons, 1979), p. 114.

[11]B. Glotov, "In Hope of a Sunday Assault," *Sotsialisticheskaya industriya*, Mar. 15, 1979, p. 2.

[12]O. Mulkibayev, "Why Are the Mines Giving Up the Positions They've Won" *Narodnoye khozyaystvo Kazakhstana*, No. 10, October 1978, pp. 38-45 in JPRS 72,902, Mar. 1, 1979, p. 98.

[13]Ibid., p. 98; Glotov, op. cit., p. 2.

[14] B. Glotov, "Arguments Instead of Action," *Sotsialisticheskaya industriya*, June 6, 1980, p. 2, in JPRS 76,242, Aug. 18, 1980, p. 43.

[15]Dienes and Shabad, op. cit., p. 116.

[16]*Soviet Geography*, March 1980.

Photo credit: TASS from ©SOVFOTO

Rotary bucket excavator in use at Ekibastuz

power stations refuse shipment whenever possible. Kansk-Achinsk lignite cakes onto boilers and has highly variable ash melting points.[17]

About 32 mmt of coal were extracted here in 1977, and the 1981 target is 36 mmt (see table 22). This basin is considered by many Soviets to be the best hope for expanded coal production (production that has been forecast as high as 1 bmt/yr nationwide[18]), and there are plans for a large fuel, energy, and industrial complex to be built in the area. However, the feasibility of this venture will rest importantly on the development of boilers suitable for the coal (see ch. 5.)

Moscow Basin (No. 3 on map)

This basin, covering 120,000 km² south of Moscow, contains some 4.8 bmt of low-quality coal, having high ash and sulfur contents and low calorific value. Production peaked in 1958 when new underground mine construction stopped,[19] and has been declining since 1971. Output fell from 36.2 mmt in 1970 to 29.5 mmt in 1977, and plan targets envisage a further reduction in 1981 to 23 mmt. Given the high cost of this coal, underground production probably would have ceased altogether were it not for the proximity to consumers. (In addition, there is relatively cheap coal suitable for surface mining in the

[17]*Ugol'*, No. 12, December 1975, p. 62; Dienes and Shabad, op. cit., p. 251; Campbell, op. cit., pp. 175-6.

[18]L. Sizov, "Fuel Base of Siberia: How to Develop Kansk-Achinsk Fuel-Energy Complex," *Trud*, June 27, 1980, p. 2.

[19]A. D. Breyterman, *The Economic Geography of Heavy Industry* (Moscow: Izd. "Vysshaya shkola," 1969), in JPRS 49,321, Nov. 26, 1969, p. 66.

southern part of the basin.) The Moscow basin's output is largely of local importance, serving the industrial regions around Moscow primarily as boiler fuel.

SUMMARY

Although the U.S.S.R. has large coal reserves, their geographic distribution is unfavorable, with most of the coal lying in little-studied basins in remote areas with adverse climates. Not only has coal production declined over the past several years, but the calorific value of a ton of Soviet coal has decreased and probably will continue to do so.

Major characteristics of the primary coal basins are summarized in table 23. Important features to note include the fact that the Donets basin, the major coal producer in the U.S.S.R., has an unfortunate geological structure. More than 80 percent of the re-

maining coal is in seams less than 1 m thick and only 19 percent of this coal is between ground level and 300 m. Coal from Ekibastuz is very high in ash content, which means that Soviet output figures for the basin overstate its contribution to the production of energy. Kansk-Achinsk coal is low in heat value and cannot be transported economically to the central industrial region in untreated form.

In short, despite large reserves, coal output has been falling and the outlook for the future is not as bright as one might expect. Increasingly, the unfavorable geographical distribution of unexploited coal reserves will have its effect on the costs of production and utilization, especially in view of continuing depletion of the Donets reserves. Most of the major underground mining basins are having difficulty meeting output goals. Soviet hopes therefore rest on coal which can be surface mined.

Table 23.—Characteristics of Major Soviet Coal Deposits

	Type of mining	Explored reserves (billion metric tons)	Average thickness of seam (meters)	Average depth of mine (meters)	Average calorific value (Btu per pound)	Moisture content (percent)	Ash content (percent)	Share of production in 1980 (percent)
Donets	Underground	40	0.9	566	10,900	6.5%	19.2%	28%
Kuznetsk	Underground and surface	60	2.5	262	9,990	10.2	19.0	22
Pechora	Underground	8	2.4	454	9,390	8.3	25.1	4
Karaganda	Underground	8	2.5	384	9,250	7.5	28.8	7
Ekibastuz	Surface	4	10-40	—	7,250	7.7	39.1	10
Kansk-Achinsk	Surface	72	—	—	6,490	33.0	10.7	6
Moscow	Underground	5	2.5	135	4,550	32.3	35.5	4

SOURCE: Office of Technology Assessment; CIA, "USSR: Coal Industry Problems and Prospects," ER 80-10154 (March 1980).

SOVIET COAL INDUSTRY TECHNOLOGY – PROBLEMS AND PROSPECTS

The technological level of Soviet-designed and produced coal mining equipment is uneven. At its best, the Soviet coal mining equipment industry has produced sturdy and well-designed equipment. But the technological level of equipment in place varies. Major basins can be expected to command

the better equipment, while some mines must make do with old, deteriorated machinery.

Soviet coal mining equipment stocks are large, greater in fact than American stocks, yet the U.S.S.R. produces less coal than the

United States. Despite the large inventory of equipment, the level of mechanization is often low, including the main extraction operations in some basins. This is due in part to failure to produce needed quantities of equipment of proper quality; failure to maintain equipment properly; lack of sufficient parts or repair crews; and neglect of maintenance schedules. Anomalies abound. In underground mines, coal may be extracted and the mine roof supported with sophisticated pieces of equipment—which operate along roadways prepared by hand. In the country's open-pit or surface mines, large capacity excavators may be teamed with trucks mismatched in size and strength. Such examples are not isolated extremes; technological inconsistencies of this type are widespread and chronic. In short, despite a seemingly abundant stock of equipment, the failure to produce an appropriate mix of machinery models for the special conditions imposed by different coal seams has led to shortages in some basins.

Production of coal mining equipment has, in the past, been of secondary priority to Soviet planners, subordinate to oil and gas development. The quality of Soviet machinery reflects this. The older equipment that makes up the bulk of the stocks is equivalent to models produced in the United States 10 to 20 years ago, smaller and less productive, although apparently mechanically reliable. This has been due in part to Soviet reluctance to adopt new technologies in coal mining, even technologies that would be readily available outside the U.S.S.R. Plants continue to produce equipment that is no longer in great demand, while production of new equipment, to mine thin seams for example, is lagging seriously.

The failure to change products has two major causes. Perhaps most important is the pervasive reluctance of plant managers to jeopardize output plan fulfillment by interrupting production to retool for a new product. A change in the product line not only means risking bonuses given for plan fulfillment, but also requires new supply arrangements, possible changes in profitability, and risk associated with new production technology and new products. Soviet managers have little incentive to incur such risks and so prefer to continue to use and produce established models, even if they are outmoded or unwanted.

An additional problem stems from the fact that the Ministry of the Coal Industry has been in a relatively weak position vis-a-vis its equipment suppliers. Responsibility for producing coal mining and transport equipment was scattered among many factories, all of which also produce a variety of other machines for other customers. Nor can the ministry participate in the research, design, and testing of new mining equipment—as does the Ministry of the Power Industry, for instance, with respect to power generation.

The current renewed interest in coal has led to some attempts to alleviate these problems. In the early 1970's an effort was made to make both manufacturers of some mining equipment and coal mine construction organizations more responsive to the needs of the industry, and administrative responsibility for these activities was transferred to the Ministry of the Coal Industry, known as Minugleprom. (Underground equipment is handled by Minugleprom, but the production of surface mining equipment is under the Ministry of Heavy and Transport Machine Building.) This has produced some improvement: production of modern equipment has increased in recent years and the quality of output has reportedly risen. For instance, between 1973 and 1977, the number of mining equipment models awarded the State Seal of Quality, the U.S.S.R.'s highest category of product quality, increased 2.4 times.[20] However, demand is still not being satisfied.

Nor does it seem that investment in productive capacity in underground coal mining machine building is sufficient to support in-

[20]"Make Decisions of the 25th Congress of the CPSU a Reality," *Ugol'*, No. 4, April 1978, pp. 3-7 in JPRS 71,340, June 22, 1978, pp. 3-7.

creased output. It was hoped that transferring coal machinery plants to Minugleprom would eliminate administrative barriers frustrating the satisfaction of demand for equipment. Minugleprom, already responsible for the fulfillment of coal output targets, would itself set the production programs of the equipment plants and oversee their fulfillment. Instead, it appears that Minugleprom may be diverting capital away from the machinery plants in an attempt to assist the fulfillment of short-term coal mining targets.

Inability to produce appropriate mining equipment in the required quantities and of required reliability is only one aspect of the equipment problem facing the industry. Perhaps even more serious in the long term is the seeming inability of miners themselves to use and maintain equipment properly. In large part, the difficulty stems from the inattention to maintenance and repair schedules, lack of spare parts, improper operations, and use of equipment inappropriate to geological conditions. As a result of poor maintenance, downtime on machinery is seriously in excess of established norms.[21] Several examples may be cited. Equipment failures in the Karaganda basin have increased by 27 percent in recent years.[22] In 1978, one Soviet journal reported that coal mining equipment idleness had reached 25 percent,[23] while another reported 350 work stoppages due to equipment failure in one Donets mine alone.[24] Poorly maintained equipment is also leading to an increased rate of accidents in the labor force. Soviet fatalities per million tons of coal mined were several times greater than the U.S. rate in the mid-1970's.[25]

[21]Glotov, op. cit., p. 2.
[22]Mulkibayev, op. cit.
[23]G. Dorofeyev, "Lost Perspective," *Sotsialisticheskaya industriya*, Dec. 3, 1978, p. 2.
[24]A. Zharkikh, "Far Behind," *Pravda Ukrainy*, Dec. 16, 1978, p. 2.
[25]"More Coal for the Country," *Ugol' Ukrainy*, No. 1, January 1979, pp. 1-4, in JPRS L/8370, Apr. 3, 1979, p. 54; Joseph J. Yancik, "Some Impressions and Observations of Soviet Coal Mining," *Society of Mining Engineers Journal*, July 1974, p. 65.

The following sections briefly describe the most important technology and equipment in both surface and underground extraction of coal, summarizing the state of the Soviet industry, and identifying the major difficulties it is encountering.

SURFACE MINING

Surface or strip mining in 1980 accounted for 37 percent of Soviet coal production. In surface mining, the rock and earth above the coal seam, called overburden, is removed to expose the coal, which is then broken up, loaded onto transport, and hauled away. Surface mining equipment ranges from construction bulldozers or front-end loaders, to enormous draglines, the largest moving land machines in the world. Large shovels and draglines remove the overburden so the coal can be picked up by smaller shovels and front-end loaders, although the latter are relatively little used in the U.S.S.R. Large power shovels work from the floor of the coal pit, taking bits of earth from one wall of the pit, pivoting and dropping the load on the other side. Larger draglines work from the surface at the edge of the pit. Buckets are dropped from the ends of long booms and are filled by dragging them back toward the machine. Draglines then turn, reach across the pit, and drop their loads. Shovels and draglines are generally preceded by vertical or horizontal drills that bore holes for explosives that can shatter the earth and rock for easier digging.

From an engineering standpoint, the Soviet Union is generally capable of fulfilling its surface mining needs. (The exception to this rule is equipment for operating in extremely cold temperatures, as in the Yakutsk basin. Such equipment has yet to be developed anywhere in the world.) But equipment is slow in reaching production and supplies are chronically short. In 1980, draglines accounted for about 33 percent of stripping work, but there have been production shortages and the demand for draglines is not being met.[26] Lack of haulage capacity is a

[26]*Ugol'*, No. 1, 1981, p. 5.

serious problem, and there is also a need for excavators of greater bucket capacity and cutting force. With the development and introduction of 120- to 180-ton trucks, the importance of large bucket capacities increases, because of the need to match equipment productivities and achieve more efficient (i.e., uninterrupted) operations. Although some giant (5,000 m³/hr) rotary excavators exist—two are in operation at the Bogatyr Mine in Ekibastuz and one is operating at the Irsha-Borodino pit in Kansk-Achinsk—there remains a general deficit in their supply and capacity.

Climate plays a special role in contributing to downtime and constraining the efficiency of Soviet equipment. Cold climates require special design features. Electrical systems are adversely affected by the cold, and when the temperature drops below −40°, the conveyer belts on rotary excavators become virtually inoperable. These effects are very serious in Siberia where much of the U.S.S.R.'s surface mining is carried on. For example, around 70 percent of the coal in the Kansk-Achinsk basin is produced through these methods.[27]

As in other energy industries, part of the problem of supply relates to the inadequacy of facilities to produce large pieces of equipment. In an effort to alleviate this problem, work has begun on the development of a machine building industry in Siberia. Construction of the Krasnoyarsk Heavy Excavator Plant is now underway and is scheduled to be completed by 1984.[28] The plant is to produce mechanical shovels with bucket capacities of 12.5 m³, walking excavators with 40 m³ buckets, and rotary bucket equipment with capacities of 5,000 to 12,500 m³/hr.

The success of these and other attempts to improve the quality and quantity of surface mining equipment are crucial to the coal industry as a whole. As mentioned above, Soviet plans to increase coal output rest ultimately on the expansion of surface mining. To a large degree, therefore, the fate of the Eleventh FYP for coal will depend on the availability of sufficient and appropriate surface mining equipment with adequate capacities to deal with increased output.

UNDERGROUND MINING

The most common technique for mining coal underground in the U.S.S.R. is by continuous mining machines built into longwall systems.[29] Longwall mining utilizes a steel plow or rotating cutting drum that moves back and forth across a coal face several hundred feet long. As the machinery moves, it cuts the coal, which falls onto a conveyer. Broad steel beams set a few feet apart provide ceiling support. These supports are moved by self-advancing hydraulic jacks that change their position during or after each pass of the cutting machine along the coal face. This change in position is accomplished by releasing the pressure exerted on the roof and moving the machinery forward one beam at a time. The unsupported portion of the roof then collapses. A continuous mining machine tears the coal from the face and loads it for transportation in one operation.

Only about 4 percent of U.S. underground production comes from longwall mining. In contrast, it is the predominant method in the U.S.S.R. (as in Western Europe), accounting in 1979 for over 65 percent of total Soviet underground output.[30] In some basins, this percentage is even higher, (reaching 96 percent in Moscow, 90 percent in Pechora, and 86 percent in Karaganda), the national average being held down by a relatively low level of longwall mining in the Donets basin. This

[27]B. Pichugin, "Coal Made Ready During the Summer," *Sotsialisticheskaya industriya*, July 17, 1980, p. 2.

[28]V. Lisin, "A Second 'Uralmash' on the Banks of the Yenisey River," *Trud*, Feb. 25, 1979, p. 1.

[29]See Simeons, op. cit., pp. 94-95; Environmental and Natural Resources Policy Division, Congressional Research Service, Library of Congress, *The Coal Industry: Problems and Prospects*, a background study prepared for the Permanent Subcommittee on Investigations of the Senate Committee on Governmental Affairs, 1978, p. 26.

[30]"Make Decisions of the 25th Congress . . . . ," op. cit., pp. 12-20.

is probably due to the lack of longwall miners designed for work on thin seams.

The Soviet equipment stock in underground mining is large and increasing, but its quality is declining. Although the number of mechanized complexes in use over the 4 years from 1975 to 1978 (inclusive) went up by 24.4 percent, the amount of equipment recorded as nonoperational went up by 73.7 percent.[31] This can be explained by the age of the equipment in use, the use of equipment unsuited to worsened geological conditions, and equipment repair and servicing practices.

Soviet sources point repeatedly to underground equipment requirements that are not being met. In particular, miners in the Donets basin are faced with the increasingly pressing need for equipment suited to new geological conditions. Sixty percent of Donets coal is being mined from thin seams less than 1.2 m thick, 50.7 percent of which are gently sloping, and 9.3 percent of which are steep.[32] At the beginning of 1978 thin seams already constituted 83 percent of the commercially recoverable coal reserves. Yet of a total of 50 working faces at one Donets mine, only 12 were being worked with appropriate equipment.[33] Shortfalls in production by the machine building industry are blamed for

this problem. Past emphasis on production of machinery for excavation of thicker and more productive seams had relegated thin seam equipment to a secondary, nonpriority role. Despite official recognition now of the need for thin seam excavators, equipment for thicker seams has continued to be developed.[34]

In addition to not meeting the present equipment needs of the coal mining industry, machine builders are criticized for a lack of attention to quality, reliability, and ease of repair.[35] Their seemingly slow response to changing needs in the industry is a function of a mix of operational constraints: a shortage of labor, insufficient production space in factories, pressures of shortrun production targets, and the fact that the coal industry is not the sole (nor even, in some cases, the primary) customer for their products.

Other deficiencies that continue to be cited include a shortage of equipment for the transport of support materials and personnel, drills of insufficient power and productivity, highly labor-intensive timbering techniques, low mechanization of tunneling operations, and ventilating systems of inadequate power and efficiency. The claim is made that while technical solutions for these problems have been developed, the necessary equipment for implementing change is not yet being produced.[36]

[31]Ye. N. Rozhchenko, "On Some Problems of the Development of Underground Coal Mining," *Ugol'*, No. 8, August 1979, p. 6.

[32]A. I. Bashov, "Dongiprouglemash's Developers Are Working for Donbass Miners," *Ugol'*, No. 10, October 1979, pp. 41-46, in JPRS 75,145, Feb. 15, 1978, p. 25.

[33]V. Deshko, "Equipment for Thin Seams," *Rabochaya gazeta*, Dec. 14, 1979, p. 1.

[34]*Rabochaya gazeta*, May 27, 1980, p. 1.

[35]Rozchenko, op. cit., p. 7.

[36]Ibid., p. 9.

SOVIET COAL INDUSTRY INFRASTRUCTURE AND RELATED AREAS: PROBLEMS AND PROSPECTS

Aside from the quantity and quality of mining equipment, the major problems facing the Soviet coal industry lie in labor supply and productivity; in the construction of

new mines; in the transport of coal; and in the amount of capital investment available to the industry. The following sections deal with these issues.

LABOR

The labor force employed in the Soviet coal industry is enormous. It has been estimated that in the early 1970's there were more than 1 million workers involved in the production of coal.[37] In comparison, the U.S. coal mining industry required only 159,000 people in 1972.[38] And despite the high absolute level of employment, Soviet labor shortages in the coal industry are becoming increasingly serious.

The major reason for the coal industry's voracious requirements is the low level of mechanization. Over 50 percent of those employed are still engaged in manual labor. Even in the more highly mechanized longwall mines, one-third of the work performed is manual. Much of this labor relates to auxiliary operations. Mine repair, roof control, and even some coal and rock loading is done manually.[39]

Labor shortages affect both coal extraction and coal mining machine building. *Pravda* noted in 1979 that in the Kuznetzk basin, the work force was 5,000 short in the underground mines alone. Labor shortages are also reported for the Karaganda basin. The director of the Gorlovka Machine Building Plant, a major coal mining equipment producer, recently complained that production targets cannot be met because the plant lacks workers. In October 1978, M. I. Shchadov, a deputy minister of Minugleprom, indicated that the industry as a whole was facing labor shortages and that the shortages were impeding output.[40]

These shortages may be exacerbated by the progressive reductions in the length of the workweek. Before 1956, mines operated 7 days a week. Between 1956 and 1958, extraction and development work began to shut down 1 day a week and the workday was reduced to 6 hours for some workers. In 1967, a 2-day weekend was introduced and miners doing heavy labor underground were given a 30-hour workweek. These reductions have created a demand for additional labor that is not likely to be met in the next few years, for the industry is experiencing difficulty in recruiting and keeping workers. At one time, coal miners were among the highest paid workers in the U.S.S.R., but now the difference between coal miners' wages and those of the average industrial worker is decreasing. Housing for coal miners is in short supply and this does little to attract workers. Shortages of labor are especially acute in the eastern regions of the country[41] and are affecting mine construction as well as coal output there. Labor turnover is also a substantial problem. In early 1980, turnover ran at about 20 percent of the total work force per year.[42]

Problems of this kind are not unique to the coal sector; they pervade all Soviet industries and the situation is likely to grow worse in the years ahead. The probability that the coal industry will have sufficient labor to solve its problems without other reforms is low. Thus, in coming years solution to the labor problem will rest on increases in labor productivity.

Table 24 gives official Soviet productivity figures for 1971 to 1977, the last year for which they are available. The amounts shown here are inflated by the fact that, like output data, Soviet productivity statistics are in terms of "raw" (i.e., uncleaned) coal mined per "production" worker, a category that excludes workers who would be counted in the West. Nevertheless, several trends are clear. Labor productivity for the industry as

---

[37]Stephen Rapawy, "Estimates and Projections of the Labor Force and Civilian Employment in the U.S.S.R., 1950 to 1990," (Washington, D.C.: U.S. Department of Commerce, Bureau of Economic Affairs, September 1976), p. 31; Strishkov, Markon, and Murphy, "Soviet Coal Productivity . . . ," op. cit., p. 48.

[38]Campbell, op. cit., p. 132.

[39]V. P. Podgurskiy and A. S. Minevich, "Reserves of Labor Productivity Growth," *Ugol'*, No. 7, July 1980, p. 43.

[40]Bogachuk, op. cit.; Mulkibayev, op. cit., p. 99; V. Vylgin, "In Every Column—A Minus," *Rabochaya gazeta*, May 27, 1980, p. 1; M. I. Shchadov, "Coal: Increase Extraction, Accelerate Deliveries," *Gudok*, Oct. 12, 1978, pp. 1-2 in JPRS 72,821, Feb. 14, 1979, p. 66.

[41]Kurnosov, op. cit.

[42]Podgurskiy, op. cit.

Table 24.—Labor Productivity in Soviet Coal Mining
(metric tons mined per person per month)

	1971	1972	1973	1974	1975	1976	1977	1979
Minugleprom	62.3	66.3	69.7	73.1	75.4	75.1	75.3	70.2
Underground mining..........	48.0	40.5	52.6	54.3	55.2	54.6	53.7	48.6
Surface mining	310.0	335.1	362.5	391.2	428.3	435.5	454.0	448.0
By basin:								
Donetsk (underground)	39.9	41.7	43.3	43.9	43.7	42.5	41.4	—
Kuznetsk	74.1	78.6	82.7	87.0	92.8	95.1	96.3	—
Underground	—	—	—	70.4	74.2	75.7	75.9	—
Surface	—	—	—	231.3	253.5	260.0	271.9	—
Karaganda	73.5	79.4	84.5	91.2	96.2	98.6	98.9	—
Underground	—	—	—	86.7	91.3	93.7	93.9	—
Surface	—	—	—	295.1	316.4	328.2	338.4	—
Moscow....................	74.0	78.4	82.7	87.2	90.5	87.4	86.0	—
Underground	—	—	—	80.2	83.4	80.4	78.8	—
Surface	—	—	—	306.5	303.4	283.6	272.2	—
Pechora (underground)	61.0	64.4	67.5	70.6	75.0	77.8	79.1	—
Kansk-Achinsk (surface)	—	—	—	—	—	929.9	909.3	—

SOURCES: April issues of *Ugol'*.

a whole rose through 1975, but since then appears to have stalled at around 75 mt per person month. Labor productivity in underground mining has decreased since 1975, although this decline has been offset by gains in surface mining, where productivity is 8.5 times higher. Continued gains in surface labor productivity must be counted upon to offset underground declines such as those apparent in the Donets and Moscow basins. Labor productivity in Donets is not even one-half as great as in the other major basins, due largely to a relatively low level of mechanization of mining operations. Since the Donets basin employs about 55 percent of the industry's labor force,[43] improvements in labor productivity here are particularly important.

MINE CONSTRUCTION

Coal mine construction organizations, like underground mining equipment manufacturers, were transferred to the administration of Minugleprom in the early 1970's. But here too, there have been complaints that no improvements have resulted. Instead, the construction firms have been cut off from

their old ministries and suppliers, and do not have production capacities of their own.[44] They are inhibited by lack of resources and labor shortages. In addition they must still contend with all the traditional impediments to the conduct of business in the Soviet economy: lack of cooperation from other organizations, poor plan development, shortages of labor and funds, and improper work practices.

The Central Intelligence Agency (CIA) has reported that additions of new coal mining capacity between 1976 and 1979 fell to the lowest level in nearly a decade. At the same time the rate of mine depletion has risen, and it has been estimated that over three-quarters of new mine capacity now merely offsets mine depletion.[45] Estimates of recent yearly mine depletions are given in figure 7. The depletions shown for 1978 through 1980 are substantial and may pose a serious impediment to coal output growth in the next 5 years. Since new Soviet mine capacity is taking 10 to 11 years to introduce, the Soviets will be largely dependent, for many years to come, on assimilating mines currently under construction.

[43]Central Intelligence Agency, "U.S.S.R.: Coal Industry Problems and Prospects," ER 80-10154, March 1980, p. 8.

[44]N. Klunduk, "Is This Really Economic?" *Pravda*, Dec. 11, 1978, p. 2.
[45]CIA, "USSR: Coal Industry . . . ," op. cit., p. 3.

Figure 7.—Estimated Mine Depletion, 1975-80
(million metric tons)

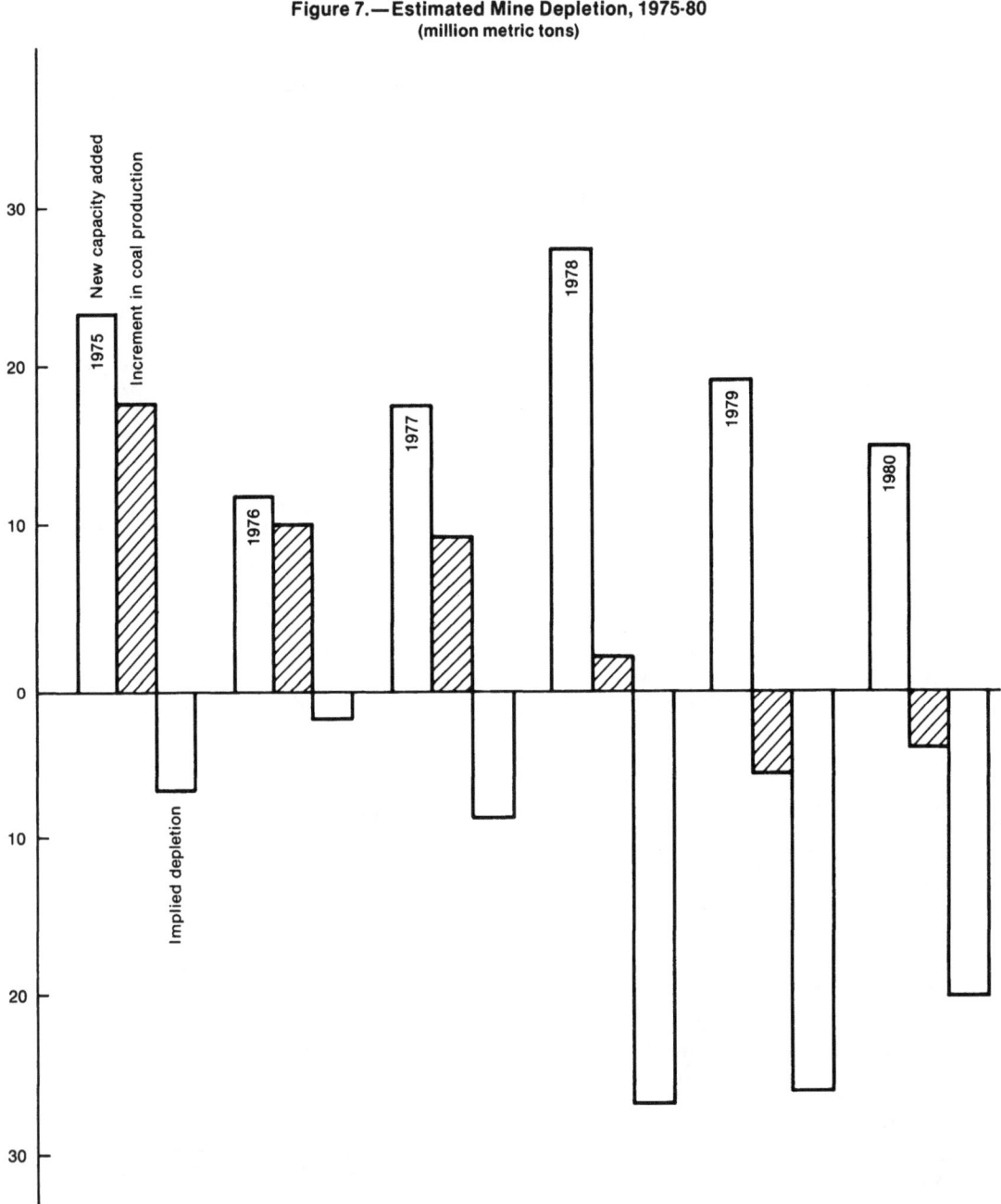

SOURCE: Methodology taken from CIA, "U.S.S.R.: Coal Industry. . .," op. cit., p. 5.
Data for 1975-1979: *Narodnoye khozyaystvo, op. cit.* various years.
1980: *Trud* (December 21, 1980), p. 2.

Mine reconstruction is also lagging. According to Soviet norms, a mine with a capacity of up to 3 mmt per year should require 5 to 7 years for reconstruction. In practice, reconstruction of many mines takes three times longer.[46] Due to shortages of appropriate new capacity, mine operators who must fulfill their output plan targets mine whatever coal is available, often damaging longrun development plans in the process. In their attempts to maintain output, they push mining into ill-studied coal seams and use machines in conditions for which they were not designed.

The severity of the problem may be suggested by the following example from Kuznetsk. In this leading Soviet coal region only one new mine has gone into operation in the past 18 years in two of the biggest production units—the Kuzbassugol and Leninskugol mining associations. At present, not one mine is under construction in the province that includes much of the basin.[47]

TRANSPORT

The bulk of growth in coal production is coming from Siberian basins. The limitations posed by transport conditions were officially recognized by the November 1979 Plenum of the Communist Party Central Committee in the emphasis it placed on solving transport problems associated with the growing flow and volume of freight. Basically, there are three choices open to the U.S.S.R. for the transport of coal from remote regions: rail, slurry pipeline, and the conversion to electricity at source and transmission by wire. This latter option is discussed in chapter 5.

Rail

At present, coal is transported almost entirely by rail and a number of factors hinder its delivery. These include losses of coal during shipment, and more important, the inefficient management and insufficient capacity of the rail system.

[46]S. Zayganov, et al., "On Mining Technology," *Pravda*, Oct. 14, 1979, p. 2.
[47]Shumkin, op. cit., p. 1.

First, some Soviet coal is difficult to transport economically. Ekibastuz and Kuznetsk coals, for instance, can be transported in run-of-mine form, but Kansk-Achinsk coal, which tends to self-ignite when dried, can be shipped only 1,500 to 2,000 km in untreated form. Longer distances will not be practical until beneficiation technologies have been developed and put in place. This is not likely to occur until the latter part of the 1980's.

Second, loss of coal from rail cars is a major problem. Some rail cars leaving the Donets basin arrive at the power station with only one-half their cargo remaining, and about 4 percent of the coal shipped from the Karaganda basin is lost during transport. Aside from theft, coal is lost in two ways. It either leaks out of the bottoms and sides of the cars or it is blown out of the open cars by the wind. The introduction of continuous mining machinery has led to shipments containing finer coal than was the case previously. Consequently, during transport in open rail cars 1 ton of coal may be blown away in each car for every 1,000 km traveled. The

[48]V. Sonin, "The Tracks Are Sown With Coal," *Pravda*, Nov. 1, 1979, p. 3; "Russians Plan Surface Mine Complex for Power Generation," *Coal Age*, October 1978, p. 47.

Photo credit: TASS from ©SOVFOTO

Coal-loaded trains leave Karaganda

Ninth FYP (1971-75), called for Minugle-prom to construct eight facilities in the Donets basin to coal the coal in the rail cars with a protective film, but as of late 1979 none had been built.[49]

Third, there is substantial evidence of mismanagement in the rail transport of coal. While there is a shortage of rail cars in Siberia to transport coal to the Urals, rail cars are standing idle on sidings in other regions. In 1978, the Soviet paper *Izvestiya* reported that coal from Uzbekistan was being shipped to electric power stations in Kirgiziya, while coal from Kirgiziya was being shipped to electric power stations in Uzbekistan. This was because coal had to be shipped to the Angren electric power station to maintain operations during the peakload period. When the station returned to normal operation the coal shipments continued (and this despite the fact that the coal had already damaged station equipment). Coal continued to arrive from Kirgiziya for the next 2 years and piled up at the station.[50] Again, such anecdotes are not isolated instances. They reflect deep systemic problems.

The most important constraints in the ability of the existing rail system to handle coal shipments, however, lie in factors related to rail management and engineering. These difficulties are not grounded in the Soviet Union's capability to solve technical problems in rail transport. The technology for electrification, double-tracking, and locomotive and freight car design and construction is well-established. The real constraint lies in a past heritage of mismanagement and in the inadequacy of capital investment funds allotted to the rail system, manifest in the railroad's poor economic performance in recent years.[51]

The engineering and technical improvements needed to increase coal's share of total rail freight would include:[52]

- increasing the length of jointless track on improved roadbeds;
- increasing track capacity at arrival and departure points;
- increasing the share of eight-axle rail cars in the rolling stock fleet in combination with locomotives of increased power;
- continuation of ongoing and planned double-tracking and electrification; and
- improvements in freight car and track repair and maintenance practices to curb coal losses en route.

The Eleventh FYP calls for construction of only 3,500 km of new mainline and 5,000 km of secondary rail line.[53] Much of this relates to the Baikal-Amur Mainline (BAM) railroad in the Far East, and will contribute little to facilitating the transport of Siberian coal to the European U.S.S.R. Thus, while marginal improvements in the system may continue, the improvements necessary to support an increased flow of Siberian coal are not likely to be met in the near future. Expansion of the rail system to allow for greater coal freight may be hindered by a reluctance on the part of the Soviet leadership to commit itself to Siberian coal development (see ch. 8).

Slurry Pipelines

Coal slurry pipelines are a possible solution to the transport problem posed by the substantial distance separating primary sites of energy consumption and the country's largest fuel reserves. Although the initial capital investment for such pipelines is high, operational costs relative to rail trans-

[49]Sonin, op. cit., p. 3.
[50]G. Dimov, "Why Take Coal to Coal?" *Izvestiya*, Sept. 5, 1978, p. 2.
[51]Central Intelligence Agency, "The Soviet Economy in 1978-79 and Prospects for 1980," June 1980, p. 8.

[52]T. M. Borisenko and V. P. Vodyanitskiy, "Evaluation of Possible Ways of Increasing the Economic Effectiveness of Systems of Long-Distance Hydraulic Transport of Coal," *Izvestiya an SSSR: Energetika i transport*, No. 4, April 1979, p. 44.
[53]*Izvestiya*, Dec. 2, 1980, p. 5.

port are low.[54] Pipeline transport of coal would also circumvent the problem of stepping-up use rates on an already intensively operated rail system.

While underground slurry pipelines hold certain advantages over rail for long-distance coal transport (no loss of transported material en route; reduction in noise and pollution; increase in land made available for alternative uses; and greater process automation) several problems remain to be resolved. There has been little study of the physical-technical processes of pumping coal slurry through the large diameter pipes required for efficient pipeline transport over long distances.[55] Moreover, the Soviet Union does not now produce the basic equipment needed for slurry preparation plants, pumping stations, and end-of-line installations for preparation of the coal for burning. These include high-capacity slurry pumping units, centrifugal pumps capable of handling large amounts of slurry, and dependable wear-resistant fittings (flush ball cocks, reflux valves, etc.).[56]

Two industrial coal slurry pipelines have been built and are in operation in the Kuznetsk basin. The large particle size of the coal being pumped through these lines (up to 50 mm) has led to significant wear in the pipes and has required that they be periodically turned and replaced.[57] These pipelines are short (not over 10 to 11 km) and lie above ground, facts that facilitate maintenance. Nevertheless, pipeline erosion of this type would have significant impact on the cost of operating coal slurry pipelines of greater lengths.

Construction of a 250-km pipeline, connecting the hydraulic underground mine Inskaya in the Kuznetsk basin to a thermal electric power and heat station in Krasnsyarsk, was scheduled to begin in 1982, but there are now indications that the project will be delayed until 1984.[58] This line is to serve as the prototype for 2,000- to 4,000-km pipelines connecting the Kuznetsk basin to the western U.S.S.R. The earliest date given for creation of long-distance coal slurry pipelines is 1990. Additional constraints on development of long-distance coal slurry pipelines arise from the need to prevent the slurry from freezing in cold weather, and the high-volume water requirements associated with hydrotransport. These caveats are particularly important for pipelines originating in Siberia.

In short, there is a disparity between Soviet progress in slurry theory and practice. On the one hand, the Soviet Union has led the world in development of hydrotransport theory. "Soviet insight into the structure of fine coal slurries is of the highest order,"[59] even in the absence of a long-distance pipeline of a length comparable to the U.S. Black Mesa coal slurry pipeline. Yet Soviet domestic capabilities for construction of pumping equipment remain undeveloped.

CAPITAL INVESTMENT

Annual investment in the coal industry rose from 1.4 billion rubles in 1965 to 2.0 billion rubles in 1979 (see table 25), but coal's share in total investment in industry fell from 6.9 to 4.5 percent. In comparison, investment in the petroleum industry rose from 10.0 percent of all industry investment in 1965 to 12.9 percent in 1979, while investment in the gas industry rose from 3.0 to 4.5 percent. The coal industry has thus failed to keep pace with the other fuel sectors, and the current state of the industry reflects this fact. Its recent poor performance supports the judgment that past investment has been too small and too irregular to maintain, let

[54]"Development of New Types of Transport," *Stroitelstvo truboprovodov*, No. 11, November 1978, pp. 36-37.

[55]Borisenko and Vodyanetskiy, op. cit., p. 37.

[56]Ye. Olofinskiy, *Planovoye khozyaystvo*, No. 8, August 1980, p. 95, in JPRS 76,585, Aug. 8, 1980, p. 36.

[57]Ibid.

[58]"Kuzbass-Novosibirsk Coal Pipeline," *Sotsialisticheskaya industriya* Sept. 12, 1980, p. 2 in JPRS 76,654, Oct. 20, 1980, p. 50; *Planovoe khozyaystvo*, No. 5, May 1981, p. 25.

[59]John W. Kiser, III, "Report on the Potential for Technology Transfer from the Soviet Union to the United States," (Santa Monica, Calif.: Rand, August 1974), p. 40.

**Table 25.—Capital Investment in
Leading Fuel Industries**

Year	Coal	Oil	Gas
(million rubles, constant prices)			
1965	1,426	2,070	615
1970	1,541	2,527	1,041
1975	1,759	3,853	1,798
1976	1,747	4,066	1,835
1977	1,848	4,503	2,031
1978	2,035	5,270	2,210
1979	2,020	5,860	2,020
(percent of total investment in industry)			
1965	6.9%	10.0%	3.0%
1970	5.3	8.8	3.6
1975	4.4	9.7	4.5
1976	4.3	10.0	4.5
1977	4.5	10.6	4.8
1978	4.5	11.6	4.9
1979	4.5	12.9	4.5

SOURCE: *Narodnoye khozyaystvo*, various years.

alone expand, productive capacity. The situation is aggravated by constantly rising costs of mining and mine construction in the European basins.

Evidence of insufficient investment in the coal industry can be seen in a number of areas: 1) the extremely low level of introduction of new mine capacity at many basins; 2) the insufficient productive capacity at plants producing mining equipment; 3) the lack of repair facilities at many basins and the attendent rise in machinery downtimes; 4) the shortage of enrichment facilities; 5) the ill-repair of rail cars; 6) the shortages of locomotives and rail cars; 7) the short supplies of spare parts; 8) the lack of equipment suitable for working thin or pitching seams; 9) the low level of mechanization of many basins, including the Donets; and 10) the lack of large capacity trucks and rail cars at surface mines. Probably the two most important of these areas are mine capacity and transport.

Mine Capacity

One of the most serious effects of the present low level of investment in the coal industry has been the failure to prevent a decline in productive mine capacity. The

Soviets hope to achieve greater increases in productive capacity per ruble of investment by switching from underground to surface mining. At present, the share of surface mining in total mining is substantially lower for coal production than for the mining of ferrous and nonferrous metal ores, and for chemicals. Soviet industry officials believe that a 1-percent increase in the share of surface mining in total output (accompanied by a 1-percent decrease in underground mining) could save 80 million rubles per year, lower the capital intensity of the industry by 1 percent, and raise labor productivity by 1.4 percent.[60] It must be noted that increases in surface mining would have to be substantial before such an effect would result in decreased expenditures by the industry, but increases here would help the Soviets to achieve greater increases in capacity per ruble spent.

The differences between surface and underground mining have an important geographic dimension. Coal production from the underground mines in the European U.S.S.R. is stagnating or falling. At the same time, production is becoming increasingly costly as mines go deeper and the quality of mined coal falls, and as thin or steep seams account for greater shares of output. The cost of coal mined in these basins is now as much as several times higher than the cost of coal mined at Kansk-Achinsk and Ekibastuz, even when compared on a calorific basis.

In the mid-1970's, the required capital investment for introducing new mine capacity in the Donets basin was 64.3 rubles/mt of new capacity and in the Moscow basin, 89.7 rubles/mt. By comparison, the required capital investments in the Ekibastuz and Kansk-Achinsk basins were 8.2 rubles/mt and 9.6 rubles/mt, respectively. The cost per ton of coal mined from new capacity was correspondingly high in the Donets and Moscow

---

[60]M. I. Shchadov, "Improving Equipment and Technology for Surface Mining of Coal Deposits," *Ugol'*, No. 1, January 1981, p. 1.

basins (17.0 rubles/mt and 24.1 rubles/mt, respectively), and low in the Ekibastuz and Kansk-Achinsk basins (2.5 rubles/mt and 2.4 rubles/mt). Extraction of Kansk-Achinsk and Ekibastuz coal is thus economically more attractive than extraction of any other coal in the U.S.S.R.[61]

The relatively large investments that would be required to maintain Donets and Moscow basin production are a serious deterrent to production there. This fact is largely responsible for the turn to eastern coal in Soviet economic planning.

Coal Transport

A recent Soviet source has given rough cost estimates for various options for transporting coal from Siberia to the Urals or farther west. These options include: 1) rail transport of coal not requiring beneficiation before transport (presumably Kuznetsk coal); 2) transport of coal requiring beneficiation, i.e., Kansk-Achinsk coal; and 3) transport of coal as electricity. This assumes a volume of traffic on the order of 250 to 300 mmt of coal per year.[62]

Rail transport of Kuznetsk coal to the Urals or the European U.S.S.R. is an attractive option. Coke from Kuznetsk is significantly less expensive to produce than coke from Donets, and is competitive with the latter anywhere in the U.S.S.R. However, expansion of rail traffic from the

Kuznetsk basin would call for substantial investments in the rail system, and would increase the rail sector's demand for heavy-duty steel rails, labor, and improved freight cars and locomotives. This would place additional stresses on the already strained steel and machine building industries, and on an increasingly tight labor market.

Building a railroad, even in the more favorable terrain of the area, might cost up to 1 million rubles/km.[63] (It probably would not be necessary to build completely new rail lines, since junction with existing railroads would be possible at certain points.) Increased traffic on the railroads will also lead to faster depreciation of the track. Present rails handling 100 to 120 mmt of traffic a year wear out in 4 to 5 years. Their life would decrease to 2 to 3 years if 320 to 350 mmt of traffic were carried,[64] and steel rail demand would therefore remain high after construction of the initial rail line was completed.

Transport of 250 to 300 mmt of coal per year from Kansk-Achinsk would entail not only upgrading or expansion of the rail system, but also the expenditure of huge sums on coal beneficiation facilities. Since the coal is of such low quality, about 600 to 700 mmt would have to be mined in order to obtain 250 to 350 mmt of upgraded coal. No less than 25 treatment plants would have to be built at a total cost of 10 billion to 12 billion rubles. The investment in the treatment plants alone is enormous and is on a level with the required investment for the proposed Siberian natural gas pipeline to Western Europe.[65]

[61]These costs are termed *privedennyye zatraty,* a cost that includes direct costs, plus a capital charge at an interest rate appropriate for the given industry. Ya. Mazover, "Perspectives of the Kansk-Achinsk Coal Basin," *Planovoye khozyaystvo,* No. 5, May 1976, p. 66.

[62]V. E. Popov (ed.), *Siberian Fuel-Energy Complexes,* (Novosibirsk: Izd. "Nauka," 1978), p. 207; See also A. Probst, "Ways of Developing the Fuel Economy of the U.S.S.R.," *Voprosy ekonomiki,* No. 6, June 1971, p. 57; Ya. Gantman, "Structural Changes in the Fuel Balance of the U.S.S.R.," *Planovoye khozyaystvo,* No. 11, November 1971, pp. 88-91.

[63]B. S. Filippov, "The Effectiveness of Transporting Kuznetsk Coals and Coke to the Center of the European U.S.S.R.," *Koks i khimiya,* No. 3, March 1977, p. 51.

[64]Popov, op. cit., p. 207.

[65]David Brand, "Soviet Slip-Up," *Wall Street Journal,* Jan. 23, 1981, p. 1.

WESTERN TECHNOLOGY IN THE SOVIET COAL INDUSTRY

As the above analysis has indicated, the Soviet coal mining machinery industry is capable of producing—and does produce—equipment of sufficient quality to meet the needs of the industry. The power and capacity of this machinery often tends to be below the best Western models, but the Soviets are improving in this respect. For all the shortcomings described above, the production of many types of machinery increased at substantial rates between 1970 and 1979. Continuous miner production was up 5 percent; heading machines were up 78 percent; and loaders were up 28 percent. In addition, the U.S.S.R. actually exports mining equipment, and the share of annual production of mining machinery exported has risen. For example, the export of continuous miners rose from 5 percent of production in 1970 to 9 percent in 1979. Overall, in 1979, the Soviets imported about 74 million rubles worth of mining equipment (all types), while they exported about 211 million rubles worth.[66] This reinforces the impression that the Soviets suffer not so much from an

overall equipment shortage as from the lack of capacity to produce specific models. These areas of need—in which thin-seam miners and surface mining equipment (power shovels, draglines, drilling equipment, bucket excavators, trucks, and rail cars) figure prominently—have been supported by modest imports.

Table 26 shows official Soviet foreign trade statistics for imports of all types of mining machinery. These statistics should be treated with some caution. First, they are certainly incomplete. For instance, Japan is not listed as a source of imported machinery, but significant amounts of Japanese equipment are known to be in use in the South Yakutian basin. Second, the figures are highly aggregated; they do not break out coal from other mining equipment.

The volume of Soviet mining equipment imports rose dramatically during the 1970's, peaking at about 92.6 million rubles in 1978. In 1979, 38.3 percent of these were from the West (8 percent of total imports were from the United States); 47.5 percent were from Eastern Europe; and 14.2 percent were unidentified. It would appear, therefore, that purchases from the United States are neg-

[66]U.S.S.R. Ministry of Foreign Trade, *Vneshnyaya torgovlya SSSR v 1979 g.* (Moscow: Izd. "Statistika," 1980), pp. 21, 34.

Table 26.—Soviet Imports of Equipment for Underground and Surface Mining of Minerals
(thousand rubles)

Year	Source							Total
	U.S.A.	Poland	G.D.R.	F.R.G.	France	Sweden	Czechoslovakia	
1970	2,045	—	7,987	556	969	1,515	6,394	19,857
1971	338	—	3,383	1,792	1,868	1,925	1,343	10,768
1972	353	—	2,026	1,552	2,250	404	1,033	7,623
1973	377	—	2,606	2,022	2,370	2,117	2,687	12,356
1974	438	—	3,449	10,344	1,068	7,432	—	26,719
1975	7,858	—	4,017	20,515	3,973	10,108	—	58,181
1976	7,287	—	5,461	6,964	9,999	7,961	—	49,976
1977	674	—	8,826	8,676	851	1,419	—	36,303
1978	2,836	11,426	33,098	20,710	1,898	9,923	—	92,636
1979	9,984	10,624	24,504	14,104	449	3,798		73,935

SOURCE: U.S.S.R. Ministry of Foreign Trade, *Vneshnyaya torgovlya SSSR* (Moscow: Izd. "Statistika") for various years.

ligible and that Eastern Europe figures more prominently as a supplier of Soviet mining equipment than does the West. Poland, for instance, is an important producer of underground equipment. The U.S.S.R. has also purchased excavators from East Germany.

Western trade statistics and trade journals provide more information on the precise nature of imports from the West. The two major areas here are equipment purchased from Japan for development of South Yakutian coal, and trucks for use in surface mining operations. (Meanwhile, the U.S.S.R. is attempting to increase its own capacity for the production of large—up to 180 tons—dump trucks.) In general, compared to the oil and gas industries, the Soviet coal industry shows little reliance on Western technology and equipment. The Soviets have opted largely for domestic development and production of equipment, despite the fact that superior models may be available in the West.

Nor is there evidence of much use of Western computers in the Soviet coal industry. Underground and surface mining are not particularly amenable to the application of computers. The Soviets would only be likely to turn to the West in these areas if they could acquire breakthroughs in automated mining. Given the low level of mechanization in the industry and its secondary priority after petroleum and nuclear power, such a development is unlikely. Computerization in Soviet coal mines is, therefore, not expected to be important in the next 10 years, although it could contribute to the rationalization and management of the industry. Given the pervasive systemic problems described above, the coal industry would at best benefit slowly and indirectly from transfers of software. It has not sought such technology itself, and is not likely to do so in the near future.

PROSPECTS FOR THE SOVIET COAL INDUSTRY

The Eleventh FYP calls for the production of 770 to 800 mmt of coal per year by 1985. Achievement of output in this range, which is lower than the original 1980 goal (790 to 810 mmt), would represent the reversal of previous trends of declining production and restore a modest rate of growth for the industry as a whole. Most of this growth would be achieved by expanding surface mining. The FYP targets envisage surface mining constituting 39 to 40 percent of total output (300 to 320 mmt), leaving 470 to 480 mmt of underground production. 1980 underground output was 451 mmt. Thus, the intention is to at least hold underground production stable.

Soviet targets for the Twelfth FYP (1986-90), if they exist, have not been published. However, the literature does support the qualitative judgment that the Eleventh FYP period is intended to be a time of preparation for a period of more intense growth to follow.

The 1981-85 respite will hopefully allow time to permit the expansion and upgrading of surface mine capacity, coal processing capacity, the stock of surface mining equipment, and the rail transport system.

In this section, OTA has attempted to evaluate these goals and to provide estimates of plausible levels of production in 1985 and 1990. As with all of the projections in this report, **the figures provided here are not predictions**. Rather, they are projections based on OTA's judgments of likely outcomes, given explicit accompanying assumptions. These estimates, together with 1980 production figures, are given in table 27.

1981-85

OTA believes that the production range for the Soviet coal industry specified in the plan is unrealistically high, and that the best

Table 27.—Estimated Soviet Coal Production
(million metric tons)

		Years		
	1980	1985 (plan)	1985 (best case projection)	Percentage change
U.S.S.R. total	716	(770-800)	765	+ 7
Major basin[a]:				
Donets	203		195	– 3
Kuznetsk	154	(167)	160	+ 4
Ekibastuz	67	(85)	85	+ 20
Kansk-Achinsk . . .	40		50	+ 25
Karaganda	49	(49)	49	0
Pechora	30		26	– 13
Moscow	26		25	– 4
South Yakutia. . . .	3		11	+ 267
By surface mining . . .	264	(300-320)	320	+ 21

[a]Estimates.

SOURCE: Office of Technology Assessment; *Soviet Geography*, April 1981, p. 280.

that could be expected is production of 765 mmt. This growth of roughly 7 percent is short of the low end of the range specified in the plan, but even this figure should be regarded as a highly optimistic best case, which might be possible if the U.S.S.R. could fulfill announced plan targets for surface mining and halt the decline in underground output. Some experts believe that the latter is impossible and that a more realistic projection would be some 20 mmt lower.

OTA's most optimistic scenario corresponds closely to recent CIA projections,[67] and is based on the following assumptions:

- No dramatic changes in the present organization of the economic system as it affects coal production.
- No dramatic change in the priority to be accorded to the coal industry; i.e., OTA assumes that coal will retain its "second-class" status, at least for the next 5 years, while attention is concentrated on nuclear power and gas development. This subject is discussed in more detail in chapter 8.
- No major labor shortages. Growth in coal output will come almost exclusively from Siberian surface mines that

have a labor productivity nearly nine times as great as underground mines. The shift to surface mining, coupled with continued mechanization and automation of underground operations, should, therefore, help to alleviate labor shortages.

- Few, if any, new measures taken to provide greater protection of the environment. OTA assumes that despite official rhetoric affirming the need for greater environmental protection, only those measures that would not lead to significant sacrifices in output will be instituted.
- Investment resources increased sufficiently to provide for a low level of growth.
- Expansion of coal mining equipment production in the following areas: larger capacity power shovels and draglines; excavating equipment and electrical systems for Siberian climate; special subcomponents, lubricants, ventilation and other systems for excavators; large diameter steel cables and rolled metal for excavators; spare parts; drilling equipment with improved productivity; larger capacity mine trucks; conveyer belts with improved strength; and equipment for mining coal from thin and steeply pitched seams.
- Continuation of present levels of equipment and technology imports, i.e., of the policy of relying heavily on domestic technology. Specifically, the projections assume that the U.S.S.R. continues to import Japanese equipment for development of Yakutia; to purchase little or no Western underground equipment; and to expand imports of surface mining equipment, mainly from East Germany and Poland.

A major industry concern will be the maintenance of coking coal production. Virtually all Soviet coking coal is mined underground, and underground mining will decline in many basins. Yet, even with substantial declines, four factors suggest that industry need not suffer from a lack of coke:

---

[67]The CIA in "U.S.S.R.: Coal Industry . . . ," op. cit., posits a range of 765-785 mmt.

1) there is reason to believe that domestic consumption of coke will rise by only 4.5 mmt between 1977 and 1985, and by only 0.5 mmt between 1986 and 1990;[68] 2) about 44 to 56 mmt of coking coal per year are burned at electrical power generating plants (due to insufficient enrichment capacity to render these coals suitable for coking);[69] 3) some coking coal being mined is improperly categorized as steam coal;[70] and 4) of total coking coal mined each year, about 4 percent may be lost in transport.[71] The latter "use" represents a potential source of coking coal for productive domestic consumption if the Soviets are willing to make necessary but expensive improvements in the transportation system.

The best case growth in output projected here would be largely supported by a growth in surface mined coal from about 264 mmt in 1980 to about 320 mmt in 1985, in accordance with the FYP target. The share of surface mining in total output would therefore rise from 37 percent to about 40 percent. These coal increments could come almost exclusively from Siberia. The following developments seem likely in individual basins:

Donets

Mine depletions here will probably exceed the introduction of new capacity. The share of coal mined from deep, thin seams will increase, resulting in slower rates of coal extraction. At the same time, the cost of coal mined will continue to rise, and this will promote increased substitution in consumption of cheaper eastern coals. Production in the basin will probably fall to 195 mmt or less by 1985.

Kuznetsk

This basin has vast reserves, and coal mined here is cost competitive with Donets coal in many regions of the European U.S.S.R. It is likely to become the U.S.S.R.'s leading producer of coking coal before 1985. However, the growth in new mine capacity has been slow and more coal enrichment capacity is needed. Only a small increase in output—from 154 to 160 mmt—can be expected by 1985.

Karaganda

Introduction of new mining capacity has lagged seriously. Much of the coking coal cannot be coked without prior enrichment and, due to insufficient enrichment capacity, is not being mined. As in the Kuznetsk basin, only a small increase in production can be expected by 1985. Output may rise to 49 mmt by 1985, but stagnation or a decline in production at the basin cannot be ruled out.

Moscow

Possibilities for increased coal production here are virtually nonexistent. The coal has a high ash and sulphur content and is becoming more and more expensive to mine. Annual production should fall by at least 1 mmt—and perhaps by as much as 5 mmt—by 1985.

Pechora

The Pechora basin contains large reserves of high-quality coking coal and production could have been substantially expanded, but little new mine capacity has been added in the last 15 years. The basin probably will lose 4 to 5 mmt of yearly capacity by 1985.

Urals

The coalfields in the Urals are being depleted and production will decline. Current production is not sufficient to meet even local needs.

---

[68]*Bakinskiy rabotnyy*, Apr. 19, 1981, p. 1.

[69]The slowing of growth in coke consumption is due to likely reductions in the requirements for coke per ton of pig iron produced. This conclusion is based on the finding of a 1980 Battelle report on energy efficiency in the Soviet iron and steel industry. See ch. 7.

[70]M. V. Golitsyn and V. F. Cherepovskiy, "Analysis of U.S.S.R. Coal Reserves and Main Directions of Geological Prospecting Works," *Sovetskaya geologiya*, No. 4, April 1980, pp. 25-28.

[71]Ibid., pp. 27-28.

Kansk-Achinsk

Production could grow rapidly, but problems of transport and use will remain. The contribution this coal can make to the energy supply of the central regions or the Urals before 1985 therefore is highly questionable. In any event, annual production could increase by 10 mmt by 1985.

Ekibastuz

Plans for Ekibastuz production have been announced and call for increases of some 15 mmt by 1985. However, the quality of Ekibastuz coal is extremely low.

South Yakutia

Development of this basin is behind schedule, but production could grow to 11 mmt by 1985. However, a large share of output will be exported to Japan as compensation for developing the basin (see ch. 11.)

It is at least possible that the Soviet Union could come close to reaching its 1985 coal output targets. But the significance of this growth in output should not be overestimated. First, as noted above, the calorific content of Soviet coal has been falling steadily. If past trends continue, it will probably fall by roughly 1 percent per year between 1980 and 1985. Gains in run-of-mine output will therefore be largely offset by declines in calorific content. Fuel output, if calculated in tons of standard fuel, could actually decline unless output at high-quality coal basins remains at least relatively stable.

Second, the fact that much of the increase in coal output is to come from Kansk-Achinsk and Ekibastuz puts severe limitations on the use to which the coal can be put. Kansk-Achinsk coal cannot at present be transported to the Urals or the European U.S.S.R., let alone to export markets. Production is soon likely to exceed local demand and, as chapter 5 discusses in detail, generation of electricity at the mine site creates a number of other problems for Soviet plan-

ners. Ekibastuz coal is also of very poor quality, some of it nearly half ash.

In sum, even an increase in production of coal to 765 mmt per year, very close to plan targets, may mean an absolute decline in standard fuel produced. Moreover, much of what is mined cannot at present contribute to fuel supplies in consuming centers of the European U.S.S.R. because it is uneconomical to transport in untreated form. Achievement of 1985 plan targets, therefore, will contribute little to efforts to substitute coal for oil in existing powerplants or the few new ones to be constructed in the European part of the country.

1986-90

Projections for the Twelfth FYP period are necessarily highly speculative and must remain sketchy. In general, however, if present trends continue, and if the U.S.S.R. can come close to realization of 1985 targets, output could continue to grow. The amount of this increase would depend on the success of surface mining operations, although gains in surface mining would continue to be offset by declines in underground production.

The most likely areas in which to expect high rates of growth in the latter half of the decade are the Kansk-Achinsk, Kuznetsk, Ekibastuz, and South Yakutian basins. Stable or declining production can be expected in the Moscow and Donets basins, but the Soviet literature hints at investment plans that could lead to a small growth in output at Karaganda and a recovery at Pechora to about the 1980 level.[72]

Even assuming very high rates of growth in surface mining basins, however, (50 percent in Kansk-Achinsk, 25 percent in Kuznetsk, 35 percent in Ekibastuz, and 35 percent in South Yakutia)—a highly optimistic assumption—it is difficult to imagine coal

---

[72]"Russians Plan Surface Mine Complex for Power Generation," *Coal Age*, October 1978, p. 47.

output rising over 1985 levels by more than 100 mmt. Surface-mined coal would thus have to constitute about one-half of all coal. The significance of this level of output for the Soviet economy would depend upon success in constructing the coal treatment plants necessary for making use of Kansk-Achinsk and Ekibastuz coal, and on the fate of plans for long-distance electricity transmission.

SUMMARY AND CONCLUSIONS

The Soviet coal industry has encountered serious problems in the past few years for which no solution is yet in sight. These have to do with the declining output of underground mines located near centers of consumption; the fact that new deposits lie in remote areas of Siberia; and the declining quality of the coal that is being produced.

The Eleventh FYP establishes goals that are dramatically less ambitious than those of previous plans, a fact that may reflect a realization and acceptance by planners of the real limits placed on growth of output by the combination of problems facing the industry. Even so, these targets are probably excessively optimistic, and even gains in overall coal production will be offset to some degree by the fact that the quality of much of the new coal being mined is low. In fact, coal output could increase and its standard fuel equivalent actually decline.

The Soviet coal industry suffers from many of the same ailments afflicting most sectors of the economy. The problems are to a large degree systemic and have no permanent solutions short of major reforms of the system itself. The time has come for the coal industry to "fine tune" its operations. Unfortunately, the Soviet economy is ill-suited to such a task. The situation here has been aggravated by the low priority assigned to the coal industry in the past, and the fact that in order to achieve meaningful increases in output, a number of problems must be simultaneously addressed. These include labor productivity, additions to mine capacity, increasing the quality and quantity of mining equipment, resolving coal transport problems, and devising ways to use the low-quality coals that are making up an increasing share of production.

It is the combination of these difficulties that has led to the declining performance of the coal industry as a whole. There is little reason to expect that such obstacles will be overcome in the present decade. Nor is it clear that even massive improvements in one or several of these areas (e.g., labor productivity) could do more than increase coal industry efficiency, without necessarily significantly affecting output.

At present, there is no evidence to suggest that extensive Western participation in Soviet coal development would greatly boost production. Aside from the South Yakutian basin, which is being developed with the assistance of the Japanese, the Soviets have made little use of Western coal mining equipment and technology. Most such imports have been in the area of surface mining, especially large capacity mining trucks. The cessation of these supplies could have an impact on the efficiency of Soviet surface mines, but it is unlikely that the converse would hold, i.e., that more Western trucks would alone lead to increased coal output. The Soviets, moreover, have recourse to their own truck industry. If sufficient resources are allocated to production of such domestic models (something that cannot be taken for granted), the Soviets could satisfy demand for large-capacity trucks themselves. The Soviets are constructing a plant near Kansk-Achinsk to manufacture heavy excavators. Its successful completion would be another step towards independence from Western surface mining technology.

An embargo now, therefore, of all Western trade with the U.S.S.R. would inconvenience it—but would not seriously impair coal production. Similarly, Western assistance alone is unlikely to be able to boost coal production. Possibilities for expanded domestic production of equipment and for imports from Poland and other East European countries would compensate for losses of Western equipment. The longer run impact of such an embargo is more difficult to predict. If the U.S.S.R. places priority on expanded coal output, this growth will have to come largely from surface mining, since it appears that underground mining capacity has irreversibly peaked. If later in the decade bottlenecks in surface excavation and haulage equipment become troublesome, it is possible that the U.S.S.R. would look to the West for significant amounts of this equipment. However, these imports would have to be accompanied by serious efforts to solve a much wider array of coal industry problems.

CHAPTER 4

The Soviet Nuclear Power Industry

CONTENTS

LIST OF TABLES

LIST OF FIGURES

The Soviet Nuclear Power Industry

Soviet planners see nuclear energy as an increasingly important source of electricity. Current plans call for nuclear power to generate most of the incremental electricity provided to the European U.S.S.R. in this decade. The Soviet nuclear program has logged impressive gains. The portion of electricity supplied by nuclear power rose from 0.5 percent in 1970 to more than 5 percent in 1980; and production increased more than twentyfold, from 3.5 billion kilowatt hours (kWh) to as much as 73 billion kWh, over the same period.[1] Moreover, the current Five Year Plan (FYP) calls for nuclear's share of electricity production to more than double to 14 percent by 1985 and estimates for the year 2000 have ranged as high as 33 percent.[2] The ease with which the U.S.S.R. is able to adjust to its problems in the coal and oil industries will depend in part on its ability to fulfill—or at least approach—these targets.

This chapter summarizes Soviet policy toward nuclear power, describes the present state of the Soviet nuclear power industry, and evaluates Soviet planners' goals for the contribution of nuclear-generated electricity to the energy balance in 1985 and 1990. It also examines the past and potential contribution of Western equipment and technology to that industry.

---

[1] *Ekonomicheskaya gazeta,* No. 12, 1981, p. 1, gives a figure of 73 billion kWh. Earlier estimates were 70.5 billion kWh.

[2] W. Lepkowski, "U.S.S.R. Reaches Takeoff in Nuclear Power," *Chemical and Engineering News,* Nov. 6, 1978, pp. 31-36.

SOVIET POLICY CONSIDERATIONS

The Soviet Union has generated electricity from nuclear power since 1954, when its first nuclear power station (NPS) came online at Obninsk, near Moscow. The Soviets are proud of the fact that the U.S.S.R. was the first country in the world to produce commercial nuclear-powered electricity. At the same time, development of the nuclear industry was slow, and Soviet-installed nuclear capacity at the end of 1980 was about one-fourth that of the United States in the beginning of that year—13,460 megawatts (MW) as opposed to 52,300 MW.[3] The two industries have also experienced different patterns of growth. While nuclear capacity in the United States expanded rapidly in the 1960's and early 1970's, the pace of the Soviet industry did not begin to accelerate until middecade, just the time, in fact, that the program in the United States was beginning to slow.

There are several factors that have contributed to the Soviet Union's policy to expand its nuclear industry. Its growth is related to the recognition that fossil fuel resources have become increasingly difficult and expensive to exploit. As the oil, coal, and gas of the European U.S.S.R. have been depleted, production has shifted northward and eastward to Siberia, causing both extraction costs and the cost of transporting energy to the consumer to rise greatly. Nuclear power has therefore become an economically viable option, particularly since power stations are largely located in the more densely populated European part of the country.

---

[3] Eric Morgenthaler, "Eastern Energy: Soviet Bloc is Pushing Nuclear Powerplants Even as U.S. Pulls Back," *Wall Street Journal,* Jan. 4, 1980, pp. 1, 4, op. cit.

Soviet nuclear policy is also marked by relatively little anxiety over safety and environmental issues compared to the West. Soviet ideology has characteristically reflected a boundless faith in technology and the beneficial effects of technological developments on the welfare of mankind. Accidents such as the one that occurred at Three Mile Island are attributed by the Soviet press to the irresponsible behavior of profit-seeking private firms, rather than to any dangers inherent in nuclear power. The press assumes that in the U.S.S.R.—where private enterprise does not officially exist, and production is carried out by planners armed with what are considered rational, scientific methods—the welfare of the citizenry and of the environment will be carefully considered.[4] The official Soviet position is that concerns over the safety of nuclear power are founded on ignorance. This point is illustrated in a 1971 statement of A. M. Petrosyants, Chairman of the State Committee on the Utilization of Atomic Energy:

> It can be stated that nuclear power stations are no more dangerous than any other industrial type plant, and can be sited in any densely populated area and even within the confines of large cities . . . Widespread publicity on the safety of nuclear power stations, explanations of the facts with demonstrations of how nuclear power stations operate, are indispensable measures in sweeping away the skepticism and lack of confidence observed in some parts of the population in some instances.[5]

Despite recent publicity in the West over alleged large-scale nuclear accidents in the Soviet Union,[6] there is no evidence that this position has changed.

Soviet nuclear policy may also be at least partly driven by a desire to demonstrate conspicuous technological achievement in a large-scale program that the U.S.S.R. itself regards as necessary to its role as a world superpower. A successful nuclear program is seen as a means of enhancing the prestige of the nation and providing visible evidence of the superiority of socialism. The power station at Obninsk, for example, has been hailed as "a triumph of advanced Soviet science and technology. It confirmed with new strength the indisputable advantages and the richest creative potentialities of the socialist society."[7] This is not so much a reflection of any Soviet world lead in terms of installed nuclear capacity—in 1980, the U.S.S.R. lagged behind the United States, Japan, and France in this respect—as much as an affirmation of the fact that Soviet advances in this area have proceeded with little direct technical help from the West. This technical independence is underscored by the fact that the Soviet Union has had recent successes exporting its nuclear reactors (e.g., to Finland); that breeder technology is more advanced in the U.S.S.R. than in the West (with the possible exception of France); and that Western scientists are quite impressed with Soviet fusion research.

Factors such as these have contributed to the rapid growth of the nuclear sector in the U.S.S.R. and to the formulation of ambitious plans for the next decade. The following sections describe the present state of the atomic power industry and discuss and evaluate these targets.

[4]Gloria Duffy, *Soviet Nuclear Energy: Domestic and International Policies* (Santa Monica, Calif.: Rand Corp., December 1979), p. 38.

[5]A. M. Petrosyants, "Nuclear Power in the Soviet Union," *Soviet Atomic Energy*, No. 3, March 1971, pp. 297-302.

[6]See, for example, John R. Trabalka, L. Dean Eyman, and Stanley I. Auerbach, "Analysis of the 1957-1958 Soviet Nuclear Accident," *Science*, July 1, 1980, pp. 345-353.

[7]Oleg Kazachkovskiy, "Condition and Outlook for Work To Create AES With Fast Neutron Reactors," *Ekonomicheskoye sotrudnichestvo stran-chlenov SEV*, No. 2, 1980, p. 1, in JPRS 76,135, July 30, 1980, pp. 1-8.

PRESENT NUCLEAR POWER CAPACITY AND PRODUCTION IN THE U.S.S.R.

In 1976, the 25th Party Congress adopted the Tenth FYP (1976-80), which called for the production of 1,340 billion to 1,380 billion kWh of electricity in 1980. Of this, 80 billion (about 5.8 percent) was to be provided by nuclear power stations.[8] Although production failed to meet this ambitious target, Soviet accomplishments in the area of nuclear electrification over the plan period were impressive.

1980 CAPACITY AND PRODUCTION

As table 28 demonstrates, Soviet nuclear powerplants produced 73 billion kWh (or 70.5 billion kWh, depending on which Soviet source is used) in 1980, nearly 3½ times as much as in 1975 and over 25 percent more

[8]E. E. Jack, J. R. Lee, and H. H. Lent, "Outlook for Soviet Energy," in Joint Economic Committee, U.S. Congress, *Soviet Economy in a New Perspective* (Washington, D.C.: U.S. Government Printing Office, 1976), p. 466.

than in 1979. Installed capacity at the end of 1980 reached 13,460 MW, having nearly doubled in 3 years. Nuclear powerplants accounted for 5.6 (or 5.4) percent of all electricity produced in 1980 and 5.1 percent of installed electrical capacity as of December 31, 1980.[9] On a Btu basis, it is estimated that nuclear energy accounted for a little over 1 percent of Soviet primary energy output in 1980.[10]

This energy is currently being produced from at least 29 online reactors that are distributed among 13 sites. These sites are listed in tables 29 and 30, and shown in

[9]Total installed capacity (all sources) as of Jan. 1, 1981, was taken from *Ekonomicheskaya gazeta*, No. 12, March 1981, p. 2. The installed nuclear capacity (13.5 million kW) was then divided by this total (267 million kW) to derive the indicated percentage.
[10]L. Melentyev and A. Makarov, "Future Development of the Fuel-Energy Complex," *Planovoye khozyaystvo*, No. 4, April 1980, pp. 87-94, in JPRS 75,903, June 19, 1980, pp. 13-22.

Photo credit: TASS from ©SOVFOTO

The Novovoronezh NPS

Table 28.—Production of Electricity by Soviet Nuclear Power Stations

Year	Production (billion kWh)	Percent of total production	Installed nuclear capacity (end of year, million kW)
1960	Negligible[a]	—	—
1965	1.4[b]	0.3[g]	0.3[c]
1970	3.5[b]	0.5[g]	0.9[b]
1975	20.2[b]	1.9[g]	4.7[b]
1976	26.4[c]	2.2[g]	5.7[c]
1977	34.8[d]	3.0[g]	7.1[c]
1978	44.8[d]	3.7[g]	9.1[e]
1979	54.8[f]	4.4[h]	11.4[i]
1980 (plan)	80.0[b]	5.8[b]	18.4[b]
1980 (actual) ..	70.5[j]	5.4[j]	13.5
	(73.0)[k]	(5.6)[k]	—

SOURCES: [a]E.E. Jack, J.R. Lee, and H.H. Lent, "Outlook for Soviet Energy," in Joint Economic Committee, U.S. Congress, *Soviet Economy in a New Perspective* (Washington, DC: U.S. Government Printing Office, 1976), p. 462.

[b]*Energetika SSSR v. 1976-1980 g.*, A.M. Nekrasov and M.G. Pervukhin, (eds.), (Moscow: Izd. "Energiya," 1977), pp. 11, 61 and 62.

[c]L. Dienes and T. Shabad, *The Soviet Energy System* (Washington, D.C.: V.H. Winston and Sons, 1979), p. 153.

[d]*Ekonomicheskaya gazeta*, No. 7 (February 1980), p. 1.

[e]*Elektricheskiye stantsii* (January 1979), pp. 2 and 3.

[f]*Elektricheskiye stantsii* (April 1980), p. 7.

[g]Calculated by dividing the figure in column 2 by total electricity produced in the given year, as reported in the Soviet statistical yearbook, *Narodnoye khozyaystvo SSSR v. 1978 g.* (Moscow: Izd. "Statistika," 1979), p. 142.

[h]Calculated by dividing the figure in column 2 for 1979 by the total electricity produced in that year, as reported in *The U.S.S.R. in Figures* for 1979 (Moscow: Izd. "Statistika," 1980), p. 107.

[i]*The Columbus Dispatch* (Columbus, Ohio), Jan. 29, 1980, p. A-6.

[j]*Elektricheskiye stantsii* (January 1981), p. 2.

[k]*Ekonomicheskaya gazeta*, No. 12 (1981), p. 1.

figure 8. Most Soviet nuclear power stations are located in the European portion of the country, primarily to the west of the Volga River, where electricity demand is concentrated, but there is also some demand for nuclear plants in remote regions (e.g., the Bilibino NPS), apparently because of the high cost of other energy sources.

In both regions, the Soviets are interested in using nuclear energy to provide district heating as well as for generating electricity. The first heating plants may already be in operation, and it is claimed that six atomic heating and cogeneration plants will be built during the Eleventh FYP. Because of Soviet confidence in the safety of nuclear power, there is little written about the environmental implications of locating nuclear power stations in populated areas. In fact, in the case of nuclear district heating (where heat losses are highly sensitive to transmission distances), the plants are actually to be sited within urban areas.

REACTOR TYPES

The Soviets have experimented with a number of types and sizes of reactors, but they are now concentrating on two models that will be standardized to facilitate mass production: large capacity channel reactors and pressurized water reactors. The former are the most commonly used. Called RBMKs (the initials stand for the Russian words for "large capacity channel reactors"), they supply approximately 8,000 MW or 61.6 percent of the total capacity of operational nuclear power stations. RBMKs are boiling water reactors; they are light-water cooled and graphite moderated. The reactor consists of a large pile of graphite with small tubed channels running throughout. Some channels house the fuel rods while others allow for coolant flow.

There are several advantages to the RBMK. Its modular design allows the reactor to be almost entirely assembled on the station site, with only a few components requiring preassembly at the manufacturing plant; it is capable of online refueling; and it provides the capacity to detect a failed fuel element in a given pressure tube during operation. This latter characteristic permits immediate removal and reinstallation of the fuel element without shutdown.[11] The RBMK also uses lightly enriched uranium and produces more plutonium than the other most common model. The U.S.S.R. is now the only country actively engaged in the construction of this type of reactor, the United States having abandoned plans for its commercial development some years ago.

---

[11]Joseph D. Lafleur and Victor Stello, "NRC Team Visit to U.S.S.R., Feb. 5-18, 1978," information report No. SECY-78-11B, Mar. 21, 1978, p. 11.

Table 29.—Soviet Nuclear Power Stations in Operation

Station name	Location	Reactor No.	Year of initial operation	Reactor designation[j]	Rated reactor capacity, MW(e)	Reactor type[j]
Obninsk	Obninsk	1	1954	—[a]	5	BWR
Siberian	Troitsk[h]	1	1958	—	600[b]	BWR
Dimitrovgrad	Dimitrovgrad	1	1965	VK-50	50[c]	BWR (Vessel)
	(Formerly Melekess)	2	1969	BOR-60	12	Breeder
Beloyarskiy	Zarechnyy	1	1964	AMB-100	100	BWR
		2	1967	AMB-200	200	BWR
		3	1980	BN-600	600	Breeder
Novovoronezhskiy	Novovoronezhskiy	1	1964	VVER-210	210[d]	PWR
		2	1969	VVER-365	365	PWR
		3	1971	VVER-440	440	PWR
		4	1972	VVER-440	440	PWR
		5	1980	VVER-1000	1,000	PWR
Shevchenko	Shevchenko	1	1973	BN-350	350[e]	Breeder
Kola	Polyarnyye Zori	1	1973	VVER-440	440[f]	PWR
		2	1974	VVER-440	440[f]	PWR
Bilibino[g]	Bilibino	1	1974	EGP-6	12	BWR
		2	1974	EGP-6	12	BWR
		3	1975	EGP-6	12	BWR
		4	1976	EGP-6	12	BWR
Leningrad	Sosnovyy Bor	1	1973	RBMK-1000	1,000	BWR
		2	1975	RBMK-1000	1,000	BWR
		3	1979	RBMK-1000	1,000	BWR
Kursk	Kurchatov	1	1976	RBMK-1000	1,000	BWR
		2	1979	RBMK-1000	1,000	BWR
Armenian	Metsamor	1	1976	VVER-440	405[i]	PWR
		2	1979	VVER-440	410[i]	PWR
Chernobyl	Pripyat	1	1977	RBMK-1000	1,000	BWR
		2	1978	RBMK-1000	1,000	BWR
Rovno	Kuznetsovsk	1	1980	VVER-440	440	PWR

[a]In U.S. sources, this reactor has been designated as the AM-1 (Dienes and Shabad, op. cit., p. 156) and the VAM-1 (Sutton, op. cit., p. 243).

[b]The station reportedly has a total capacity of 100 MW in 1958. At present, the station is said to have a capacity which "significantly exceeds 600 MW" (A. M. Petrosyants, op. cit., p. 123.

[c]According to Petrosyants, op. cit., p. 171, the VK-50 reactor was upgraded to 65 MW(e) in 1974.

[d]This capacity reportedly was raised to 240 MW in February 1965 and, briefly, to 280 MW in January 1969.

[e]If the Schevchenko reactor were used exclusively to produce electricity, its capacity would be 350 MW(e). In fact, only 150 MW(e) of capacity are devoted to generation of electricity, while the balance is used to produce 120,000 metric tons of desalinated sea water per day.

[f]These two reactors of the station's first phase reportedly have been operating at capacities as high as 470 MW each (940 MW total) since December 1978.

[g]This station generates commercial heat as well as electricity.

[h]According to Dienes and Shabad, op. cit., p. 153, this location is given in U.S. lists of foreign reactors; Soviet sources have not identified the Siberian station's location.

[i]It is not clear why these two reactors are rated at lower capacities than other VVER-440's; one source [Atomnaya energiya (May 1977), p. 419], relates the lower capacities to "cooling conditions" of the reactors.

[j]PWR = pressurized water reactor (vessel type), which is designated VVER in Russian, BWR = boiling water reactor, one series of which is designated RBMK in Russian; in the West the RBMK often is described as a light-water-cooled, graphite-moderated reactor (LGR or LWGR).

SOURCES: A. M. Petrosyants, Sovremennyye problemy atomnoy nauki i tekhniki v SSSR (Moscow: Atomizdat, 1976), pp. 118-192). Sovetskaya atomnaya nauka i tekhnika (Moscow: Atomizdat, 1967), pp. 91-110. Izvestiya akademii nauk SSSR: energetika i transport, No. 5 (1977), pp. 13-31. Elektrifikatsiya SSSR (1967-1977 gg.), P. S. Neporozhniy (ed.) (Moscow: Izd, "Energiya," 1977), p. 50. Elektricheskiye stantsii, No. 2 (February 1978), pp. 8-13; No. 2 (February 1980), pp. 7-11; and No. 6 (June 1980), pp. 2-8. Kommunist (Jan. 1, 1980), p. 1; and Feb. 29, 1980), p. 1. Atomnaya energiya, (April 1980), pp. 220-223. Stroiteinaya gazeta (Apr. 9, 1980), p. 1. L. Dienes and T. Shabad, The Soviet Energy System (Washington, D.C.: V. H. Winston & Sons, 1979), pp. 153, 156, and 157. A. C. Sutton, Western Technology and Soviet Economic Development (Stanford, Calif.: Hoover Institution Press, 1973), p. 243. Trud, (June 1, 1980), p. 1. Izvestiya (Dec. 25, 1980), p. 1.

Table 30.—Estimated Total Operating Electric Generating Capacity of Soviet NPS's as of Dec. 31, 1980[a]

Station	Total capacity, MW(e)
Obninsk	5
Siberian	600
Dimitrovgrad	77
Beloyarskiy	900
Novovoronezhskiy	2,485
Shevchenko	150
Kola	940
Bilibino	48
Leningrad	3,000
Kursk	2,000
Armenian	815
Chernobyl	2,000
Rovno	440
Grand total	13,460

[a]These are not rated capacities, but OTA's best estimates of the actual operating capacities at each station.

SOURCE: Table 29.

The next most commonly used reactors, pressurized water reactors, supply approximately 4,200 MW or 31 percent of current capacity. They are known as VVERs in the Soviet Union (for the Russian words "water-water power reactors"), and are similar to models available in the West. Light water is used in these reactors as both moderator and coolant. A large, cylindrical, steel pressure vessel houses the fuel and control rods along with other necessary internal apparatus. The high pressures involved dictate that major components, including the reactor vessel, be manufactured and tested before shipment to the station site for final assembly.

REACTOR MANUFACTURE

Thus far, the expansion of Soviet nuclear capacity in 1,000-MW increments has been based almost exclusively on the RBMK-1000 reactor, the components for which can be produced at ordinary manufacturing plants. Pressure vessels for reactors of the VVER series require specialized production facilities.

One enterprise that produces pressure vessels and equipment for reactors of both types is the Izhora Plant Production Asso-

ciation near Leningrad. Izhora began producing main power equipment for NPSs in 1964. In order to manufacture this equipment, it has had to undergo extensive expansion and retooling, and it is now the main supplier of nuclear power reactors. Izhora is producing 1,000-MW VVER reactors to be installed at NPSs under construction in the Southern Ukraine and at Kalinin, and it has begun making an RBMK-1500 unit for the Ignalina NPS.[12] However, this enterprise alone will not be able to produce all the reactors required in the next decade. Some of this burden has been shifted to Czechoslovakia's Skoda Works, which has been assigned the task of producing the smaller VVER-440 pressure-vessel reactors.[13] In the Soviet Union, a major share of the burden is to be assumed by the gigantic new Volgodonsk Heavy Machine Building Plant, better known as "Atommash."[14] As Atommash reaches full capacity (by 1990), more and more reliance will be placed on the VVER-1000.

When Atommash is fully operational, it will produce VVER-1000 pressure-vessel reactors on an assembly-line basis, at the rate of eight reactors per year. Since Atommash is designed to specialize in reactor production, this rate of output, if achieved, undoubtedly will outstrip that of a more conventional (albeit upgraded) manufacturing enterprise like Izhora.

In addition, the Soviets have long been interested in introducing commercial breeder or fast-neutron reactors (designated either

---

[12]Yu. Sobolev, "The Direction for the Search," *Sotsialisticheskaya industriya*, Nov. 3, 1979, p. 2 in JPRS 75,069, Feb. 5, 1980, pp. 1-3); "Heroes of Izhora," *Leningradskaya pravda*, Dec. 19, 1980, p. 2.

[13]Various items of largely Soviet-designed equipment for NPSs are being or will be produced in East European CMEA countries, for use both in these countries and in the Soviet Union. For a description of this "division of labor," see Vyacheslav Zorichev and Yevgeniy Fadeyev, "The Course for Accelerated Growth of Atomic Power," *Ekonomicheskoye sotrudnichestvo stran-chlenov SEV*, No. 6, 1979, pp. 51-55.

[14]Because of the growth in size and scope of operations since its inception, this plant has been redesignated as a production association. See Anatoliy Mashin, "First Phase," *Znamya*, No. 3, March 1979, pp. 190-199, in JPRS 73,863, July 19, 1979, pp. 1-13.

Figure 8.—Soviet Nuclear Reactor Sites

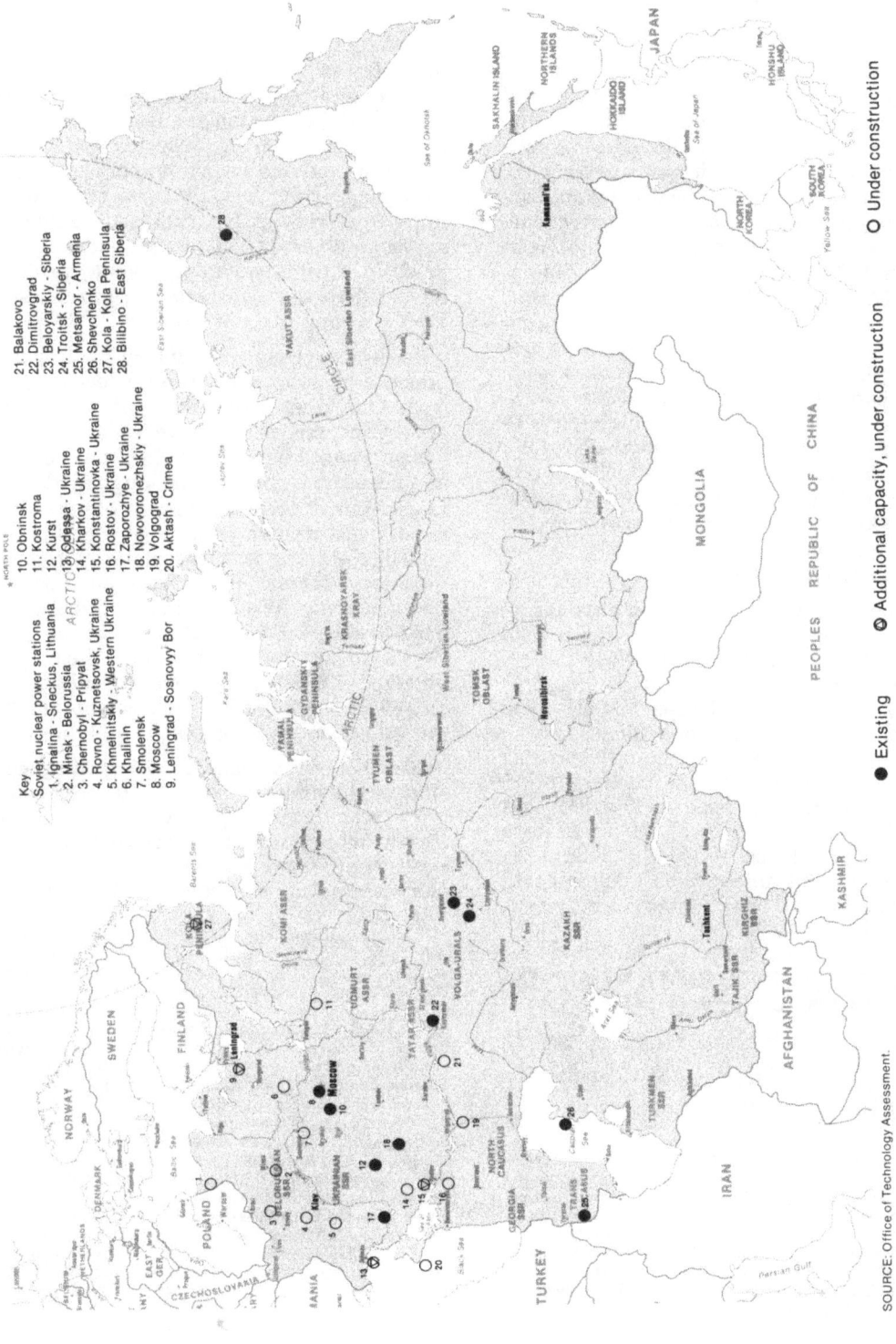

Key
Soviet nuclear power stations
1. Ignalina - Sneckus, Lithuania
2. Minsk - Belorussia
3. Chernobyl - Pripyat
4. Rovno - Kuznetsovsk, Ukraine
5. Khmelnitskiy - Western Ukraine
6. Khalinin
7. Smolensk
8. Moscow
9. Leningrad - Sosnovyy Bor

10. Obninsk
11. Kostroma
12. Kurst
13. Odessa - Ukraine
14. Kharkov - Ukraine
15. Konstantinovka - Ukraine
16. Rostov - Ukraine
17. Zaporozhye - Ukraine
18. Novovoronezhskiy - Ukraine
19. Volgograd
20. Aktash - Crimea

21. Balakovo
22. Dimitrovgrad
23. Beloyarskiy - Siberia
24. Troitsk - Siberia
25. Metsamor - Armenia
26. Shevchenko
27. Kola - Kola Peninsula
28. Bilibino - East Siberia

● Existing   ⊕ Additional capacity, under construction   ○ Under construction

SOURCE: Office of Technology Assessment.

BOR or BN in Soviet terminology) in order to make better use of available nuclear fuel supplies. The major characteristic of this type of reactor is that it "breeds" or produces more fuel than it consumes. Systematic investigation of fast-neutron reactors began in 1948, and the first Soviet reactor of this type was started up at Obninsk in 1956. In 1969 the BOR-60 went into operation in Dimitrovgrad, and in 1973 the first demonstration breeder reactor began operating in Shevchenko on the east coast of the Caspian Sea. This reactor has a generating capacity of 150 MW, as well as the ability to desalinate 120,000 cubic meters of seawater per day. In April 1980, the BN-600, with an electrical generating capacity of 600 MW, went into operation as Unit 3 at the Beloyarskiy station.[15] These three reactors currently account for 5.8 percent of Soviet nuclear generating capacity.

At one time it was expected that large-scale commercial stations equipped with breeders would be ready by the early 1980's. However, recent Soviet literature suggests that this stage will not be reached until the end of the 1990's. Technical difficulties with the use of sodium and other liquid metals as heat-transfer agents are the primary problem (no moderator is used in breeders and the high temperaures involved necessitate the use of liquid metal, usually sodium, as the coolant), but unexpectedly high costs associated with all basic processes in the external fuel cycle have also been cited.[16]

PAST CONTRIBUTION OF WESTERN EQUIPMENT AND TECHNOLOGY

Although the U.S.S.R. collaborates with its East European allies on nuclear develop-

ment, its progress has been largely autonomous. The Soviet nuclear power industry has relied heavily on domestic equipment, purchasing relatively little from the West. Western controls on the export of nuclear technology have almost certainly been an important reason for this self-sufficiency, although shortages of foreign exchange and pride in Soviet technology would probably have led to some restraint in purchases of Western equipment in any case.

Soviet purchases of primary nuclear components such as reactors and reactor parts from the West have been infrequent and poorly documented.[17] However, the U.S.S.R. has purchased equipment that could be used in nuclear powerplants. Although evidence is incomplete,[18] documented purchases consist mostly of machine tools, heavy equipment, and engineering services slated for Atommash and Izhora. These have been sold by firms in Italy, West Germany, Japan, Sweden, France, and the United States. Italy seems to be the country most heavily involved in such trade; several Italian companies have supplied equipment, mainly machine tools, to Atommash and Izhora.

Besides heavy manufacturing equipment, Atommash and Izhora are said to be receiving well-outfitted quality control laboratories, including large destructive-testing equipment, from European, Japanese, and North American sources. The Soviet nuclear

---

[15]Kazachkovskiy, op. cit, pp. 3-4: Peter Feuz, "The Nuclear Push in the U.S.S.R. and Eastern Europe," *Power Engineering*, No. 8, Aug. 1978, pp. 100-101; "Late Breeder in the Ural Mountains," *Wirtschaftswoche* (Dusseldorf), No. 18, May 2, 1980, pp. 14-16, in JPRS 75,973, July 2, 1980, pp. 8-16.

[16]N. Dollezhal and Y. Koryakin, "Nuclear Power Engineering in the Soviet Union," *The Bulletin of the Atomic Scientists*, No. 1, January 1980, pp. 33-37. (Originally published in the September 1979 issue of *Kommunist*.)

[17]U.S. Department of Commerce trade statistics report only two entries under SITC Code 7117, "nuclear reactors and parts thereof." One is a 1978 export of $448,700 from the United States; the other $329,000 from West Germany in 1976. No further unclassified information on either sale could be found. OTA has been told, however, that a West German firm was to deliver all fittings and equipment for the nuclear cycle in three Soviet powerplants with a combined generating capacity of 1,880 MW; the delivery was to be completed before the end of 1980.

[18]The remainder of this section is based on various issues of *Soviet Business and Trade* and on *Nucleonics Week*, Nov. 8, 1979, p. 12. In general, Soviet sources make no mention or only passing reference to the use of foreign equipment. For example, one Soviet source noted in passing that metal refining equipment from a Swiss firm is used at the Izhora Plant Association. See V. P. Goloviznin, "Soviet Power Equipment—The Basis for the Development of Power Engineering in Our Country," *Energomashinostroyeniye*, No. 4, April 1980, pp. 2-4 in JPRS L/9176, July 2, 1980, pp. 11-18, especially p. 14.

industry has also been acquiring valves from at least one Western source, a Canadian firm that has supplied valves for the Novovoronezhskiy NPS; and the Belgorod Power Machine Building Plant reportedly has introduced technology for automatic argon-arc welding of austenitic steel pipeline units for NPSs. The technology uses AM-11 automatic welding units manufactured by a U.S. firm.[19]

The U.S.S.R. also may be purchasing large capacity cranes, both for building NPSs and for handling heavy items at manufacturing plants. In 1976, there was speculation in the industry that foreign bids would be solicited for 1,200-ton cranes for Atommash, but OTA has been unable to confirm any completed transactions. Similarly, in 1979 the U.S.S.R. reportedly ordered three 300-metric-ton truck-mounted cranes from a West German firm. These cranes are technically well-suited for use at nuclear facilities.

It is important to note that none of these purchases appears in an area in which the Soviets lack technology. Moreover, OTA found no evidence that equipment of the sort described has been sought or purchased in massive amounts or that the Soviets have thus far sought large-scale active Western cooperation in their nuclear industry.

1980 PLAN FULFILLMENT

As table 28 shows, the Soviet nuclear power industry was originally charged with producing 80 billion kWh of electricity during 1980, and completing some 18,000 MW of installed capacity by the end of that year. By early 1980, it was clear that these targets could not be reached and in February of that year a revised production goal of 71.9 billion kWh was published. As of December 31, 1980, installed nuclear capacity was approximately 13,460 MW, a figure that takes into account the start of unit 1 (440 MW) at the Rovno NPS in December 1980.[20] This is well below the plan target of 18,400 MW.

Failure to meet the plan was largely due to delays in the installation of nuclear power stations. These delays were apparently the result of a variety of systemic problems and were more closely associated with the construction industry and suppliers of material and equipment than with technological difficulties. Although one of its production lines has been opened, Atommash is now several years behind schedule. The long lead-times required to install reactors and to make nuclear power stations fully operational make it unreasonable to blame Atommash for past failures to meet goals for installed capacity or production of nuclear electricity. However, problems in bringing Atommash online may become a major factor in any failures to meet nuclear targets in the 1980's.

Soviet literature provides numerous accounts of the difficulties encountered in the building of nuclear power stations,[21] including poor organization of labor, as well as delays on the part of ministry officials, designers, builders, and suppliers of materials, construction modules, and equipment. Similar complaints can be found even in the construction of the BN-600 breeder reactor (unit 3 at the Beloyarskiy NPS)—where the Soviets felt themselves to be in

[19]N. V. Kulin, V. G. Medalye, and I. Ye. Shulman, "The Status and Prospects for the Advancement of Welding Production at Nuclear Machine Building and Boiler Building Enterprises," *Energomashinostroyeniye*, No. 1, January 1981, pp. 28-30. The Belgorod plant is a boiler building plant that is believed to produce auxiliary equipment for NPSs.

[20]Not all sources agree with this 13,460-MW figure. A Soviet article, "Results of the Advancement of Electric Power in 1980 and Tasks for 1981," *Elektricheskiye stantsii*, No. 1, January 1981, pp. 2-4, reports the total capacity of Soviet NPSs at the end of 1980 as 12,300 MW. In part, this discrepancy can be explained by lower Soviet figures for the capacities of the Dimitrovgrad, Kola, and Novovoronezhskiy NPSs, which would account for a difference of 105 MW (see tables 29 and 30). In addition, the Soviet total may exclude the capacity of the Siberian NPS (600 MW). Finally, the Soviet figure may not take into account unit 1 of the new Rovno NPS. For purposes of this report, and on the basis of documented evidence, OTA includes the capacity increments apparently omitted from the reported Soviet figure for total nuclear capacity.

[21]G. Dolzhenko, "Sanctions? Forget It!" *Stroitelnaya gazeta*, July 16, 1980, p. 2.; Zh. Tkachenko, "You Cannot Replace Metal With Messages," *Sotsialisticheskaya industriya*, Apr. 2, 1980, p. 2, in JPRS 75,973, July 2, 1980, pp. 17-20.

the international technical spotlight.[22] In this case, the general contractor was several months late in issuing technical documentation; equipment and materials were late or defective; there was a shortage of spare parts; and when equipment did arrive at the construction site, managerial personnel and skilled workers were in such short supply that there frequently were long delays before the equipment was installed or even properly stored.

Such problems are not peculiar to the Soviet nuclear power industry, although they may be exacerbated by the high technological standards required for nuclear equipment. Similar sorts of complaints are published regularly in the Soviet press and cover a variety of industries. To a certain extent, too, these problems may be seen as "growing pains." Installed nuclear capacity rose an average of 1,760 MW per year during the period 1976-80, v. only 760 MW per year in 1971-75. As computed from table 28, the annual rate of addition continued to increase during the late 1970's, rising from 1,000 MW in 1975-76 to 2,300 MW in 1978-79. The Eleventh FYP (1981-85) projects continued rapid growth and it is clear that nuclear power's contribution to the Soviet energy balance will accelerate in the next decade. Nevertheless, it is likely that this rapid buildup in the rate of construction of nuclear power stations has itself created numerous problems for the future and that the difficulties described above will persist as the industry struggles to meet the 1985 and 1990 targets.

---

[22]"Late Breeder . . . ," op. cit.

NUCLEAR CAPACITY AND PRODUCTION: FUTURE PROSPECTS

TARGETS FOR 1985

Soviet goals for nuclear power are ambitious. The draft of the Eleventh FYP calls for the generation of 1,550 billion to 1,600 billion kWh of electricity from all sources in 1985, of which 220 billion to 225 billion kWh, or about 14 percent, are to be generated at nuclear power stations.[23] In order to achieve these goals, 24,000 to 25,000 MW of additional nuclear generating capacity are to be installed during the FYP. If these installation goals are achieved, the U.S.S.R. will have about 38,000 MW of installed nuclear capacity by the end of 1985.[24] (The United States had 52,300 MW of nuclear capacity as of January 1980.[25])

These targets assume a high utilization rate for nuclear generation facilities. The rate for 1985, calculated as the ratio of planned nuclear electricity production during 1985 to estimated midyear planned capacity in 1985, implies that the Soviets expect to operate nuclear stations for an average of about 6,250 hours per year during 1985. This rate is not impossible, but it is substantially higher than the utilization rate of 5,420 hours achieved in 1980.[26]

TARGETS FOR 1990

Current Soviet goals for 1990 are less explicit. One Western journal cited a 1990 nuclear capacity target of 80,000 MW in one issue, while another issue 6 months later reported a goal of 90,000 MW.[27] In a speech de-

---

[23]"Draft of the Main Directions of Economic and Social Development of the U.S.S.R. for 1981-1985 and for the Period of 1990," Izvestiya, Dec. 2, 1980, pp. 2-7.

[24]This assumes that the Soviet Union had 13,460 MW of effective nuclear power generating capacity, as of Dec. 31, 1980. Addition of 24,000 to 25,000 MW of capacity during the Eleventh FYP would result in total nuclear capacity at the end of 1985 of 37,460 to 38,460 MW. OTA has used the approximate midpoint of this range—38,000 MW.

[25]Morgenthaler, op cit.

[26]Calculated from table 28 using estimated capacity as of July 1, 1980, 13,020 MW.

[27]"Afghanistan Imperils Co-operation on Soviet Programme," Nuclear Engineering International, February 1980, p. 5; Richard Knox, "Progress in the U.S.S.R.," Nuclear Engineering International, August 1980, p. 14.

livered in December 1978, P. S. Neporo-zhniy, the Minister of Power and Electrification, spoke of reaching 100,000 MW of nuclear capacity in the following 10 to 12 years,[28] and a West German newspaper has quoted 1990 targets of 100,000 and 110,000 MW, concluding that one "cannot accurately determine from Soviet data what the U.S.S.R. wants to have ready or wants to build by 1980 or 1990."[29]

Soviet goals for 1990 are contingent on the achievements of the industry by 1985. Reported downward revisions in the 1985 goals make it likely that targets for 1990 also have been adjusted and that any existing plans for 100,000 to 110,000 MW have now been abandoned. Barring large-scale infusions of Western equipment (e.g., turnkey NPS projects), a possibility that will be discussed in more detail below, the published goal that OTA regards as being most reasonable is 80,000 MW. The discussions that follow assume this figure, which would require the addition of 42,000 MW of nuclear capacity during the Twelfth FYP (1986-90), given that 38,000 MW of capacity were in place at the end of 1985. If these capacity goals are achieved by 1990, the Soviets would have the ability to generate approximately 410 billion to 475 billion kWh of nuclear electricity per year, depending on one's assumptions regarding utilization rates.[30]

SOVIET ABILITY TO MEET PLAN TARGETS

The following sections evaluate the potential for success of these plans, identifying the major obstacles likely to be encountered, and noting the areas in which Western tech-nology and equipment might make the greatest contribution. This evaluation is organized around four key areas:

1. construction of the stations;
2. manufacturing reactors, steam equipment, and electrical machinery;
3. facilities for the external fuel cycle, which include supply of enriched uranium for reactor fuel and the storage and disposal or reprocessing of spent fuel; and
4. computer technology to support station operation. (Electricity transmission lines and pumped storage electricity generating capacity are discussed in ch. 5).

For purposes of this discussion, OTA has assumed a profile of yearly additions to nuclear capacity. This profile is illustrated in figure 9, which also shows the actual annual additions between 1975 and 1980. The latter increased from about 1,000 MW in 1976 to about 2,300 in 1979, falling to 2,040 in 1980. In order to add 24,000 to 25,000 MW by 1985 and 42,000 more by 1990, the Soviets could proceed at any of a number of different paces. The one shown in figure 9—successive increments of 3,000, 4,000, 5,000, 6,000, and 7,000 MW to 1985 and 7,000, 8,000, 8,000, 9,000, and 10,000 MW to 1990—is not the only possible profile, but it is a reasonable one to serve as an illustration of the demands that will be placed on the Soviet nuclear industry to achieve a total of 80,000 MW installed capacity by 1990.

Table 31 presents much of the published information on nuclear power stations that are now under construction or planned. Rated reactor capacities have not been published for all of these, but those that have been released total 41,820 MW (excluding unit 6 at Rovno)—enough to cover the entire capacity addition called for in the Eleventh FYP and roughly 40 percent of the likely addition during the Twelfth FYP.

Several features of these planned additions are noteworthy. First, only three more VVER-440 reactors are scheduled to be built

[28]"Speech of P. S. Neporozhniy," *Izvestiya*, Dec. 1, 1978, p. 4.

[29]"From Lake Baikal to the Elbe River," *Wirtschaftswoche* (Dusseldorf), No. 46, Nov. 12, 1979, pp. 64, 68, 70, 71, 74, and 76, in JPRS 674,792, Dec. 19, 1979, pp. 20-29.

[30]The low end of the range assumes an operating rate of 5,420 hr/yr, the rate achieved in 1980, while the upper assumes the operating rate apparently sought for 1985, 6,250 hours. Mid-1980 capacity is estimated as 75,300 to 76,300 MW.

Photo credit: TASS from ©SOVFOTO

NPS control room in the U.S.S.R., 1980

for domestic NPSs, two for the Kola NPS and one for Rovno. Production of this type of reactor is to shift to the Skoda Works in Czechoslovakia, and such reactors will be installed in Eastern Europe and exported elsewhere, e.g., to Cuba or non-Communist Third World countries. Second, most announced capacity additions will involve 1,000-MW reactors. Of the 31 reactors of this size that have been scheduled, nine are RBMKs and 22 are VVER units, the latter to be produced primarily by Atommash. Third, there are plans to install four RBMK- 1500 reactors at the Ignalina NPS in Lithuania,[31] and

possibly several more at the Smolensk and Kostroma NPSs.[32] Finally, there are plans to eventually build on RBMK-2400,[33] but no details on the installation of such reactors before 1990 have been published.

Construction

Time.—Plans that call for the addition of 24,000 to 25,000 MW of nuclear capacity by 1985, and an additional 42,000 MW by 1990 can only be realistic if they allow sufficient time for the construction of new plants. The Soviets themselves have maintained that the construction time norms are 6 years for

[31]*Atomic Science and Technology in the U.S.S.R.* (Moscow: Atomizdat, 1977), p. 26. The RBMK-1500 is an upgraded version of the RBMK-1000. See A. M. Petrosyants, *Problems of Atomic Science and Technology* (Moscow: Atomizdat, 1979), p. 139.

[32]F. Ovchinnikov, "By Advancing Tempos," *Sotsialisticheskaya industriya*, Jan. 30, 1981, p. 2.

[33]N. A. Dollezhal, "Atomic Power Engineering: Scientific-Technical Tasks of Development," *Vestnik akademii nauk SSSR*, No. 7, July 1978, pp. 46-61.

Figure 9.—Past and Projected Additions to Nuclear Generating Capacity

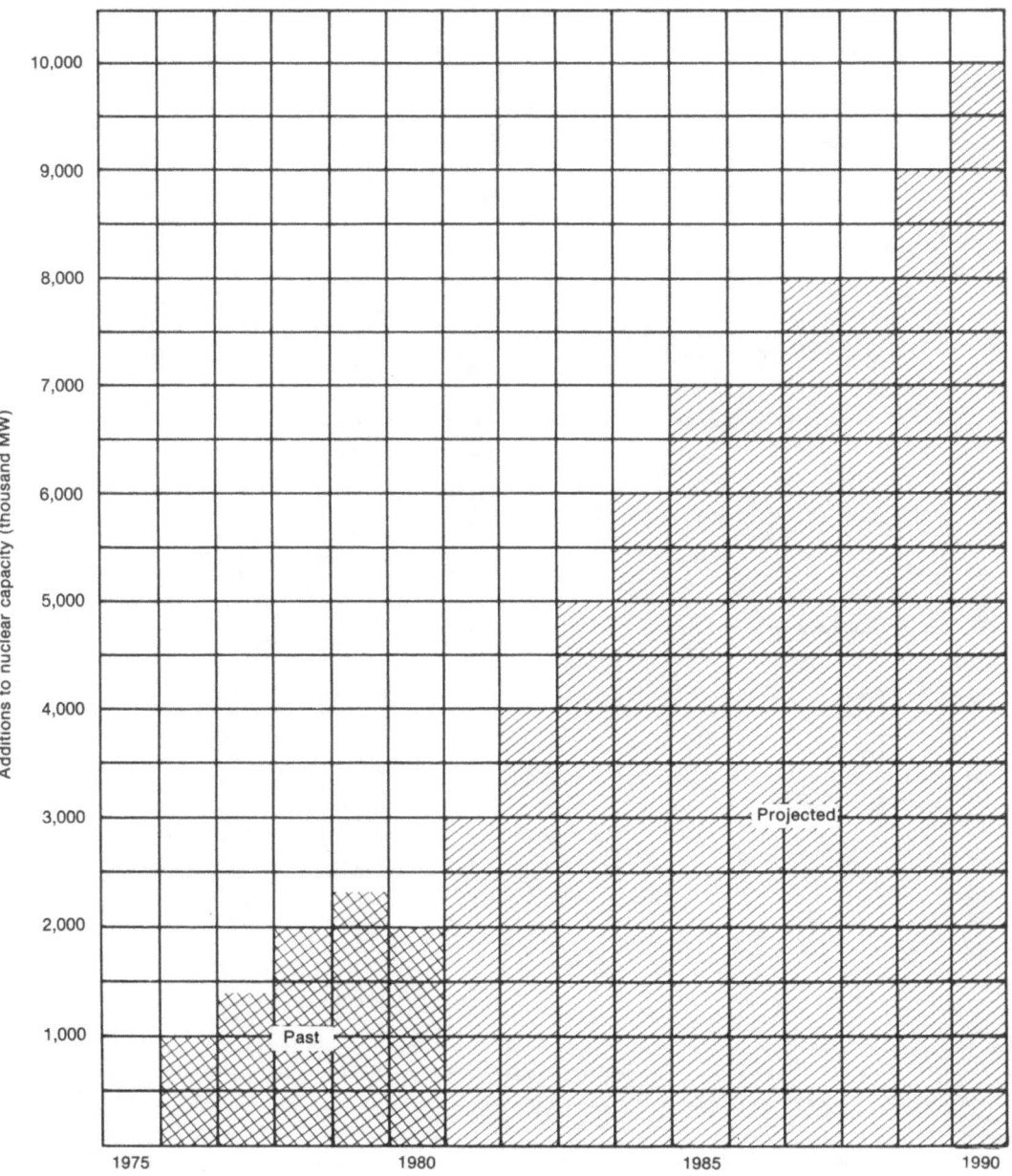

Year

SOURCE: Office of Technology Assessment.

Table 31.—Planned New Nuclear Electric Generating Capacity in the U.S.S.R., Post-1980

Station name	Location	Reactor No.	Year of initial operation	Reactor designation	Rated reactor capacity, MW(e)	Reactor type[e]
Existing Stations:						
Kola	Polyarnyye Zori	3	1981	VVER-440	440	PWR
		4	1982	VVER-440	440	PWR
Leningrad	Sosnovyy Bor	4	1981	RBMK-1000	1,000	BWR
Kursk	Kurchatov	3	1981	RBMK-1000	1,000	BWR
		4	By 1985	RBMK-1000	1,000	BWR
Chernobyl	Pripyat	3	1981	RBMK-1000	1,000	BWR
		4	By 1985	RBMK-1000	1,000	BWR
New Stations:						
Rovno[a]	Kuznetsovsk	2	1981	VVER-440	440	PWR
		3	1983	VVER-1000	1,000	PWR
		4	—	VVER-1000	1,000	PWR
		5	—	VVER-1000	1,000	PWR
		6	By 1992	VVER-1000[d]	1,000	PWR
Southern Ukrainian[a]	Konstantinovka	1	1981	VVER-1000	1,000	PWR
		2	By 1985	VVER-1000	1,000	PWR
		3	—	VVER-1000	1,000	PWR
		4	—	VVER-1000	1,000	PWR
Kalinin[a]	Udomlya	1	By 1985	VVER-1000	1,000	PWR
		2	By 1985	VVER-1000	1,000	PWR
		3	By 1985	VVER-1000	1,000	PWR
		4	By 1985	VVER-1000	1,000	PWR
Smolensk[a]	Desnogorsk	1	1981	RBMK-1000	1,000	BWR
		2	By 1985	RBMK-1000	1,000	BWR
		3	By 1985	RBMK-1000	1,000	BWR
		4	By 1985	RBMK-1000	1,000	BWR
Ignalina[a]	Snieckus	1	By 1985	RBMK-1500	1,500	BWR
		2	By 1985	RBMK-1500	1,500	BWR
		3	—	RBMK-1500	1,500	BWR
		4	—	RBMK-1500	1,500	BWR
Western Ukrainian[a]	Khmelnitskiy	1	By 1985	VVER-1000	1,000	PWR
		2	By 1985	VVER-1000	1,000	PWR
		3	—	VVER-1000	1,000	PWR
		4	—	VVER-1000	1,000	PWR
Odessa[b]	Near Odessa	1	—[c]	VVER-1000	1,000	PWR
Gorkiy[b]	Gorkiy	2	—	VVER-1000	1,000	PWR
Minsk[b]	Minsk	—	—	—	—	—
Kharkov[b]	Kharkov	—	—	—	—	—
Volgograd[b]	Volgograd	—	—	—	—	—
Zaporozhye[a]	Energodar	1	—[c]	VVER-1000	1,000	PWR
		2	—	VVER-1000	1,000	PWR
		3	—	VVER-1000	1,000	PWR
		4	—	VVER-1000	1,000	PWR
Balakovo[a]	Balakovo	—	—[c]	VVER-1000	1,000	PWR
Rostov[a]	Near Volgodonsk	—	—[c]	VVER-1000	1,000	PWR
Crimean[a]	Aktash	—	—[c]	VVER-1000	1,000	PWR
Tatar	Nizhnekamsk	—	—	—	—	—
Bashkir	Nertekamsk	—	—	—	—	—
Kostroma	—	—	—	RBMK-1500	1,500	BWR

[a]These stations reportedly are under construction.

[b]The Odessa station is to be the first of a series of large nuclear heat and power stations (NHPS) located near major urban centers. A small (48-MW) NHPS is in operation at Bilibina. Although virtually any type of reactor theoretically can be adapted to this type of station, preference reportedly is being given to PWR's, but evidence suggests that there will be at least two. Plans for the period 1981-1985 also call for NHPS's to be built in Minsk, Kharkov, and Volgograd; the reactor type for these stations is not known.

[c]Unspecified amounts of capacity are to be introduced at these stations during the period 1981-1985.

[d]The Rovno NPS is to have six units by 1992; starting with Unit 3, they are to be based on the VVER-1000 reactor.

[e]See notes to table 29.

SOURCES: A.M. Petrosyants, *Problemy atomnoy nauki i tekhniki* (Moscow: Atomizdat, 1979), p. 139. *Atomnaya nauka i tekhnika v SSSR* (Moscow: Atomizdat, 1977), p. 26. *Atomnaya energiya*, Vol. 43, No. 5 (November 1977), pp. 418-420. *Elektricheskiye stantsii*, No. 6 (June 1980), pp. 2-5; No. 10 (October 1980), pp. 10-14; and No. 12 (December 1980), pp. 60-63. *Teploenergetika*, No. 7 (July 1979), pp. 2-9; and No. 8 (August 1980), pp. 2-5. *Ekonomicheskaya gazeta*, No. 1 (January 1981), pp. 11 and 12; and No. 12 (March 1981), p. 2. *Izvestiya* (December 2, 1980), pp. 2-7. *Pravda* (Jan. 6, 1981), p. 1. *Sotsialisticheskaya industriya* (Jan. 30, 1981), p. 2. *Stroitelnaya gazeta* (July 16, 1980), p. 2. *Soviet Geography*, various issues.

an NPS equipped with two VVER-1000 reactors and 6 years, 3 months for an NPS with two RBMK-1000 reactors.[34] While only one VVER-1000 has been installed to date, experience with smaller VVER reactors and with the RBMK-1000 suggests that these estimates may be optimistic.

Unit 1 at the Novovoronezhskiy NPS (VVER-210) was begun in 1957, but did not go into operation until 1964.[35] The first phase of the Leningrad NPS (equipped with two RBMK-1,000 units) was under construction for 7 years, as was the first phase of the Chernobyl NPS (also equipped with two RBMK-1000 units).[36] Unit 1 (an RBMK-1000) of the Kursk NPS took 6 years to build and unit 2 (also an RBMK-1000) required 3 years, stretching the total time of phase 1 to 9 years.[37] In the case of the Armenian NPS, where construction began in 1971, unit 1 (VVER-440) achieved criticality in December 1976 and full-power operation in November 1977; however, unit 2 (VVER-440) did not come online until 1979—8 years after construction began.[38] These experiences would suggest that 7 years is a more realistic estimate of the time required to bring the first two units of an NPS online.

According to F. Ya. Ovchinnikov, Deputy Minister of Power and Electrification, in mid-1979 the U.S.S.R. had 20 NPSs with a total rated capacity of over 60,000 MW in operation or under construction.[39] If additions to capacity proceed at the rate suggested in figure 9, the Soviets could have 60,000 MW *installed* capacity before the end

of 1988. This allows over 9 years for completion of all of the nuclear power stations reported by Ovchinnikov. On the surface, therefore, the contemplated additions to capacity are consistent with Soviet experience and seem realistic in terms of construction time.

There is an additional dimension, however. Expenditures for construction and assembly work are not spread evenly over the time required to build an NPS. Rather, they begin at a low level, rise steadily until the fourth and fifth years of construction, and then fall off. During construction of an NPS equipped with two VVER-1000 reactors, for example, only 4 percent of total construction and assembly costs is incurred in year 1, but 24 percent of total costs is incurred in year 4 and a similar amount in year 5. In the case of an NPS equipped with two RBMK-1000 reactors, the percentages rise from 7 percent in the first year to 20 percent in year 4 and 20 percent in year 5.[40]

As a result, it may be relatively easy to begin the construction of an NPS, and it may be easy to complete one that is in the late stages of construction, but great effort on the part of the construction industry is required during years 3 to 5. Based on their past experience wth the construction of nuclear power stations, it is possible that the Soviets could succeed during the 1980's in beginning the construction of all planned NPSs, and could complete by 1990 those that are nearly finished. But, the U.S.S.R. might still reach the end of the decade with a great deal of unfinished construction because many NPSs were stalled in the demanding third to fifth construction years.

Required Investment.—Although Soviet data on per unit investment cost is notoriously unreliable, OTA has attempted to convey a rough idea of the order of magnitude of investment entailed in Soviet nuclear plants. A number of estimates of the investment cost per kilowatt of nuclear

[34]N. Ya. Turchin, et al., *Building Fuel and Atomic Power Stations* (Moscow: Stroyizdat, 1979), vol. II, p. 505.

[35]Lafleur and Stello, op. cit., app. B, p. 20.

[36]William F. Savage, "Report of the Second Meeting of the U.S.-U.S.S.R. Joint Committee for Cooperation in the Field of Energy" (Washington, D.C.: U.S. Department of Energy, Dec. 15, 1977), p. 12; V. P. Akinfiyev, A. D. Gellerman, and V. K. Bronnikov, "Experience in Assimilating the Rated Capacity of the First Phase of the Chernobyl NPS," *Elektricheskiye stantsii*, No. 2, Feb. 1980, pp. 7-11.

[37]S. Troyan, "Labor Watch of the Atom," *Izvestiya*, Feb. 1, 1979, p. 1, in JPRS 73,092, Mar. 28, 1979, pp. 39-41.

[38]U.S. Nuclear Regulatory Commission, op. cit., app. B, p. 26.

[39]F. Ya. Ovchinnikov, "Nuclear Power Engineering Is a Quarter Century Old," *Teploenergetika*, No. 7, July 1979, pp. 2-5, in JPRS L/8700, Oct. 4, 1979, pp. 1-8.

[40]V. B. Dubrovskiy, et. al., *Building Atomic Electric Power Stations* (Moscow: Izd. "Energiya," 1979), p. 187.

capacity have appeared in Soviet publications and have ranged from as low as 175 rubles/kW to as high as 413 rubles/kW, depending on the date of the estimate, the cost categories encompassed by the estimate, the assumed size of the power station, and the prospective site.[41]

The most authoritative recent estimates of specific capital investment are probably those that appeared in a February 1979 article by A. Troitskiy, a deputy head of the Department of Power and Electrification of Gosplan.[42] Troitskiy's estimate for the "Center," which best typifies the European U.S.S.R., where most plants are likely to be built, is 380 rubles/kW. This figure includes all associated costs—transmission lines, mining uranium, providing housing for construction workers, etc.

Assuming that the Soviet Union does add the 24,000 to 25,000 MW of capacity called for in the Eleventh FYP, and using Troitskiy's figure, the total cost of additions to nuclear capacity during the period 1981-85 should be 9.1 billion to 9.5 billion rubles. The Eleventh FYP target for planned capital investment in the period 1981-85 is 711 billion to 730 billion rubles. Given that industry has usually been accorded about 35 percent of the total, planned industrial investment in the period is probably about 250 billion rubles. The cost of additions to nuclear capacity therefore constitutes about 3.6 to 3.8 percent of industrial investment. However, investment in the electric power industry as a whole has been usually about 4.6 percent of the total. This means that the nuclear program, which will account for 45 to 50 percent of new electric capacity by 1985 (see ch. 5), could take up about 80 percent of all electric power investment.

This is a large burden indeed—and if even greater increases in nuclear capacity are to be realized in the 1986-90 period, the investment problem can only intensify. Nuclear plant construction on this scale will therefore place much greater demands on the Soviet economy in the coming decades. On the one hand, it is hard to argue from these figures that the aggregate investment demands cannot be met, if the U.S.S.R. is prepared to rearrange its priorities and direct a larger share of total investment funds to this sector. On the other hand, such a reallocation is likely to cause painful readjustments in the Soviet economy and "losers" in the system can be expected to resist the change.

Labor Requirements.—Although data on the construction labor requirements in the nuclear power industry are scarce, available information suggests that these requirements are heavy. An idea of their order of magnitude might be gleaned from the following calculation. In nonnuclear power stations, the lowest levels of labor expenditure achieved in construction were 2.4 to 2.7 person-days/kW of installed capacity.[43] If the Soviets were able to add nuclear capacity with a rate of labor expenditure of only 2.5 person-days/kW of capacity, addition of 24,000 to 25,000 MW during the period 1981-85 would require 60 million to 62.5 million person-days of construction and installation work, while the indicated addition of 42,000 MW during the Twelfth FYP would require about 105 million person-days.[44] If the time profile shown in figure 9 were followed, the construction labor embodied in annual commissionings would rise

[41]Turchin, et al., op. cit., vol. II, p. 894; A. Troitskiy, "Electric Power: Problems and Perspectives," *Planovoye khozyaystvo,* No. 2, February 1979, pp. 18-25. For a thorough discussion of Soviet estimates of nuclear power costs, see William J. Kelly, Hugh L. Shaffer, an J. Kenneth Thompson, "The Economics of Nuclear Power in the Soviet Union," presented at the 55th Annual Conference of the Western Economic Association, San Diego, June 1980. (Forthcoming in *Soviet Studies.*)

[42]Troitskiy, op. cit., p. 20.

[43]Turchin, et. al., op. cit., vol. II, p. 896.

[44]Interestingly, the figure of 2.5 person-days/kW is approximately correct for construction of nuclear power plants in the United States, if U.S. figures are revised to make them more compatible with Soviet practice. See L.M. Voronin (ed.), *Atomnyye elektricheskiye stantsii,* No. 2 (Moscow: Izd. "Energiya," 1979), p. 47. However, the 2.5 person-day figure (derived from Soviet fossil-fired power construction) understates labor requirements for construction of Soviet NPSs, if one can judge from figures on labor requirements *per ton* of equipment. As an example, it is estimated in Voronin that the amount of installation labor required per metric ton of equipment installed is 2 to 2.5 times greater for an NPS than for an analogous fossil-fired power station. Ibid., p. 68.

from 5.8 million person-days in 1979 to 7.5 million in 1981 and 25 million in 1990. Assuming, further, that the average construction worker provides 250 person-days of labor per year, a work force of 23,200 would have been required to deliver the 1979 total, while 30,000 workers would be required in 1981 and 100,000 in 1990.

This growing need for workers to build nuclear power stations comes at a time when the total demand for construction labor appears to be outstripping the supply. It is true that the Soviet Union has a large construction labor force. In 1979, 11.2 million workers and office employees were employed in construction, 10.1 percent of all employment in the national economy. But the rate of growth of construction employment has begun to slow,[45] and in the last 3 years the absolute increment to total construction employment has fallen successively. These declining growth rates do not signify a diminished interest in construction on the part of Soviet planners, but rather a necessary adjustment of the economy in general to slow growth in the labor force.[46] As a result, all sectors of the Soviet economy are under pressure to achieve output gains through increases in labor productivity rather than through expansion of employment.

The use of mechanization and automation as solutions to these problems has met with limited success in the construction industry. F. Sapozhnikov, Deputy Minister of Power and Electrification, has observed that "nonindustrial and technologically ineffective solutions" still are used in the construction of nuclear power stations, and that the share of manual labor remains high.[47] One reason for this seems to be that specially developed equipment and tools for installation work have not been introduced into practice fast enough to keep up with the rapid growth rate of nuclear construction projects.[48]

Other approaches to productivity improvements have been tried, including an effort to reduce onsite labor requirements by increasing the amount of assembly work carried out at factories before components are shipped to the construction site. It has also been suggested that the turnover rate for construction labor could be reduced (and labor productivity increased) if more attention were paid to the needs of construction workers for housing, health care, entertainment, and other social amenities.[49]

Despite the growing demand for workers to build nuclear power stations and the declining rate of growth of the total construction labor force and of the nonagricultural labor force, overall labor shortages need not impede nuclear power development if Soviet authorities plan ahead and allocate their available workers carefully. According to the rough estimates above, nuclear plant commissionings in 1979 would have required an aggregate amount of construction and installation labor equal to only 0.2 percent of the 11.2 million worker construction labor force available in that year. Projected commissionings in 1985 and 1990 would represent only 0.3 and 0.9 percent of the 1979 labor force, respectively. Although there appear to be limited opportunities to shift workers from construction of fossil fuel or hydroelectric power stations, the aggregate labor requirements to support NPS construction need not pose a serious problem.

While the Soviet construction labor force as a whole may be adequate, shortages of individual skills could well arise. Installation is a crucial part of the activity involved in building a nuclear power station, and when

[45]All figures from U.S.S.R., op. cit., pp. 387 and 388.
[46]See Murray Feshbach and Stephen Rapawy, "Soviet Population and Manpower Trends and Policies," in *Soviet Economy in a New Pespective,* compendium submitted to the Joint Economic Committee of the U.S. Congress, Oct. 14, 1976 (Washington, D.C.: U.S. Government Printing Office, 1976), p. 133.
[47]F. Sapozhnikov, "Energy, Heat, and Light for All," *Stroitelnaya gazeta* Dec. 21, 1979, p. 2, in JPRS 75,069, Feb. 5, 1980, pp. 18-20.

[48]Yu. S. Medvedev, "Methods of Increasing the Efficiency of TES, AES Equipment Intallation," *Energeticheskoye stroitelstvo,* No. 11, November 1979, pp. 7-11, in JPRS L/8955, Feb. 28, 1980, pp 7-16. Medvedev implies that this lag exists.
[49]Ovchinnikov, "By Advancing Tempos," op. cit., p. 2.

construction reaches the heavy installation stages at a large plant, more than 2,000 installation workers may be needed onsite for up to 1 to 1.5 years.[50] If a number of plants were to reach the installation stage at about the same time, a severe shortage of installation workers could result.

Construction Support.—Besides labor, the Soviet nuclear program will require adequate support in the way of equipment and facilities for construction and installation work.[51] An important equipment category at nuclear construction sites is large cranes and other hoisting equipment to move building materials such as steel structures and to install heavy components such as reactor vessels. This equipment will become increasingly important as the construction of nuclear plants is based more and more on standardized modular components—large, prefabricated pieces—which reduce labor costs and speed the work at the site.[52] Available hoisting equipment may already be in short supply; one source mentions the possibility of using this equipment for both construction and installation work by combining these two phases of operations wherever possible.[53]

Besides more and better construction equipment, the Soviet Union sees a need to improve the organization and management of construction operations in order to efficiently utilize both equipment and available labor. One form of improvement, as mentioned above, is the timely scheduling of construction and installation operations at an NPS site so that scarce equipment can be shared. On a larger scale, an attempt is being made to organize nuclear station construction work on a "flow-line" basis, much like that in the housing construction industry.[54] As a model for further efforts in this direction, a new type of facility called a nuclear power construction combine (NPCC) is being created in the city of Energodar, at the site of the Zaporozhye plant, now under construction. The NPCC is to manufacture metallic and reinforced-concrete structurals and, presumably, to assemble these pieces into buildings at the plant site.[55]

More NPCCs will probably be organized,[56] but it is not clear how many construction projects would be supported by each combine and, therefore, how many combines will be needed to support the entire nuclear program to 1990. Nor has the level of capital and labor investment in NPCCs been determined. While additional staff and equipment will no doubt be required, it is conceivable that the combines could be formed, in part, on the basis of existing construction organizations and equipment. The aim is to save time and money by using labor and equipment efficiently. Since construction delays have been a major source of bottlenecks in the Soviet nuclear program, these aims take on particular significance.

Plant and Equipment Requirements

This section discusses plant and equipment requirements for the U.S.S.R. to meet its nuclear targets for 1985 and 1990. These requirements include major components of NPSs—reactors, turbines, and generators.

Reactors.—As noted above, the growth of Soviet nuclear generating capacity over the next 10 years will be based on 1,000-MW reactors of the RBMK (graphite-moderated) and the VVER (water-moderated) series, and also on the RBMK-1500. The announced plans for expansion and new construction of Soviet nuclear power stations summarized in table 31 more than cover the capacity needed

[50]Medvedev, op. cit.

[51]Russian construction terminology distinguishes the work of "constructing" or erecting the buildings of a nuclear station from the work of "installing" the reactors and other equipment housed in these buildings. Installation (*montazh*) often is more specialized and intricate work and, thus, requires more highly skilled labor than does construction (*stroitelstvo*). As an industry, "construction" takes in both types of work, and it is in this sense that the word is used here.

[52]D. B. Fedorchukov, "Basic Directions for Raising the Effectiveness, Quality, and Rates of Construction of Nuclear Power Stations," in Voronin (ed.), op. cit., pp. 49-57.

[53]Medvedev, op. cit., p. 13.

[54]Fedorchukov, op. cit., p. 56.

[55]A. Podgurskiy, "NPS's—On a Flow Line!" *Stroitelnaya gazeta*, May 5, 1980, p. 1.

[56]Fedorchukov, op. cit., p. 56.

to reach the 1985 goal of 24,000 to 25,000 MW of additional capacity. At stations where the types and capacities of new reactors are known, the new capacity planned by 1985 amounts to at least 27,320 MW—some 780 MW more than needed to achieve the assumed goal of 38,000 MW by the end of 1985. Table 31 also shows at least ten 1,000-MW units (including units 3 and 4 at Rovno) and three 1,500-MW units to be added, presumably after 1985. Together with the aforementioned 780 MW, these additional units give a total of 17,280 MW of capacity to apply toward the anticipated growth in 1986-90, leaving some 24,720 MW of new capacity unaccounted for. Some flexibility exists in choosing how to cover this remainder.

From table 31, it appears that the U.S.S.R. is building nuclear power stations according to a regular pattern with either four 1,000- or 1,500-MW units per station,[57] each unit consisting of a reactor and accompanying turbine-generator sets. Typically, the Soviets construct stations in "phases" (ocheredi) of two power units each. Assuming that this pattern continues, it is logical to assume that planned stations for which total capacities have not been announced will have at least four units and, therefore, four reactors of 1,000- to 1,500-MW capacity each.[58]

Table 31 shows seven prospective stations for which total capacities have not been announced. However, the type of reactor to be installed at four of these stations is known;

[57]One exception to this pattern is the Rovno station, which is planned to have six units. However, the first two are VVER-440 reactors, both of which originally were planned to come onstream by the end of 1980; only one of them did. The four subsequent units are to be VVER-1000s, thus conforming to the general construction pattern for the 1980's. Another exception may be the nuclear heat and power station near Odessa, which apparently will have two 1,000-MW units.

[58]This strategy was outlined at a conference held in May 1980 at the site of the Zaporozhye plant. It was noted that the rated capacity of stations under construction in the next 10 years will be 4,000 to 6,000 MW, based on units with capacities of 1,000 to 1,500 MW. See "Conference on the Flow-Line Construction of NPS's and the Curtailment of Labor Expenditures and Duration of Construction," *Atomnaya energiya*, vol. 49, No. 4, October 1980, pp. 264 and 265.

OTA has therefore assigned at least one reactor of the designated type to each station and included these reactors in the figures given above for planned new capacity. Assuming that the four stations will have at least three more reactors of the same type when completed, the stations will account for additional capacity totalling 13,500 MW and consisting of nine VVER-1000 reactors at three of the stations and three RBMK-1500s at one (Kostroma).

If the remaining three stations, which are to be heat and power plants, were to have at least two VVER-1000 units each (by analogy with Odessa), all seven nuclear power stations would provide a total of 19,500 MW of new capacity, or 5,220 MW less than the 24,720 MW needed to attain the 1990 goal. One conceivable way of covering this difference would be by building one more *ochered* of two 1,000 MW units at three of the stations. This would signify a departure from current practice. It is also possible that plans for other sites have not yet come to light. Neither of these possibilities alters the crucial question: **How many 440-, 1,000-, and 1,500-MW units are needed to attain the desired goal, and can the Soviets produce and bring them online?**

In order to equip those stations for which reactors types are given in table 31 (excluding unit 6 at Rovno), the Soviet nuclear industry must be supplied with three VVER-440 reactors and at least 24 VVER-1000s, nine RBMK-1000s, and five RBMK-1500s. In addition, based on the alternatives discussed for covering the remaining capacity, as many as 21 more VVER-1000s, and at least three more RBMK-1500s may be required. In all, besides the 440-MW units, this would mean the production of some 45 VVER-1000 reactors, 9 RBMK-1000s, and 8 RBMK-1500s before 1990. One factor that may affect the choice of the reactor mix is that the RBMK-1500 is as yet unproven in practice; operating experience at the Ignalina station may determine its future.

In summary, the growth of the Soviet nuclear power industry in the next 10 years will be based largely on 1,000-MW reactors of the VVER and RBMK series and also on the RBMK-1500, an upgraded version of the 1,000.[59] The 440-MW class of VVERs is apparently being phased out; additional units of this size are to be installed only at Kola and Rovno. Although the U.S.S.R. seems committed to the larger classes of reactors and has designed many of its new stations accordingly, there is some flexibility as to which reactors to install. This flexibility is mainly in the choice between the RBMK-1000 and the RBMK-1500. If problems develop with the RBMK-1500, the Soviets could in principle substitute the RBMK-1000. If continued indefinitely, however, such substitutions would result in a capacity shortfall that would have to be covered in some way, such as by the installation of more 1,000-MW units.

It must be noted that reactor substitution is economic only between the 1,000- and 1,500-MW RBMK units and not between this reactor series and the VVER series. Unless a decision were made very early, substituting between the RBMK and VVER would be difficult and costly. Therefore, if there were a production shortfall in VVER-1000 reactors, stations designed to receive these could not employ the RBMK-1000 without a fundamental redesign. It is possible of course that construction schedules could be altered so that work on stations with RBMKs could move ahead and ones needing VVERs delayed. This would mean that more of the RBMKs could be installed than VVERs in order to keep up with capacity growth plans and to allow the solution of production problems. Eventually, in order to fulfill nuclear capacity goals either Soviet nuclear manufacturers would have to supply the needed reactors of both series, or some other measure—such as the import of

reactor equipment—would have to be adopted.

Soviet hopes for domestic reactor production rest on Izhora and Atommash. OTA has attempted to determine the demands that might be placed on these facilities during the next 10 years. The results, broken down by facility and year, are shown in table 32. **This table is not a prediction for installed capacity or facility output over the next 10 years.** Rather, it combines numerous statements from the Soviet press and open literature regarding the future of the nuclear industry with OTA's own understanding of that industry's development. It endeavors to show what the U.S.S.R. *must* accomplish if it is to achieve its goals. The most important of the assumptions that underlie table 32 are as follows:

1) The table assumes the Soviet stated goals of adding 24,000 to 25,000 MW of installed capacity by 1985 and achieving 80,000 MW by the end of 1990.

2) The table assumes that annual production of nuclear capacity at an individual facility in any given year will equal or exceed production in the previous year.

3) The table assumes that Izhora now has the ability to produce annually reactors having a total capacity of about 5,000 MW. This assumption is based on three pieces of evidence. First, representatives of the U.S. Nuclear Regulatory Commission were told during a visit to Izhora on February 10, 1978, that the plant had an annual capacity of 5,000 MW or more.[60] Second, a 1980 article lists 1980 production obligations for the plant that total at least 5,320 MW (three VVER-440s, two RBMK-1000s, and two VVER-1000s).[61] Finally, the nuclear plans incorporated in the recently adopted Eleventh FYP would be absurd if Izhora were not capable of producing about 5,000 MW of capacity per year. OTA has postu-

[59]The 50-percent increase in capacity in the RBMK-1500 is achieved by increasing the amount of coolant circulated through the core, thereby yielding more steam. For comparative specifications on the two reactors, see T. Kh. Margulova, *Atomnyye elektricheskiye stantsii* (Moscow: Izd. "Vysshaya shkola," 1978), p. 181; and *Atomnaya nauka . . .*, op. cit., p. 37.

[60]Lafleur and Stello, op. cit.
[61]V. V. Krotov, "The Contribution of the Power Machine Builders to the Fuel-Energy Complex of the Country," *Energomashinostroyeniye*, No. 1, January 1980, pp. 2-5.

**Table 32.—Projected Production of Nuclear Reactors and Export Potential
(MW)**

Year	Production capacity			Domestic requirements	Available for export
	Izhora	Atommash	Total		
1981...............	5,000	1,000	6,000	5,000	1,000
1982...............	5,000	1,000	6,000	6,000	0
1983...............	5,000	2,000	7,000	7,000	0
1984...............	5,000	2,000	7,000	7,000	0
1985...............	5,000	3,000	8,000	8,000	0
1986...............	5,000	4,000	9,000	8,000	1,000
1987...............	5,000	5,000	10,000	9,000	1,000
1988...............	6,000	5,000	11,000	10,000	1,000

SOURCE: Office of Technology Assessment.

lated that Izhora will continue to produce reactors at a rate of 5,000 MW per year until 1988, when a rise to 6,000 MW is assumed. This second assumption may well be too conservative. It is possible that Izhora's capacity will expand more rapidly.

4) The table takes into account indications that the Skoda Works will assume primary responsibility for providing VVER-440 reactors for Eastern Europe, but also assumes that the U.S.S.R. will provide some VVER-1000 reactors for Eastern Europe during the Twelfth FYP (see below and ch. 9).

5) Finally, the table discounts Soviet statements that seven VVER-1000 reactors, produced at Atommash, will be shipped by 1985; and that Atommash will be fully operational, producing 8,000 MW per year, by 1990.[62] Assuming that 2 years must be allowed for the installation of a reactor, it is unlikely that Atommash can supply seven 1,000-MW reactors for installation during the 1981-85 period. Four is a more probable total (i.e., annual production of 1,000, 1,000, and 2,000 MW in the years 1981, 1982, and 1983).

Given these assumptions, table 32 demonstrates the rate at which both Izhora and

Atommash must produce reactors if nuclear capacity—and therefore nuclear electricity production—goals are to be met. One vulnerable point in these projections is Atommash, the successful and timely completion of which will dictate the success of the Soviet nuclear program and, to some extent, the program of the Council for Mutual Economic Assistance (CMEA) countries. Construction of Atommash, which is being carried out in stages, has been underway since 1976. Originally planned for startup in 1977, the plant's first stage, with a rated output of 3,000 MW of nuclear capacity per year, was not operating until the end of 1978.[63] This delay is indicative of progress with the project as a whole. Since its inception, it has been beset with both internal and external problems that are frequently publicized in Soviet newspapers and journals. These include excessive idle time and poor organization of workers at the construction site, as well as delays in the deliveries of equipment and materials from external suppliers. The labor problem is complicated by shortages of workers in the plant and at the construction site. As a result, construction of Atommash is estimated to be at least 2 years behind schedule; although the first 1,000-MW reactor vessel was completed in

[62]"Five Year Plan Accelerates Nuclear Programme," *Nuclear Engineering International,* January 1981, p. 4; V. Pershin, "To Eliminate Disproportions," *Sotsialisticheskaya industriya,* Jan. 30, 1981, p. 2.

[63]"Building the Atommash Plant," *Culture and Life,* No. 5, 1976, p. 10; and "Chronicle of Events," *Stroitelnaya gazeta,* Dec. 12, 1980, p. 1.

February 1981, series production at the plant probably will not begin before 1983 or 1984.[64]

Even after the plant is completed, there are likely to be problems with acquiring skilled workers to operate its sophisticated machinery. As recently as 1979, efforts to train such workers—for example, operators of automatic welding equipment—were said to be highly inadequate to meet the needs of Soviet industry as a whole, let alone those of Atommash.[65] Fortunately it appears that the reconstructed Izhora plant will be able to compensate for at least some of the difficulties at Atommash.

The profile assumed here should permit the Soviets to meet domestic requirements for reactors and to supply 1,000 MW for export in the years 1981, 1986, 1987, and 1988. During the first of these years such exports would take the form of VVER-440s (two), while in the later years exports would be in the form of VVER-1000s. It must be noted, however, that it is not clear that these exports will necessarily be sufficient to support the planned Soviet role in CMEA cooperative nuclear power development.

As chapter 9 notes, Eastern Europe plans to have installed nuclear capacity totaling some 37,000 MW by 1990. Until 1985, nuclear powerplants built in CMEA countries will be based on the VVER-440; in the period 1986-90, the plan is to switch to the VVER-1000. The needed VVER-440s presumably will be supplied mainly by Skoda, with one or two reactors possibly coming from Izhora.[66] Since no plans have been announced for

building VVER-1000s outside the U.S.S.R., the intent seems to be for these units to be supplied by the Soviet Union. Thus, some undetermined portion of that plant's output after 1985 will be designated for export. Overall, according to the CMEA agreement, the Soviet Union is to supply about 50 percent of the basic equipment for nuclear plants in Eastern Europe, but the U.S.S.R. might well decide to allocate Atommash's early units to Eastern Europe to offset future hydrocarbon requirements there.[67] This could strain Soviet manufacturing facilities as they attempt to keep pace with large-scale nuclear growth both at home and abroad, particularly if Skoda is slow in building to capacity. Moreover, the U.S.S.R. has in the past exported reactors outside CMEA. Should it wish to continue these exports, facilities would be taxed even further.

In sum, although Soviet plans for expansion of nuclear power are reasonable, they undoubtedly pose a difficult task for plants that manufacture reactor equipment. In the past, the capacity of Soviet industry to meet the increasing needs for this equipment has been inadequate. More recently, Atommash has become operational, albeit with construction lagging behind schedule, and Izhora has been reconstructed and its capacity significantly upgraded. **It is therefore possible that the U.S.S.R. will be able to produce reactors fast enough to meet domestic requirements and to maintain some exports** —two 400-MW units in 1981 and at least 1,000 MW per year after 1985.

Steam Turbines.—The planned expansion of Soviet nuclear power will also necessitate increased production of steam turbines. In principle, the power generation equipment used in nuclear power stations is essentially the same as that in nonnuclear thermal plants. This is particularly true of turbines, and Soviet designs of steam turbines for nuclear stations sometimes duplicate or incorporate design features of those used in fossil-fired stations. This means that Soviet manufacturing plants have been able to ap-

[64]"Obligated by the Initiative," *Soltsialisticheskaya industriya*, Feb. 14, 1980, p. 2, in JPRS 75,741, May 2, 1980, pp. 1 and 2; "Commentary by the Industrial Construction Department, *Stroitelnaya gazeta*, Dec. 12, 1980, p. 1; "Soviets Building Nuclear Reactor Assembly Line," *The Columbus Dispatch*, Oct. 13, 1980, p. A-8. According to a report in *Izvestiya*, Feb. 21, 1981, p. 3, Atommash completed its first reactor as planned.

[65]V. Pershin, "Personnel for Atommash," *Trud*, Feb. 9, 1979, p. 2, in JPRS 73,015, Mar. 16, 1979, pp. 14-16.

[66]The VVER-440 may also be exported to non-CMEA nations. Two of these rectors are in operation at the Loviisa NPS in Finland, and at least one other country—Libya—has purchased a Soviet power reactor. For more on the Libyan deal, see Gloria Duffy, op. cit., pp. 84-86.

[67]Zorichev and Fadeyev, op. cit., p. 53.

ply some of their experience in making turbines and generators for fossil-fuel power stations to steam and electrical equipment for nuclear stations, and also that similar kinds of problems are likely to occur. By the same token, the demands of the nuclear power industry do place new requirements on these plants. Output of equipment must be increased to keep pace with nuclear growth, and the equipment itself, especially turbines, must be compatible with reactor systems.

Current Soviet practice is to install at least two turbines with each reactor. The VVER-440 reactor uses two 220-MW turbines, and the VVER-1000 and the RBMK-1000 units use two 500-MW turbines. The addition of the RBMK-1500, however, will require development of a 750-MW turbine for use with saturated steam. In addition, work reportedly is under way to produce turbines with a rated capacity of 1,000 MW, in order to couple the VVER-1000 and RBMK-1000 with a single turbine.[68]

These steam turbines can be either "low-speed" or "high-speed" units, operating at 1,500 and 3,000 rpm respectively. In the United States, for large capacity turbines (500 MW or more) operating in saturated steam, the low-speed design is considered more suitable; large high-speed turbines operate better and last longer when run on superheated steam. This is because large, high-speed turbines used with steam at low parameters require very long blades in the low-pressure sections of the turbine, and the high velocities at the outer ends involve too much stress. Such difficulties would explain reported Soviet problems with the high-speed (3,000 rpm) 500-MW turbines used with the 1,000-MW reactors. Based on U.S. experience, it is reasonable to expect that unless the Soviet nuclear industry switches to low-speed (1,500-rpm) turbines, it will probably encounter even greater problems with the 750- and 1,000-MW turbines that it plans to use in the near future.

It is interesting that Soviet experts seem to have weighed the costs and benefits and concluded that high-speed turbines are more promising, although considerable expenditures have already been made on preparing designs for powerhouses using the low-speed variety.[69] The only low-speed turbines known to be in place to date are two 500-MW turbines manufactured for the VVER-1000 reactor of unit 5 at Novovoronezhskiy.[70] At the same time, the U.S.S.R. is pursuing the development of high-speed turbines with capacities of 750 and 1,000 MW. Fabrication of the latter has begun, although there is no evidence that a 750-MW high-speed turbine is actually yet in production.

In sum, with the possible exception of the 750-MW size, the Soviet Union is producing turbines of both types and with the unit capacities that it needs, including 1,000 MW. However, only one size—500 MW—actually is in use; the large turbines remain unproven in practice. Whether problems will be encountered, particularly with the high-speed equipment, remains to be seen. Preference does seem to be for the high-speed type. More of these are in service, at stations with the RBMK-1000. Part of the reason for this preference may be that Soviet nuclear manufacturing facilities are not yet adequately equipped to produce high-capacity, low-speed turbines, which are bulkier and more metal-intensive.

Major Soviet turbine and generator manufacturers have been expanding and upgrading their facilities for purposes of turning out more and better equipment. One such plant—the Kharkov Turbine Plant Production Association—has been designated the chief enterprise for the development of low-speed turbines. Another leading manu-

---

[68] Margulova, op. cit., pp. 184, 224.

[69] B. M. Troyanovskiy, *Turbines for Atomic Electric Power Stations*, (Moscow: Izd. "Energiya," 1978); V. Krotov, "Power Machine Buildng and Scientific-Technical Progress," *Planovoe khozyaystvo*, No. 5, 1981, p. 8.

[70] See table 28 and G. I. Grigorash, "Turbogenerators of the Kharkov Plant 'Elektrotyazhmash' for NPS's," *Elektricheskiye stantsii*, No. 8, August 1980, pp. 5-8. Grigorash notes that the Kharkov Turbine Plant has been designated as the chief Soviet enterprise for designing low-speed turbines for nuclear stations.

facturer that has been undergoing expansion and remodeling is the Leningrad Metal Plant Production Association.[71] These two associations appear to be the major suppliers of turbines for the nuclear industry.

Despite such efforts, an official of the power machine building industry observed in 1979 that too little attention has been given to turbine construction, with the result that that industry cannot provide proper nuclear equipment.[72] No evidence was found to suggest that this situation has improved. Too many problems with the high-speed turbines larger than 500 MW may inhibit plans for the growth of the nuclear industry. In the case of the 750-MW turbine, the option exists, if necessary, to fall back to the 500-MW size, at least until any problems with the former are resolved. The same should be true with 1,000-MW turbines.[73] At some point, however, if the goals of its expansion plans are to be met, the Soviet nuclear industry will have to be assured of supplies of the larger turbines.

Generators.—As with steam turbines, turbogenerators are essentially the same for both fossil-fired and nuclear plants, and some generators are able to operate in either type of station.[74] Development of larger and larger turbines requires the creation of generators to match. Turbines and generators must be designed to produce alternating current power at the correct frequency. High-speed generators are designed with two magnetic poles and low-speed with four poles. Turbines and generators are direct-coupled and installed in sets, so that a generator will be required for each new turbine unit produced.

As power stations are built with larger unit capacities, generators too must be designed with higher rated capacities—1,000 MW and 2,000 MW in the case of those destined for nuclear plants. In the past, generator production has been a chronic source of bottlenecks in nuclear construction. Now responsibility for large generators is being given to plants like Elektrotyazhmash in Kharkov and the Elektrosila Electrical Machine Building Production Association in Leningrad, facilities that have extensive experience in designing generators for nonnuclear stations.[75] Elektrotyazhmash has been expanded and charged with producing four-pole generators with capacities of 500 and 1,000 MW. The first two 500-MW generators of this type were manufactured for unit 5 at Novovoronezhskiy, but present plans include 1,000-MW generators. Although no evidence was found regarding large capacity two-pole generators for high-speed turbines, expanded production facilities at Elektrotyazhmash seem to be adequate to manufacture both two- and four-pole generators with capacities as large as 2,000 MW.[76]

Summary.—If Soviet claims are accurate, the U.S.S.R. is fully technologically capable of developing the necessary equipment for its nuclear power industry, including reactors and compatible turbine-generator sets. While limited options exist for deciding which type of equipment to use, the burden of supplying this equipment rests with manufacturing plants. There is evidence that such plants are already having difficulty meeting the demands being placed on them by the nuclear program. OTA has not investigated the extent of Soviet investment in manufacturing other nuclear plant equipment—valves, tubing, etc.—but all evidence suggests that difficulties may be widespread in the nuclear industry. This is not implausible, given the problems faced by Soviet industry as a whole in fulfilling output quotas. Moreover, these difficulties undoubtedly will

[71]G. A. Shishov, "The 'Leningrad Metal Plant' Production Association in the Tenth Five-Year Plan," *Energomashinostroyeniye*, No. 9, September 1980, pp. 5-8.

[72]V. Krotov, "Prospects for Power Engineering Dictate," *Sotsialisticheskaya industriya*, Feb. 3, 1979, p. 2, in JPS 73,015, Mar. 16, 1979, pp. 8-11.

[73]It should be noted that constraints on the substitutability of turbines increase as the construction of a station proceeds. The turbine size must be decided and the order for this equipment placed prior to the start of construction of the station's turbine hall.

[74]Grigorash, op. cit.

[75]Ibid., p. 5.

[76]Ibid.

be aggravated by the increasing rate of growth of nuclear power and the demands for supporting this growth.

The External Fuel Cycle

An important part of the infrastructure for nuclear power generation is the external fuel cycle—the supply of uranium and the system for disposal or reprocessing of spent fuel and wastes. Little information is available on the first part of the fuel cycle in the Soviet nuclear industry. The size of uranium supplies is an official State secret and there is virtually no unclassified information regarding Soviet plants that manufacture the fuel elements themselves.[77]

Based on what is known about its consumption of uranium, the U.S.S.R. seems to be accumulating a substantial stockpile.[78] This may, in part, be a response to perceived worldwide scarcity of this element. OTA is unable to judge whether stockpiling will provide the U.S.S.R. with uranium adequate to support its nuclear program, but on the evidence of its announced plans, the U.S.S.R. seems to be confident that it has or can get the uranium it will need in the years ahead. The emphasis on breeders presumably is at least partly based on ensuring adequate uranium supplies.[79] Moreover, it apparently has adequate enrichment capacity to convert the uranium to usable fuel for its reactors. The demand for Soviet enriched fuel will increase as more and more VVER reactors are installed in the U.S.S.R., in Eastern Europe, and in Third World countries. These reactors require more highly enriched fuel than do the RBMK models.

The growing demand for this fuel may put a strain on the enrichment industry in the U.S.S.R., but it must be noted that this industry already exports its services abroad, including to Western Europe and more recently to the United States.[80]

The other end of the external fuel cycle—disposal or reprocessing of spent fuel and other waste material—is discussed somewhat more openly in the Soviet literature, although little quantitative information is available. Although Soviet experts recognize the increasing importance of dealing with wastes from NPSs, current disposal methods should be adequate for some years. Despite its rapid growth, the total nuclear generating capacity of the U.S.S.R. is still relatively small compared, for example, to that of the United States. The amount of spent fuel produced is therefore correspondingly small.

Commercial reactors in the U.S.S.R., like those in other countries, operate on a "once-through" fuel cycle. After the fuel has been burned up in the reactor, it is discharged to interim storage.[81] At Soviet nuclear power stations, the spent fuel elements are placed in water-filled storage basins that cool the fuel and allow it to decay over time.[82] Eventually, this spent fuel, its radioactivity significantly lessened during interim storage, will be transported to special plants for reprocessing in order to recover reusable fissile materials, particularly uranium and plutonium.

Technology for reprocessing spent fuel has been available in the Soviet Union since about 1950,[83] and there is reason to believe that reprocessing has also been in practical

---

[77]Peter Feuz, "The Nuclear Push in the U.S.S.R. and Eastern Europe," *Power Engineering,* No. 8, August 1978, pp. 100 and 101. See also Dienes and Shabad, op. cit., for information on the location of the uranium industry.

[78]Duffy, op. cit., p. 68.

[79]Jean A. Briggs, "Soviet Nuclear Power: Tortoise and Hare?" *Forbes,* vol. 122, No. 9, Oct. 30, 1978, pp. 123-126, reported that the U.S.S.R. imports uranium from East Germany and Czechoslovakia. Regarding the need to seek outside sources of uranium, Petrosyants, op. cit., p. 268, declared that the development of the nuclear industry and nuclear power in the "socialist countries" (including the U.S.S.R.) "in no way depends on supplies of uranium from the capitalist world."

[80]Duffy, op. cit., p. v; Theodore Shabad, "Russian Uranium Exported to the United States," in *The New York Times,* Aug. 17, 1981.

[81]For a discussion of nuclear fuel cycles in non-Communist countries, see I. Spiewak and J. N. Barkenbus, "Nuclear Proliferation and Nuclear Power: A Review of the NASAP and INFCE Studies," *Nuclear Safety,* No. 6, November-December 1980, pp. 691-702.

[82]*Atomic Science . . .,* op. cit., p. 154.

[83]Ibid., p. 153. For a description of a process designed by Soviet scientists for what they call "regeneration" *(regeneratsiya)* of spent fuel, see pp. 153-159 of this source.

use there for some time. Indeed, American scientists suspect that the U.S.S.R. was extensively involved in chemical separation of nuclear wastes as early as the mid-1950's.[84] If this is true, the Soviet Union may already have practical experience with the technology it will need to eventually perform full-scale commercial reprocessing. It has been estimated that full-scale reprocessing of spent fuel will not be economical in the U.S.S.R. until Soviet nuclear stations are producing 1,500 tons of spent fuel per year—about as much as the United States was producing from 60 reactors in 1978. At that time, total U.S. nuclear capacity was some 47,000 MW.[85] Based on OTA's projection in figure 9, the U.S.S.R. will not reach this level until after 1985. This reasoning is supported by a 1978 Soviet source that maintained that while no reprocessing plant had yet been built in the U.S.S.R., that country would have reprocessing in the 1980's.[86]

Besides spent fuel, the operation of NPSs produces other waste products, primarily liquid wastes, of high, medium, and low radioactivity. Highly radioactive liquid waste results mainly from reprocessing, and the Soviet nuclear industry, therefore, will probably not have to handle large amounts of these products until after 1985. By and large, nuclear plants produce wastes of medium and low radioactivity.[87]

As a nuclear industry grows, so does the problem of liquid-waste disposal. In response, Soviet scientists reportedly are developing various permanent disposal methods, from deep underground burial in natural geological formations to different ways of concentrating and solidifying waste material (depending on its level of radioactivity), and at least one Soviet source considers the problem of deep burial of low- and medium-level waste to be solved.[88] Concentrated low-level wastes from research facilities are already being stored underground at Zagorsk, near Moscow, and at Dimitrovgrad. In the future, high-level wastes are to be compacted, solidified, and encapsulated for storage in abandoned salt or coal mines.[89]

Computers

Computers have been used more intensively for process control applications in nuclear facilities than elsewhere in the power generation industry, but the U.S.S.R. evidently recognizes a need for wider use of this technology.[90] This need will grow as the number of NPSs and the demands placed on them increase.

The Soviet Union reportedly began developing computerized control systems for fossil-fired power stations as early as 1961, and development of these systems for nuclear stations began in 1971.[91] Nevertheless, there is evidence that Soviet NPSs lack components such as microprocessors and other control equipment to support their functions. For example, Western observers have noted that Soviet NPSs employ manually operated plumbing in reactor cooling systems. The computer systems themselves are said to be relatively primitive,[92] and new systems are introduced into plants slowly. If the Soviets do not meet their goals for the construction of NPSs in the Eleventh and Twelfth FYPs, it will not be due primarily to deficiencies in computer control systems.

[84]Duffy, op. cit., p. 64.

[85]Biggs, op. cit., pp. 126, 124.

[86]John J. Fialka, "Soviets Think They've Solved Atom Safety Problems," *Washington Star,* Oct. 1, 1978, pp. A-1 and A-10.

[87]Petrosyants, op. cit., p. 356, notes that relatively small amounts of highly radioactive waste also are produced in the operation of NPSs. At the same time, V. B. Dubrovskiy, et al., op. cit., p. 33, states that such waste products are not formed at NPSs, Small amounts of highly radioactive waste undoubtedly are produced by industrial and research reactors.

[88]Petrosyants, op. cit., pp. 356-366.

[89]Briggs, op. cit., p. 126.

[90]Ye. P. Stefani and V. I. Gritskov, "The Status and Prospects for the Development of Computerized Technological Process Control Systems for Power Units of Thermal and Nuclear Power Stations," *Teploenergetika,* No. 7, July 1980, pp. 2-7.

[91]Ibid., pp. 2 and 3.

[92]Fialka, "Russia's Nuclear Program . . .," op. cit.

REQUIREMENTS FOR NEW EQUIPMENT AND TECHNOLOGY

NUCLEAR CAPACITY AND PRODUCTION: LIKELY ACHIEVEMENTS

Based on the foregoing analysis of the problems facing the Soviet nuclear power industry and its surrounding infrastructure, OTA has developed a set of projections for likely levels of achievement of nuclear capacity and electricity production.

As noted above, the Soviet Union has set a goal of adding 24,000 to 25,000 MW of nuclear capacity during the Eleventh FYP period. If this goal is achieved, total installed nuclear capacity at the end of 1985 would be about 38,000 MW. If the 1985 target is reached, or not seriously underfulfilled, there is good reason to believe that the Soviets would set a goal of 80,000 MW of installed capacity by the end of 1990. Figure 9 presented a time profile of capacity additions that would permit the U.S.S.R. to meet these goals. This profile hypothesized successive increments of 3,000, 4,000, 5,000, 6,000, and 7,000 MW during the period 1981-85, followed by additions of 7,000, 8,000, 8,000, 9,000, and 10,000 MW during the Twelfth FYP.

The above discussion has indicated that the U.S.S.R. should be capable of producing reactors fast enough to meet domestic requirements, to export 1,000 MW in 1981, and at least 1,000 MW per year after 1985. Therefore, reactor production itself is not a particularly weak link in the Soviet nuclear power industry. The picture is less sanguine with respect to other requirements for meeting the capacity targets assumed for 1985 and 1990.

Bottlenecks will probably develop in construction and installation work. The contemplated rate of commissioning of nuclear capacity implies a need to triple the size of the construction labor force. If the rate of addition of generating capacity (of all forms) were expected to moderate in the 1980's, it might be possible to meet this need for labor by drawing workers from nonnuclear power projects. But no such moderation is in prospect. Gross additions to generating capacity (all forms) are to rise from 52,000 MW in the Tenth FYP to about 64,000 MW in the Eleventh FYP, and about 85,000 MW in the Twelfth FYP (see ch. 5). Furthermore, since there appears to be no prospect that the absolute rate of construction of fossil-fired and hydroelectric powerplants will decrease during the coming two FYP periods, it may not be possible for nuclear projects to draw any workers (on net) from these competing projects unless there are substantial increases in the productivity of construction labor. Power station construction overall probably will require large numbers of additional workers, while NPS construction in particular will require a great deal of additional specialized construction and installation labor (e.g., workers to assemble and install reactors and associated equipment).

A second area in which bottlenecks may develop is the production of turbines and generators. The Soviets have not yet mastered the production of low-speed 500-MW turbines, high-speed 750-MW turbines, and high-speed 1,000-MW turbines. All of these will be needed in the present decade. If technical problems arise, the Soviets may be forced to fall back on the more tested high-speed 500-MW turbine that has found widespread application in conjunction with the RBMK-1000. Generator problems may develop in the production of the four-pole 500-MW models needed to go with the new low-speed 500-MW turbines. A more serious problem may be a general shortage of capacity to manufacture generators. As in the case of labor, no help will come from a reduced burden of fossil-fired and hydroelectric capacity additions. The U.S.S.R. could reduce its exports of turbines and generators, but it is not clear that this action would increase its ability to produce the types of turbines and generators needed for nuclear power stations.

As a result of these factors—likely delays in construction of NPSs and delays in installation work at NPSs, plus shortfalls in the production of turbines, generators, and other equipment required for NPSs—OTA expects the Soviet nuclear program to fall behind schedule. Although estimates of developments 5 and 10 years into the future are necessarily speculative, **it seems reasonable to adopt as an optimistic or best-case projection the expection that the Soviets will have installed 36,000 MW by the end of 1985** (v. 38,000 MW planned) and 75,000 MW by the end of 1990 (v. 80,000 MW planned). If the U.S.S.R. achieves 36,000 MW of capacity by the end of 1985, its nuclear power stations will generate about 190 billion to 210 billion kWh of electricity during that year.[93] Achievement of 75,000 MW by the end of 1990 should lead to generation of 400 billion to 445 billion kWh in 1990.[94]

[93]Estimated by multiplying projected **capacity for** mid-1985 (33,800 MW) by operating rates of 5,663 to 6,259 **hr/yr.** The former rate is the estimated 1980 rate, while the **latter** is the 1985 operating rate implied by targets announced for the Eleventh FYP.

[94]Estimated capacity for mid-1990 (71,000 MW) operated for 5,663 to 6,259 hr/yr.

THE POTENTIAL ROLE OF WESTERN EQUIPMENT AND TECHNOLOGY

Despite the achievements justly claimed by the U.S.S.R. in developing and upgrading its nuclear technology and equipment, there is room for new technology and equipment available in the West. For example, while some spent fuel reprocessing apparently is being carried out, the U.S.S.R. has not yet built a commercial-size plant for this purpose. Such a plant will require large-scale production equipment that may still be lacking despite the long-standing availability and use of basic reprocessing technology. In addition, there may be alternative processes that have not been developed or explored thoroughly in the U.S.S.R..

Another area that probably could benefit from new technology and equipment is the manufacture of metal parts, such as turbines and reactor tubing. Soviet metallurgical processes and fabrication techniques for these parts seem to be less advanced than similar processes and techniques used in the United States. A third area of technological need is in control equipment, particularly microprocessors.

Soviet experts have themselves noted some areas where new or improved equipment is needed. According to one article, the horizontal drum-type steam separators currently used with RBMK reactors should be replaced with more efficient vertical separators, a process that requires more research.[95] A new type of reactor vessel made of prestressed steel-reinforced concrete is being developed for a 500-MW reactor for nuclear heat and powerplants,[96] and according to another source, work is being done and more is needed to find better materials and designs for the motors of main circulatory pumps of nuclear stations.[97] In the area of steel pressure vessel fabrication, Soviet industry is said to lack progressive forging technology for making reactor vessels. There are reports, however, that such technology is being introduced at Atommash.[98] Finally, it has been suggested that one way to improve the American uranium

[95]G. V. Yermakov, "Scientific and Technical Tasks for the Advancement of Atomic Power in the U.S.S.R.," *Elektricheskiye stantsii*, No. 7, July 1979, pp. 5-9.

[96]Ibid., p. 8.

[97]O. L. Verber, et. al., "High-Power Asynchronous Electric Motors for Main Circulatory Pumps of Nuclear Power Stations," *Elektricheskiye stantsii*, No. 9, September 1980, pp. 5-9.

[98]Ye. N. Moshnin and S. A. Yeletskiy, "Directions of the Advancement of Forging and Pressing Production Technology of Atomic Machine Building," *Energomashinostroyeniye*, No. 1, January 1980, pp. 36-38.

enrichment industry is through the use of advanced isotope separation processes such as laser enrichment.[99] Conceivably, such processes could also be applied in the U.S.S.R.

While the acquisition of these technologies would undoubtedly benefit the Soviet nuclear industry, it may be that more mundane needs in the area of construction equipment and manufacturing of parts for nuclear reactors, turbines, and generators would be more critical for the fulfillment of 1985 and 1990 plan targets. These needs include heavy machinery for manufacturing plants; large capacity cranes for construction and installation work; small parts such as valves and circulatory pump motors; and small computer components and other electronic control equipment.

One way for the U.S.S.R. to fill both kinds of needs is to purchase technology and equipment from the West. There are two possible motives for making such purchases, depending on whether the need is technical or economic. In the first instance, the Soviet Union is motivated to make foreign purchases because its own scientists and engineers have not yet developed or will not be able to develop the desired equipment or technology. OTA believes that this need is slight, if it exists at all, in the Soviet nuclear industry. Soviet nuclear power R&D capabilities apparently have been and will continue to be adequate to meet virtually all the requirements of the nuclear industry for new equipment and technology.

Economic motives are far more likely to animate Soviet trade with the West in this area, largely because of the widespread systemic problems that have caused bottlenecks in the nuclear program as well as in other areas of the Soviet economy. As a result of such problems, domestic equipment and technology cannot be introduced or manufactured fast enough to keep pace with planned nuclear expansion. Such delays have apparently led the Soviets to seek outside

sources for additional supplies of equipment and technology to augment domestic sources, and there is no reason to believe that such purchases will not continue or indeed increase—if development of these items in the U.S.S.R. remains costly in comparison.

If the Soviets fail to meet their plan targets for installed nuclear capacity, they should have continued interest in acquiring equipment from the West. Under these circumstances, the U.S.S.R. might well attempt to purchase an entire nuclear power station from the West on a turnkey basis. This option would be attractive from several points of view. Purchase of one 4,000- to 5,000-MW NPS would fill the gap between the assumed target capacity for 1990 and OTA's projected actual capacity. In addition, although the Soviets have a highly developed nuclear technology, purchase of such a plant from an advanced Western nation probably would have substantial technology transfer benefits. It might also be possible to reduce the immediate financing and foreign exchange cost of the project by negotiating a barter arrangement in which the Soviets would pay for the plant in part by supplying electricity to Western countries (e.g., to West Germany).[100]

Alternatively, the U.S.S.R. might choose to purchase selected components for nuclear powerplants in order to compensate for particular shortfalls in domestic equipment production. Since the Soviets seem likely to have an adequate supply of reactors, their purchase seems unlikely unless the Soviets foresee technology transfer benefits. However, shortfalls in domestic production of turbines, generators, and other components could occur and could lead to Soviet interest in Western imports.

Nuclear power is a high-priority sector in the Soviet Union and probably has been able

---

[99]Spiewak and Barkenbus, op. cit., p. 695.

[100]At least one such barter deal has already been suggested by the Soviet Union. The deal, whereby West Germany would have built a nuclear station in Kaliningrad in exchange for electric power, fell through in 1976 for economic and political reasons.

to command foreign exchange to finance imports. It seems likely that the volume of Soviet imports for this industry has been constrained by Western export restrictions and by the adequacy of Soviet-made equipment, rather than by the availability of foreign exchange to this sector or the availability of Western credits.

If trade restrictions are relaxed, Soviet purchases of nuclear equipment are likely to rise even if purchases must be made on a cash basis. The availability of Western credit on liberal terms probably would lead to larger volumes of imports. A barter deal in which the Soviets acquired a turnkey NPS and paid for it over a number of years with exports of electricity would almost certainly be the most attractive alternative from the Soviet perspective because the project would be self-amortizing.

In short, even though the Soviet Union has developed a substantial nuclear power sector relying largely on its own efforts, the major surge planned for the 1980's is likely to encounter problems, and progress will probably not be as fast as the Soviets hope. This gap between expectations and results could create interest in the possibility of importing Western nuclear components. If Western trade restrictions were relaxed, Soviet needs for equipment and traditional Soviet interest in Western technology could lead to substantial imports for this industry. Liberal credit terms are not essential to such trade, but might result in greater volumes. However, Soviet imports are likely to be limited to amounts needed to meet plan targets for NPS capacities. The prospects of the United States in competing for shares in this hypothetical market are discussed in chapter 6.

SUMMARY AND CONCLUSIONS

This chapter has described the recent growth of the Soviet nuclear industry, including officially announced plans for adding 24,000 to 25,000 MW of nuclear generating capacity by 1985 and another 42,000 MW by 1990. While these plans are feasible in terms of the number and capacity of planned power stations and required levels of investment, OTA has identified a number of potential problems.

Many of these problems relate to the construction of NPSs. Past experience suggests that completion of facilities will lag behind plan targets. Although nuclear construction need not necessarily be hampered by insufficient labor in general, highly skilled installation workers may be in short supply. Shortages of materials and equipment, poor organization of available resources and equipment, and lack of experience in installing and bringing online the relatively untried VVER-1000 reactor may also contribute to delays.

Another potentially important source of bottlenecks in the Soviet nuclear program is

inadequate capacity to manufacture turbines and generators. In addition, Soviet industry seems to lack experience with, and perhaps manufacturing capacity for, producing low-speed turbines, which the U.S. nuclear industry has found more suitable than high-speed turbines for running on the low-parameter steam generated by nuclear reactors. A similar situation may exist with respect to four-pole generators used with the slower turbines. Evidence suggests that the U.S.S.R. intends to install low-speed turbines with VVER-1000 reactors, as is the case with unit 5 at the Novovoronezhskiy NPS. If so, turbines and generators needed for new NPSs with these reactors could be in short supply if manufacturing plants are unable to turn out this equipment in sufficient quantities.

Thus, despite a record of self-sufficiency in the nuclear industry, shortages of equipment and materials at home could force the U.S.S.R. to seek these products abroad. The CMEA countries have adopted a long-range plan for cooperative nuclear development that will support the U.S.S.R.'s program by providing additional sources of needed products. In addition, the Soviet Union is known to be engaged in trade with Western countries for this purpose. Although such trade so far has been modest and for the benefit of Soviet plants manufacturing nuclear equipment, it is possible that the situation could change as the demands of the industry grow. (See ch. 12 for a discussion of potential deals in Italy and West Germany. The role of the United States in this trade is discussed in ch. 6.)

In summary, OTA believes that while announced and estimated Soviet goals for nuclear growth to 1990 are attainable in principle, given the systemic problems that continue to hamper Soviet economic growth in general and that seem to be aggravated by the qualitative and quantitative demands of the nuclear industry for materials, equipment, and labor, the U.S.S.R. will probably fall *at least* 5,000 MW short of its targets. The demands of the nuclear industry will place great strain on other already overburdened areas of the Soviet economy, particularly the construction industry and nuclear manufacturing plants. To relieve some of this strain, the U.S.S.R. might wish to rely on foreign trade, both with CMEA countries and with the West, as it has done in the past. If shortfalls in domestic production are coupled with delays in or curtailment of supplies of needed equipment and technology from the West, the U.S.S.R.'s planned rate of nuclear growth will probably slow.

CHAPTER 5

The Soviet Electric Power Industry

CONTENTS

LIST OF TABLES

FIGURE

The Soviet Electric Power Industry

Ever since Lenin articulated the goal of electrification of the entire country, the electric power industry has been considered fundamental to the task of Soviet economic development. Although realization of this goal still belongs to the future, the use of electricity has been promoted throughout the Soviet economy, and the construction of generating stations and an integrated power transmission and distribution system have been given high priority in State plans. As a result, electricity consumption in all sectors of the Soviet economy has grown considerably.

Electric power is produced in the U.S.S.R. through the conversion of nuclear or hydropower, or through burning fossil fuels—coal and liquid hydrocarbons. The status of and prospects for the Soviet nuclear industry are treated separately in chapter 4; and OTA has not studied Soviet hydropower. This chapter, therefore, concentrates on the technological and other problems facing the U.S.S.R. in the conversion of coal, oil, and gas to electricity. These problems fall into three major categories: problems of electricity generation, problems relating to the con-

struction of electricity transmission lines, and problems associated with the development of integrated electricity networks.

The chapter is concerned with the difficulties encountered in, and the prospects for, generation of electricity at powerplants fired by fossil fuels, with the ability of the U.S.S.R. to construct high voltage (HV) power transmission lines, and with the formations of power systems and the problems associated with managing these systems. It analyzes the present and prospective role of electric power in supplying energy to the Soviet economy, and the changes—in generating capacity and output, in location of generating stations, and in technology and equipment—which must take place for this role to expand. It goes on to describe present activities in, and plans for, electricity transmission lines and integrated networks, to discuss the present and potential contributions of the West in each of these areas, and to evaluate Soviet prospects for meeting growing electricity demands and for fulfilling existing plans for the addition of installed electricity capacity and growth in electricity production.

THE FUTURE FOR ELECTRICITY GENERATION IN THE U.S.S.R.

Table 33 shows that in 1980 the U.S.S.R. generated 1,295 billion kilowatt hours (kWh) of electricity, a 4.5 percent increase over 1979. Approximately 80 percent of this electricity came from fossil fuel or conventional plants, and about 5.5 percent from nuclear power. But the U.S.S.R. is planning a sharp shift from fossil-fired to nuclear and hydro generating capacity. In the next 5 years, the contribution of the nuclear industry will triple, while the amount of electricity provided

by fossil-fuel stations is expected to grow only 5 to 9 percent for the entire period. This shift is further demonstrated in table 34, which shows that while installed fossil-fuel generating capacity is expected to grow about 15 percent by 1985, nuclear capacity is slated to nearly triple and hydropower to rise some 23 percent. This planned relative growth in hydropower's share of installed capacity is much higher than in previous years.

Table 33.—Soviet Electricity Production
(billion kWh)

	1975	1979	1980[2]	1985 (plan)
Total.............	1,039[1]	1,238[1]	1,295	1,550-1,600[5]
Fossil-Fired....	893[4]	1,011[4]	1,038	1,100-1,135[6]
Nuclear........	20.2[7]	54.8[3]	73	220-225[5]
Hydro[a]	126.0[1]	172.0[1]	184	230-235[5]

[a]Includes production at pumped-storage stations.

SOURCES: [1]U.S.S.R. Central Statistical Administration, *Narodnoye khozyaystvo SSSR v 1979 g.*, (Moscow: Izd. "Statistika," 1980), p. 168.
[2]*Ekonomicheskaya gazeta*, No. 12 (1981), p. 2.
[3]*Ekonomicheskaya gazeta*, No. 7 (1980), p. 1.
[4]Residual.
[5]*Izyestiya* (Dec. 2, 1980), p. 3.
[6]Fossil-Fired generation to account for 71 percent of total generation in 1985 [*Ekonomicheskaya gazeta*, No. 12, (1981), p. 2].
[7]L. Dienes and T. Shabad, *The Soviet Energy System* (Washington, D.C.: V. H. Winston & Sons, 1979), p. 153.

Table 34.—Installed Soviet Electrical
Generating Capacity
(thousand MW end of year)

	1975[10]	1978[1]	1979	1980	1985 (plan)
Total	217	246	255[2]	267[5]	332[6,8]
Fossil-Fired[a]....	171	190	194[4]	201[4]	230[6,8]
Nuclear.........	4.7	8.4	11.4[3]	13.4[6]	38[9]
Hydro[b]	41.5	47.5	50.0[2]	52.3[7]	64[5]

[a]Includes about 76,000 MW at heat and power stations (TETs) in 1980[5]
[b]Includes pumped-storage stations

SOURCES: [1]*Elektricheskiye stantsii*, No. 8 (1979), p. 6.
[2]*Narodnoye khozyaystvo, SSSR b 1979 g.*, p. 168.
[3]See ch. 4.
[4]Residual.
[5]*Ekonomicheskaya gazeta*, No. 12 (1981), p. 2.
[6]*Teploenergetika*, No. 1 (1981), p. 2.
[7]*Elektricheskiye stantsii*, No. 1 (1981), p. 3.
[8]*Planovoye khozyaystvo*, No. 1 (1981), p. 7, reports a planned gross addition of 71,000 MW between 1981 and 1985. OTA has subtracted 6,000 MW to represent retirement of depreciated capacity in the period.
[9]*Izvestiya* (Dec. 2, 1980), p. 3, reports 24,000-25,000 MW of new capacity to be added between 1981 and 1985.
[10]*ACES Bulletin*, No. 1 (spring 1978), p. 41,

Despite this change in emphasis, however, fossil-fired plants still make up the bulk of Soviet generating capacity and will account for 44 percent of the new capacity called for in 1985. This section describes the present status of fossil-fired generating capacity, the ways in which the U.S.S.R. expects this capacity to change, and the demands that will be placed on the electric power and related industries if these plans are to be met.

One notable trend in fossil-fired electric power generation has been the substantial reduction in the relative importance of coal in power station supply over the past 15 years. Between 1960 and 1975, coal's share in the fuel structure of powerplants fell from 70.9 to 41.3 percent (see table 35). Coal was replaced largely by liquid fuels, the use of which has risen from 7.5 to 28.8 percent of the total; and by natural gas, which rose from 12.3 to 25.7 percent. The shift away from coal was due largely to the relatively low cost to the Soviets of petroleum in this period.

Now there are important incentives to return to coal for that increment to installed capacity that is not to come from nuclear or hydropower stations. Figure 10 shows the location of major sites for the construction of new fossil-fuel generating plants and of plants where plans exist to increase installed capacity. Table 36 summarizes the known characteristics of these new plants and additions. It is obvious that the Soviets hope that much of the increment in fossil-fired installed capacity and electricity production in the next Five Year Plan (FYP) period will come from coal produced in remote regions of the U.S.S.R. Seven of the nine new stations shown in figure 10 will be built in Kazakhstan, central and eastern Siberia, and

Table 35.—Structure of Fuel Use in Fossil-Fired Electrical Power Generation (percent)

	1960	1965	1970	1975	1980[a]
Gas	12.3%	25.6%	26.0%	25.7%	25.1%
Liquid fuel	7.5	12.8	22.5	28.8	28.0
Coal	70.9	54.6	46.1	41.3	42.5
Peat	7.0	4.5	3.1	2.0	2.6
Shale	1.0	1.5	1.7	1.7	1.4
Other	1.3	1.0	0.6	0.5	0.4
Total	100.0%	100.0%	100.0%	100.0%	100.0%

[a]Planned structure.

SOURCE: A. M. Nekrasov and M. G. Pervukhin (eds.), *Energetika SSSR v. 1976-1980 godakh*, (Moscow: Izd. "Energiya," 1977), p. 151.

Figure 10.—Major Sites for the Construction of New Fossil-Fuel Burning Plants or Additions to Installed Capacity, 1981-85

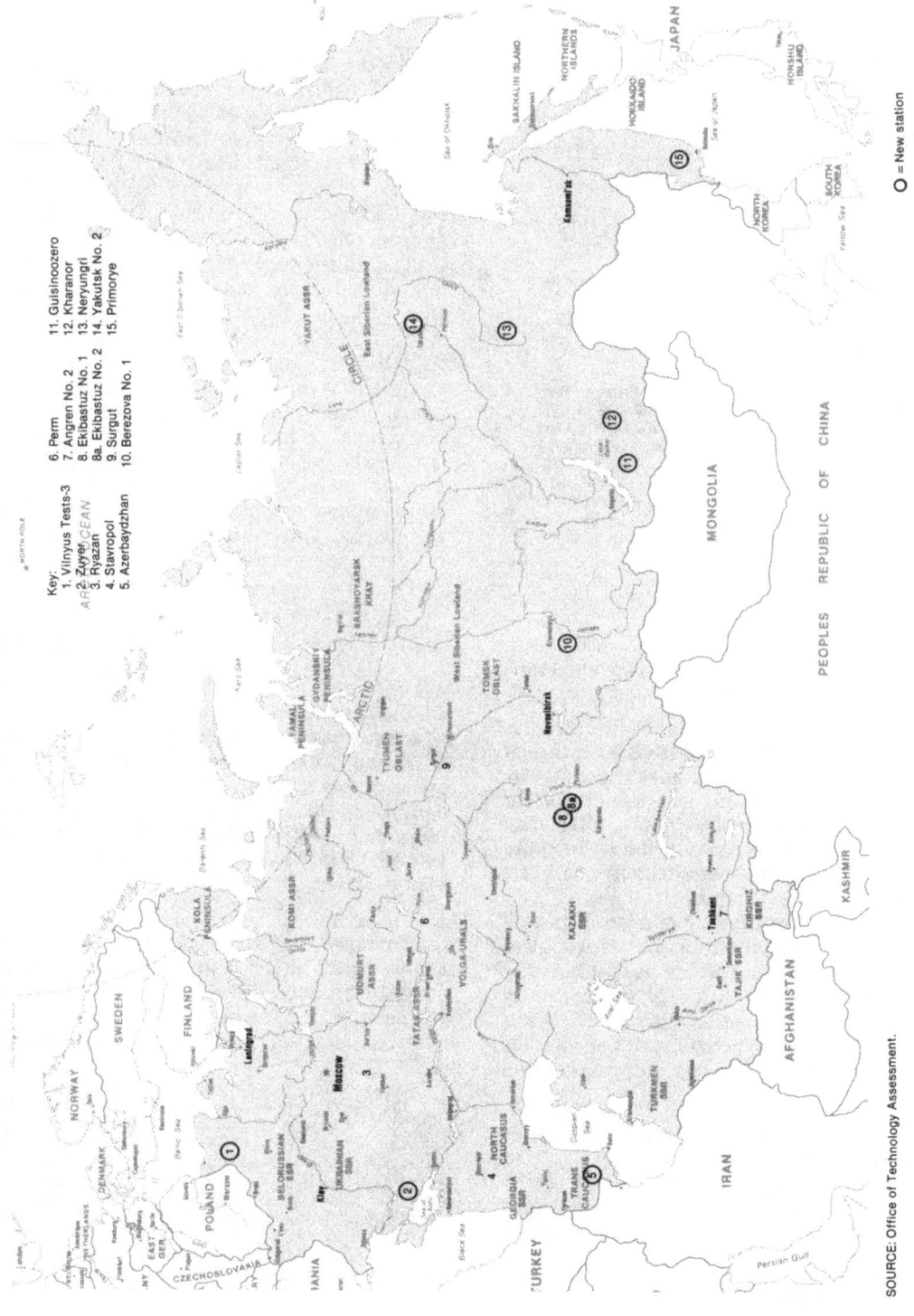

Key:
1. Vilnyus Tests-3
2. Zuyev
3. Ryazan
4. Stavropol
5. Azerbaydzhan

6. Perm
7. Angren No. 2
8. Ekibastuz No. 1
8a. Ekibastuz No. 2
9. Surgut
10. Berezova No. 1

11. Gusinoozero
12. Kharanor
13. Neryungri
14. Yakutsk No. 2
15. Primorye

O = New station

SOURCE: Office of Technology Assessment.

Table 36.—Planned Expansion of Fossil-Fired Powerplant Capacity, 1981-85

Location	Fuel	Comments
New stations		
Azerbaydzhan	Unknown	
Berezovskoye No. 1	Lignite	Boiler system not
Gusinoozero	(Coal)	developed
Kharanor	(Coal)	
Neryungri	(Coal)	
Yakutsk No. 2	(Coal/gas)	
Primorye	(Coal)	
Ekibastuz No. 2	Coal	
Additions to existing stations		
Zuyevka	(Coal)	300 MW in 1981
Stavropol	Gas/mazut	300 MW in 1981
Ryazan	Lignite	800 MW in 1981
Perm	(Coal)	
Surgut	(Gas)	
Ekibastuz No. 1	Coal	2 500 MW generators installed as of 1980
Angren No. 2	(Coal)	

NOTE: (   ) - indicates probable fuel.
SOURCES: *Ekonomicheskaya gazeta*, No. 1 (January 1981), p. 11, and *Izvestiya* (December 2, 1980), p. 6.

the Far East. The plants will all be coal fired, although one at Yakutsk may also burn natural gas.

In fact, these plans not withstanding, it may be natural gas—not coal—which shows the most significant growth as a fuel for power generation in the 1980's. Recently published Soviet figures show that the planned structure of fuel use shown in table 35 was not fulfilled.[1] During the Tenth FYP there was a further jump in the share as well as the quantity of oil and gas used in power stations. Given the problems facing the Soviet coal industry (see ch. 3) and the enormous planned increases in gas output, it is not unreasonable to expect an appreciable surge in the share of gas as a power station fuel.

This outcome is made even more plausible by the nature of Soviet plans to utilize coal in power generation. Nearly all the increment in coal production will now come from Kazakhstan, Siberia, and the Far East.

Given the difficulties in coal transport, one way of utilizing this coal is to use it to generate electricity at the mine itself. Large electric power complexes intended to supply local and regional needs and fired by local coal are already under construction in both the Ekibastuz and Kansk-Achinsk basins. Indeed, one Soviet source claims that 77 percent of all the new fossil-fired capacity to be introduced in the present FYP will be mineside plants at Ekibastuz and Kansk-Achinsk.[2] Eventually long-distance power transmission lines are expected to make this electricity available to the Urals and European U.S.S.R. These complexes, therefore, are central to known existing plans for electric power generation in the coming decade. The following sections describe the difficulties that are most likely to inhibit their completion and consequent growth in coal-fired power generation.

THE GENERATION OF ELECTRICITY IN MINE-MOUTH POWER COMPLEXES

The first Ekibastuz State Regional Power Station was to go online and construction of a second to begin during the Tenth FYP (1976-80). Eventually, four of these stations are to be built in the Ekibastuz region, each equipped with eight 500-MW generator blocks. Over 600 million rubles were to be invested in the Ekibastuz power and fuel complex between 1977 and 1980.[3]

Similarly, 8 to 10 power stations are to be built as part of the Kansk-Achinsk Fuel and Power Complex, development of which was formally called for in a 1979 decree.[4] Power stations erected near the eastern Kansk-

[1]*Elektricheskiye stantsii*, No. 5, 1981.

[2]*Planovoye khozyaystvo*, No. 1, 1981, p. 7.
[3]"The Pavlodar-Ekibastuz Complex," *Ekonomicheskaya gazeta*, No. 22, May 1978, p. 2; Ye. I. Borisov, "Energy in the Anniversary Year 1977 and Problems in 1978," *Teploenergetika*, No. 1, January 1978, p. 3; "A Giant at Power Engineering," *Izvestiya*, Feb. 25, 1978, p. 2; O. Mulkibayev, "Problems at Ekibastuz," *Narodnoye khozyaystvo Kazakhstana*, No. 12, December 1977, pp. 59-65, in JPRS 71,029, Apr. 28, 1978, pp. 44, 50.
[4]A. Sergeyev, "KATEK: The Forecasts and the Present," *Tekhnika i nauka*, No. 1, January 1977, pp. 4-6, 8; "A Constellation of Industry," *Izvestiya*, Apr. 7, 1979, p. 3.

Achinsk deposits are intended for local and regional power generation only, while those near the Western deposits are to supply power to the Urals and European U.S.S.R. Long-range plans envisage the investment of 13 billion to 14 billion rubles on development of mining, power generation, and coal treatment facilities.[5]

There are four major problems confronting these complexes. The first two—construction and supply of power equipment—are common to all forms of powerplant construction. The third—the development of boilers—relates to the intended use of low-grade coal from Siberia, especially Kansk-Achinsk lignites. The fourth factor is the development of appropriate electricity transmission technology and is discussed in a subsequent section of this chapter. Such technology is critical to the ability of the Ekibastuz and Kansk-Achinsk complexes to supply electricity to the Western U.S.S.R.

Construction

A major obstacle to the introduction of planned electrical generating capacity is the low quality of construction operations at many plant sites. Problems here are similar to those found in other sectors of the construction industry. They include labor shortages, supply problems, plant design errors, planning inefficiencies, and long construction times.

Like many other Soviet industries, the electrical powerplant construction industry faces labor shortages, resulting in part from a high degree of usage of manual labor (40 percent of the work in building powerplants).[6] This is caused largely by the low level of mechanization of auxiliary operations and insufficient use of prefabricated building elements. The situation is aggravated by the frequent need to "disassemble"

completed work because of errors in construction or changes in plans. Such disassembly is not highly mechanized. In addition, labor shortages are exacerbated in Eastern regions where poor living conditions promote high labor turnover.

The Soviet press carries numerous articles on the problems of equipment and materials supply to powerplant construction sites. Producers of boilers or turbines often ship equipment in installments at their own convenience, and builders must store this machinery awaiting other needed parts. Often components arrive in insufficient quantities or in unsuitable grades or types. The supply system itself seems to be straining to maintain the flow of materials to a growing number of construction sites. It is increasingly common for materials to pile up at one site, while another runs short.

A frequent complaint is that the blueprints for powerplants contain errors, for which no one will take responsibility and which no one will correct. Often construction is well underway when the errors are discovered and it becomes necessary to rebuild part of the plant. Or, modifications are added to the initial designs during construction, and again, construction must be halted. Designers out of touch with construction problems may incorporate unobtainable parts or equipment in their designs. Problems of this sort are endemic to the Soviet system and are caused largely by the absence of a single point of responsibility for all phases of a project.

Attempts to fulfill construction plans often result in surges in new capacity start-ups during the fourth quarter of the year. This leads to a practice called "storming" in which intense efforts are made to finish work in a short time and projects of lower priority are abandoned in order to divert resources to others. Resources are increasingly dispersed among too many projects. This results in supply breakdowns, lower labor productivity, and increasing volumes of unfinished plant construction at the end of each year. Plant construction often consumes 1.5 to 2

[5]G. S. Ageyev, et al., "Basic Directions of Formation of the Kansk-Achinsk Fuel and Power Complex," *Teploenergetika*, No. 4, April 1974, pp. 31-34.
[6]P. P. Falaleyev, "Basic Directions of Increasing the Effectiveness of Power Engineering Construction," *Energeticheskoye stroitelstvo*, No. 6, June 1979, pp. 2-6.

times as long as called for in plan norms.[7] The situation is aggravated by increasing downtime of construction equipment, due in large part to the poor quality of maintenance work.

Supply and Quality of Power Machinery

Important problems here are the apparent difficulties of the power machine building and electrical equipment industries in meeting delivery schedules, and the unreliability of equipment. The power machine building industry produces, among other things, boilers, turbines, and generators for hydroelectric, fossil-fired, and nuclear powerplants. The performance of this industry generally reflects the high priority accorded electrification. Production of electrical equipment grew by over 21 percent between 1975 and 1978, and industry labor productivity grew by over 14 percent. Its share of products in the highest product quality category nearly doubled between 1975 and 1978, rising from 12 to 22 percent.[8] The Soviets claim that the latest products of this industry are on a par with the best Western equipment. Indeed, Soviet hydroelectric turbines have found a market in the West where few other Soviet industrial products are competitive, and the efficiency of Soviet oil- and gas-fired boilers is said to be 0.5 to 1 percent higher than that of foreign analogs.[9]

Yet the performance of the industry is uneven, and not all its power machinery is up to the technical level of the export-worthy models. The industry as a whole seems not to have an integrated plan for solution of its quality problems and only faces those which can no longer be ignored.[10] Low quality here can be traced directly to the economic incentive system where, output being the prime

goal, much may be sacrificed to achieve it. In the power machinery industry, such sacrifices can take the form of inadequate testing of new equipment before the start of serial production. In addition, finished equipment is not "debugged" before delivery; rather it is left to the engineers of the powerplants to correct the defects. A turbogenerator destined for the Nazarovo regional power station, for example, was not tested before delivery. Vibration problems surfaced in operation and over a 5-year period resulted in 62 shutdowns—equivalent to nearly 3 years of idle time.[11] Such problems will persist until the incentive system is reoriented toward rewarding producers for production of "quality equipment" instead of merely "equipment."

Boiler Development

A major problem of coal-fired power stations is declining coal quality. The Soviet Ministry of the Coal Industry (Minugleprom) is required to monitor coal quality and to deliver suitable supplies to power stations. Minugleprom's plan targets, however, are expressed in terms of the quantity, rather than the quality of the coal shipped. Indeed, it has been known to falsify records to hide the low quality of its coal.[12] The Ministry of Power and Electrification (Minenergo) frequently complains about the coal it receives, and promotes the idea of either moving quality control outside the coal industry or of setting up an independent agency to perform this function. Meanwhile, poor quality coal—particularly coal with high ash content—poses serious problems for electricity generation.

Both Kansk-Achinsk and Ekibastuz coal tend to form sediment on the convective surfaces of boilers. The Soviets have reportedly succeeded in designing a boiler which suits the particular properties of coal from Ekibastuz, and this is now being burned at

[7]Ibid., p. 35.

[8]*Ekonomicheskaya gazeta*, No. 22, May 1979, pp. 1-2, in JPRS 73,859, July 18, 1979, pp. 9, 11.

[9]V. P. Goloviznin, "Soviet Power Machine Building—Base of Development of Power Engineering of the Country," *Energomash-inostroyeniye*, No. 4, April 1980, p. 3.

[10]V. Krotov, "Complex Approach to Management," *Trud*, Mar. 15, 1979, p. 2, in JPRS 73,380, May 4, 1979, p. 62.

[11]I. Sidanov and A. Zarnadze, "Effectiveness of Introducing New Technology," *Voprosy ekonomiki*, No. 2, February 1980, p. 128.

[12]V. Levin, "Padding the Load," *Sotsialisticheskaya industriya*, May 11, 1980, p. 2.

15 thermal power stations. Kansk-Achinsk coal has proved less tractable. The Soviets claim to have modified a large boiler system so that it can burn some types of Kansk-Achinsk coal; however, coal from at least one of the basin's large deposits has not yet been sufficiently studied to permit development of boilers which can be fired by it.[13] Until this problem is solved, Kansk-Achinsk coal will be of limited utility in electricity generation.

[13]"The Problem of the Combustion of Kansk-Achinsk Coal," *Teploenergetika*, No. 7, July 1975, p. 92.

ELECTRIC TRANSMISSION

A major purpose in creating the Kansk-Achinsk and Ekibastuz complexes is to provide electricity to the Urals and the European U.S.S.R. Development of appropriate ultrahigh voltage (UHV) transmission technology is necessary if this goal is to be realized. (Lines at voltages between 250 and 1,000 kilovolt (kV) are considered extra-high voltage (EHV) and voltages above this are UHV.) The Ekibastuz complex is to be linked to the Urals by an 1,150-kV alternating current (AC) line and to the central regions by a ±750-kV direct current (DC) line. The Kansk-Achinsk complex is to be linked to the Urals or the central regions by a ±1,100 to ±1,200-kV DC line. This section examines the current status of UHV technology and prospects for its development by 1990.

Transmission of large amounts of power over very long distances is expensive. The amount of power that an electrical transmission line can carry increases as the square of the voltage, i.e., if the voltage of a line is doubled, it carries four times the original power. Thus, HV transmission lines mean that power can be more economically transmitted over longer distances than lower voltage lines. But the task of bringing electricity from the East to the European part of the country requires the construction and operation of UHV lines at unprecedented voltages. In this respect, the U.S.S.R. will be entering relatively uncharted territory.

In the Soviet Union, as in the United States, AC is the most common method of transmitting electric power, allowing high voltages to be transmitted and then easily reduced to lower voltages at the point of utilization. On the other hand, HV DC transmission requires less insulation, and when the same size cables and insulation are used, a DC circuit will carry considerably more power than an AC circuit. In addition, because no alternating magnetic field exists inside the wires carrying DC, energy losses and the problem of synchronizing systems are reduced. But the cost of DC wires is raised by the necessity of placing converters at both ends of the line. For this reason, DC transmission is not economical over short distances. The U.S.S.R. considers DC more economic than AC for transmitting power over distances in excess of 1,500 to 2,000 km,[14] and it is pioneering the use of direct current in UHV from power stations in Kazakhstan and Siberia to the European part of the country.

The U.S.S.R. has thus far built only two DC transmission lines. The newer and larger of these is a ±400-kV line between Volgograd and the Donets basin. Commissioned in stages between 1962 and 1965, this line is scheduled to be overhauled within the next 5 years in order to upgrade its equipment.[15] Experience gained in the construction and operation of the ±400-kV line is being used to develop DC lines of higher voltages.

[14]Zhimerin, op. cit., p. 82. For more on AC v. DC power transmission, see Ronald Amann, Julian Cooper, and R. W. Davies (eds.), *The Technological Level of Soviet Industry* (New Haven and London: Yale University Press, 1977), pp. 202-204 and 23.

[15]"On Reconstruction of the Volgograd-Donbass Direct-Current Power Transmission Lines," *Energetik*, No. 1, January 1981, p. 37. The first DC line, with a voltage of ±200 kV, runs from the Kashira Power Station to Moscow. See D. G. Zhimerin, *Energetika: nastoyashcheye i buduschcheye (Moscow: Izd. "Znaniye," 1978)*, p. 82.

Increasing line voltages has been a basic trend in the development of Soviet power engineering. At present, the voltage in Soviet trunklines has reached 500- to 750-kV AC and ± 400-kV DC, and there have been plans for lines of 1,150-kV AC and ± 750-kV and higher DC. The attainment of these voltage levels is based on many years' experience with powerline development and construction, which in the past has earned the U.S.S.R. a leading position worldwide in high-voltage transmission.[16]

The construction of UHV powerlines of 1,150-kV AC and ± 750-kV DC signifies a qualitatively new level in Soviet power engineering—a transition to what is still largely an experimental technology, both in Soviet and world practice. In the case of the ± 750-kV DC line, all equipment reportedly has been developed in the U.S.S.R. and will

be produced at Soviet plants.[17] Nevertheless, some technical problems apparently remain. For example, at least one Soviet expert sees the need to hasten the development of new reactive-power compensation devices to maintain voltage levels and reduce energy losses in AC lines of 1,150 kV (and also 750 kV).[18] In the development of UHV DC transmission, the major problems have centered around circuit breakers and, especially, converter equipment.[19] While Soviet experts seem to be confident that these problems have been or will be solved,[20] Western experts are less certain.

---

[16]K. D. Lavrenenko, "Soviet Electric Power in the Past 60 Years," *Teploenergetika*, No. 11, November 1977, pp. 2-8; P. S. Neporozhniy (ed.), *Elektrifikaysiya SSSR (1967-1977 gg.)* (Moscow: Izd. "Energiya," 1977), pp. 260-261; Amann, Cooper, and Davies, op. cit., pp. 222-224.

[17]M. Pchelin, "A River of Energy Will Start to Flow," *Stroitelnaya gazeta*, Jan. 23, 1980, p. 3.
[18]Peterson, op. cit., p. 66.
[19]Amann, Cooper, and Davies (eds.), op. cit., pp. 215-220.
[20]A major reason for the planned overhaul of the ± 400-kV Volgograd-Donets line is to replace less efficient mercury-arc converter equipment with more advanced thyristor devices. See "On Reconstruction . . .," op. cit. Similar devices reportedly by have been developed in the U.S.S.R. for the ± 750-kV Ekibastuz-Tambov line. See V. P. Fotin, "Development of a Complex of Equipment for the 1,500-kV Ekibastuz-Center Direct-Current Power Transmission Line," *Elektrotekhnika*, No. 6, June 1978, pp. 1-6.

SYSTEM INTEGRATION

Soviet efforts to extend electricity supply to all sectors of the economy have aggravated the problem of "maneuverability," i.e., meeting widely varying demands for electricity, increasing the need for reserve capacity and maneuverable equipment at power generating stations. This section describes the ways in which the Soviet load pattern is changing, and Soviet problems and plans for responding to these changes. It deals first with peaking problems—including programs for creating equipment for this purpose and the difficulties associated with introducing large amounts of baseload nuclear capacity—and then with plans for integrating the electricity system through a nationwide power grid.

PROSPECTS FOR COPING WITH DEMAND VARIATIONS

Table 37 illustrates the growth in electricity generation and consumption in the U.S.S.R. from 1960 to 1980. From this table, it can be calculated that total electricity consumption increased from 292 billion kWh in 1960 to about 736 billion kWh in 1970 and about 1,276 billion in 1980. (The difference between total production and total consumption in the latter 2 years is due to exported electricity.) Electricity consumption is clearly growing rapidly and demand must be met by the construction of adequate amounts of generating capacity. The table also shows dramatic changes in electricity consumption

Table 37.—Electricity Generation and Consumption in the U.S.S.R.
(billion kWh)

	1960	1970	1980[a]
Generation of electricity	292	740.9	1293
Consumption of electricity:			
Industry....................	188.7	438	696.2
Construction	8.9	15	23.3
Transportation	17.6	54.4	102.5
Agriculture.................	9.9	38.5	109
Municipal services and households	30.5	81.1	155
Electricity generation and transport[b]...........	36.4	108.7	189.7

[a]1980 figures are preliminary.

[b]Includes electricity consumed at power stations (approximately 6 percent of the total) and grid losses (between 8.5 and 9 percent of the total).

SOURCE: *Elektricheskiye stantsii*, No. 12 (December 1980), p. 44; and No. 1 (January 1981), p. 2.

by economic sector. The relative share of consumption by industry—the heaviest user—has decreased, while shares of agriculture and municipal services and households have increased.

As demand has grown in the agricultural, urban services, and household sectors, the power system has been confronted with increasingly irregular load curves, with more pronounced periods of moderate to high demand—so-called semipeakloads and peakloads—during certain hours of the day. These alternate with periods of sharply reduced demand.[21] (The maximum continuous demand throughout all periods is called the baseload).

The Soviet power industry continues to have difficulty covering semipeak and peak loads, primarily because of a lack of generating equipment designed for this purpose. Soviet convention distinguishes three types of generating capacity: 1) baseload units, 2) semibaseload (or semipeakload) units, and 3) peakload units.[22] The equipment stock

now consists mainly of the baseload type, in both fossil-fired units and nuclear units.

The U.S.S.R. lacks adequate gas-turbine technology, pumped-storage facilities, and hydro units for handling sharp, short-time peaks. Moreover, the problem of coping with semipeak fluctuations is aggravated by the increasing importance of large (300 MW or more) generating units which are technologically unsuited for this purpose.[23] The power machine building industry is aware of these deficiencies and is being called on to step up the development and construction of peak and semipeak equipment, including 150- to 200-MW gas-turbine units and hydroturbines for pumped-storage plants.[24]

The Soviets have tended to build generating stations and units with larger and larger capacities in order to reap benefits of economies of scale in power generation. This has created a problem, because the market for electricity from baseload capacity is limited. Even with the growing overall demand for electricity, particularly in the European part of the country, there may even now be a surplus of available baseload capacity. The lack of highly maneuverable equipment has already forced power stations to use ill-suited baseload equipment to cover peak and semipeak periods.[25]

The installation of more baseload equipment will increasingly raise both technical and economic problems. Both fossil-fuel and nuclear units are slow to start up and to reach rated capacity. They, therefore, cannot respond to sudden sharp load fluctuations. Indeed, such fluctuations can even damage the equipment.[26] The equipment may be used to cover moderate load fluctuations, but this practice is economically inadvisable, especially in the case of nuclear units.

[21]Loads also vary on other bases, including weekly, monthly, and seasonally; however, the shorter and more frequent daily variations seem to pose the greatest difficulties for power stations.

[22]V. N. Gusev and V. I. Rozova, "On the Possibility of Operating Nuclear Power Stations Under Variable Loads," *Elektricheskiye stantsii*, No. 9, September 1977, pp. 9-11.

[23]Leslie Dienes and Theodore Shabad, *The Soviet Energy System* (Washington, D.C.: V. H. Winston & Sons, 1979), p. 191.

[24]Goloviznin, op. cit.; Ye. Borisov, "High Tension," *Sotsialisticheskaya industriya*, Dec. 21, 1980, p. 1; and V. Boldyrev, "For the Needs of Heat Supply," *Sotsialisticheskaya industriya*, Jan. 30, 1981, p. 2.

[25]Dienes and Shabad, op. cit., pp. 189-192.

[26]Neporozhniy, op. cit., p. 215.

Besides being technically suited to baseload operation, nuclear stations must be operated for a large number of hours per year. This is because nuclear stations have low operating costs but high fixed costs (e.g., high construction costs per kW of capacity). Only by producing large volumes of electricity per year can the fixed cost per kW be brought low enough to make the average cost of nuclear electricity competitive. Soviet planners are well aware of the need to balance the advantages of nuclear power stations (NPSs) in conserving fossil fuels against their high investment costs.[27]

Soviet experts recognize the importance of using nonnuclear capacity wherever possible to compensate for load fluctuations. It is now recommended that fuel-intensive fossil-fuel stations be unloaded before nuclear ones.[28] In such a case, nuclear capacity is used as a substitute for less economical fossil-fuel capacity. As the proportion of nuclear capacity increases, it will eventually become necessary to operate both fossil-fuel and nuclear units under variable loads.[29] OTA's information does not permit it to determine the point at which this problem will become acute in the U.S.S.R., but the evidence does allow some observations on this subject.

Much has been written about the maneuverability of nuclear generating units.[30] Until recently, nuclear plants were designed to operate only under baseloads. An all-union conference was held in 1977 to discuss results of research on this problem, and trials have been conducted to determine the feasibility of running nuclear stations under variable loads. In order for them to perform well under such conditions, several technical problems must be solved, including removal of limitations on the number of start-stop cycles for the equipment; choice of the best fuel, fuel cladding, and designs of fuel elements; and optimization of reactor control and protection systems. Moreover, operating conditions themselves will have to be improved. Stations presently operating under variable loads are very inefficient.

According to one Soviet source, baseloading of nuclear capacity will be possible as long as the following conditions pertain: 1) NPSs account for no more than 22 to 24 percent of total generating capacity; 2) other types of capacity are unloaded to the degree possible, as needed, including complete weekly shutdowns of one or two units at regional fossil-fuel stations; and 3) maneuverable equipment (hydraulic, pumped-storage, and gas-turbine units) accounts for at least 18 to 19 percent of total capacity (in the European part of the U.S.S.R.).[31]

OTA has estimated that Soviet NPSs will account for approximately 11 percent of total installed capacity by 1985, and for no more than 18.5 percent by 1990 (see below). This suggests that baseloading of Soviet NPSs should present no problems until after 1990. However, if NPSs account for as much as 18.5 percent of installed capacity nationwide by 1990, their proportion could exceed 24 percent in the European U.S.S.R. This would force NPSs there to operate under variable loads. In fact, the Soviet source cited above anticipates some unloading of nuclear capacity, mainly on weekends, even before the 24-percent level is exceeded.[32] The likelihood that this will happen depends, in part, on how successfully the U.S.S.R. exercises its options for coping with load fluctuations.

One such option is building new, flexible heat and power (cogeneration) stations, designed to operate under either base or varying loads. This would obviate the construc-

[27]I. M. Volkenau and Ye. A. Volkova, "Operating Conditions of Nuclear Power Stations in Power Systems," *Elektricheskiye stantsii*, No. 3, March 1978, pp. 7-9.

[28]S. Ye. Shitsman, "The Effectiveness of NPS's Under Daily Unloading," *Elektricheskiye stantsii*, No. 8, August 1980, p. 11.

[29]Ibid.

[30]N. A. Dollezhal, "Nuclear Power and Scientific-Technical Tasks of Its Advancement," *Atomnaya energiya*, vol. 44, No. 3, March 1978, pp. 203-212.

[31]Volkenau and Volkova, op. cit., pp. 8 and 9. According to this source, in 1975 the share of maneuverable equipment in the European part of the Unified Power System was approximately 22 percent, but this share is expected to decrease to 18 to 19 percent in the future.

[32]Ibid., p. 9.

tion of specialized semipeak condensing stations, which would generate only electricity and consume fuel at a higher rate than would cogeneration stations operating under variable loads. Proponents of this option contend that the expansion of cogeneration capacity in the European U.S.S.R. will be necessary despite the growth of nuclear capacity in that region.[33]

This position is controversial, however. A 1979 article by A. Troitskiy, Deputy Head of the Department of Power and Electrification of Gosplan U.S.S.R., argues that construction of cogeneration stations should be "drastically limited" so that these stations will not displace generating capacity at NPSs.[34] Troitskiy recommends: 1) the retention of obsolete units which are not physically worn out to serve as reserve capacity for short-term peakloading, 2) the improvement of the load-following characteristics of large fossil-fuel units, and 3) the construction of pumped-storage stations. Pumped-storage stations (PSSs) are a form of hydroelectric capacity.

Hydropower is highly maneuverable. The Soviet power industry is well aware of this option and is striving to maximize its value.[35] Unfortunately, the availability of hydraulic capacity is affected by water levels in the rivers and reservoirs that feed hydroelectric stations. In the European part of the U.S.S.R., where the load-variation problem is at its worst and where most NPSs are being built, suitable water resources are much more limited than in remote regions such as Siberia.[36] Since as much as 70 percent of the suitable hydraulic resources in the European U.S.S.R. have already been de-

veloped,[37] the construction of conventional hydrostations alone will not satisfy the growing need for maneuverable capacity in areas where it is most needed. Therefore, the U.S.S.R. is turning more and more to the construction of PSSs for peakload coverage.[38] The specially designed reversible hydraulic turbines of pumped-storage units serve a dual purpose: during offpeak hours, excess generating capacity from other units is used to run the turbines to pump water up into a reservoir; in peak hours, this water is released to generate electricity by turning the turbines in the opposite direction. The use of pumped-storage capacity, consequently, can help to maintain baseload operation of nuclear or other stations by providing coverage of peakloads and also additional consumption during offpeak hours. For this reason, the construction of PSSs is considered an inseparable part of Soviet plans for growth in nuclear power production.[39]

Despite such plans and the expressed need for pumped-storage capacity, progress with the design and construction of PSSs in the U.S.S.R. is said to be slow, mainly because Soviet designers have neglected these stations, which are expensive to build. Only one small PSS near Kiev is presently in operation.

The first PSS to be built in conjunction with an NPS is, however, underway. This is the Southern Ukrainian Power Complex, which includes the Southern Ukrainian NPS, the Tashlyk Hydroelectric Station, and the Konstantinovka Hydroelectric and Pumped-Storage Station. When completed, the complex will have a total capacity of more than 6,000 MW, nearly two-thirds of which will be nuclear.[40] Another PSS has been under construction at Zagorsk, near Moscow, since 1976, but its completion is apparently not yet in sight.[41] The Kayshyadoris PSS in

[33]V. P. Korytnikov, "Basic Tasks for Heightening the Effectiveness and Reliability of Heat Supply to the Country's Economy," *Teploenergetika*, No. 8, August 1980, pp. 2-5.

[34]Troitskiy, op. cit., p. 22. Both Troitskiy's and Korytnikov's arguments are aimed at lowering fuel costs and conserving fossil fuel. To cover growing heat demand, which would ordinarily be met with cogeneration capacity, Troitskiy calls for construction of large boiler houses, presumably in conjunction with conventional fossil-fired and nuclear electric stations. Korytnikov, op. cit., p. 3, points out, however, that this arrangement would result in greater fuel expenditures than those incurred at cogeneration stations.

[35]Dienes and Shabad, op. cit., pp. 133-136.

[36]Peterson, op. cit., p. 65.

[37]Dienes and Shabad, op. cit., p. 133.

[38]Neporozhniy, *Elektrifikatsiya . . .*, op. cit., p. 216.

[39]Peterson, op. cit., p. 65.

[40]P. S. Neporozhniy, "Lenin's GOELRO Plan Is 60 Years Old," *Elektricheskiye stantsii*, No. 12, December 1980, pp. 2-8, especially p. 6.

[41]V. Vennikov, "Contemplating the Future," *Sotsialisticheskaya industriya*, Jan. 30, 1981, p. 2.

Lithuania is to be commissioned during the Eleventh FYP, and plans have been drafted for at least two other stations—the Dnestrovsk and the Leningrad PSSs.[42]

But even the timely construction of PSSs will not completely solve the problem of covering sharp load fluctuations in the European U.S.S.R. Pumped-storage capacity must be augmented with other highly maneuverable equipment, particularly gas-turbine units, which may be used alone or as part of combined "steam-and-gas" units.[43] The U.S.S.R. is reported to be working on the practical use of gas-turbine units with a capacity of 100-MW and also of 250-MW steam and gas units.[44] However, there is no evidence that these units are being used extensively in the Soviet power industry.[45]

A final option for coping with load fluctuation is capacity substitution through the creation of large-scale, interconnected power systems or grids. Such systems permit generating capacities to be shared by shifting their output from one grid to another. This is particularly advantageous to the U.S.S.R., with territory that spans 11 time zones. When a grid in one time zone is experiencing peak demand for electricity, it can borrow power from an interconnected grid in another time zone. The supplier also benefits by utilizing capacity that would otherwise be idle. Predictable load variations allow capacity exchange schedules to be worked out in advance, and this has reportedly been done for Soviet power systems. On the other hand, unplanned variations require more immediate response. This situation is said to be covered through the intervention of dis-

patcher personnel and the operation of the automatic frequency and power regulating system.[46] The effectiveness of response to unplanned loads by many power stations probably is reduced, however, by the shortage of maneuverable generating equipment.

In sum, if the U.S.S.R. carries out its plans: 1) for building peakload capacity, including hydroelectric and pumped-storage stations as well as gas-turbine and steam-and-gas units, 2) for improving the maneuverability of fossil-fuel stations, and 3) for expanding its unified power grid to facilitate capacity sharing and substitution, the baseloading of NPSs should be feasible until 1990. Delays in these plans could force some unloading of nuclear capacity during offpeak hours. This situation could present technical problems for the Soviet nuclear industry; as recently as 1978, the ability of conventional reactors to withstand repeated load variations was still in question.[47]

THE UNIFIED POWER SYSTEM

The Soviet Union is attempting to take full advantage of large-scale grids by the formation of a nationwide Unified Power System (UPS). When complete, UPS will incorporate 11 smaller joint power systems ranging across the entire U.S.S.R. In addition, UPS will be tied into the unified system of the East European countries.

The core of the unified system was formed in the 1950's in the European U.S.S.R. The "European" UPS presently takes in eight joint systems in the Northwest, the Center, the South, the Middle Volga region, the North Caucasus, Transcaucasus, the Urals, and North Kazakhstan. In 1978, a 500-kV line was strung linking the European UPS with the Joint Power System (JPS) of Siberia. Other joint systems are in Central Asia and the Far East,[48] and plans exist for

[42]*Ekonomicheskaya gazeta*, 1981:2, p. 2; Neporozhniy, *Elektrifikatsiya...*, op. cit., p. 217; Borisov, "High Tension," op. cit.

[43]Peterson, op. cit., p. 65.

[44]Neporozhniy, "Lenin's GOELRO...," op. cit., p. 8, uses the passive form *osvaivayutsya*, which literally means that the new types of equipment "are being mastered."

[45]Workers of the "Kharkov Turbine Plant" Production Association reportedly have begun work on adjusting and putting into operation a gas turbine designated the GT-35. The turbine is part of the U.S.S.R.'s first steam-and-gas unit, the PGU-250, which is installed at the Moldavian Thermal Power Station. See V. M. Velichko, "The Labor Contribution of Power Machine Builders," *Energomashinostroyeniye*, No. 1, January 1981, pp. 2-5.

[46]L. G. Mamikoyants, et. al., "The Development of Power Engineering in the U.S.S.R. and the Control of Electrical Power Generation and Distribution," presented at the Control Data Corp. Seminar on Power Industry Development, Washington, D.C., Dec. 6 and 7, 1979.

[47]Volkenau and Volkova, op. cit., p. 9.

[48]Fotin, op. cit., pp. 184 and 185.

these to be linked to the European UPS in the 1980's, thus completing the formation of the nationwide system.[49] The smallest units of the unified system are the so-called Regional Power Systems, which can cover several administrative regions or *oblasts*.[50]

To form the unified system, JPSs and regional systems are tied together with HV transmission lines of 220- to 750-kV AC and ± 400-kV DC. In the future, higher voltages are to be used—1,150-kV AC and ± 750- to ± 1,125-kV DC and above.[51] The main AC voltage level for system interties in the UPS is 500 kV. In the South and Northwest JPSs, 330-kV interties have been used in the past, but a network of 750-kV lines is being developed. At present, a 750-kV system intertie connects Leningrad and Moscow, and a second line of this voltage runs from the Donets basin to the Western Ukrainian substation and on into Hungary. Plans also call for the construction of a ring of 750-kV lines around the Moscow region to transmit and distribute power from nuclear stations, and 1,150-kV lines linking Ekibastuz to the Urals. Construction of the first of the latter lines, which will be nearly 1,500-km long, reportedly is already underway.[52]

The Soviet Union claims that the JPSs presently tied into the UPS encompass an area of 10 million km² with a population of nearly 220 million people; that UPS unites 88 of the 97 power systems in the U.S.S.R.; that only two JPSs and several power systems "in remote regions" remain isolated from UPS; and that in 1979, power stations of UPS accounted for 82 percent of the installed capacity and 88 percent of the electricity generated in the U.S.S.R.[53]

It is difficult to evaluate these claims, however. While the Soviet literature stresses the achievements of UPS in providing connections beween grids and tends to convey the impression of a sophisticated system, Western electrical engineers who have visited the U.S.S.R. report that these connections are tenuous and that the entire system is run from a single, underequipped dispatching office in Moscow.[54]

The latter point is particularly important. The coordination and management of a power system covering a large territory requires a well-organized system of control centers and effective control equipment. In theory, overall management of the Soviet UPS is assigned to the system's central dispatching department (CDD), which oversees the work of dispatching departments of the 11 joint systems. The latter departments, in turn, supervise the work of regional control centers and power stations.[55] The CDD's primary responsibility is to ensure the stable, efficient operation of UPS and its components and, thus, the delivery of reliable, quality electric power to consumers. In addition, the CDD takes part or assists in research, development, and planning aimed at maintaining and improving UPS.

To accomplish all of this would require constant upgrading of control facilities and equipment, including the introduction of new communication and data transmission techniques. The rapid transmission and processing of information on all aspects of grid performance and operating conditions are necessary to effect timely shifts of power from one system to another. It is not clear that the U.S.S.R. has as yet acquired these capabilities.

[49]Mamikoyants, op. cit.

[50]W. G. Allinson, "High Voltage Electric Power Transmission," ch. 5 in Amann, Cooper, Davies, op. cit., pp. 199- 224. It should be noted that not all sectors of the Soviet economy are served by the Unified System and joint systems. Until recently, agriculture and certain other sectors were excluded, giving rise to the spread of small, unconnected electric power stations. This problem of a "dual economy" in the electric power industry is discussed by Dienes and Shabad, op. cit., pp. 185-187.

[51]A. Pivenko, "High-Capacity Transformers," *Izvestiya* June 1, 1980, p. 1; and V. Ganzha, (title unknown), *Sotsialisticheskaya industriya*, June 14, 1980, p. 2.

[52]L. L. Peterson, "The Development of Power Systems," *Elektricheskiye stantsii*, No. 12, December 1980, pp. 63-66.

[53]Ibid.

[54] Discussion with Val Lava and Frank Young, members of the Joint American-Soviet Committee on Cooperation in the Field of Energy.

[55]Mamikoyants, et al., op. cit.

Computer technology is also important to UPS management, and there is evidence that the Soviet Union recognizes a need to use computers more extensively for this purpose.[56] For example, the automatic monitoring of frequency (which is supposed to be the same throughout UPS) and of active power, is reportedly done using minicomputers at

all levels of control, from the CDD down to local power systems. Voltage levels in basic networks will also eventually be placed under centralized automatic control, which likely will require the use of computers, but this work is said to be only at the preliminary development stage. Such control equipment is the basis for the formation of a hierarchical computerized system for managing UPS—a goal which is still to be achieved.

---

[56] Ibid. See also the section below on *Western Technology in the Soviet Electric Power Industry.*

WESTERN TECHNOLOGY AND THE SOVIET ELECTRIC POWER INDUSTRY

The Soviet Union has so far been successful in the development and implementation of HV (up to 750-kV AC and ± 400-kV DC) transmission, apparently with little or no assistance from the West.[57] Moreover, at least on the evidence of Soviet literature, there seems no reason to doubt the U.S.S.R.'s ability to continue to progress and to make the transition to UHV transmission. Certainly given the great distances over which electric power is to be transported in the U.S.S.R., it probably has greater economic motivation than any other country in the world to employ UHV.

But, from a technical standpoint, the fact remains that UHV is still a relatively new and experimental technology. There is evidence that the U.S.S.R., which in the 1960's emerged as a world leader in HV power transmission, has since lagged behind the West in some aspects of UHV, particularly in the development of thyristor converters for DC lines.[58] These observations cast some doubt on Soviet claims about UHV power-line construction and suggest that the Soviet program might benefit from foreign experience and equipment.

OTA found no direct evidence that the U.S.S.R. is purchasing or intends to purchase UHV equipment and technology from

the West, yet the possibility of future purchases cannot be ruled out. In the case of thyristor converters, for example, the U.S.S.R. has developed its own equipment. But this may be inferior to that available from Western countries; Sweden has at least one firm that is actively engaged in commercial applications of this technology.[59] While the U.S.S.R. can and does achieve the same effect as one foreign thryistor by using several of its own, it is possible that a decision could be made to purchase foreign models if large numbers were required. Soviet industry may also be unable to manufacture enough cable for power distribution. A 1976 source reported that the U.S.S.R. had been placing large orders for cable for small (10-kV) powerlines with suppliers in West Germany and Finland, and that even larger orders could be expected in the future.[60] Conceivably, the U.S.S.R. could also turn to the West for assistance with the development or supply of compensation devices for UHV AC lines.

Finally, despite its gains in the automation and computerization of UPS, the Soviets have a long way to go before they can possibly realize centralized control of the whole system. A considerable amount of computerization has been applied at the

---

[57] Amann, Cooper, and Davies, op. cit.
[58] Ibid., p. 220.

[59] Ibid.
[60] "Power Lines," *Soviet Business and Trade,* No. 9, 1976, pp. 5-6.

regional and power-system levels for economic and grid management. But very little closed-loop control has been implemented and the software lags that of Western systems. Only limited computerization exists at regional control points and powerplants. These functions tend to be limited to accounting, monitoring, and short-term planning. Some form of closed-loop control is not planned until the early 1980's, and the Soviet goal of automating the whole system over the next 10 years does not seem realistic.

The power industry has had access to and made use of most of the major computers produced in the U.S.S.R. over the last 15 years. Conversely, it has not relied extensively on the West for computer equipment although, as has generally been the case with economic management, the indirect influence of U.S. computing has been great.

The addition of inflexible nuclear powerplants, the need for more power generating capacity, and more reliance on Siberian plants now make management tasks considerably more difficult, increasing the need for more sophisticated control of the whole power system. The more the Soviets try to tie the network together, the greater will be their requirements for real-time control systems which can model a wide variety of situations and optimize operational economics.

In addition, very large computers may allow the Soviets to do more extensive modeling, as opposed to field testing, of various network configurations. The U.S.S.R. has spent millions of dollars on test generators and other testing equipment. In the United States, such testing is performed by simulations on computers. Soviet facilities for computer modeling are only now being created. As the grid becomes more complex, substantial savings could be realized here.

Building a multilevel hierarchical process control system of the magnitude of UPS raises enormous software engineering problems. The Soviets lack sophisticated software design tools and experience. Their usual practice is to farm out the development of separate pieces of large systems to various institutes. Without rigorous specifications of interfaces and frequent communications, the overall system is unlikely to work correctly. Consequently, the U.S.S.R. will probably have to settle for considerably less during the eighties than UPS envisaged: greater manual intervention at each level, slower system response, greater cost, and less reliability.

The West could supply integrated software tools, data base management systems, and other software which would help the Soviet software industry and trickle down to this application. Joint ventures, long-term contacts, training, and other transfer mechanisms would also help to build up overall software engineering capabilities. It is unlikely that the Soviets would seek to purchase a large computer for modeling. The new large Soviet-made computers are sufficient for this purpose, provided they are available to the power industry over the next few years. This is another area in which general help in software would play the most decisive role. The same can be said for software for management of construction.

In sum, the Soviets have introduced a large number of computers at various levels of the electric power generation hierarchy. Most of these are concerned with economic management tasks and the overall level of closed-loop control of the power grid is not great. As new atomic and peakload capacities are added and more electricity is generated in Siberia, the management problems will become even more complex. Building this system requires software abilities that the Soviets probably do not yet possess. The U.S.S.R. may also encourage technology transfer from the West in this area. This would be a departure from past experience, for the electric power industry has not previously made extensive purchases of hardware and software. Despite Soviet claims about domestic developments in power transmission technology, the

U.S.S.R.'s apparent loss of supremacy in this field in recent years suggests the potential for Western assistance, particularly in the areas of UHV transmission and computer equipment, software, and software engineering tools and techniques for power system management and control.

THE FUTURE OF THE SOVIET ELECTRIC POWER INDUSTRY

ALTERNATIVES FOR MEETING INCREASED ELECTRICITY DEMAND

This chapter has discussed two ways in which the U.S.S.R. can generate and transmit more electricity to meet a still growing demand in the European part of the country. It can build power stations in the European U.S.S.R. itself, or it can transmit power there over long-distance lines which originate at remote coal basins. Soviet planners are pursuing both approaches, but it seems that some preference is being accorded the first, which is based mainly on the construction of nuclear power stations and the curtailment of fossil-fuel generation in this region. The second approach involves the construction of UHV power transmission lines from coal-fired stations in Kazakhstan and Siberia. As noted above, this project has begun, although the ultimate fate of the program may be delayed, pending the outcome of the nuclear program.

Soviet preference for localized power generation at NPSs can be understood by comparing the costs of electricity supplied by each approach. Troitskiy has compared the costs of electricity generated at a baseloaded NPS in the European U.S.S.R. and electricity transmitted there over a 1,500-kV DC line from a coal-fired station in Ekibastuz in Kazakhstan.[61] According to his figures, the total cost of 1 kWh of electricity, including the costs of extracting, producing, and transporting fuel (coal or uranium) and electricity, is 6 percent higher at the point of consumption for electricity from an Ekibastuz power station than for electricity from a local NPS (1.22 kopecks/kWh v. 1.15 kopecks/kWh, respectively).[62] Given the possibility of error in the calculations, actual costs could be roughly the same, or nuclear electricity could be even less expensive than indicated. In either case, one may question the U.S.S.R.'s decision to build long-distance transmission lines at all if nuclear electricity costs about the same or is cheaper to produce locally. Indeed, Troitskiy himself argues that power transmission westward from Ekibastuz and Kansk-Achinsk is advisable only if the demand for baseload capacity in the European U.S.S.R. cannot be covered with nuclear stations.[63]

One answer may be to substitute fossil-fired generation for nuclear, particularly if nuclear growth falls short of planned targets. In such a case, a shortfall in nuclear generating capacity might be covered with coal-fired capacity in Ekibastuz. Troitskiy also gives figures for capital investment costs per kilowatt of capacity required to deliver electricity to the European U.S.S.R. (including transmission lines) via either

---

[61]A. Troitskiy, "Electric Power: Problems and Prospects," *Planovoye khozyaystvo*, No. 2, February 1979, pp. 18-25. OTA believes that Troitskiy's figures for the center are indicative of the European U.S.S.R. as a whole.

[62]Ibid., p. 20. Troitskiy also considers the option of building gas-fired condenser stations in the Center and supplying them with natural gas from the Tyumen region. Although this option is cheaper (1.08 kopecks/kWh) than the other two options described, Troitskiy points out that gas-fired stations "cannot be recommended" for the Center for two reasons: 1) possibilities for long-distance transport of natural gas in the future are still limited, and 2) additional gas resources are necessary, first of all, in order to replace residual fuel oil (*mazut*) as a fuel at power stations.

[63]Ibid., p. 21. Troitskiy's estimate of the cost of electricity from Kansk-Achinsk is based on transmission over a 2,250-kV DC line. Because this voltage level is not likely to be reached before 1990, if at all, the Ekibastuz-Center option is considered in the present discussion.

nuclear capacity there or coal-fired capacity in Ekibastuz.[64] To cover a shortfall in nuclear capacity of, for example, 5,000 MW with an equivalent amount of coal capacity in Ekibastuz would require only about 25 million rubles of capital investment less than the investment in the equivalent nuclear capacity, a difference of about 1 percent.[65] Again, given the likely margin for error in Troitskiy's estimates and the magnitude of the investment costs (billions of rubles), this difference is insignificant.

In terms of capital investment, then, using low-grade Ekibastuz coal to generate electricity for the European U.S.S.R. appears to be at least as good as, and maybe slightly better than, building nuclear stations near consumers. Nevertheless, the fact remains that the estimated total cost per kilowatt-hour of electricity (as opposed to capacity) is lower for the nuclear than for the Ekibastuz coal option.

Furthermore, from a technical standpoint, there are at least two major risks associated with coal-generated electric power from Kazakhstan or Siberia. First, Soviet equipment for burning low-grade coal at power stations, particularly boilers for burning Kansk-Achinsk coal, remains to be perfected. Second, as noted above, technology for transmitting coal-generated electricity, especially UHV DC technology, is unproven in practice. Although the first ± 750-kV DC line from Ekibastuz apparently is under construction, the U.S.S.R. has not demonstrated practical mastery of this voltage level. Even more uncertain is the possibility of practical application of ±1,125-kV DC, which is the minimum voltage planned for use in UHV lines from Kansk-Achinsk.

Nor are the economics of using AC lines entirely clear. Troitskiy views 1,150-kV AC lines as an effective means of supplying power to the Urals from Ekibastuz, but he does not mention the possibility of further extending such lines.[66] Conceivably, announced plans to do this could be questioned, since UHV AC lines from Ekibastuz to the area around Moscow presumably would create problems similar to those discussed above in connection with UHV DC transmission v. nuclear power generation. In any case, as with UHV DC, Soviet success with UHV AC will depend on the solution to whatever technical problems exist. Here, again, Soviet plans reflect a confidence that such problems have been or can be solved.

Nuclear technology, on the other hand, is proven. NPSs have been operating successfully in the European U.S.S.R. for years, and the voltage level for transmission lines from these stations—750-kV AC—apparently has been mastered. Moreover, the cost of building these lines is lower than that for UHV lines. If Troitskiy's estimates of capital investment costs are accurate, the share of powerlines in total investment costs is greater for coal-fired stations in Ekibastuz than for NPSs around Moscow.[67]

In sum, while the option of supplying electricity by building NPSs in the western part of the country requires slightly higher capital investment than building coal-fired stations in Ekibastuz for the same purpose, the delivered cost of electricity from NPSs is slightly lower than for coal-generated electricity from Ekibastuz. Moreover, a large portion of the investment costs for the coal-fired stations, and the higher cost of Ekibastuz electricity, can be attributed to the

[64]Ibid., p. 20. The figures are 380 rubles/kW of nuclear capacity and nearly 375 rubles/kW of fossil-fired capacity—a difference of only about 1 percent.

[65]OTA's projection in chapter 4 on the "base case" scenario for the Soviet nuclear power industry included an estimated shortfall in nuclear capacity of 5,000 MW by 1990, relative to the assumed target for that year. Using Troitskiy's investment figures, this amount of nuclear capacity in the European U.S.S.R. would cost about 1.90 billion rubles, while the same amount of coal-fired capacity in Ekibastuz would cost about 1.88 billion rubles.

[66]Troitskiy, op. cit., p. 21.

[67]Based on figures in Troitskiy, op. cit., p. 20, OTA estimates that UHV powerlines account for roughly one-fourth of the cost of building a coal-fired power station in Ekibastuz to supply power to the Center. This estimate was derived by comparing the cost per kW of Ekibastuz capacity serving the European U.S.S.R. (374.8 rubles) to the difference between this figure and the cost per kW of Ekibastuz capacity serving Siberia (278.2 rubles). Presumably, this difference (96.6 rubles/kW) can be largely attributed to the cost of UHV lines to the Center.

cost of UHV DC transmission lines—technology which is unproven in practice. These factors, plus unsolved problems of burning low-grade coal at power stations and successful experience with nuclear power, make the nuclear option seem more desirable than the coal option. Technological advancements in UHV and coal power generation, together with practical experience, may make coal use more viable in the future. At least for the present, however, OTA believes that Soviet planners will downplay coal, particularly plans to use Kansk-Achinsk coal to generate electricity, and place more emphasis on nuclear power.[68]

PROSPECTS FOR MEETING PLAN TARGETS

1981-85

Tables 33 and 34 above showed Soviet plans for commissioning of new generating capacity and for electricity production between 1981 and 1985. Achievement of this program could be jeopardized in at least two important ways. First, Soviet plans call for the commissioning of only 10,000 MW of new capacity in 1981,[69] leaving over 15,200 MW to be added each year from 1982 through 1985. In the past, introductions of new capacity have averaged about 10,000 MW per year (in 1980, however, the increment was some 13,000 MW), although production of turbines and generators has been running at 18,000 to 20,000 MW per year.[70] Construction must be expanded in the last 4 years of the FYP if the goal of installing

71,000 MW of new capacity is to be met. Second a substantial share of this new capacity is to be built at Kansk-Achinsk (the Berezovskoye No. 1 Plant) (see table 35), but there is no evidence that a suitable boiler has yet been developed. It is known that trials using a 2,650 tons of steam/hour boiler have failed to solve the problems.

Taking these factors into account, OTA estimates that lags in construction at fossil-fired plants make it likely that the projected growth of 29,000 MW will not be attained and that net growth might more probably be around only 24,000 MW. If the estimated "best-case" shortfalls projected in chapter 4 (2,000 MW) and the plan targets for hydropower are factored in, the result is the achievement of 325,000 MW by 1985, 7,000 MW short of the plan. These projections are summarized in table 38.

The Eleventh FYP calls for the generation of 1,550 billion to 1,600 billion kWh of electricity by 1985.[71] As table 39 demonstrates, OTA estimates that actual generation will be

[71]"Draft of the Main Directions of Economic and Social Development of the U.S.S.R. for 1981-1985 and for the Period to 1990," *Izvestiya*, Dec. 2, 1980, p. 2.

---

[68]Besides greater technical risks, the option of using Siberian coal to supply electric power to the European U.S.S.R. involves higher costs than the Ekibastuz option. According to Troitskiy (Ibid.), supply of the Center from generating capacity in Kansk-Achinsk would cost nearly 390 rubles/kW, presumably because of the greater distance from the European U.S.S.R. and the use of higher voltage (2,250-kV DC v. 1,500-kV DC) in transmission lines. Overall, electricity from Kansk-Achinsk would cost an estimated 1.28 kopecks/kWh.

[69]"Results of Development of Electric Power Engineering in 1980 and Tasks for 1981," *Elektricheskiye stantsii*, No. 1, January 1981, p. 181.

[70]U.S.S.R. Central Statistical Administration, *Narodnoye khozyaystvo SSSR v 1979g.* (Moscow: Izd. "Statistika," 1980), p. 181.

Table 38.—Estimated Soviet Electrical Generating Capacity, 1985 and 1990 (thousand MW)

	1985 Planned (from table 34)	1985 Projected	1990
Total	332	325[a]	405[a]
Fossil-Fired	230	255[b]	255[e]
Nuclear	38	36[c]	75[c]
Hydro	64	64	75[d]

[a]Sum of fossil-fired, nuclear, and hydro capacities.

[b]Estimated. Of 71,000 MW to be added in the period, 35,000 MW are assumed to be fossil-fired. This figure has been adjusted downward by 11,000 MW to account for retirements and underfulfillment of plan targets.

[c]Estimated.

[d]Extrapolated trend. V. S. Serkov (ed.), *Ekspluatatsiya gidroelektrostantsiy*, (Moscow: Izd. "Energiya," 1977), p. 18, indicates that hydroelectric capacity in 1990 is to be 82,000 MW. This appears ambitious, but indicates that sites for new capacity are not exhausted.

[e]Represents a net addition of 30,000 MW. (A gross increase of 35,000 MW, less 5,000 MW of retirements.)

SOURCE: Office of Technology Assessment.

Table 39.—Estimated Soviet Electricity Production, 1985 and 1990

(billion kWh)

	1985		1990
	Planned (from table 33)	Projected	
Total.........	1,550-1,600	1,515-1,625[a]	1,900-2,040[a]
Fossil-Fired	1,100-1,135	1,095-1,180[b]	1,235-1,330[d]
Nuclear	220-225	190-210[c]	400-445
Hydro......	230-235	230-235	265[e]

[a]Sum of fossil-fired, nuclear, and hydro capacities.

[b]Mid-1985 capacity of 223,000 MW times operating rate range of 4,900 to 5,291 hr/yr. The latter is the 1980 rate.

[c]Estimated.

[d]Mid-1990 capacity of 252,000 MW times operating rate range of 4,900-5,291 hr/yr. The latter is the 1980 rate.

[e]Estimated mid-1990 capacity of 73,900 MW used at a rate of 3,600 hr/yr, that is, at roughly the 1979 rate of utilization. See Narodnoye khozyaystvo, op. cit. (1980), p. 169.

SOURCE: Office of Technology Assessment.

1,515 billion to 1,625 billion kWh in 1985, well within the scope of the plan. This estimate too uses the optimistic nuclear projections in chapter 4 and adopts the goal set for hydroelectric generation; OTA estimates that fossil-fired electricity generation will reach 1,095 billion to 1,180 billion kWh in 1985. Combining these three components yields a result that should equal the plan target even if capacity introductions fall short of the goal. This outcome depends on the assumed rate of utilization of installed capacity. If the 1980 rate is maintained, fossil-fired generation could reach 1,180 billion kWh, but if rates continue to decline—say to 4,900 hr/yr—then generation will reach only 1,095 billion kWh, still within the plan target range.

It must be noted that opportunities to increase the amount of electricity produced from coal may be limited by the availability of coal. Although *planned* coal growth is commensurate with *planned* growth in fossil-fired electricity generation over the FYP (8 to 12 percent v. 5 to 9 percent, respectively), chapter 3 estimates that at most coal output will actually increase by only 7 percent, and even this is a highly optimistic projection. Given probable growth of demand for coal in other industrial sectors, notably in ferrous metallurgy, even a 7-percent growth in coal

production is insufficient to achieve the upper end of the range for electricity growth. In addition, if growth in Kansk-Achinsk output is excluded owing to boiler problems, growth in coal production to 1985 falls to 5 to 6 percent. When likely declines in the average calorific value of other coals are taken into account, there is a possibility that coal production on a Btu-basis will fall from the 1980 level. There is, then, a substantial probability that coal's contribution to total fuel consumption in electrical power generation will decline. As a consequence, some plans for adding new coal-burning capacity may have to be scrapped or at least postponed until after 1985.

In sum, present rates of power machinery production suggest that the Soviets are capable of producing the 35,000 MW of power generating equipment needed to achieve the planned gross addition of fossil-fired capacity by 1985. But unless greater resources are allocated to construction of powerplants, there will be insufficient finished plant to house much of this equipment at the end of 1985. In addition, it is possible that growth in coal production will not be sufficient to support a 5- to 9-percent growth of fossil-fired electricity production. As a consequence, it is difficult to see how coal's share of the fuel balance in electricity generation can be expanded between 1981 and 1985.

1986-90

Estimates for Soviet electricity generating capacity in the Twelfth FYP are highly speculative. The figures shown in table 38 indicated that this capacity could exceed 400,000 MW by 1990, possibly amounting to 405,000 to 415,000 MW. Fossil-fired capacity will grow between 1986 and 1990, although the extent of the growth is hard to judge. During this period, additional plants will be built at the Ekibastuz and Kansk-Achinsk basins and probably at the South Yakutian basin. Other plants may be built in the Far East and in the Urals, In all, OTA has assumed a net addition of 30,000 MW of capacity between 1986 and 1990. If so, fossil-fired capacity in 1990 could reach 255,000 MW (see table 38). Total electricity production could reach 1,900 billion to 2,040 billion kWh by 1990, with 1,235 billion to 1,330 billion kWh being generated by fossil-fired stations (see table 39).

SUMMARY AND CONCLUSIONS

The U.S.S.R. has amassed great experience in power transmission, including long-distance HV transmission. This experience has proved valuable in and has been enhanced by the formation of the nationwide Unified Power System. At the same time, Soviet power engineering is moving into a qualitatively new field—UHV transmission. The move to UHV, at least initially, will involve substantially higher investment and operating costs; but the Soviets are confident that these costs will be offset by the great savings to be gained from long-distance UHV transmission. The success of this move may determine the outcome of Soviet plans to complete the UPS. The chances for success will depend, in large part, on the applicability to UHV of past Soviet experience in power transmission and, perhaps, the availability of assistance from the West.

Plans for UHV DC transmission of coal-generated electricity to the center of the European U.S.S.R. from Kazakhstan and West Siberia seem to be viewed as a way to supplement nuclear power, particularly in the event of a shortfall in planned nuclear capacity. However, the technical risks involved in UHV transmission and power generation using low-grade coal support the conclusion that Soviet planners may be downplaying coal in favor of nuclear power, at least for the present. This may mean that construction of UHV DC lines other than the ±750-kV line now being built will be delayed, pending success of the nuclear program and the solution of technical problems connected with developing a boiler to burn Kansk-Achinsk coal and technology for DC transmission at voltages of ±1,125-kV and more. Gas may also come to play a more important role in power generation.

There are no economic constraints to building the planned 1,150-kV AC lines to supply power to the Urals, assuming that technical problems are or can be solved. Presumably, however, plans to extend these lines to the Western U.S.S.R. would raise economic questions similar to those entailed in proposed UHV DC transmission vis a vis nuclear power.

To some extent, the generation of electricity by fossil-fired plants in the 1980's will be tied to the fate of the nuclear electrification program. But if nuclear power falls behind schedule, fossil-fired equipment will be called on to cover the shortfall.

On the face of it, there would appear to be a substantial amount of flexibility in the system, at least in handling the baseload. Fossil-fired capacity at the end of 1980 (201,000 MW) would satisfy and even exceed the 1985 generation targets of 1,100 billion to 1,135 billion kWh if utilization rates of 5,475 to 5,650 hr/yr could be achieved. If the 1985 goal for fossil-fired capacity (230,000 MW) were met, operation of this capacity at the 1979 utilization rate (5,651 hr/yr) would permit the generation of 1,300 billion kWh in 1985—almost enough to cover the 1985 targets for both fossil-fired and nuclear plants.

There is sufficient evidence to suggest that the U.S.S.R. will have substantial reserve capacity in its power generation system in the 1980's and will be able to compensate to some degree for shortfalls in the nuclear program by increasing the rate of utilization of fossil-fired capacity, provided that the needed fossil fuel supplies are available. Nor does the availability of generating equipment appear to be a problem. Present annual production of turbines and generators seems more than sufficient to support a gross addition of 65,000 MW of fossil-fired capacity by 1990 (54,000 MW net).

Examination of the available literature, both Soviet and Western, revealed virtually no imports of Western technology and equipment in nonnuclear power generation. In fact, the Soviets have opted for domestic development of this industry and have largely succeeded in achieving a high technological level in their equipment, comparable in some cases with the best available in the West.

One equipment problem—the development of large boilers for Kansk-Achinsk coal—probably must be solved domestically. Boilers are custom designed for specific types of coal and suitable boilers would not be available from the West. The Soviets have been working on the development of such boilers for years, but to date appear to have had little success. This is not unusual, however. Boiler development can take decades, and there is no assurance that an efficient boiler can be developed for a given type of coal. In any event, Soviet capabilities in boiler design are comparable with Western

capabilities. In short, Soviet reliance on domestically produced power generating equipment will probably continue into the foreseeable future and the availability of Western technology and credits should be of small concern in this area.

Similarly, the U.S.S.R. appears to be largely self-reliant in the construction of power transmission lines. This self-reliance may be reduced if Soviet plans for UHV transmission are carried out; the U.S.S.R. may be forced to turn to the West to help it supply the large body of technology and equipment which would be required. At the same time, the need for the U.S.S.R. to move ahead at full speed with plans for UHV development, particularly UHV DC, is questionable, given past success with and future plans for nuclear power. Postponement or abandonment of Soviet UHV plans would limit the potential impact of Western technology in this field.

Although reliance on the West for computer technology for UPS has been very limited, the U.S.S.R. could profit substantially from using U.S. software engineering techniques to build the network's computer control system. Continued indirect acquisition of these techniques is at least as likely, however, as direct acquisition.

Western Energy Equipment and Technology Trade With the U.S.S.R.

CONTENTS

LIST OF TABLES

LIST OF FIGURES

Western Energy Equipment and Technology Trade With the U.S.S.R.

The previous four chapters have discussed the impact of Western energy technology and equipment on Soviet energy production in qualitative terms. This chapter examines the nature and extent of Soviet energy-related purchases from the West. Using available trade data, it seeks to ascertain the magnitude and sources of this trade, to analyze identifiable patterns and trends, and to illuminate the role of the United States in providing material assistance to Soviet energy industries. In the latter context, the chapter addresses the issue of "foreign availability," i.e., the extent to which the United States is the sole or preferred supplier of energy equipment and technologies that the Soviet Union has purchased in the last 5 years or is likely to seek during the present decade.

The methodology employed in this chapter is as follows: OTA used the surveys of the Soviet oil, gas, coal, nuclear, and electricity transmission industries which appear in chapters 2 through 5, together with information on common Western practice, to compile a broad list of important energy technologies, items of equipment, and services.[1] This list was refined by using trade statistics to identify those areas in which the U.S.S.R. has made purchases from the West in the past 5 years. Major items were then subjected to a foreign availability analysis in which OTA attempted to ascertain which West European and Japanese firms manufacture similar equipment or possess expertise comparable to those companies that actually supplied the U.S.S.R.

[1] No attempt has been made here to distinguish between "technology" and "products" or "equipment". For a discussion of this subject, see OTA, *Technology and East-West Trade* (Washington, D.C.: U.S. Government Printing Office, 1979), ch. VI.

Foreign availability is a highly subjective concept. As chapter 13 discusses in detail, there is no universal agreement on the parameters that define the degree of equivalency necessary to constitute foreign availability, on how the parameters should be weighted, or indeed on how equivalency itself can and should be measured. Given the limitations of this study, no attempt was made to conduct an exhaustive worldwide search for all alternative sources for each item; in most cases identification of two or three suppliers was deemed sufficient.

Nor did OTA construct a rigorous framework for defining and measuring foreign availability. Information on energy technologies and equipment was collected from industry sources in a number of countries and evaluated by independent technicians. These evaluations were supplemented by interviews with representatives from energy-related companies in the United States and with members of the intelligence community. The criteria of comparability were quality, price, and technical capabilities (i.e., speed, capacity, precision). OTA did not investigate the potential manufacturing capacity of alternative suppliers or evaluate the willingness of firms to sell to the U.S.S.R. The latter are both issues that would have to be taken into account in an exhaustive foreign availability analysis.

Although the results of the analysis carried out here are limited, they can be used to indicate those areas in which the United States enjoys a significant technological edge over other Western nations in an energy-related process, system, or piece of equipment important to the oil, gas, coal, nuclear, or electric power industries of the U.S.S.R.

A variety of statistical systems were examined in the course of this study, no one of which provided an ideal data base. OTA has dealt elsewhere with the problems associated with measuring and reporting trade,[2] but a brief review of the data sources employed here can provide a sense of the limitations of the analysis that follows and of the strength of the generalizations that may legitimately be drawn from it.

This chapter relies most heavily on the United Nations' Standard International Trade Classification (SITC). This scheme summarizes trade information for thousands of different items by organizing them into commodity groupings of up to seven digits. The SITC system is the only readily available statistical source that reports data for all of the Western countries examined during the course of this assessment. The disadvantage of SITC lies in the fact that it is highly aggregative, i.e., its codes encompass items that are not specifically energy related. Consequently, **many of the values shown are inflated, and should be understood to represent relative orders of magnitude rather than precise amounts of energy-related exports.** OTA has collected data for those SITC codes believed to consist largely, albeit not exclusively, of energy-related items. These codes are shown in table 40. The data are best used for comparative purposes—to identify the relative importance of suppliers of particular items to the U.S.S.R., for example.

In some cases a more discrete analysis than that possible from SITC data seemed warranted. Here three other data bases were employed: Schedules B and E from the U.S. Department of Commerce,[3] and the European Economic Community (EEC)'s Nomenclature of Goods for the External Trade Statistics of the Community and Statistics of Trade Between Member States

—————————
[2]Ibid.
[3]DOC's Schedule B is based on the U.N. SITC and also employs seven digits. It was replaced in 1978 with Schedule E, a system designed to reflect greater precision in the codes. DOC data are, therefore, Schedule B through 1977 and Schedule E thereafter.

Table 40.—UN SITC Energy-Related Equipment Codes

SITC code	Description
571.1	Propellent powders and other prepared explosives
571.2, 571.2	Safety and detonating fuses, percussion and detonating caps, igniters, etc.
629.4	Transmission, conveyor or elevator belts of vulcanized rubber
642.93	Gummed or adhesive paper in strips or rolls
655.92	Transmission, conveyor or elevator belts of textiled material
678, 672.9	Tubes, pipes and fittings of iron and steel
695.24	Rock drilling bits; tools and bits for assorted hand tools
711.1	Steam and other vapor generating boilers and parts, n.e.s.
711.2	Auxiliary plant for use with steam and other vapor generating boilers
711.7	Nuclear reactors and parts thereof n.e.s.
714.3	Automatic data processing machines and units thereof; magnetic and optical readers and machines for processing data
714.92	Parts, n.e.s. of and accessories for ADP and other calculating machines
718.42	Self-propelled shovels and excavators, self and nonself propelled levelling, tamping, boring, etc., machinery and parts thereof
718.51	Machinery for sorting, screening, separating, washing, crushing, etc., for earth, stone, ores and other minerals
719.21	Pumps for liquids and parts thereof
719.22	Air and vacuum pumps and air or gas compressors and parts thereof
719.23, 712.31	Filtering and purifying machinery and apparatus for liquids and gases
719.31	Other lifting, handling, loading and unloading machinery, n.e.s.
722.1	Rotating electric plant and parts thereof; transformers, converters, rectifiers, inductors and parts.
722.2	Electrical apparatus for making or breaking electrical circuits
723.1	Insulated electric wire, cable, bars, strip and the like
723.21	Electrical insulators of other materials
729.52	Electrical measuring, checking, analysing or controlling instruments, n.e.s.
729.92	Electric welding, brazing, soldering and cutting machines and apparatus and parts thereof, n.e.s.
732.4	Special purpose motor lorries, vans, crane lorries, etc.
735.92	Light vessels, floating cranes and other special purpose vessels, floating docks
735.93	Floating structures other than vessels
861.81, 729.51	Gas, liquid, and electricity supply or production meters.
861.91	Surveying, hydrographic, etc., and geophysical instruments (nonelectrical)
861.99	Parts, n.e.s., of meters and counters; nonelectrical and electrical measuring, checking, etc., instruments of SITC 729.52, 861.8 and 861.97

SOURCE: Office of Technology Assessment.

(NIMEXE). The former were useful in providing a clearer sense of the role of energy-related items in U.S. exports; the latter fulfilled the same role for West European countries. Cross-checking Western export with Soviet import data proved impossible. This phenomenon is not unique to the energy sector; it reflects different classification systems and data reporting criteria.

It must be noted that U.S. export statistics do not include the value of Western technology and equipment sales that originate from U.S. subsidiaries or licensees abroad; nor will the category of U.S. exports to the U.S.S.R. include items that are first destined for third countries. Corporations that export energy technology and equipment are often multinationals or at least have international corporate affiliations in other countries. Some sense of the international nature of this industry can be gleaned from appendix A, which shows a partial list of energy corporation affiliations worldwide. The complexity of these relationships makes it extremely difficult, if not impossible, to always attribute technology sales to the true country of origin. The data collected here, therefore, very likely understate the contribution of American technology to the U.S.S.R.

Finally, export statistics are ill-suited to identifying such technology transfers as the sale of licenses or turnkey plants. For this reason, OTA supplemented its statistical data bases with a comprehensive search of *Soviet Business and Trade,* a biweekly publication that reports major trade deals between the U.S.S.R. and the West.[4] Tables B-1 and B-2 in appendix B summarize energy-related transactions reported in this publication between 1975 and 1980.

---

[4]*Soviet Business and Trade* is published in Washington, D.C. by Welt Publishing Co.

EAST-WEST TRADE IN ENERGY TECHNOLOGY AND EQUIPMENT

Although Soviet trade with the West has grown markedly in the past decade, it has remained a relatively small part of world trade as a whole. Except for sales of agricultural commodities (i.e., grain), the United States has captured relatively small market shares of this trade compared to other nations in the Industrial West (IW, defined here as Canada, France, Italy, Japan, Netherlands, Norway, Sweden, Switzerland, the United Kingdom, the United States, and West Germany).[5] Table 41, which shows total Soviet imports from the IW, demonstrates that when agricultural commodities are excluded, in 1979 the United States lagged behind West Germany, Japan, and France in industrial exports to the U.S.S.R.

Table 42 shows the value of Soviet imports of items in the energy-related SITC codes listed in table 40. From these data,

OTA estimates that energy-related items constituted about 25 percent of total Soviet imports from the IW in 1975, and about 22 percent in 1979. When total imports are adjusted to omit agricultural commodities, the relative importance of energy-related trade increases slightly. In 1979, about 28 percent of Soviet nonagricultural imports from the West consisted of energy equipment and technology.

It is clear from table 42 that the vast preponderance of the U.S.S.R.'s energy-related imports are destined for its oil and gas industries, a fact that is demonstrated graphically in figure 11. The oil and gas sector in 1979 took up 77 percent of such imports. Assuming that a similar proportion of multiarea items—i.e., those that could be employed in more than one energy industry—were destined for the oil and gas industries, approximately 81 percent of all Soviet purchases of energy equipment and

---

[5]See OTA, op cit., ch. III.

Table 41.—Industrial West Exports to the U.S.S.R., Selected Countries[a] 1975-1979 (million U.S. dollars)

Year	Total industrial West[a]	United States	Japan	France	West Germany	Italy	United Kingdom	Other
1979	$15,255	$3,604	$2,461.0	$2,007	$3,619	$1,220	$694	$1650.4
Industrial.........	12,151	1,319	2,461	1,791	3.551	1,191	655	1,183.1
Agricultural	3,104	2,285	0.2	216	68	29	39	467.3
1978	12,419	2,249	2,502	1,455	3,141	1,133	665	1,274
Industrial.........	10,485	804	2,501	1,408	3,128	1,084	615	945
Agricultural	1,934	1,445	1	47	13	49	50	329
1977	10,788	1,624	1,934	1,496	2,789	1,228	607	1,110.1
Industrial.........	9,411	747	1,934	1,369	2,765	1,218	602	776
Agricultural	1,377	877	0	127	24	10	5	334
1976	11,051	2,306	2,252	1,118	2,685	981	432	1,277
Industrial.........	9,005	946	2,252	977	2,659	974	427	770
Agricultural	2,047	1,360	0	141	26	8	5	507
1975	10,092	1,834	1,625	1,147	2,824	1,020	464	1,178
Industrial.........	8,469	720	1,625	1,059	2,816	996	459	794
Agricultural	$1,623	$1,114	$ 0	$ 88	$ 8	$ 24	$ 5	$ 384

[a]Industrialized Western nations include Canada, France, Italy, Japan, the Netherlands, Norway, Sweden, Switzerland, the United Kingdom, the United States, and West Germany.

SOURCE: OECD, *Statistics of Foreign Trade, Series B*, Paris (Annual).

Table 42.—Soviet Energy-Related Imports From Selected Western Countries[a]
(million U.S. dollars)

Year	Oil/gas	Coal	Nuclear	Electric power	Multiarea commodities	Total
1979....................	$2,652.5	$255.9	$84.1	$279.3	$155.1	$3,427
1978....................	2,321.0	237.0	70.4	331.9	134.8	3,095
1977....................	1,991.7	174.4	98.8	252.6	103.4	2,621
1976....................	2,250.0	202.8	70.2	157.0	86.5	2,767
1975....................	$1,989.5	$212.5	$90.5	$117.6	$90.3	$2,500

[a]Includes Canada, France, Italy, Japan, the Netherlands, Norway, Sweden, Switzerland, the United Kingdom, the United States, and West Germany.

NOTE: Data here is for the SITC codes listed in table 40.

SOURCE: United Nations SITC data.

technology were used to find, produce, and deliver oil and gas. The share of the nuclear power sector was less than 3 percent; that of coal 7 percent; electric power, 8 percent; and other multiarea items, 1 percent.

The role of the United States in this energy-related trade has been small, both in terms of its share in U.S.-Soviet trade as a whole, and in comparison to the relative shares of energy-related equipment and technology in the Soviet trade of America's allies. As table 43 displays, the value of U.S. exports to the U.S.S.R. in energy-related SITC codes for 1979 was $237.6 million. This constituted some 6.6 percent of America's total and about 18 percent of its industrial exports to the Soviet Union in that year. Table 43 also shows that in 1979, the value of U.S. energy-related exports to the

Figure 11.—Soviet Imports of Western Energy Equipment and Technology, by Industry

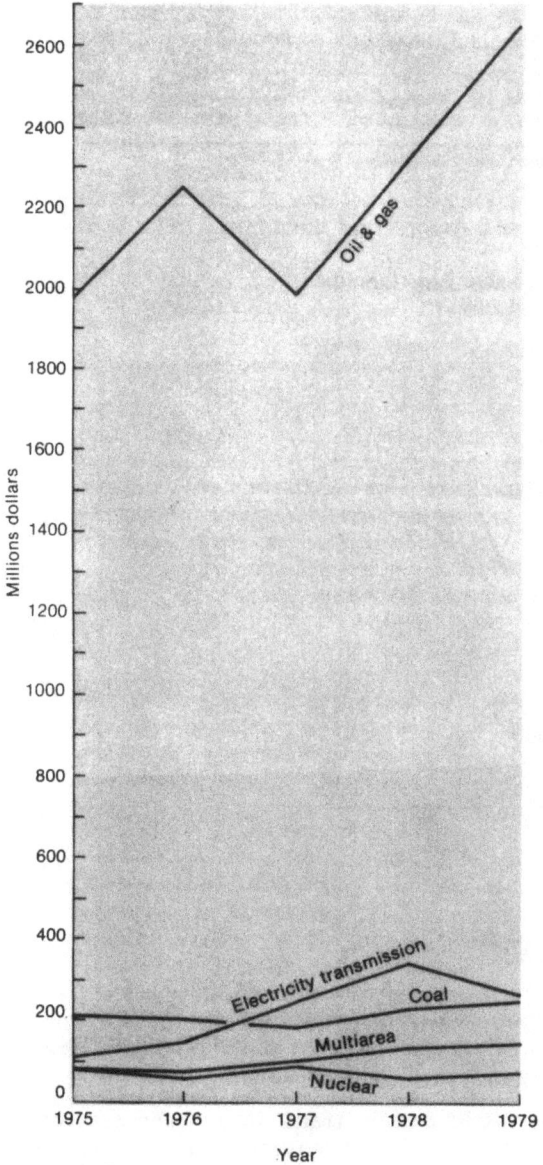

SOURCE: Table 42.

U.S.S.R. was lower than those of Japan, West Germany, France, and Italy. Japan was by far the largest energy equipment supplier to the U.S.S.R., its exports totalling about one-third of all Soviet purchases in this area. West Germany was a close second. France and Italy both recorded exports about double those of the United States. In 1979, the United States accounted for under 7 percent of energy-related exports to the U.S.S.R. of the Western countries examined. Moreover, given the fact that the SITC data on which this analysis is based seriously inflates the export figures by including items not destined for the Soviet energy sector, this percentage may actually be even smaller. U.S. trade statistics as of this writing are available only through October 1980, but these suggest that U.S. energy-related exports plummeted in 1980, the result of the post-Afghanistan technology embargo.

The composition of the Soviet Union's purchases from the United States reflects the same pattern as its energy-related imports from the West as a whole, i.e., they are largely composed of items destined for the oil and gas industries. This pattern is shown in figure 12. But although U.S. sales have not been high in dollar amounts, or particularly impressive as a percentage of total Soviet energy-related imports, it has been contended that their importance to the U.S.S.R. is greatly magnified by the fact that they have been composed of critical items, some of which are unavailable elsewhere in the world. The remainder of this chapter will investigate this assertion, examining sector by sector the composition and magnitude of Soviet imports in each of the five energy industries under consideration, and identifying the sources of these imports. Where U.S. firms have been active traders, an attempt has been made to ascertain whether or not comparable items are available elsewhere in the West.

Table 43.—Energy-Related Exports to the U.S.S.R. by Selected Countries[a] (million U.S. dollars)

Year	United States	Japan	France	West Germany	Italy	United Kingdom	Other	Total energy exports
1979	$237.6	$1,097.1	$474.4	$906.1	$408.2	$90.5	$213.1	$3,427
1978	159.8	1,067.5	391.4	839.3	390.6	113.8	133.0	3,095
1977	211.7	599.9	418.6	745.4	447.5	49.7	148.2	2,621
1976	284.7	904.4	308.3	627.8	438.4	46.3	156.7	2,700
1975	$218.3	$479.5	$334.4	$854.8	$433.8	$38.4	$140.3	$2,500

[a]Includes Canada, France, Italy, Japan, Netherlands, Norway, Sweden, Switzerland, United Kingdom, United States, and West Germany.

SOURCE: United Nations SITC Data.

Figure 12.—U.S. Energy-Related Exports to the U.S.S.R., by Industry

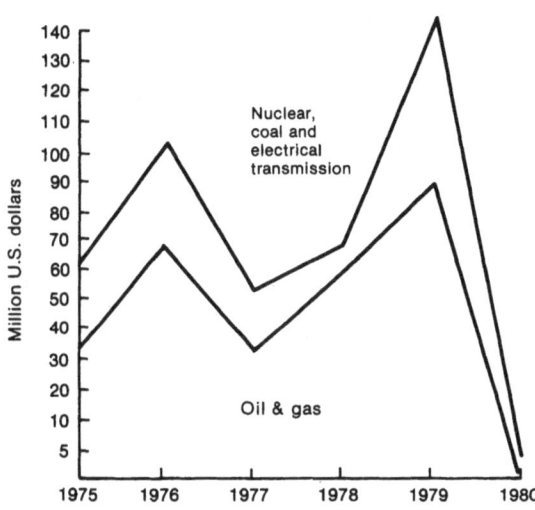

SOURCE: U.S. Department of Commerce.

NUCLEAR

WESTERN EXPORTS

Two SITC codes relate particularly to the nuclear industry:

711.7   Nuclear reactors and parts thereof.

729.92   Electric welding, brazing soldering and cutting machines and apparatus and parts thereof.

The latter code clearly incorporates a broad range of equipment that can also be utilized by other industries.

Little Soviet trade in nuclear reactors and parts has been recorded—exports of $448,700 from the United States in 1978 and $329,000 from West Germany in 1976. OTA has been unable to find any unclassified accounts that provide further information on either of these figures. There have been report that the U.S.S.R. may be considering the purchase of reactors from Italy, but as yet no such deal has been completed. (It must be noted that the U.S.S.R. has also exported nuclear reactors to the West.) However, the limitations of using and difficulties in interpreting trade statistics may be illustrated by the fact that a $47 million contract purportedly signed with the Italian firm Breda Termomeccanica in 1976 for the design and building of reactor manufacturing facilities at Atommash and Izhora does not appear in this SITC code.[6] It is possible that this deal was not consummated or that it is reflected elsewhere in the trade data.

SITC code 729.92 covers electric welding, brazing, soldering, and cutting machines and apparatus and parts. IW trade in this area amounted to over $90 million in 1975 and over $84 million in 1979, but as figure 13 indicates, U.S. shares have been falling since

[6] *Soviet Business and Trade,* Nov. 24, 1976, p. 4.

Figure 13.—Western Energy Trade With U.S.S.R.—Welding Equipment (SITC 729.92)

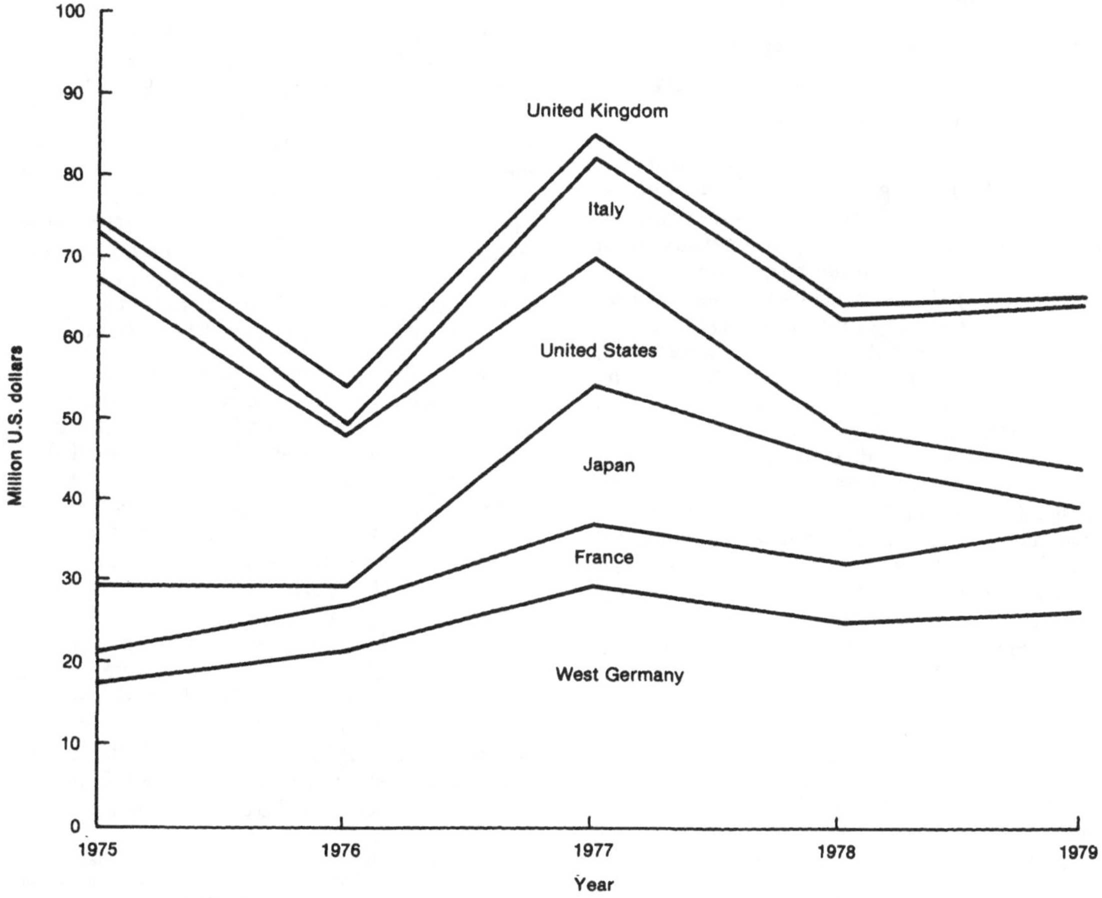

SOURCE: United Nations SITC Data.

1975. In 1979, less than 10 percent of the welding equipment supplied to the U.S.S.R. by the IW came from the United States, while West Germany and Italy together supplied nearly half.

In addition to the items contained in these two codes, several purchases that may have been destined for the nuclear industry have been noted in *Soviet Business and Trade*. These include a hydraulic press purchased from Japan;[7] pumps from a United Kingdom firm,[8] and various valves, shutters, and plugs from West Germany and Canada.[9] Velan Engineering Co. of Canada is a major supplier of valves for the Soviet oil, gas, chemical, and nuclear industries.

FOREIGN AVAILABILITY OF NUCLEAR EQUIPMENT AND TECHNOLOGY

Aside from complete reactors, chapter 4 has identified some areas in which, should they decide to step up their nuclear-related purchases, the Soviets might usefully turn to the West to supplement their own manufacturing capabilities. These areas include nuclear grade tubing and pipes, welding and brazing equipment, steam generators, pumps and casings (to circulate coolant through the reactor), nuclear valves, and computers, software, and automatic control. It must be stressed that at present the U.S.S.R. imports very little, if any, such equipment. Indeed, analysis of the trade data has confirmed chapter 4's generalization that the Soviet nuclear power industry has been virtually self-sufficient. Assuming for purposes of argument that the U.S.S.R. might reverse its policy, OTA sought to determine in which, if any, of these areas the

United States might be considered a sole or preferred supplier to the U.S.S.R.

OTA's "foreign availability analysis" for the nuclear industry was based on information from industrial trade journals and interviews with representatives of U.S. firms (Westinghouse, Rollmet, Ransome), Oak Ridge National Laboratories, the Nuclear Regulatory Commission, and the U.S. Department of Energy. The results of this analysis are reported in tables C-1 through C-7 of appendix C. It can be seen from these tables that firms in a variety of countries could potentially supply the U.S.S.R. with the kind of nuclear equipment and technology it is most likely to seek.

There is one area in which the United States could be a strongly preferred supplier. The United States is the acknowledged world leader in many areas of computer technology. Computers are normally used at nuclear power stations for rather mundane tasks—data acquisition and simple process control—and the U.S.S.R. presently relies heavily on domestically produced computers at its nuclear stations. If greater power station automation is planned, however, more sophisticated systems might be necessary.

The United States is also a world leader in developing and mass-producing exotic, high-strength materials (zircalloy, and high nickel content stainless steel). These are useful in the nuclear industry, particularly for advanced breeder reactors. Such technology has been fairly widely diffused. Indeed, the means by which nuclear technology is spread throughout the world may be illustrated by several examples: The U.S.-based Westinghouse Electric Co. exports nuclear grade tubing to Mitsubishi in Japan, Framatone in France, and ENSA in Spain, and each of these companies constructs nuclear reactors under a Westinghouse license. Toshiba and Hitachi of Japan are General Electric licensees.[10] The reactor in the state-owned plant at

---

[7]*Soviet Business and Trade*, Nov. 24, 1976. This, together with Italian large boring and milling machines, German machine tools, and U.S. welding and X-ray equipment, was destined for Atommash. See also *Nucleonics Week*, Nov. 8, 1979, p. 12.

[8]Hayward Taylor & Co., Ltd. *Soviet Business and Trade*, Feb. 17, 1975.

[9]*Soviet Business and Trade*, Feb. 17, 1975; Apr. 28, 1975; and Dec. 6, 1978.

[10]Mans Lonnroth and William Walker, "The Viability of the Civil Nuclear Industry," International Consultative Group on Nuclear Energy (The Rockefeller Foundation, 1979).

Krsko, Yugoslavia, was purchased from Westinghouse.

SUMMARY AND CONCLUSIONS

The U.S.S.R. is presently self-sufficient in the design, manufacture, and operation of its nuclear plants. Should it reverse its policy and begin seeking Western imports in this area, it would find numerous potential suppliers in several nations.

COAL

WESTERN EXPORTS

The U.N. SITC codes that contain equipment applicable to coal mining are as follows:[11]

571.1   Propellent powers and other prepared explosives.

571.2   Safety and detonating fuses, percussion, and detonating caps, igniters, etc.

629.4   Transmission, conveyor and elevator belts of vulcanized rubber.

655.92 Transmission, conveyor and elevator belts of textile materials.

695.24 Rock drilling bits; tools and bits for assorted hand tools.

718.42 Self-propelled shovels and excavators, self and non-self-propelled leveling, tamping, boring, etc., machinery and parts thereof.

718.51 Machinery for sorting, screening, separating, washing, crushing, etc., for earth, stone, ores, and other minerals.

Figure 14 summarizes trade in each of these categories for the period 1975-79. It shows that the U.S.S.R. has purchased no fuses or explosives from any IW nation (SITC 571.1 and 571.2) since 1977. During 1975 and 1976, U.S. exports in these categories were very small ($5,000 to $28,000). Japan and West Germany led the sale of

transmission, conveyor, and elevator belts of vulcanized rubber (SITC 629.4) to the U.S.S.R. in 1975-79; Soviet purchases from the United States were a weak third or fourth (behind Italy). These sales totaled about $48.7 million for all countries during the entire period. Sales in the category 655.92 were negligible.

In sales of rock drilling bits and tools, and bits for assorted hand tools (SITC 695.24) to the U.S.S.R., the United States lagged West Germany, Japan, France, Italy, and the United Kingdom. Indeed, the U.S. market share here has fallen precipitously since 1975 when sales of $1.3 million were recorded—in 1979 only $13,000 of such goods were purchased from the United States. Because this equipment might equally well be destined for the oil and gas sector, it is discussed in more detail below.

The Soviet Union buys considerable amounts of self-propelled shovels and excavators, leveling, tamping, and boring machinery (SITC 718.42). However, the finer breakdowns available from U.S. Scheduled B and E and NIMEXE data reveal that most of the purchases here are for items probably destined for the oil and gas industries. Coal-relevant subcategories from Schedule B and Schedule E include draglines, dragline buckets, coal cutting machines, continuous mining machines, longwall mining machines, and excavating machines (including attachments). No exports from the United States to the U.S.S.R. whatsoever have been reported in any of these subcategories.

In 1979, the U.S.S.R. bought $81.8 million worth of machinery for sorting, screening,

---

[11]SITC 719.3l (other lifting, handling, loading, and unloading machinery) proved on further examination using DOC and NIMEXE data to consist mostly of items relating to oil and gas.

Figure 14.—Coal-Related Equipment Exports to the U.S.S.R. (million U.S. dollars)

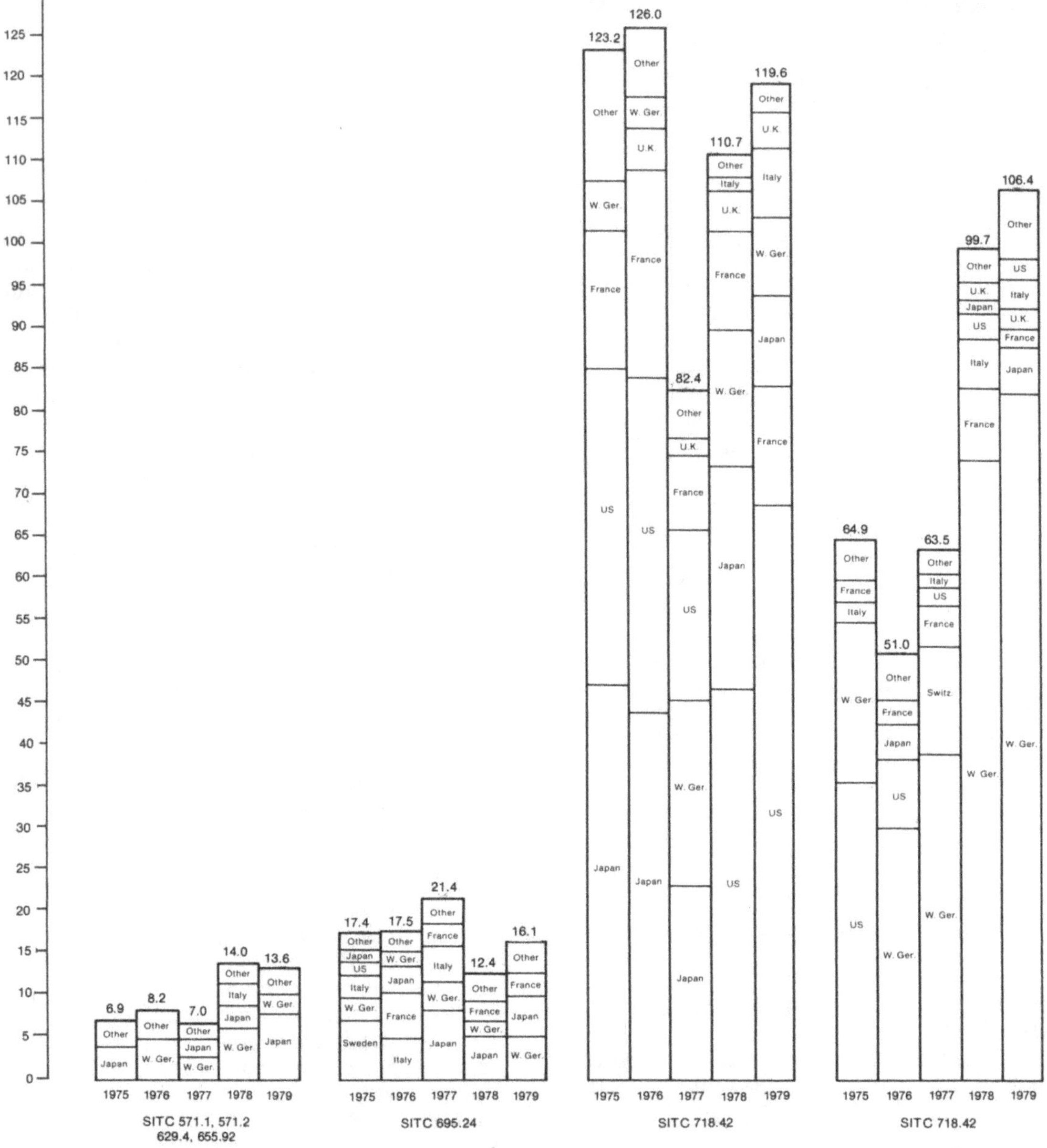

SOURCE: United Nations, SITC data.

separating, washing, and crushing (SITC 718.51) from West Germany. Imports from the United States in the same category came to less than $2.4 million, mostly for crushing, pulverizing, and grinding machines and parts. West Germany has carried most of this market since 1976.

In sum, it would appear that Soviet purchases from the West for its coal industry have been relatively modest and that between 1975 and 1979 little of these came from American firms. Soviet import statistics corroborate this broad conclusion. Although the Soviet method of reporting commodity trade differs from that employed in the West—in commodity groupings, valuation, coverage, and the method used for identifying trade partners—data from the U.S.S.R. reflect the same patterns as Western export statistics.

Table 44 shows official Soviet import data on items that may have an impact on coal-related activities. These data indicate that in recent years, Soviet purchases from West Germany and Japan have been much larger than those from the United States. For example, in 1978, the figures show imports of $115.4 million of sorting machinery from West Germany, but only $6 million from the United States. West Germany ($30.4 million) also led the United States ($4.2 million) in exports of mechanical shovels and excavators. Although the only Soviet imports of ships, derricks, and cranes reported in 1977 and 1978 come from the United States, the dollar amounts ($891,000 and $5.7 million) are relatively small. The U.S.S.R. purchased $58.6 million and $92.4 million worth of these goods from Japan in 1976 and 1975, respectively.

A search of *Soviet Business and Trade* for transactions that might not have appeared in the SITC codes indicates that the most important category of U.S. exports likely destined for the Soviet coal industry was transportation. Approximately 100 trucks (ranging in size from 100 to 200 tons) purchased from the Unit Rig & Equipment Co. of Tulsa, Okla., for about $70 million were

Table 44.—Soviet Imports From Selected Western Nations[a]

	1978	1977	1976	1975
U.S.S.R. imports of machinery for sorting				
United States ..	6,096	26,287	32,900	24,110
West Germany .	115,423	30,780	7,573	4,555

	1978	1977	1976	1975
U.S.S.R. imports of mechanical shovels and excavators				
United States ..	4,169	917	9,692	10,923
West Germany .	30,444	11,799	9,262	28,516
France	2,790	1,157	13,299	5,522

	1978	1977	1976	1975
U.S.S.R. imports of ships, derricks and cranes				
United States ..	5,677	891	266	10,424
Italy	—	—	3,693	16,068
United Kingdom	110	037	2,758	3,700
Japan	—	—	58,646	92,403

[a]OTA collected Soviet data for France, Italy, Japan, the United Kingdom, United States, West Germany, and West Berlin. Only those areas that showed trade in a given category are presented here.

SOURCE: CIA, based on Soviet trade data.

for use in coal-related activities.[12] Other American deals have included $14 million worth of slurry pumps from Ingersoll-Rand between 1974 and 1976;[13] and front-end loaders contracted for $1 million from the Clark Equipment Co. in 1978, and for $2.9 million from Dart Division of Paccar, Inc. in 1979.[14]

Between 1974 and 1980 Japanese firms involved in the South Yakutian Development Corp. were responsible for sales totaling approximately $450 million,[15] most of which consisted of shovels and other surface mining equipment. In addition, in 1976 and 1975, respectively, the U.S.S.R. reportedly contracted for $500,000 worth of equipment for excavation in underground mines from

[12]*Soviet Business and Trade*, Mar. 1, 1978, p. 3.
[13]*Soviet Business and Trade*, Aug. 1, 1975, pp. 1-2.
[14]*Soviet Business and Trade*, June 6, 1979, p. 7.
[15]*Soviet Business and Trade*, June 6, 1979; Nov. 22, 1978; May 26, 1976.

the United Kingdom, and a similar amount in front tunneling machines and loaders from West Germany.[16]

FOREIGN AVAILABILITY OF COAL TECHNOLOGY AND EQUIPMENT

As chapter 3 has indicated, the U.S.S.R. is well able to design, test, and manufacture its own coal mining equipment. Soviet equipment is heavier and somewhat less sophisticated than U.S. equipment, but it is adequate. A common Soviet practice has been to buy items applicable to a specific phase of coal mining and reproduce them. Soviet-made continuous miners, for example, are copies of West German, English, and French models. Drill bits that were formerly purchased from Western Europe are now domestically produced. Equipment for Siberian surface mining was originally purchased from Marion in Japan.

In short, like the nuclear industry, the Soviet coal mining industry has been essentially self-sufficient. In an attempt to ascertain whether a reversal of past Soviet practice with respect to coal-industry equipment imports would focus on items in which the United States is a sole or preferred supplier, OTA assembled a list of essential equipment for coal mining operations, and attempted to locate suppliers of this equipment in Western Europe and Japan. The results may be found in tables C-8 and C-9 of appendix C. Outside of the few deals discussed above, there is no evidence that any of the companies listed here have actually supplied or intend to supply equipment to the U.S.S.R. Nevertheless, it is clear from table C-9 that there exist many European and Japanese suppliers of coal mining equipment.

Comparison of the items available from these companies and those produced in the United States reveals substantial differences between underground and surface mining capabilities. The majority of Soviet underground mining utilizes longwall techniques not widely employed in the United States. Longwall mining was invented in the 1950's in West Germany; in fact, the United States is heavily dependent on Britain and West Germany for longwall research and development, and it also imports substantial amounts of European longwall equipment. (The U.S.S.R. has attempted to export its own longwall systems to the United States.) Much of the underground coal mining equipment manufactured in large quantities in the United States is therefore of little or no use to the U.S.S.R. Undercutters are available only from the United States, but these are not necessary for longwall mining operations. There are also differences in geologic formations that render much U.S. equipment inapplicable in the U.S.S.R.; e.g., the narrow seams in many Soviet mines do not easily lend themselves to mechanization.

A different situation pertains with respect to surface mining, which, as chapter 3 has pointed out, will become increasingly crucial to Soviet coal production as the decade progresses. The United States is a world leader in surface mining equipment and technology, and produces items that could be of great use to the U.S.S.R. One example is the dragline. This is the only piece of surface mining equipment in continuous operation, and coal output is heavily affected by its speed in removing overburden. U.S. draglines and excavators have the largest capacities currently available. Increased excavation capacity would allow Soviet surface mining to become more productive, and bigger draglines would facilitate deeper surface mining. U.S. firms also produce trucks, power shovels, and excavators with capacities much larger than any available from Western Europe or Japan. These items, though not vital to the ability to mine coal, could help to increase output.

SUMMARY AND CONCLUSIONS

Both Soviet and Western trade data show that the U.S.S.R. has purchased relatively little coal mining equipment and technology

---

[16]*Soviet Business and Trade,* Nov. 22, 1978; June 6, 1979.

from the West, perhaps less than 10 percent of its total equipment needs. With the possible exception of transport vehicles, American market shares in this trade are smaller than those of other Western countries, especially West Germany and Japan. The only coal mining technology that OTA could establish as unique to the United States is undercutters, but these are not used in the U.S.S.R. American underground mining technology is unlikely to be particularly attractive to the Soviet Union.

While no surface mining technologies seem to be unique to the United States, U.S. firms do produce the largest capacity trucks, draglines, and excavators in the world. Chapter 3 has maintained that the future of Soviet coal production rests on expanding its surface mining operations. Should the U.S.S.R. depart from past practice and begin to import large quantities of Western surface mining equipment, the United States would—all other things being equal—be the preferred supplier.

ELECTRIC POWER

WESTERN EXPORTS

Figure 15 shows export statistics for the following SITC codes containing equipment used for the generation and transmission of electric power:

711.1 Steam generating boilers.
711.2 Auxiliary plant for use with steam and other vapor generating boilers.
722.1 Rotating electric plant and parts thereof; transformers, converters, rectifiers, inductors and parts.
722.2 Electrical apparatus for making or breaking electric circuits.
723.1 Insulated electric wire, cable, bars, strip and the like.
723.21 Electrical insulators and other materials.

1979 statistics for SITC 722.1 indicate that Japanese, French, and West German firms supplied the U.S.S.R. with $15.2 million, $11.9 million, and $11.1 million worth of transformers, converters, rectifiers, inductors, and parts thereof, respectively. Purchases from the United States in this category amounted to only $1.35 million. Much the same situation has prevailed since 1977. Similar patterns hold for electrical apparatus used in making or breaking electrical circuits (SITC 722.2). From 1975 to 1979, Soviet imports from the United States ($3

million) have been dwarfed by imports from France ($59.2 million), Japan ($51.7 million), West German ($11.3 million), and Italy ($13.6 million). The Soviet Union does not purchase many electrical insulators (SITC 723.21). Most of its 1979 purchases came from the United States, but these amounted to only $666,000. Even this was an anomaly. Between 1975 and 1979, no Soviet purchases from any country exceeded $48,000.

A similar aberration can be seen in the data for SITC 711.1, steam generating boilers. The $19 million recorded in this category for U.S. exports in 1979, although not a large amount in absolute terms, represented a departure from Soviet practice over the past 5 years in two ways. The United States had not before been the largest Western supplier of this equipment, and the amount was significantly greater than any recorded previously for a single country in a single year. OTA has been unable to discover any details of the transaction or transactions that accounted for these exports. SITC 711.2 contains auxiliary plant for such boilers. Except for large U.S. sales in 1976 and 1977 ($9.4 and $22.7 million), for which OTA has been unable to obtain more information, Soviet purchases in this category have come almost exclusively from France in amounts ranging between about $2 and $5 million per year.

Figure 15.—Electric Power Equipment Exports to the U.S.S.R. (million U.S. dollars)

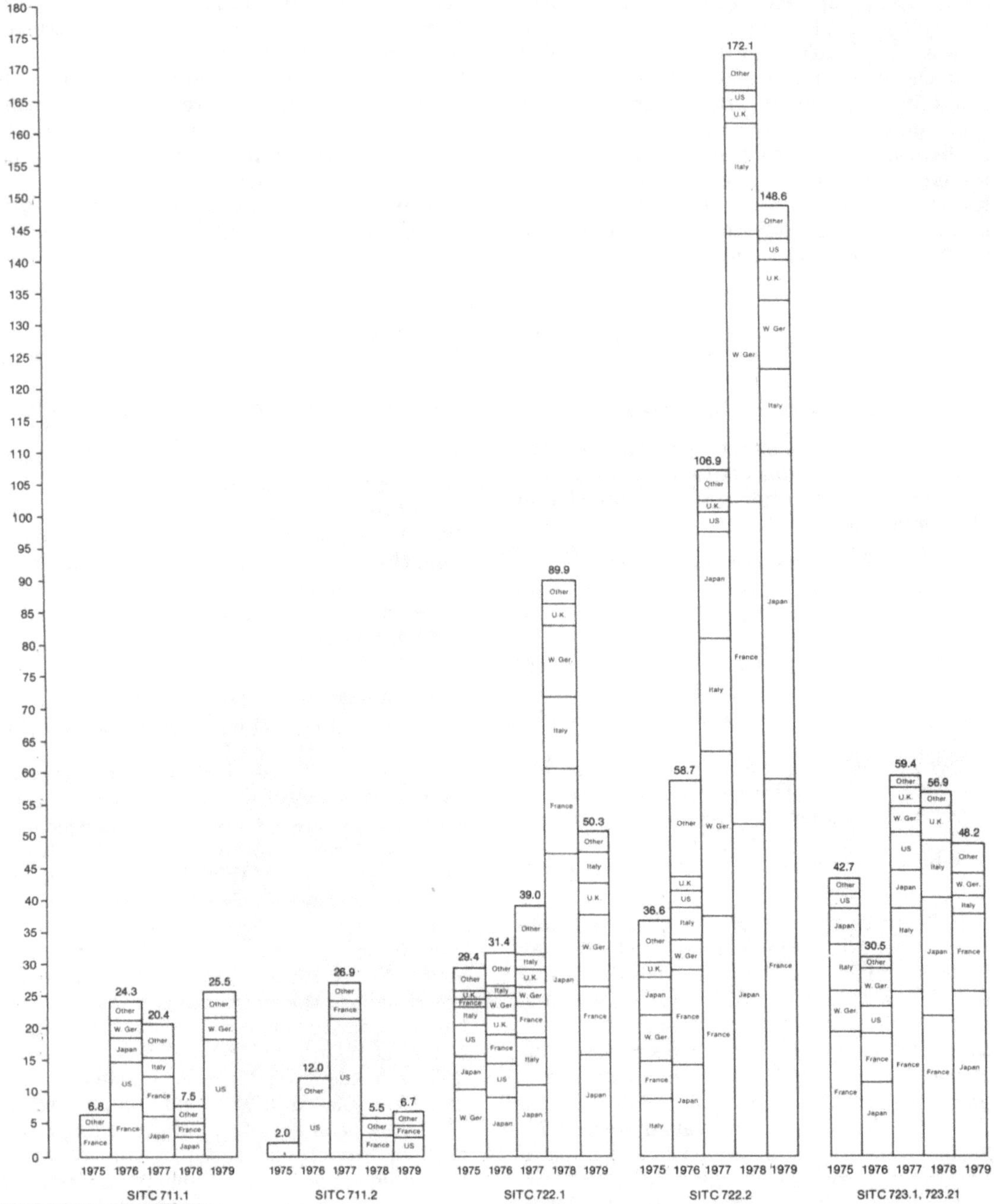

SOURCE: United Nations, SITC data.

It was not possible to determine the proportion of the equipment in these codes that was actually destined for the Soviet electric power industry. *Soviet Business and Trade* noted only one relevant deal over the past 5 years, a Soviet purchase of cable for the Siberian power grid from Siemans AG of India and West Germany.[17]

FOREIGN AVAILABILITY OF ELECTRIC POWER EQUIPMENT AND TECHNOLOGY

The U.S.S.R. does not purchase very large amounts of electric generation or transmission equipment from the West. As in the coal and nuclear sectors, however, Soviet policy could change. OTA, therefore, attempted to determine whether competing firms in Europe and Japan could supply the U.S.S.R. with electrical transmission technology.

American industry representatives have maintained that technology for the production of most of the necessary equipment for electricity generation and transmission is widespread and available in the open literature. Interviews with General Electric (GE), Westinghouse, the Electric Power Research Institute (EPRI), and the Bonneville Power Administration produced a consensus view that European-produced equipment typically costs less than American equipment, and that the quality and capacity of the equipment produced by West European and Japanese companies compare favorably with U.S. items. A representative from GE claimed that the West Europeans were at

---

[17]*Soviet Business and Trade*, Sept. 29, 1976.

least on par with, and possibly ahead of, the United States in high voltage transmission, and a representative from the Bonneville Power Administration told OTA that West Germany, Sweden, and Japan equal the United States in producing high capacity underground cable technology and equipment.

West European and Japanese firms that produce electric generation and transmission equipment are listed in tables C-3 and C-10 of appendix C. Licensing agreements between Mitsubishi and Siemens exist for many of the items used in transmission. Moreover, the U.S.S.R. itself produces some transmission components for export. The Soviet trading organization, Energomash, has sales agents in Australia, Austria, Argentina, Belgium, Brazil, Canada, West Germany, and the United States. Among the products it markets are coupling capacitors for high voltage transmission lines, three-phase power transformers, and circuit breakers for use in substations.

SUMMARY AND CONCLUSIONS

While the Soviets buy some electrical equipment from the West, these purchases are relatively modest and no corroborating evidence is available to link them to electric transmission. The sole exception is purchases of cable (from Siemens of West Germany). International sales representatives for General Electric and Westinghouse, and electric transmission experts who have visited the U.S.S.R. agree that the Soviets design and produce virtually all their own transmission equipment.

OIL AND GAS

WESTERN EXPORTS

Figure 16 shows United Nations data for the most important oil and gas related SITC codes:

Exploration

714.3   Automatic data processing (ADP) machines and units thereof; magnetic and optical readers and machines for processing data.

714.92   Parts and accessories for ADP and other calculating machines.

861.91   Surveying, hydrographic, etc., and geophysical instruments (nonelectric).

861.99   Parts of meters and counters; nonelectric and electrical measuring, checking, etc., instruments.

Drilling

695.24   Rock drilling bits; tools and bits for assorted hand tools.

732.4   Special purpose motor lorries, vans, crane lorries, etc. (includes rigs).

Production/completion/transportation

678, 679.2   Tubes, pipes, and fittings of iron and steel.

642.93   Gummed or adhesive paper in strips or rolls (for pipe insulation).

719.21   Pumps for liquids and parts thereof.

719.22   Air and vacuum pumps and air or gas compressors and parts thereof.

719.23, 712.31   Filtering and purifying machinery and apparatus for liquids and gases.

719.31   Ships, derricks, cranes and mobile lifting cranes and parts; other lifting, handling, and loading and unloading machinery.

Offshore

735.92   Light vessels, floating cranes and other special purpose vessels, floating docks.

735.93   Floating structures other than vessels.

Multiarea

729.52   Electrical measuring, checking, analyzing, or controlling instruments.

861.81, 729.51   Gas, liquid, and electricity supply or production meters.

The codes presented here are clearly unsuitable for any precise analysis of Soviet oil and gas industry imports. They include many items that may have been destined for other energy sectors or for another part of the Soviet economy altogether, and they fail to reflect known important transactions—the U.S. sale by Dresser Industries of a drill bit plant, for instance. For this reason, OTA has supplemented the SITC data with Department of Commerce Schedules B and E statistics—which provide detailed information about U.S. exports—the EEC NIMEXE system, and information about specific transactions gleaned from *Soviet Business and Trade*. These sources have allowed a rather more detailed, albeit sometimes qualitative, discussion of the nature, extent, and source of Western exports in the Soviet oil and gas industry.

The Department of Commerce data shown in table 45 indicate that U.S. oil and gas equipment trade with the U.S.S.R. nearly tripled between 1975 and 1979, from about $31 million to about $90 million. This growth has been largely due to increases in the value of computers and parts, drill rigs and parts, and pumps. The data also show that while the share of Soviet purchases of exploration equipment has declined slightly, that of drilling-related equipment has grown enormously, largely at the expense of well completion and production items. This may partly reflect a shift in Soviet emphasis away from

Figure 16.—U.S.S.R. Imports of Oil and Gas Equipment (million U.S. dollars)

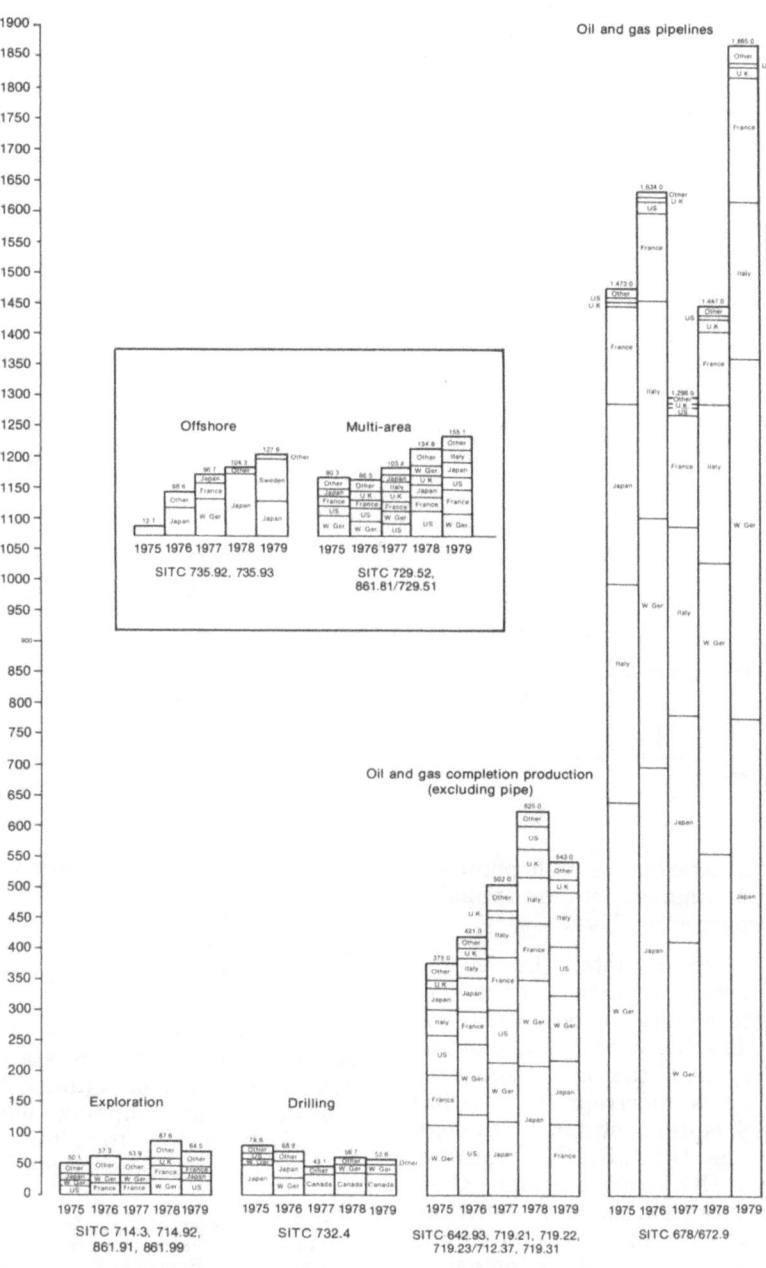

SOURCE: United Nations, SITC data.

Table 45.—U.S. Oil and Gas Equipment Trade With U.S.S.R. Relative Percentage of Each Technology Area (thousand U.S. dollars)

	1975	1976	1977	1978	1979
Total	30,818.0	68,418.8	33,882.8	59,652.7	89,741.0
Exploration					
Geophysical	2,282.0	983.1	480.7	862.5	2,022.6
Computers & pts	8,282.0	16,215.0	3,643.4	17,578.3	22,311.0
Total exploration	10,569.8	17,198.1	4,124.1	18,440.8	24,333.6
Percent of total	34%	25%	12.2%	31%	27%
Pipe	—	2,821.0	1.9	404.7	579.2
Bits	219.9	—	147.2	—	—
Rigs and pts	1,477.0	6,513.0	8,272.0	33,247.9	37,235.3
Total drilling	1,696.9	9,334.0	8,421.1	33,652.6	37,814.5
Percent of total	5%	13.6%	24.8%	56.4%	42%
Well completion/ production					
Pumps	11,657.4	22,220.1	3,674.3	1,241.9	9,979.7
Pump parts	5,928.8	8,235.6	11,743.9	242.9	17,613.2
Gas compressors	765.3	10,445.6	5,432.3	5,937.8	—
Oil and gas sep.	199.8	685.4	487.1	136.7	—
Total comp/prod	18,551.3	41,886.7	21,337.6	7,559.3	27,592.9
Percent of total	60%	61%	63%	12.7%	30.7%

SOURCE: Department of Commerce.

the use of U.S. electric submersible pumps in favor of gaslift techniques that are available from non-U.S. sources (see below).

According to these data, the United States captured only 3.3 percent of the 1979 estimated Western sales in the oil and gas sector reported in table 42 above. It must be noted, however, that the Department of Commerce statistics underrepresent the full value of the U.S. equipment and technology purchased by the U.S.S.R. For instance, neither SITC nor DOC trade statistics show any U.S. contributions in the area of offshore equipment. Yet, a recent survey of Soviet offshore rigs revealed drilling equipment of

U.S. origin.[18] This apparent contradiction may be attributed to the fact that although the U.S.S.R. has purchased its rigs from other nations, these suppliers themselves have imported U.S. drilling equipment for installation on the rigs. The U.S. sale will appear as an export to the third country.[19] Similarly, although U.S. statistics show no sales in the area of refining, U.S. firms are known to have supplied engineering and technical services to West European and Japanese companies engaged in the con-

---

[18] 1980-81 Directory of Marine Drilling Rigs, pp. 19-172.
[19] American technology reexported from third countries is still subject to U.S. export control laws.

struction of petroleum refineries in the U.S.S.R.[20] These transfers of know-how— like the sale of the Dresser drill bit plant— are not recorded in Schedule B/E data and are thus not reflected in table 45 or figure 17.

These data problems limit the precision of any conclusions that can be drawn from Western export statistics. Nonetheless, some generalizations about Western exports to the U.S.S.R. in a number of oil and gas industry sectors are possible.

Exploration

Most of the exploration equipment exported to the U.S.S.R. has consisted of computers and geophysical equipment. Figure 18, for instance, illustrates Soviet purchases of automated data processing equipment from selected Western countries. This figure shows that the primary exporters of this exploration-related computer equipment were the United States, France, and West Germany. Through 1978, the United States tended to supply the majority of the hard-

[20]*Soviet Business and Trade,* Jan. 1, 1979.

Figure 17.—U.S. Oil and Gas Equipment Trade With U.S.S.R. by Technology Area

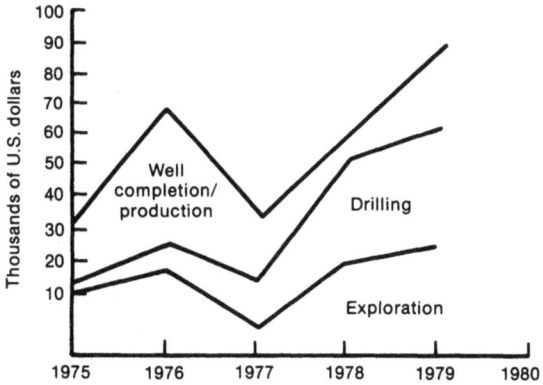

SOURCE: Table 45.

ware and the French to specialize in software.[21] The post-1978 decline in such exports may be due to tighter multilateral export controls on computers, or to improvements in the Soviet hardware base.

The Control Data Corp. (CDC) has been by far the leading exporter of American computers capable of processing the large quantities of data associated with seismic surveying. In recent years, CDC has sold three Cyber 172s and a Cyber 73 to various Soviet ministries engaged in seismic exploration. IBM has also sold two S/370-148 computers for geophysical applications. Another American firm, Geosource, Inc. of Houston, had by 1979 sold 13 Command II field processing systems that are used to preprocess seismic data in the field.[22] This sale alone, valued at $6 million, accounted for nearly 30 percent of total U.S. sales of computers and computer parts in 1979. The French firms Ferney-Voltaire and CIE Generale Geophysique (CGG) have provided the majority of the specialized software used with the CDC computers.[23]

The United States and France are also the U.S.S.R.'s leading Western suppliers of geophysical equipment. NIMEXE and DOC data show that between 1975 and 1979 French sales in this area have increased, while those from the United States have been erratic, ranging from as high as $2.2 million to as low as $480,000. CGG was a major supplier of geophysical equipment, in 1976 alone selling to the U.S.S.R. $14 million worth of digital seismographic recorders, magnetometers, gravity meters, and hydrophones. The equipment was used to equip two geophysical ships that were built for the Soviet Union by Mitsubishi of Japan.[24]

Geosource is the largest American supplier of geophysical equipment, and it sold approximately $30 million worth of equipment to the U.S.S.R. between 1975 and

[21]*Soviet Business and Trade,* Apr. 11, 1979.
[22]Ibid.
[23]Ibid., and Mar. 18, 1979.
[24]*Soviet Business and Trade,* Sept. 29, 1976.

Figure 18.—Western Energy Trade With U.S.S.R. Automated Data Processing Equipment

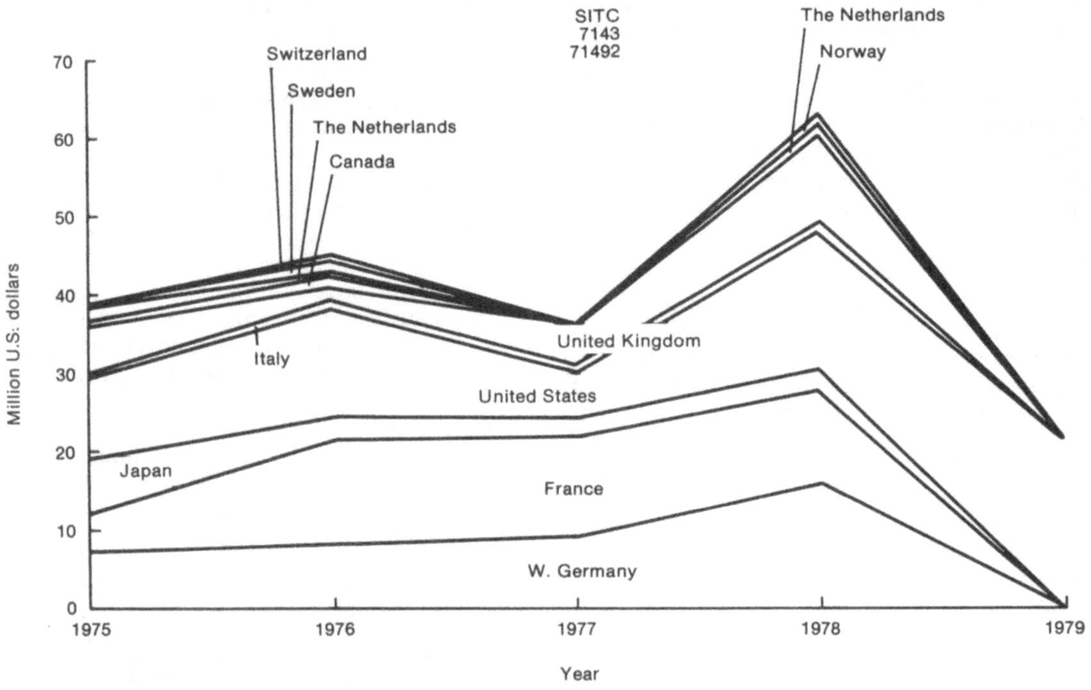

SOURCE: United Nations SITC Data.

1979.[25] Geosource has supplied the U.S.S.R. with a wide array of geophysical prospecting equipment, including 13 photodot automated digital display systems that are used in conjunction with the Command II system.

Drilling

The major Soviet imports in this area have been drill pipe and casing. Soviet imports of drill pipe and casing have come predominantly from Japan, West Germany, France, and Italy. Within these countries, important suppliers have been Mannesmann (West Germany); Vallourec (France); and Finsider (Italy). The U.S.S.R. has also sporadically purchased packers, mud additives, power tongs, and heavy drilling equipment from the West. The Soviets have purchased a number of packers from both Technip in France and Lynes International in the United States.[26]

U.S. exports of drill pipe and casing to the U.S.S.R. totaled less than $1 million over the last 3 years. These relatively low levels are at least partially due to a rapid increase in U.S. drilling activity, which caused U.S. demand for drill pipe and casing to exceed domestic supply. The American shortfall has largely been made up with pipe and casing imported from Japan. The majority of U.S. drilling-related exports to the U.S.S.R. are drilling rigs and parts for drilling rigs. While the

---

[25]*Soviet Business and Trade,* Feb. 14, 1979.

[26]*Soviet Business and Trade,* Jan. 2, 1980.

United States has not been a major supplier of complete rigs—approximately 12 U.S. rigs have been sold to the Soviet Union over the last 10 years—the Soviets are purchasing U.S. drilling rig parts in significant dollar values.

U.S. firms account for most of the non-Communist world's production of drill bits, and U.S. bits are of significantly higher quality than Soviet counterparts. The Soviets, however, have purchased fewer than 100 U.S. drill bits over the last 5 years. Instead they have opted to purchase the design and equipment for a drill bit plant from Dresser Industries. Nor have drill bit imports from Western Europe been large. The NIMEXE system classifies drill bits into both bits made of base metal and metal carbide. EEC exports of both types have been inconsequential.

Well Completion/Production

Figure 19 shows a steady growth in Soviet imports of pumps for liquids. The leading exporters of these items have been France, West Germany, and the United States. While it is not clear that West German and French pumps are used in the production of oil and gas, Schedule B/E data show that almost all of the U.S. trade is in oil well and oilfield pumps. This is confirmed by articles in *Soviet Business and Trade* that indicate that U.S. companies such as TRW-Reda, Centrilift, and Oil Dynamics have been major exporters of electric submersible pumps for Soviet oilfields.

Finally, Western trade activity in pipe handlers and gas lift equipment was not ascertainable from the trade data due to problems of aggregation. It was apparent from

Figure 19.—Western Energy Trade With U.S.S.R. Pumps for Liquids

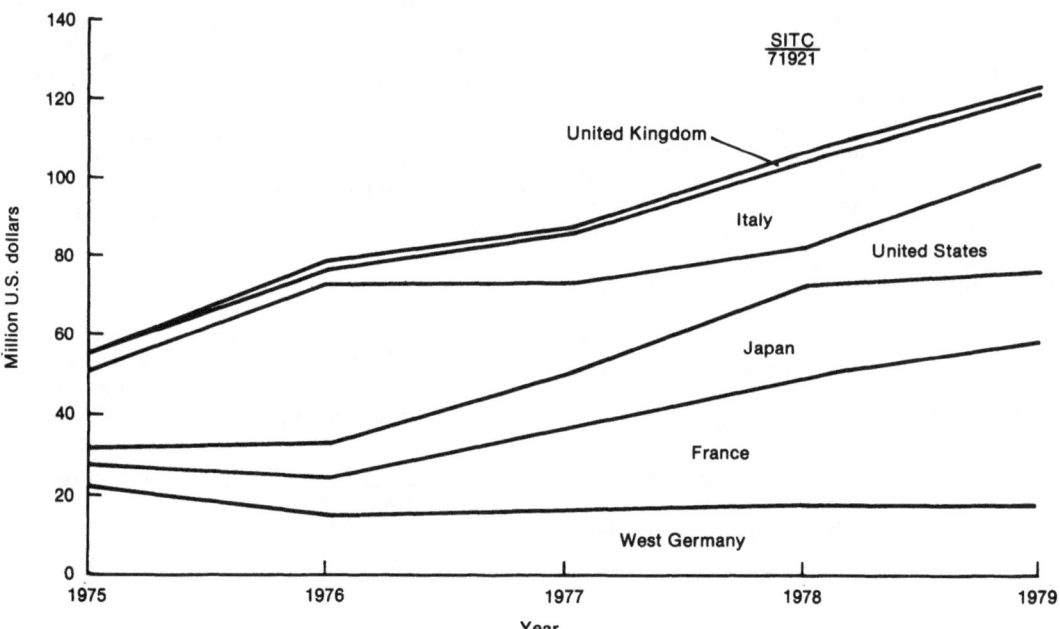

SOURCE: United Nations SITC Data.

articles in *Soviet Business and Trade,* however, that these items have been exported to the Soviets by the United States, Japan, and France: gas lift equipment predominantly by the French, and pipelayers primarily by the United States and Japan. Technip of France has sold gas lift equipment for over 2,000 wells. Pipelayers have been sold to the U.S.S.R. by Caterpillar and International Harvester in the United States and Komatsu in Japan.

The Soviet Union has made sporadic purchases of enhanced recovery equipment and technology. Among the transactions that have appeared in *Soviet Business and Trade* have been the sale of two carbon dioxide (CO_2) plants by Borsig of West Germany, two surfactant plants by Pressindustria of Italy, a surfactant plant as well as chemicals from Sanyo Chemical Industries of Japan, and an alpha-olefin plant from Davy International in Great Britain.[27]

Transportation

The most important commodities in this sector have been large diameter pipe and gas pipeline compressor stations. SITC codes 678 and 672.9 contain a number of subcategories and cover a wide variety of tubes and pipes. Figure 16 shows that in each category, Japan, West Germany, and France are the major Soviet suppliers, and the United States is by far the smallest. In the category of tubes, pipes, and fittings of iron and steel, for instance, 1979 Japanese exports were worth approximately $0.75 billion and West Germany's over $0.5 billion, while U.S. sales amounted to a little over $1 million. Some of this pipe may have been used in the nuclear industry, but it is probably safe to assume that a large portion of it went to the oil and gas sector. The principal companies supplying the pipe are Sumitomo (Japan), Mannesmann (West Germany), Vallourec (France), and Finsider (Italy). **The United States does**

---

[27]*Soviet Business and Trade,* Aug. 15, 1979.

not produce the 56-inch diameter pipe that the Soviets use to construct gas pipelines.[28]

Compressor stations for gas pipelines are another active commodity. The largest exporters of compressor stations to the U.S.S.R. have been Italy, West Germany, and the United States, with Nuovo Pignone of Italy and GE of the United States the major suppliers.[29]

Refining

The U.S.S.R. has purchased refineries and refinery equipment from Japan, Italy, and France, the tendency being to import entire refineries rather than component parts. The primary contribution of the United States in this area has been through Fluor Corp., which provided design and engineering services to Italian and Japanese construction firms.[30]

Offshore

Trade in this area has consisted primarily of sales of offshore drilling rigs and auxiliary vessels and equipment, and the principal suppliers have been Japan and the Netherlands.[31] Rauma-Repola Oy of Finland was recently granted a contract to build three dynamically positioned drill ships for the Soviet Union, to be delivered in 1981 and 1982. The United States has supplied auxiliary equipment for rigs sold to the Soviets, but Soviet purchases in this area have been both moderate and sporadic.[32]

Conclusions

Examination of trade data reveals that the U.S.S.R. has been very selective in the kinds of Western equipment and technology it has purchased to supplement its domestic

---

[28]Interview with J. Brougher, Bureau of East-West Affairs, Department of Commerce.
[29]*Soviet Business and Trade,* Nov. 10, 1976.
[30]*Soviet Business and Trade,* Jan. 31, 1979.
[31]*Soviet Business and Trade,* July 15, 1980; May 21, 1980.
[32]1980-81 Directory of Marine Drilling Rigs, pp. 19-172.

oil and gas equipment. Indeed, the relatively modest imports of may items lead one to suspect that the U.S.S.R. has been supplementing domestic equipment stocks at times of peak demand and/or purchasing the best available product for particularly difficult application. It is equally probable that some items have been procured for laboratory examination and duplication, or to serve as guides to correct specific problems.[33] **The most prominent exception here is Soviet imports of Japanese and West European large diameter pipe, which have been consistently large and which seem to be required because of insufficient domestic production capacity.**

Interestingly, the U.S.S.R. has not purchased many items basic to the petroleum industry. These include magnetometers, gravimeters, mud-pumps, drilling mud, casing cement, engines, pipe insulation, separation equipment, and offshore floating production platforms. These omissions or gaps in trade with Western countries can be interpreted in a number of ways: that the U.S.S.R. and its CMEA partners have an adequate industrial base to supply their needs, even if the result is inefficient by Western standards; that insufficient hard currency has forced priority-setting among Western imports; or that the U.S.S.R. has made a policy decision to be as independent as possible of supplies from the Western countries in certain critical segments of the oil and gas industries. It is most likely that a combination of such factors is at work.

Be that as it may, the following generalizations seem warranted by the data:

- In value terms, by far the largest Soviet purchases from the West have been in the area of iron or steel seamless pipes and tubes (including the large diameter pipe used in Soviet oil and gas lines). Purchases in this area from the United

States have been negligible. By far the largest suppliers have been Japan and West Germany.
- The U.S.S.R. has also purchased substantial amounts of various pumps and gas compression equipment. Here, the United States has had larger market shares. The U.S.S.R. has made only a few large purchases in the area of light vessels, floating docks, etc., which includes offshore drilling rigs. None of the vessels themselves have come from the United States. In 1979, Japan and Sweden were the only large exporters in this category.
- The U.S.S.R. has purchased very few drill bits from the West, apparently preferring to acquire its own additional manufacturing capacity in the form of an entire plant.

FOREIGN AVAILABILITY OF OIL AND GAS INDUSTRY EQUIPMENT AND TECHNOLOGY[34]

Much oil and gas technology originated in the United States, but that technology has lost its American identity over the years through licensed production, wholly owned subsidiaries overseas, and employment of U.S. commodities and expertise worldwide. Other sophisticated technology was developed elsewhere. For example, Schlumberger of France first developed electric well log-

---

[33]There is substantial evidence of duplication. In an interview with OTA, a Vice President of TRW-Reda Pump, Inc., asserted that when one of his technicians toured a Soviet pump plant in 1979, he saw 20-year-old Reda models being produced.

[34]This section is based on trade journals, industry catalogues, *Soviet Business and Trade, The Composite Catalogue of Oil Field Equipment and Services* (Houston, Tex.: Gulf Publishing Co., 1980), and interviews with representatives of the following firms: Gulf Oil Exploration & Production Co., Geosource, Inc., Dresser Industries, BWT World Trade, Cameron Iron Works, Inc., Hughes Tool Co., TRW-Reda Pump, Inc., Brown & Root, Inc., and Williams Bros. Engineering Co. Representatives of these firms provided candid, forthright observations of their past dealings with the U.S.S.R., insight gained during country visits and an appraisal of their foreign competitors. These visits, coupled with the other reported U.S.S.R./Western trade deals, provided a basis on which to judge the availability of Western oil and gas technology. While these sources could not provide complete identification of all possible suppliers, OTA believes that it was acquainted with the most significant suppliers and the strongest competitors to U.S. firms.

ging equipment that is recognized today as the world's standard. Likewise, the steel industries in Western Europe and Japan generally produce products that are as good as, if not better than, those available from the United States—and at lower prices. Technological leads are perishable with time. Licensed manufacturers frequently improve upon designs or manufacturing processes based on local conditions and equipment. Wherever the original development work and design may have been done, ideas soon become general knowledge. The following sections discuss the foreign availability of energy technology that would be useful to the U.S.S.R. in the various phases of oil and gas production and delivery.

Exploration

As noted above, the American firm Geosource has been very active in Soviet trade. Sercel of France has been a strong competitor to Geosource for sales of field data collection and preprocessing centers. Table 46 compares basic parameters of Geosource, Texas Instruments (another U.S. firm), and Sercel products used in seismic work. The table shows that the equipment, although not identical, is similar.

Table 47 identifies the major items of seismic surveying equipment and suppliers around the world. Most items are produced by firms in Western Europe and Japan, and many are available in the Eastern Bloc, although the quality of the latter is questionable and Western equipment tends to be more advanced. The United States may lead technically in one or two items, but the general consensus is that products from Sercel, for example, are capable of performing similar functions. On the other hand, only the United States is able to supply the full range of equipment.

Table 46.—Comparison of U.S. and French Seismic Equipment

	Geosource MDS-10			Texas Instruments DFSV			Sercel 338 B
	Data channels	Sample interval (MS)	Packing density	Data channels	Auxiliary channels	Sample rates	24 Channels @ 1, 2, 4 ms
Data	24	½	1,600	SEG-B			
	24	1	800 or 1,600[a]	To 24	4	1, 2 or 4 ms	
	24	2	800 or 1,600[a]	28	2	1, 2 or 4 ms	
	24	4	800				
Channels	48	1	1,600	48	4	1[b], 2 or 4 ms	48 Channels @ 2, 4 ms
	48	2	800 or 1,600[a]	60	2	2 or 4 ms	
	48	4	800 or 1,600[a]	96	4	2 or 4 ms	
	96	2	1,600	120	4	2 or 4 ms	96 Channels @ 4 ms
	96	4	800 or 1,600[a]	240	4	4 ms	
	Solid state stacking available at all sample rates.			Packing density 800 or 1,600 bpi except (I) 1,600 bpi only.			
Packing density maximum BPI	1,600			1,600			1,600 6,250 for 338IR (IBM recorder)
Number of bits	14 bits plus sign bit 4 bit gain word			14 bits plus sign bit 3 bit gain word			14 bits plus sign bit
Frequency response	2 to 1,000 Hz			3 to 256 Hz			Unknown
Distortion	0.1% maximum @ 0.53V RMS input			0.05%			Less than 0.1% @ .05V input
Tape speed range	Unknown			10 to 120 ips			20 to 92 ips

[a]800 BPI NRZI optional.
[b]1 ms at extra cost to 56 channels.

SOURCE: Office of Technology Assessment.

Table 47.—Manufacture of Seismic Equipment by Country

	U.S.A.	France	W. Ger.	Japan	Canada	Brazil	U.S.S.R.	Bulgaria	E. Ger.	Hungary	Poland	China	U.K.	Israel
Vibrator	X		X											
Vibrator control	X													
Shooter explosive	X													
Recorder field	X	X		X	X		X			X				
Tape transport	X	X					X			X				
Camera CRT	X			X						X				
Cables	X	X	X	X	X	X	X				X			
Connector	X	X			X	X	X							
Geophone	X			X	X		X				X	X		
Airgun (marine)	X													
Marine streamer	X	X	X										X	
Marine positioning	X	X											X	
Seismic computer	X	X	X		X		X	X	X	X				
Array transform processor	X	X												
Plotter	X	X	X							X				X

NOTE: No inferences as to quality or comparability can be made from this table, which merely shows the existence of commercial manufacturers.

SOURCE: Office of Technology Assessment.

Seismic survey data must ultimately be processed by large mainframe, third generation computers with floating point and array processors. The United States has approved the sale of six large computers, with somewhat restricted array processors, for use in the major hydrocarbon producing regions in the U.S.S.R. Two French firms, Ferney-Voltaire and CGG, are known to have supplied sophisticated geophysical software (computer programs) used on the U.S. machines to analyze seismic survey results. An IBM sale included American software.

In sum, the equipment to perform seismic surveys and record seismic data are generally available worldwide. U.S. firms are unique, however, in being able to provide systems displaying the full range of equipment and know-how. U.S. firms also lead in the accuracy of some equipment; and the United States has a substantial lead in computers that process the seismic data. Soviet capabilities in this area are generally 5 to 15 years behind the West and purchases of such equipment would certainly enhance the U.S.S.R.'s seismic work. The degree to which this would necessarily lead to increased oil production in the present decade is unclear, however.

Drilling

The U.S.S.R. has purchased 15 portable drilling rigs from Tamrock Oy of Finland. Canadian sales of $12 million to $32 million each year between 1975 and 1979 were probably also portable rigs, which are known to be produced by the Canadian firm Foremost. Mobile equipment capable of drilling to 20,000 ft is also available from Romania, although it may not perform to advertised specifications.

The U.S.S.R. has imported drill pipe almost entirely from firms in Western Europe and Japan. Prominent suppliers have been Vallourec and Creusôt-Loire of France; Mannesmann of West Germany; Italsider and EFIM of Italy; and Mitsubishi, Mitsui, Sumitomo, Japan Steel Works, and Nippon Kokan of Japan. Japanese pipe is generally considered to be equal to, if not better than, U.S. pipe and is available at a significantly lower price. It is manufactured by the latest methods including inertial welding of the tool joints on the ends of the pipe, and U.S. drillers are buying Japanese pipe to supplement U.S. production capacity.

The United States is the predominant producer of drill bits in the West, with the American firms Hughes, Dresser, Smith, and Reed supplying the vast majority of bits used outside the Communist world. These firms produce the greatest diversity of high-quality bit types for varying underground rock strata. A few diamond bits and core bits are produced by Diament-Boart of Belgium and Tsukimoto Seiki in France, but these cannot substitute directly for rock drill bits. While Tsukimoto produces both diamond and metal bits, its total annual production is very small, approximately 5,000 bits. European bits have a more limited operating capability than their U.S. counterparts and their quality does not match U.S. standards. Creusôt-Loire, SMF Division of France, has recently been purchased by Hughes Tool Co. and the drill bit plant is being modernized to U.S. standards.

The Soviet Union is itself a prodigous producer of drill bits, and it has not purchased Western bits in large quantities. A great deal of publicity has accompanied the sale by Dresser Industries of a tungsen carbide journal bearing drill bit plant to the U.S.S.R., the capabilities of which are discussed in chapter 2. The export license for this plant was recently revoked, but all the technology relating to production machinery, manufacturing processes and metallurgical specifications has already been transferred. The revocation mainly prohibits Dresser from providing onsite training of Soviet technicians once the manufacturing plant is completed. The U.S.S.R. will therefore be forced to resort to trial and error to duplicate Dresser-achieved quality. In sum, the United States enjoys a significant lead in both quality and quantity of rock drill bits. But the U.S.S.R. does not purchase significant quantities of such bits and the sale of the American advantage—if the plant achieves its rated capacity of high-quality bits.

Most of the well-logging equipment currently employed in the U.S.S.R. is copied from U.S. Halliburton "Jeep" single conductor logging tools acquired as part of lend-lease equipment after World War II. The current technology in the West employs multiple conductors (up to seven) to obtain all the desired information on a single pass in the borehole. These multiple conductors significantly expedite complete logging operations. The Soviets have purchased well-logging tools from several U.S. firms (Halliburton, Dresser, Gearhart-Owens) but have not allowed experienced Western firms to enter the U.S.S.R. to provide logging service. The world's leading logging firm, Schlumberger, has a policy of selling only services, not equipment, and it performs 80 percent of the logging services outside the Communist world. Other logging services exist in France, United Kingdom, and West Germany. These firms are generally small, however, without Schlumberger's reputation for quality of service. In sum, the U.S.S.R. substantially lags in logging equipment, but the technology is available outside the United States.

Well Completion/Production

The process of completing a well entails the installation of equipment necessary to isolate the producing zones in the well, extract, and contain the crude oil or gas—well head assemblies (christmas trees, chokes, valves), downhole packers (for both single and multiple zone completion in a single well), and artificial lift equipment (sucker rod pumps, electrical submersible pumps, and gas lift equipment).

The literature reveals few exports of well completion equipment to the U.S.S.R. Sales of well head assemblies have been made by Hübner-Vamag AG in Austria; FMC Europe (Luceat), Cameron de France and Creusot-Loire of France; EFIM of Italy; producers in Romania; and BWT World Trade, CAMCO, Otis Engineering, Cameron Iron Works, FMC Petroleum Equipment Co., and Baker Oil Tools, Inc. of the United States. U.S. industrial representatives generally agree that equipment available overseas provides satisfactory service except under severe conditions, i.e., high pressure and corrosive atmospheres. These problems are usually best served by U.S.-supplied equipment. But such conditions are found infrequently in the major Soviet oil and gas producing regions—less than 5 percent of the time and then principally only in the North Caucasus, the Caspian, and Sakhalin.

Artificial lift equipment is less generally available outside the United States than wellhead equipment. The Soviets are known to produce their own sucker rod pumps and electric submersible pumps. U.S. technicians who have seen Soviet submersible pumps report that they appear to be exact copies of pumps produced by Reda in the United States shortly after World War II. None of the pumps observed were estimated at greater than 200 horsepower (hp). This may be compared to the up to 1,000 hp pumps available in the United States. Soviet pumps also have a considerably shorter life in the well than their U.S. counterparts. Within the U.S.S.R., Soviet pumps reportedly operate 30 to 90 days in the hole while pumps imported from the United States last 120 to 360 days. (American pumps routinely operate in excess of 1 year in U.S. wells before they require service.) U.S. pumps that fail in the U.S.S.R. are often not returned to service. The Soviet Union has consistently refused to allow American service technicians into the field, and the Soviets themselves have insufficient trained personnel and supplies of replacement parts. In the West, a specific pump is "fine tuned" by the manufacturer at the site to optimize usage. Since this has been made impossible in the U.S.S.R., the pumps probably operate inefficiently.

Excluding the U.S.S.R., the world supply of submersible pumps is provided by four U.S. firms. They are TRW-Reda, Hughes-Centrilift (formerly Borg-Warner/Byron-Jackson), Baker-Kobe (formerly FMC), and Oil Dynamics, Inc. Prices of U.S. pumps range from approximately $10,000 for those with small diameters and low power, to $200,000 for the largest and most powerful. Soviet purchases have averaged aproximately $100,000 per unit, suggesting that they are supplementing their own production with the larger units available only in the United States. The U.S.S.R. purchased about 1,500 pumps from the United States between 1974 and 1978, but none have been imported since. This suggests that the Soviets are now supplying their own needs or using other techniques to remove fluid from wells.

One such technique is gas lift, which the U.S.S.R. has in fact used to augment its submersible pumps. The Soviets made a major purchase of gas lift equipment—enough to equip almost 2,400 wells—in 1978 from Technip of France. They have also purchased gas compressors from Dresser Industries and gas lift equipment from CAMCO in the United States. Gas lift is more expensive than pumps per unit volume of oil produced because it requires complicated compression equipment to handle the large volumes of gas it employs. The gas distribution valves and their proper sequencing are the most critical technology required in this technique. These are generally available outside the United States.

Sucker rod pumps, such as those seen dotting the Midwestern and Western United States, are also used to lift oil. Until about 15 years ago, Soviet-made models were commonly beset with bearing failures and cracking of the rods. Through improved metallurgy, the U.S.S.R. seems to have solved these problems and it does not import in this area.

Transportation

The expansion of Soviet pipelines for both oil and gas has benefitted *extensively* from imports from the West. The Soviets have purchased a seamless pipe manufacturing plant with a capacity of 170,000 metric tons per year from Creusôt-Loire in France and a West German group composed of Mannesmann-Demag-Meev. The plant uses the French Vallourec process. The U.S.S.R. has also purchased extensive quantities of finished pipe from Mannesmann and Kloeckner of West Germany; Cie de St. Gobain-Point-a-Mousson in France; Finsider in Italy; and Mitsui, Sumitomo, Nippon, Seiko, Nippon Kokan, Kawasaki, and Itoh of Japan. Japanese steel plate is also used for rolling into pipe in the U.S.S.R. The Soviet Union has purchased pipeline valves from Hübner-Vamag in Austria; Honeywell Gmbh, Borsig Gmbh, and Klaus Union in West Germany; Petrovalves, WAGI SpA, and Grove Italia in Italy; and Kobe Steel and Japan Steel Works in Japan. Clearly this technology is available worldwide.

Pipeline booster pumping stations for oil and gas compressor stations and their related components have been supplied to the U.S.S.R. by Honeywell-Austria Gmbh in Austria; AEG-Kanis Turbinenfabrik, Klaus Union, CooperVulkan Compressor Gmbh in West Germany; Kongsberg Turbinfabrik in Norway; Nuovo Pignone and Worthington SpA of Italy; Sumitomo and Hitachi of Japan; Thomassen of the Netherlands; and John Brown Engineering of Scotland. Several U.S. firms, including Ingersoll-Rand, Dresser, GE, Cooper Industries, and International Harvester, have also exported compression and pumping equipment. GE was selected as a major supplier of compression equipment for the Orenberg gas pipeline, but 75 percent of GE's Orenberg order was filled by firms outside the United States under subcontract to GE. Compression equipment is manufactured worldwide. Some of the more modern designs are derivatives of jet aircraft engines, but the technology is not advanced and is available in Western Europe and Japan.

The U.S.S.R. has also purchased pipeline laying equipment from the West. This usually consists of a crawler tractor with side-mounted support to lower the pipe into a prepared trench. Fiat-Allis Construction Machinery, Inc., of Finland has supplied spare parts for both bulldozers and pipe-

Photo credit: Oil and Gas Journal

Equipment for work on large diameter pipelines

layers. Caterpillar-Mitsubishi and Komatsu of Japan have also exported pipelayers, as has Bunsar in Poland (an International Harvester licensee). International Harvester and Caterpillar have been the major U.S. suppliers of similar equipment. Foremost of Canada has also supplied heavy pipe carrying vehicles. The technology requirements for these vehicles are not advanced and are generally available outside the United States. Although American firms may produce the largest machines and American models may be better suited to work in cold climates, alternative models, especially from Japan, could fulfill Soviet requirements. No information was collected on production capacities in any country. The Soviets seem to be buying these commodities due to shortfalls in their own production.

U.S. firms, namely CRC International and Perry Equipment Corp., have won contracts to supply the U.S.S.R. with pipeline inspection robots, or "pigs," but competitive bidding to supply this type equipment for the Orenberg gas pipeline included Mannesmann and Prenatechnik of West Germany; Primaberg and OeMV of Austria; General Descaling of the United Kingdom; Nippon Kokan of Japan; and Aveary Lawrence of Singapore. Both Prenatechnik and General Descaling have previously sold pipeline inspection pigs to the U.S.S.R. OTA's assessment of the general capabilities of these pigs indicated that foreign equipment is comparable to U.S. models.

In sum, well completion/production equipment, blowout preventers and wellhead assemblies (Christmas trees) designed for very high pressure and/or highly corrosive conditions are available only from U.S. firms. But these types of equipment would be required for only a small percentage of the wells drilled in the U.S.S.R. The United States does maintain a monopoly on quality electric submersible pumps, but the U.S.S.R. has not purchased these for the past 2 years. It *is* purchasing large quanities of pipe and pipeline equipment—which are available in Western Europe and Japan.

Secondary Tertiary Recovery

The Soviets have been experimenting with several enhanced recovery techniques. As noted above, the U.S.S.R. has purchased two CO_2 recuperation/liquefaction plants with a combined capacity of 400,000 tons/year from Borsig Gmbh, a subsidiary of Deutsche Babcock AG, and a chemical surfactant plant (alkyl phenol) with a 100,000-ton/yr capacity from Fried Uhde Gmbh, both of West Germany. A plant capable of producing 250,000 tons of surfactant per year was obtained from Pressindustria in Italy, and other deals have been broached with firms in Japan and England. It is not clear when these facilities will be brought online. In any event, the contribution to overall production will be negligible. The benefits of tertiary recovery techniques are still being explored through testing and experimentation in the West as well as in the U.S.S.R.

More enhanced recovery experience resides in the major oil and service companies operating in the United States than anywhere else in the world. U.S. firms could probably aid the U.S.S.R., if it would allow foreigners to provide technical services. There has as yet been no sign of Soviet interest in such services.

Offshore

After the initiation of the Soviet-Japanese cooperative project on Sakhalin Island (see ch. 11), the U.S.S.R. approached the Gulf Corp. regarding the use of its highly sophisticated survey ship, the *Hollis Hedberg*. The Soviet Union, however, prohibited the use of an American crew in Soviet waters, and Gulf declined to participate. Instead, the U.S.S.R. leased a French survey ship with crew from CGG for 6 months during the summer of 1976. The Soviets also procured from CGG sufficient geophysical equipment to completely outfit two geophysical ships. In the same year, they purchased two ships from Mitsubishi of Japan, and another completely equipped geophysical survey ship has reportedly been bought from Serete Engineering of France. GECO of Norway was hired in 1978 to conduct an offshore survey using 48-channel equipment in the Baltic Sea off the East German-Polish coast. A similar ship, ordered from GECO in 1977, performed surveys during the summer of 1978 in the Barents Sea. These transactions suggest that the U.S.S.R. has been able to acquire substantial Western expertise to develop its offshore fields, with little to no direct participation from the United States.

The United States has, however, provided a third-generation main frame computer, a CDC-Cyber 172, suitable for seismic analysis. This is installed in a computer center on Sakhalin Island. The software for the CDC computer, and for another installed elsewhere in the U.S.S.R., was purchased from the French firms CGG and Ferney-Voltaire. The Sakhalin Island computer facility is used to analyze the marine seismic data acquired by at least two of the Soviet geophysical ships equipped with CGG instrumentation and equipment.

Offshore exploratory drilling in the Sakhalin region was initially performed in 1977 with a semisubmersible rig leased from a Norwegian firm, Fred Olsen & Co. This was subsequently replaced by a Mitsubishi-built semi, the *Hakuryu II.* Additionally, the Japanese consortium has provided several jackup rigs for exploratory drilling off Sakhalin. The drilling rigs are operated by Japanese-trained Soviets, and a Japanese drilling supervisor remains with each rig.

The U.S.S.R. obtained its first mobile offshore rig in 1966 from IHC in the Netherlands. This rig, which was for use in the Caspian Sea, has become the prototype for Soviet domestically produced rigs. Equipment used on Soviet domestically produced offshore rigs is also reported to be of Soviet origin.

In 1976, Armco Steel (U.S.) was granted an export license to provide Rauma-Repola Oy of Finland the necessary technical data to produce three semisubmersible drilling rigs that were to be sold to the U.S.S.R. and assembled at the Astrakhan shipyards on the Caspian Sea. The first semi, the *Kaspmorneft,* was completed for sea trials in August 1979, but is not yet operational. The second, the *Shelf-1,* was ready for sea trials in 1980. The third semi is being modified at the yard based on experience with *Kaspmorneft.* The Soviets have now ordered three dynamically positioned drill ships, also from Rauma-Repola Oy, for exploratory drilling in the Barents and Kara Seas. The dynamic positioning systems are being provided by Kongsberg Vaapenfabriken of Norway. Other competitors included Simrad A/S of Norway and Honeywell of the United States.

The drill ships, as well as many of the other assembled offshore rigs supplied to the U.S.S.R., are largely outfitted with drawworks, prime power, rotary tables, subsea blowout preventors, mud pumps, and cranes made by U.S. firms and their overseas subsidiaries and licensees. Dominant U.S. suppliers are National Supply, Ideco, Continental Emsco, Oilwell, and Gardner-Denver. The main structural platforms for these rigs are made in many shipyards around the world, but U.S. firms produce the majority of the mobile offshore designs. Major U.S. firms here are Bethlehem Steel, Marathon LaTourneau, Levingston, Avondale, Todd, McDermott, and Ingalls. Significant quantities of rigs are also produced in the Netherlands by Verolme, IHC, Rhine-Schelde-Veroime; in Canada by Davie, Halifax and Scott-Lithgow; in Finland by Rauma-Repola Oy; in Japan by Mitsubishi, Mitsui, Hitachi, Sumitomo, IHI, and Nippon Kokan; in Norway by Aker, Nylands, Trosvik and Normarig; and in France by CFEM. Lesser suppliers may be found in Taiwan, Italy, United Kingdom, West Germany, Venezuela, Scotland, Hong Kong, Singapore, Australia, Sweden, Korea, and Spain. In the Communist world, rigs have been constructed by the People's Republic of China, Romania, and the U.S.S.R. These are usually copies of Western rigs.

The prime power used on the rigs is usually diesel-electric, and the suppliers include all the world's major diesel manufacturers:

General Electric/EMD, Caterpillar, Fairbanks-Morse, Detroit Diesel Allison, SACM and Alsthom-Atlantique/SEMT Pielstick (France), and Paxman and MTU (West Germany). Dynamic positioning control systems have been supplied to the U.S.S.R. by Honeywell and Delco in the United States, Simrad A/S and Kongsberg Vaapenfabriken in Norway and CIT Alcatel in France. Mechanical anchoring systems and cranes have been provided by U.S., Japanese, West German, Norwegian, and French firms. Diving equipment on the rigs has been most frequently provided by COMEX in France, but Ocean Systems (U.S.) has also exported in this area. Subsea blowout preventors appear to be available only from U.S. suppliers, including Cameron Iron Works, Hydril, and NL Industries. These firms are also the sole suppliers of subsea well completion stacks. A leading U.S. supplier indicated that his firm had provided subsea blowout preventors for all Soviet offshore rigs and ships.

The offshore oil and gas industry is a classic example of the worldwide nature of this technology. While the earliest offshore activities were concentrated off the Gulf and California coasts of the United States, the industry is now active in other parts of the Caribbean, off Brazil, West Africa, the North Sea, Asia, and the North Slope of Alaska and Canada. The most stringent requirements for offshore technology are represented by North Sea and North Slope activities, and both U.S. and European firms are benefitting from this experience. In sum, while the U.S.S.R. sorely needs offshore equipment, with the exception of drawworks, rotary tables, mud pumps, subsea blowout preventors, and well completion stacks—the narrow range of items in which the United States still maintains a monopoly or lead—it can acquire quality items in Western Europe and Japan.

Engineering firms are perhaps the most critical element in successful offshore operations. U.S. firms clearly have the greatest breadth of experience in this area, but numerous foreign companies can supply most individual aspects of the know-how. Table 48 lists major foreign offshore engineering firms that are able to perform all or part of the engineering design required in defining and establishing a new offshore producing field and providing associated equipment.

Teamed together, the firms listed in table 48 could provide the same capability that is resident in U.S. firms like Brown & Root, Inc., and J. Ray McDermott. In fact, there has been substantial teaming for the North Sea and the Beaufort Sea.

Refining

Although current capacity seems to supply current needs, improved refining technology and equipment may well be required in the U.S.S.R. during this decade. Increased natural gas production will require processing of vast amounts of natural gas. Gas processing complexes have been sold to the U.S.S.R. by Technip and Construcion Metalliques de Provence in France and the Japan Steel Works, Nichiman Jitsugyo & Co., Ltd., and Mitsubishi in Japan. The Fluor Corp. of the United States has pro-

Table 48.—Non-U.S. Offshore Engineering Firms

Norway:
    Aker, Kvaener
Netherlands:
    Herrema
United Kingdom
    Worley
    Adkins
    Halgrove-Eubank
    Matthew-Hall
    Davey Powergas
    Willey
    Lawrence & Allison
France:
    E.T.P.M.
    U.I.E.
    Serete
Italy:
    Technomare
    Saipem
    Snamergetti
Mexico:
    Protectors
Spain:
    Initel

SOURCE: Brown and Root, Inc.

vided engineering services and technical assistance on at least two Japanese sales.

Mitsubishi has sold an oil refinery to the U.S.S.R. Competitors for that sale were reported to be Linde AG in West Germany and C-E Lummus in the United States. Other Soviet imports of oil refining equipment during the period 1975-78 have included deals worth over 190 million rubles from Japan, 165 million rubles from East Germany, 76 million rubles from France, 43 million rubles from Czechoslovakia, 1.5 million rubles from Italy, and 1 million rubles from the United Kingdom. The U.S.S.R. will probably continue to seek assistance in this area, but the technology to produce and operate an adequate refinery is not advanced and is available on a worldwide basis. (This includes the use of hydro and catalytic cracking to break up the heavy hydrocarbon molecules to form the lighter molecules in motor fuels and aviation gasoline.) Modern U.S. and Western European refineries now have sophisticated computer controls, which the Soviets lack. These controls improve efficiency but are not generally integral to the basic technology. In short, the technology required for refining crude oil is available in several Western countries, and the U.S.S.R. is certainly not dependent on the United States for refining technology to meet its near-term needs.

TECHNOLOGY TRANSFER AND "FOREIGN AVAILABILITY"

Until recently, the United States had been the sole source of "state-of-the-art" technology in virtually all technological areas.

Separation installations at a West Siberian gas compression station

America's technological lead was largely attributable to two factors: it outspent most other countries in research and development and the wealth of technological "know-how;" and equipment produced by U.S. R&D had remained resident in U.S. corporations. The rise of the multinational corporations during the 1960's altered this state of affairs.

In their quest for expanded markets and higher profit margins, the multinationals have transferred significant quantities of advanced technological know-how and equipment. This process is nowhere more evident than in the oil industry, where the first true multinationals emerged. The international nature of oil and gas exploration and production provided a natural incentive for oil industries to adopt a global approach to the dispersion of know-how and equipment. As far back as the 1940's, oil companies perceived a need for local sources of equipment and technology. Transfers of technology between U.S. firms and other Western concerns have taken place in nearly all of the key technological areas of the oil and gas industry. The data also show several transfers of American technological know-how directly to the Soviets; i.e., the Dresser drill bit plant and the Armco licensing of offshore rigs. The result of these transfers has been to significantly reduce the number of areas in which the United States is a sole source of supply.

The three principal vehicles for transfer of technology are wholly owned subsidiaries, affiliates, and licensing of production processes and know-how. Parent corporations have availed themselves of all three methods. Each provides varying levels of technology transfer, and differing amounts of control which the parent organization retains over the end use of technology.

Transfers of technology have affected the position of the United States as sole source in two ways. An initial technology transfer spreads U.S. know-how throughout the world. Once a foreign concern acquires a technological base, it can expand upon this base and develop similar product lines on its own. One example of this process in the case of GE's licensing of compressor technology to Nuovo Pignone of Italy. Shortly after acquiring the technology, Nuovo Pignone was producing its own gas pipeline compressor stations in competition with GE's line. In 1976 Nuovo Pignone won a large Soviet contract for pipeline compressors over a competing bid from GE. Nuovo Pignone is now an important supplier of this type of equipment.

The United States still leads in some areas, however. These are discussed in the following sections.

Exploration

In exploration, the United States holds a unique position in that it is able to provide the complete set of equipment, computers, and software needed to model subsurface structures and locate oil and gas. The lead of American firms in this area may be attributed to the fact that their international subsidiaries, affiliates, and licensees do the vast majority of exploration in the West. Many of the components of these systems are available elsewhere, especially from the Japanese, but the most advanced expertise resides in the United States. The United States also has a slight edge in hydrophone and geophone accuracy.

Although other sources exist for the latest technology in integrated circuits, the United States currently is the only source of minicomputers used to rapidly process and initially analyze seismic data in the field. This capability allows up to 24-hour turnaround for initial seismic results (v. a more normal 90-day turnaround of complete results from a central data processing center) to alert the field crew to particularly promising locations or to inadequate data that should be repeated. This is "state-of-the-art" technology, however, and is still used by only a small number of firms, even in the United States.

Several U.S. firms manufacture advanced geophones and hydrophones that exhibit small, incremental advances over items available in the United Kingdom, France, or Germany. The foreign models, however, can certainly perform the necessary tasks.

Computers

In the United States, computers have become an integral part of the business of finding, extracting, processing, and delivering energy. Computers do not play as pervasive a role in the U.S.S.R. both because their value was recognized later and because of systemic problems in organizing and realizing the production of hardware and software. Soviet energy industries have relied extensively on indirect transfers, although a number of direct transfers have played an appreciable role in selected areas, most notably in geophysical processing.

OTA has isolated a number of key areas where the Soviets lag behind the United States in computing. These are summarized in figure 20. The first is hardware. Although other Western nations, notably Japan, can supply equivalents, the United States still leads in supplying integrated systems for these applications. The United States also holds a commanding lead in software and software development techniques, although the French have supplied the U.S.S.R. with some geophysical software. It is in the development of integrated systems and software that Soviet systemic problems have had the greatest impact.[35] Because of the improving Soviet hardware base, there will be less overt pressure to buy from the West, but indirect Soviet reliance on Western, and in particular, American developments will very likely continue.

Other Oil and Gas Equipment

U.S. drill bit manufacturers have the most extensive variety of bits in the world and a near monopoly on bit sales outside the Communist countries. Only a few small bit suppliers exist outside the United States. Most of these specialize in diamond-coated bits that have a relatively narrow range of application. In any case, the experience base of U.S. firms and the proven quality and durability of their products clearly establish them as world leaders. Even with the sale of a U.S. bit manufacturing plant to the U.S.S.R., it is doubtful that the Soviets can produce comparable quality bits without extensive one-on-one training by U.S. technicians in the manufacturing steps and quality assurance provisions. Sustaining high-quality metallurgical raw materials will also be necessary to achieve a capability equal to that of the United States. The required knowledge and experience can be gained through trial and error, but several years may be required to achieve the capability that a few months of onsite training might provide.

The world's major purveyor of well logging services is Schlumberger, a French firm, and logging equipment was first developed in France. Nevertheless, U.S. firms do excel in the electronic technology and interpretation experience necessary to obtain high-quality well-logging survey results, and improvements made in the United States, with U.S. technology and based on the extensive U.S. drilling and logging experience, have been important.

In well completion and production, several items appear to be unique to the United States: blowout preventors and wellhead assemblies designed for either very high-pressure service (above 10,000 psi) or for use in highly corrosive hydrogen sulfide environments, and electric submersible pumps. Although firms outside the United States can provide less capable units, oil field specialists everywhere recognize U.S. blowout preventors and Christmas trees as the ultimate in quality. The electric submersible pumps needed to produce high volume wells are exclusively available in the United States. U.S. pumps have proven down-hole longevity when properly tailored to the well and served with reliable power. The size range of 25 to 1,000 horsepower exceeds by a

---

[35]S. E. Goodman, "Soviet Software: Progress and Problems," *Advances in Computers*, vol. 18, 1979.

Figure 20.—Relative Importance to Soviet Energy Industries of Computer-Related Technologies in Which the U.S.S.R. Lags the United States

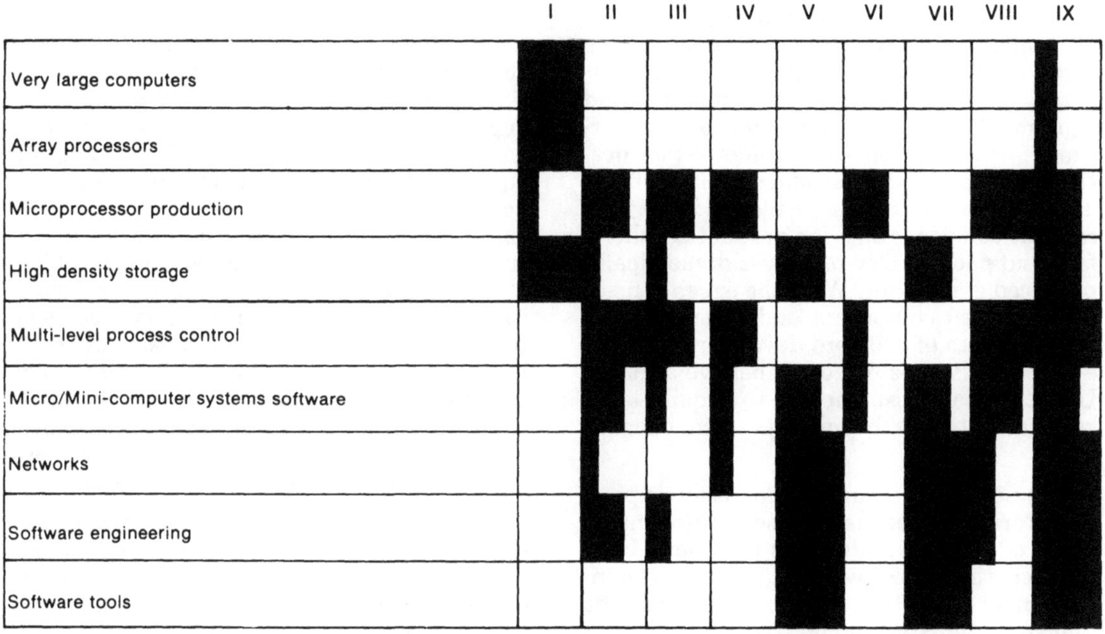

	I	II	III	IV	V	VI	VII	VIII	IX
Very large computers	■								
Array processors	■								
Microprocessor production		■	■	■		■		■	■
High density storage	■		■		■	■	■		■
Multi-level process control			■	■	■			■	■
Micro/Mini-computer systems software				■	■	■	■	■	■
Networks			■	■	■		■	■	■
Software engineering					■	■	■	■	■
Software tools					■	■	■	■	■

Legend: I. Geophysical Processing
II. Oil and Gas Production
III. Refining and Petrochemicals
IV. Pipelines
V. Economic Management (Oil and Gas)
VI. Coal Mining
VII. Economic Management (Coal)
VIII. Electric Power Process Control
IX. Electric Power and Grid Management

wide margin the range of pumps produced in the U.S.S.R.

The area of enhanced recovery equipment and know-how is difficult to evaluate. The Soviet Union has purchased entire plants to produce chemical surfactants and CO_2 to aid in the extraction of heavy oil or oil tightly bound in rock. Soviet literature has also reported experimentation with hot water and steam injection, fire flooding, and even underground nuclear explosions to achieve improved recovery of oil from a reservoir. These techniques are still not in widespread commercial use even in the United States. Nevertheless, the United States has had more experience with technical approaches to achieve improved recovery of crude oil than any other nation. It is reasonable to conclude, therefore, that the United States is the sole source of substantial experience in tertiary recovery methods—if the U.S.S.R.

were to seek that type of service. Thus far, it has not.

The final area where the United States is a sole source supplier is in subsea blowout preventors, marine draw-works, mud pumps, rotary tables, and wellhead completion assemblies used in offshore operations. Any such items with significant capacity ratings are available only from U.S. manufacturers. But the current proliferation of licensees for the manufacture of platform drilling equipment; i.e., draw-works, mud pumps, rotary tables, etc., may soon effectively remove those items from the sole source list.

In sum, of the thousands of pieces of equipment used to find, extract, and produce oil and gas, only a handful are unique to the United States. This finding reflects the dispersal of technology that occurs with multinational companies and the worldwide nature of the petroleum business.

SUMMARY AND CONCLUSIONS

In 1979, the Soviet Union devoted some $3.4 billion, approximately 22 percent of its trade with its major Western trading partners, to energy-related technology and equipment. The vast majority of its purchases—worth about $2.7 billion—was destined for the Soviet oil and gas industries. These imports clearly have played important roles in compensating for production shortfalls and poor quality of Soviet domestically produced equipment. With the exceptions of sophisticated computers and software, and some aspects of offshore development, however, there is little reason to believe that the U.S.S.R. uses these imports to acquire technologies hitherto beyond its own capabilities.

The Soviet coal, nuclear, and electric power sectors have been largely self-sufficient. It is the oil and gas industries that have been most characterized over the years by the involvement of the West, and there is no doubt that the industry would benefit substantially from Western imports on a massive scale. Whether from lack of hard currency, a deliberate policy of self-reliance, the reluctance or inability of Western firms to sell, or all three, Soviet purchases have generally remained relatively modest and strategically targeted. An important exception is large diameter pipe, an item that will be crucial to energy development in the present decade. Here, the U.S.S.R. is quite dependent on firms in Japan and West Germany. **The United States does not produce pipe in the diameter required by the U.S.S.R.**

Indeed, **the United States is not the predominant supplier of most energy-related items recently imported by the Soviet Union.** The foreign availability sections of this chapter have identified numerous foreign firms supplying oil and gas equipment to the U.S.S.R., reinforcing the theme of the international nature of the major oil and gas companies. Newly developed technology has generally been diffused throughout the world through an extensive network of subsidiaries, affiliates, and licensees.

There are a few items of oil and gas equipment which are either solely available from the United States or for which the United States is generally considered a preferred supplier: integrated computer systems and software; rock drill bits; electric well logging equipment; blowout preventors; and wellhead completion assemblies for high pressure, corrosive or subsea applications; marine draw-works; mud pumps; rotary tables; electric submersible pumps; and a substantial experience base in tertiary recovery techniques. **With the exception of computers, however, the U.S.S.R. is either not purchasing these items, is on its way to acquiring the capacity to produce them domestically, or has demonstrated that they are not essential to oil and gas production.**

This study reinforces the international extent of the oil and gas industry. The spread of technology that was originally developed in the United States has been enhanced through the growth of multinational companies that supply equipment to all users. This results in relatively few items that remain exclusively available from U.S. sources. The United States continues to represent the ultimate in quality or capability in some equipment, but the extent of that lead is diminishing. The United States still leads in exploration, drilling, offshore, well completion, enhanced recovery, and operations in extreme geologic conditions. But the item most badly needed by the U.S.S.R.— large diameter pipe—is available from Japan, West Germany, France, and Italy. The United States retains the best reputation as the supplier of pipeline pumping and compressor stations, and, in particular, for the turbine drive units that power them. But **the Soviet Union can and does obtain most of what it needs for continued development of its oil and gas resources from sources outside the United States. In short, U.S. industry could assist the U.S.S.R., but to make a significant impact the assistance would have to be massive—and unprecedented.**

Appendix A. – Energy Corporation Affiliations

Parent company	Country	Products	Divisions	Subsidiaries	Licenses
ITT	United States	Oil seal equipment		Gallino (Italy)	
Texas International	United States	Oil seal equipment, High speed presses		Rockwell Machine Tool Ltd. (Britain) Matrix Engineering Ltd. (Scotland)	
Marion Power Shovel Co.	United States	Crawler cranes, pile drivers, Mining shovels, large diameter pipes, steel piping, compressor equipment			Sumitomo (Japan)
Deutsche Babcock & Wilcox	West Germany	Ball valves for pipelines	Borsig GmbH		
Studebaker-Worthington, Inc.	United States	Booster pumps, concrete pumps, high pressure pumps	Worthington SpA (Italy) Division of Worthington Pump Inc. (U.S.)		
Int'l Systems & Controls Corp.	United States	Natural gas filtration equipment	Black, Syvallo and Bryson		
George Kent Group	United Kingdom	Turbine meters, readout instruments		Kent France S.A.	
Harnischfeger	United States	Crawler cranes, pile drivers			Kobe Steel (Japan)
Bucyrus-Erie	United States	Crawler cranes, pile drivers			Komatsu (Japan)
Baker Trading Corp.	United States	Wellhead equipment, testing equipment for wildcat wellheads, workover rigs, drilling test equipment	Baker Division	Lynes Inc.	
TRW	United States	Submersible pumping units, pumps, cable	TRW-Reda Pump Inc. TRW Crescent Wire and Cable Division		
Creusôt-Loire S.A.	France	Seamless pipe plant		Creusôt-Loire Enterprises (Licensee of SMF International Member)	
Joy Manufacturing Co.	United States	Power tongs	Hillman-Kelly		
ASEA	Sweden	Welding line (automatic)		ESAB	

Parent company	Country	Products	Divisions	Subsidiaries	Licenses
J. Ray McDermott & Co.	United States	Propane coolers, pipeline coolers		Hudson Products Corp. Licensee of Creusot-Loire (France), Hudson Italiana SpA (Italy)	
Cooper-Besemer Co.	United States			Chiyoda Co. Ltd. (Japan)	
Finmeccanica SpA	Italy	Valves		WAGI International SpA (Italy)	
Borg-Warner Corp.	United States	Submersible pumps		Centrilift Inc.	
W-K-M Valve Group	United States	Gate valves, wellheads			Hübner-Vamag AG & Co. (Austria)
Dresser Industries	United States	Mining shovels and blast hole drills; compressors for pipe line	Marion Power Shovel, Clark Division		Sumitomo Heavy Industries
Grove Valve & Regulator Co.	United States	Gate and ball valves			Japan Steel Works
Armco Steel	United States	Machine for semisubmersible offshore drilling rigs, semisubmersible rigs		National Supply	U.S.S.R. Ministry of Shipbuilding
U.S. Steel	United States	Oilwell cementing	Western Rock Bit and Oilwell Supply		
International Harvester Co.	United States	Standby power generating equipment turbines	Solar		
VOP Inc.	United States	Large dry gas scrubbers		VOP Ltd., U.K.	
Perry Equipment Co.	United States	Pig launching/receiving station			Sirtech (Italy)
Rockwell International	United States	Pipeline metering stations		Robsa (Neth.)	
Westinghouse Elec. Co.	United States	Compressor stations			Mitsubishi
Big Three Industries of Houston	United States	Welding positioners	Ransome Co.		
Stewart & Stevenson	United States	Blow-out prevention controls	Koomey Division		
Cameron Iron Works	United States	Christmas trees, stainless steel wellhead equipment	Cameron DeFrance		
FMC Corp.	United States	Christmas trees, stainless steel wellhead equipment	FMC Europe (Luceat)		
Schenck GmbH	West Germany	Screens for coal washing plant			Japan Kurimoto Iron Works Co.
McNally-Pittsburgh Manuf. Corp.	United States	Flo-driers			Sumitomo Heavy Industry

Parent company	Country	Products	Divisions	Subsidiaries	Licenses
General Electric	United States	Compressors, gas pipeline, turbines, automation equipment, compressor stations	General Electric of Britain		Nuovo Pignone (Italy) Mitsubishi Heavy Ind. AEG-Kanis (FRG) John Brown Engineering Ltd. (Scotland) Thomassen Holland (Neth.) AEG-Telefunken (FRG) Mannesmann (FRG) Hitachi (Japan)

All these associates have worked with G.E. on gas turbines. Under the agreement, G.E. supplied rotating parts and the associates supplied stationary parts and compressor to G.E. specs.

Mannesmann supplied engineering and design, procurement, installation and training services. |
Smith Int'l Inc.	United States	Vertical drill	Caldwell Division		
Kendavis Industries Int'l, Inc.	United States	Triple-joining plants for wide-diameter pipe mining dump trucks M-200s	Mid-Continental Equipment Co. Unit Rig and Equipment Co.		
Cooper Industries Inc.	United States	Centrifugal compressors	Cooper Energy Services	Cooper-Vulkan Compressor GmbH (W. Germany Joint Venture)	Creusot-Loire (France) Kawasaki Heavy Industries, Ltd. (Japan)
Crutcher Resources Corp.	United States	Spare parts - large diameter pipeline and welding equipment leases welding systems	CRC Cross CRC Automatic Welding CRC Int'l		
Honeywell-Bull, Inc.	United States	Control and measuring devices	Honeywell Austria GmbH (Austria)		
Fiat Group	Italy	Flexible hose, flexible hose expansion joints			Gilardini SpA (Member)
Geosource Inc.	United States	Digital display system seismic field recorder photo dot digital plotting system	ETL/Mandrell Products Division Petty Ray Geophysical		

Parent company	Country	Products	Divisions	Subsidiaries	Licenses
Grove Valve & Regulator Co.	United States	Flex-flo valves		Italian Affiliate	
Caterpillar Tractor Co.	United States	Spare parts - bulldozers and pipelayers	Fiat-Allis Construction Machinery, Inc. (Witractor) (Finland)		

Appendix B. – Trade by Company

The following table presents a sampling of the dealings of various Western companies involved in the export of energy-related equipment to the U.S.S.R. The data was obtained from a search of the bimonthly publication Soviet Business and Trade (SB&T) for the period 1975-80.

SB&T draws on a variety of sources, including the Soviet News Agency TASS, to gather information on Soviet trade. The staff ensures the accuracy of the reported deals through a system of cross-checking sources and phone verification with U.S. companies.

According to its publisher, fully one-half of the subscribers of SB&T are found in countries outside the United States, especially Western Europe and the Soviet Union. The subscribers provide feedback as to the accuracy of the reporting. A concerted effort is made by the editorial staff to provide a representative sample of Soviet purchases across the spectrum of both technological areas and supplying countries.

U.S. Government agencies, such as the Department of Commerce, the Central Intelligence Agency, and the Department of Energy, also subscribe to SB&T. U.S. industry representatives have indicated that information about their firms' activities is generally accurate, and they are usually contacted in advance of publication regarding the accuracy of a reported item.

No attempt has been made to validate total authenticity or completeness of the trade data contained in SB&T. OTA is confident that the major trade deals from SB&T referred to in the course of this study are factual and accurately represent the availability of energy technology and commodities from sources outside the United States. It is possible, however, that transactions recorded here may not have been consummated, or that their terms may have changed.

Table B-1.—Trade by Company—Oil and Gas

Equipment area *Exploration*

Suppliers	Country	Product	Year	Value
Foremost Industries	Canada	73 tracked geophysical survey vehicles [under subcontract to Geosource (U.S.)]	1979	
Potter Test	Canada	Portable production testing equipment		
Serete Engineering	France	1 geological survey ship	1980	$113 million
Ferneg - Voltaire	France	Geophysical software to be used on CDCs Cyber 172s	1979	
CIE Generale Geophysique	France	Special geophysical software used on the CDC Cyber 172s	1979	
CIE Generale Geophysique	France	6 month lease of a complete geophysical ship and crew	1976	
CIE Generale Geophysique	France	Digital seismographic recorders, magnetometers, gravity meters, cables and hydrophones for 2 geophysical ships	1976	$14 million
Stere	France	Underwater prospecting equipment	1975	
Comex	France	Deep sea diving equipment		
Mitsubishi Corp.	Japan	2 geophysical ships (using the geophysical equipment bought from CIE above)	1976	$2.5 million
Geosource Inc.	United States	Command II field processing systems	1979	$6 million
Control Data Corp.	United States	2 Cyber 172-4 computers	1979	$12.1 million
Control Data Corp.	United States	1 Cyber 73; 1 Cyber 172 computer		

Table B-1.—Trade by Company—Oil and Gas (Continued)

Equipment area **Exploration**

Suppliers	Country	Product	Year	Value
Geosource Inc.	United States	13 photodot automated digital display	1979	$7 million
Gearhart-Owen, Inc.	United States	Cooperative agreement on the production of direct digital well logging equipment	1978	
Magnavox Labs	United States	5 navigation systems for satellite pinpointing of geological teams	1975	
Geosource Inc.	United States	Seismic field recorder (manufactured by Geosource, Inc., ETL/Mandrell Products Division)	1975	$370,000
Petty Ray Geophysical Division of Geosource	United States	Photodot digital plotting system	1975	$300,000
Geospace Corp.	United States	Seismic plotting system with geophones	1975	$250,000
Lynes Inc.	United States	Testing equipment for wildcat wellheads	1975	$2.5 million
Mertz Inc.	United States	24 servo-hydraulic vibrator systems		$6 million
IBM Trade Development S.A. with Western Geophysical	United States France	IBM 370-148 and an array processor	1979	
Schlumberger S.A. Halliburton Services Inc. Dresser Industries	France United States	Help locate hydrocarbon reserves	1978	

Drilling

Suppliers	Country	Product	Year	Value
Maschinenfabrik Heid	Austria	Machine tools for making couplings and adapters for oilwell casing and drill pipe	1979	Sch 150 million
Tamrock Oy	Finland	15 crawler mounted drilling rigs	1978	
Airan	Finland	Drilling equipment	1975	$100,000
SMF International	France	400 kellies	1978	
Vallourec Export	France	Well head casing	1975	
Wotan Werke	West Germany	Heavy drilling equipment	1980	R4 million
Japanese Consortium	Japan	200,000 tons of seamless pipe for oil wells; to be delivered between October 1980 and March 1981	1980	
Sodeco	Japan	Casing, drill pipe, bits and clay	1976	$2 million
Baker Trading Co.	United States	Drilling test equipment	1979	$1.6 million
Farr International	United States	10 power tongs and a diesel/hydraulic power system for each	1979	$1 million
Halliburton Services	United States	Cementing systems	1978	$3 million
Joy Manufacturing	United States	Power tongs	1978	
Ekel Manufacturing Co.	United States	20 power tongs competitors: Farr International (U.S.) Joy Manufacturing Co. (U.S.)	1978	

Table B-1.—Trade by Company—Oil and Gas (Continued)

Equipment area **Drilling**

Suppliers	Country	Product	Year	Value
Dresser Industries Inc.	United States	Equip a new addition to an existing rock drill bit plant	1978	$147 million
Hercules Inc.	United States	Mud additives	1978	
Stewart & Stevenson's Koomey Division	United States	6 blow-out prevention controls	1975	
Drilco	United States	Degasser and pipe inspector	1975	
Well Completion/Production				
Hübner-Vamag AG	Austria	130 complete oil well head assemblies	1979	$14.5 million
Hübner-Vamag AG	Austria	155 natural gas drill hole plugs and production vanes	1979	$11.5 million
Hübner-Vamag AG	Austria	1,000 single slab gate valves for pipelines and well heads	1978	Sch 100 million
Hübner-Vamag AG	Austria	80 well heads	1978	
Hübner-Vamag AG	Austria	70 2.5 meter diameter ball valves for duty down to –55°C	1976	
Hübner-Vamag	Austria	Ball cock and tilt check valves for gas pipelines	1976	$5.2 million
Honeywell Austria GmbH	Austria	129 units of control and measuring devices for Orenberg line equipment built by British Sereck Controls Ltd.; UK)	1976	$9 million
Dresser Industries Ltd.	Canada	42 compressor units 21 - used for gas lift 21 - used in fire flooding	1978	$30 million
Foremost Industries Ltd.	Canada	30 metric ton payload husky 8 vehicles (pipe carriers)	1976	$4 million
Fiat-Allis Construction Machinery, Inc. (Witractor)	Finland	Spare parts for bulldozers and pipelayers	1978	R241,000
Vallourec	France	152,000 tons of large diameter pipe in 1980	1980	
Entrepose S.A.	France	Line pipe (actually supplied by Vallourec Export S.A.)	1978	
CIE Francaise d'Etudes de Constructions (Technip)	France	Gas lift equipment for 2,371 wells	1978	Fr 835 million
Cie de St. Gobain-Point-a-Mousson & Vallourec Export S.A.	France	Steel line pipe	1977	$70 million
FMC Europe (Luceat) and Cameron de France	France	Christmas trees and stainless steel wellhead equipment	1975	$6 million
Honeywell GmbH	West Germany	Large diameter pipeline valves	1980	
Borsig GmbH	West Germany	Large diameter pipeline valves	1980	
Klaus Union	West Germany	Pipeline fittings and ball valves	1979	DM 1.3 million
AEG-Kanis Turbinenfabrik	West Germany	17 gas compressor stations	1978	

Table B-1.—Trade by Company—Oil and Gas (Continued)

Equipment area **Well Completion/Production**

Suppliers	Country	Product	Year	Value
Klaus Union	West Germany	20 multipurpose pumps	1977	$200,000
Borsig GmbH	West Germany	200 ball valves for pipelines	1976	$8.8 million
Mannesmann Rohrwerke AG	West Germany	3.5 million tons of wide diameter steel pipe	1976	
Cooper-Vulkan Compressor GmbH	West Germany	15, RF2BB-30 centrifugal compressors (a joint venture with Bremer Vulkan Schiffbau und Maschinenfabrik)	1976	
Gebr Windhosst	West Germany	Fire prevention gear for Orenberg line 123 units	1976	$3.2 million
Kloeckner	West Germany	32,000 tons of large diameter pipe	n.a.	
Hudson Italiana SpA	Italy	32 propane coolers for severe climatic conditions	1978	
Nuovo Pignone	Italy	1. Automation equipment; gas compression plant	1978	
		2. Remote control equipment for gas gathering and transmission system	1978	$1.5 million
		3. 5 compressor stations	1978	$150 million
Petrovalves SpA	Italy	400 check valves for oil and gas pipelines, with diameters from 1,000 to 700 MM	1978	$6 million
WAGI SpA	Italy	Pipeline valves	1978	
Worthington SpA	Italy	20 booster pumps for pipeline	1976	
Nuovo Pignone	Italy	35 compressors for Orenberg line	1976	
Finsider	Italy	2.5 million tons of large diameter pipe	1975	
Finsider	Italy	Large diameter pipe	1974	$1.5 billion
Grove Italia	Italy	Ball valves for oil and gas pipelines	1979	
Mitsui & Co.	Japan	200,000 tons of 70 kg/mm² grade steel plates for production of wide diameter pipe	1979	Y20 billion
Kobe Steel, Ltd.	Japan	230 large diameter ball valves for gas pipelines	1979	$6.8 million
Sumitomo Corp.	Japan	Steel piping, pumping and compressor equipment, large diameter pipe	1978	$130 million
Sumitomo Metal Industries Nippon Seiko Nippon Kokorn Kawasaki Steel	Japan	500,000 tons 1,400 mm steel pipe for pipelines	1976	
C. Itoh & Co., Ltd.	Japan	Wide diameter steel pipe	1976	
Yamamoto Suiatsu Kogyosho, Ltd.	Japan	Pipe binding equipment	1976	$4.3 million
Hitachi Ltd.	Japan	Five 10,000 kW gas turbine compressor units	1976	
Japan Steel Works	Japan	56" valves for Orenberg line (subcontract to Perry)	1976	$9 million
Caterpillar-Mitsubishi	Japan	193 D-6 bulldozers/pipelayers	1975	$16 million
Sumitomo Corp.	Japan	400,000 metric tons of pipes, both large diameter and seamless	1980	

Table B-1.—Trade by Company—Oil and Gas (Continued)

Equipment area   ***Well Completion/Production***

Suppliers	Country	Product	Year	Value
Robsa	Netherlands	Pipeline metering stations, using Perry (U.S.) flow measuring equipment (Robsa is a subsidiary of Rockwell International; U.S.)	1976	$5 million
Thomassen	Netherlands	10 turbines for Orenberg (under subcontract to G.E.)	1975	$10 million
Kongsberg Turbinfabrik	Norway	Standby turbine generators for Orenberg line (a subcontract for Nuovo Pignone - Italy)	1976	
Bunsar	Poland	34 pipe layers (International Harvester Co. supplied dimensions, metallurgical specs, tooling and machinery techniques, quality control and assembly methods. International Harvester receives a royalty on each vehicle sold to a third party.)	1976	
John Brown Engineering	Scotland	33 gas turbine compressor units for Orenberg line (J.B.E. as a manufacturing associate of G.E.)	1976	$47 million
TRW-Reda Pump, Inc.	United States	90 submersible pumps	1979	$10.5 million
Ingersoll-Rand Co.	United States	Gas pipeline compressors	1979	
Baker World Trade	United States	Down-hole completion equipment, including wire-line, packers, safety valves and primary cementing equipment for 31 gas wells. About half are designed for extreme cold weather operations (–65°) for large diameter, large volume production.	1979	$2.5 million
Camco	United States	2,216 gas lift and completion units and 80 wire line units	1979	$36.1 million
Otis Engineering	United States	Down-hole completion and wire line equipment for 101 gas wells	1979	$7 million
ODI Inc.	United States	Submersible oil pumps	1978	$2 million
TRW-Reda Pump, Inc. and Borg Warner	United States	Submersible oil pumps	1978	$33.5 million
Oil Dynamics	United States	20 submersible pumps	1978	$2 million
TRW-Reda Pump, Inc.	United States	90 submersible pumps	1978	$10.5 million
Centrilift Inc.	United States	188 submersible pumps (built in the firm's Toronto plant). Note: This sale brought total number of centrilift pumps in the U.S.S.R. to 600	1978	$23 million
Occidental Int'l Eng. Co.	United States	Design and construction of a pipeline	1978	$300 millon
CRC International	United States	Internal line-up clamps, clean and wrap machines, pipe benders; pigs	1978	
Otis Engineering	United States	Completion equipment	1978	$3 million
Cameron Iron Works	United States	40 well heads	1978	

Table B-1.—Trade by Company—Oil and Gas (Continued)

Equipment area **Well Completion/Production**

Suppliers	Country	Product	Year	Value
FMC Petroleum Equipment Division & Cameron Iron Works	United States	Christmas trees	1978	
TRW-Reda Pump, Inc.	United States	350 pumps	1976	$64 million
International Harvester	United States	500 TD-25C bulldozers and pipelayers	1976	
General Electric	United States	Hot gas rotating components of the compressors for the Orenberg line	1976	
Crutcher Resources Corp.	United States	Spare parts for large diameter pipeline welding equipment	1976	$200,000
Roscoe Brown Sales Co. Inc.	United States	Pipeline augers	1976	$250,000
Cooper Industries	United States	73 RF2BB-30 centrifugal compressors	1976	
International Harvester	United States	Standby turbine generators for Orenberg line	1976	
Grove Valve & Regulator Co.	United States	The smaller valves for Orenberg line (a subcontractor for Perry)	1976	
F. H. Maloney Co.	United States	Pig signalers (subcontract to Perry)	1976	$250,000
Health Consultants Inc.	United States	Pig locaters (subcontract to Perry)	1976	$250,000
E. H. Wachs Co.	United States	Pipe cutters	1976	$300,000

Perry Equipment Corp.	United States	Entire complement of pig launching/receiving stations for Orenberg pipeline
		Competitors: Mannesmann - FRG
		  Prenatechnik - FRG*
		  Primaberg - Austria
		  OeMV - Austria
		  General Descaling - U.K.*
		  N.K.K. - Japan
		  Avery Lawrence - Singapore
		  *Have sold pigs to U.S.S.R. before.
			1976	$27,650,000
				$9 million from Perry

Baker Oil Tools Inc.	United States	20 sets of gas well completion equipment, 2 wire line units, and inflatable packers	1976	
Merrick Engineering	United States	Equipment for welding 12 to 25mm diameter pipe	1975	$700,000
CRC Automatic Welding	United States	Lease of its proprietary automatic welding system to Hungary and Poland for Orenberg line	1975	
Mid-Continent Pipeline Equipment Co.	United States	2 triple jointing plants for 42, 48, and 56 inch pipe; mandrills, clamps, cleaning, lining and coating equipment	1975	$5 million
Ransome Co.	United States	9 welding positioners	1975	$200,000
Deuma	United States	4 welding positioners	1975	

Table B-1.—Trade by Company—Oil and Gas (Continued)

Equipment area **Well Completion/Production**

Suppliers	Country	Product	Year	Value
International Harvester	United States	Bulldozers and pipe layers	1975	
General Electric Co.	United States	Gas turbines for pipelines	1975	
TRW-Reda Pump, Inc.	United States	137 electrodynamic submersible pumping units	1975	$17 million
TRW-Reda Pump, Inc.	United States	Submersible pumps	1975	$20 million
Borg-Warner	United States	120 submersible pumps	1975	
Mission Manufacturing	United States	Submersible pumps	1975	
General Electric	United States	65 MS3002, two-shaft turbines rated at 14,500 hp (all are being built under GE license in 6 different countries)	1974	$250 million
C-E Lummus & Co.	United States	Pipeline coolers	1978	
FMC Petroleum	United States	Stainless steel wellhead equipment	1978	$1 million
Creusôt-Loire Enterprises with Mannesmann-Demag-Meev Group	France West Germany	A seamless pipe plant using the Vallourec process with a capability of 170,000 metric tons per annum	1979	$230 million
Creusôt-Loire Hudson Italiana SpA	France Italy	Pipeline coolers Pipeline coolers	1978	
Caterpillar Tractor Co. and Caterpillar-Mitsubishi	United States Japan	50 Caterpillar pipelayers and 2 years of spare parts	1976	

Secondary/Tertiary Recovery

Fried Uhde GmbH	West Germany	100,000 ton per year alkyl phenol plant (used as a surfactant)	1978	DM 50 million
Borsig GmbH	West Germany	CO_2 recuperation plant	1979	

Offshore

Rauma-Repola Oy	Finland	Build the hull for the semi below	1976	$15 million
UIE & ETPM	France	An offshore oil platform fabrication yard at Baku	1980	
Serete	France	Floating drilling platforms	1975	
Blohm & Voss	West Germany	Self-propelled crane to position offshore rigs	1976	$40 million
Blohm & Voss AG	West Germany	Rebuild a shipyard at Astrakhan for assembling jackups and semi-submersible rigs	1979	
Modec	Japan	Class III 3-legged jackup rig; design by Levingston Shipbuilding (U.S.); drilling equipment by National Supply (U.S.); blowout preventor stacks and controls from N. L. Petroleum (U.S.)	1979	$35 million

Table B-1.—Trade by Company—Oil and Gas (Continued)

Equipment area *Offshore*

Suppliers	Country	Product	Year	Value
Sanwa Kizai Co. Ltd.	Japan	14 augers used for pile driving	1976	
IHC	Netherlands	Seagoing pipe layer	1971	
IHC	Netherlands	Jackup rig	1967	
Ulstein Hatlo A/S	Norway	3 ships to tow exploration and production drilling platforms	1976	
Kongsberg Vaapen Fabrikk A/S	Norway	3 dynamic positioning systems for the 3-drill ships built by Rauma-Repola Oy Competitors: Simrad A/S (Norway) Honeywell (U.S.)	1979	
Simrad A/S	Norway	Dynamic positioning system for the Armco semi built for U.S.S.R.	1979	
Armco, Inc.	United States	Equipment for a jackup being built by Mitsui Ocean Development & Engineering Co. Ltd.	1978	
Lynes International Inc.	United States	11 strings of drill stem testing equipment for offshore facilities	1979	$3.8 million
Armco, Inc.	United States	License for semisubmersible rigs built by Rauma-Repola Oy for Soviets	1978	
National Supply	United States	Provide most of the machinery for a semisubmerisble rig being built by Rauma-Repola Oy	1976	$25 million
Armco, Inc.	United States	License for production of semi's in U.S.S.R.	1976	
Refining				
Technip	France	15,000 cubic meter per annum natural gas	1975	$230 million
Constructions Metalliques de Provence	France	28 natural gas purification stations	1975	
Walworth Aloyco Grove International	Italy	1,580 ball valves for oil refineries	1975	
Japan Steel Works	Japan	Manufacturing for the above plant	1979	
Nichimen Jitsugyo	Japan	3 plant gas processing complex	1976	$250 million
Mitsubishi Corp.	Japan	Primary oil refining equipment Competitors: Linde AG - FRG C-E Lummus - U.S.	1978	
Fluor Corp.	United States	Designs, engineering, procurement, and field technical advisory services for plants to convert natural gas into ethane, methane, pentane, liquid propane, gasoline and other products	1979	
Fluor Corp.	United States	Provide engineering and technology assistance for the three plants above	1976	
Mitsubishi Heavy Industries	Japan United States	Gas processing complex	1978	
Japan Steel Works & Fluor Corp.	Japan United States	Gas processing complex	1978	$250 million

Table B-2.—Trade by Company—Coal

Equipment area *Exploration*				
Suppliers	Country	Product	Year	Value
Plategods Co.	Norway	Rock drilling equipment	1975	$2.5 million
Preparation				
South Yakutian Coal Development Corp.	Japan	Coal preparation equipment	1978	
Marubeni Corp.	Japan	33 large screens for coal washing plant Note: built by Kurimoto Iron Works Co. under a license from Schenck GmbH (FRG)	1978	Y1.2 billion
Sumitomo Heavy Industries	Japan	Four 600 ton/hour flo-driers (under license from McNally Pittsburgh Manufacturing Corp. (United States)	1978	
Transportation from Mine				
Komatsu Ltd.	Japan	30 120 ton capacity heavy mining trucks	1979	$30 million
Unit Rig & Equipment Co.	United States	30 M-200 vehicles for use in coalfields	1976	$40 million
Unit Rig & Equipment Co.	United States	Heavy duty dump trucks	1975	$13 million
Ingersoll-Rand	United States	Slurry pumps powered by a 3,000 HP engine	1975	$14 million
Ingersoll-Rand	United States	Slurry pumps	1974	
Unit Rig & Equipment Co.	United States	54 M-200s. Note: To be built by the Canadian Division.	1979	
Surface Mining Excavation				
Kent France S.A.	France	10 electric mining shovels	1975	
Orensteim und Koppel	West Germany	3 large and 1 small excavator for open cast lignite mines	1980	DM 220 million
Sumitomo Heavy Industries	Japan	20 cubic meter bucket, hydraulic mining dhovels	1978	
Sumitomo Corp.	Japan	10 self-propelled 20 cubic meter bucket, mining Shovels	1976	
Sumitomo Heavy Industries	Japan	10 "Super Front" mining shovels used in strip mining	1975	
South Yakutian Coal Development Corp.	Japan	Coal development equipment	1974	$450 million
Sumitomo Heavy Industries	Japan	10 crawler-mounted blast hole drills Note: Built under a Marion Power Shovel license		
Sumitomo Heavy Industries	Japan	5 "Super Front" mining shovels Note: Built under license from Marion Power Shovel, Division of Dresser Industries		
Paccar Inc.	United States	7 Dart D-600 15 cubic yard front-end loaders	1979	$2.86 million
Clark Equipment Co.	United States	3 Model 475 B front-end loaders	1978	$1 million

Table B-2.—Trade by Company—Coal (Continued)

Equipment area   *Transportation at Underground Mine Site*

Suppliers	Country	Product	Year	Value
Ohlemann GmbH	West Germany	36 underground mining vehicles	1979	
Komatsu Ltd.	Japan	30 120-ton capacity heavy mining trucks	1979	$30 million
Linden Alimak	Sweden	Mine shaft hoists	1979	SKr 5 million

Appendix C. – Suppliers of Essential Equipment

Table C-1.—Suppliers of Nuclear Grade Pipes and Tubes Outside the United States

Austria	Austriatom	Japan	IHI
	Vereinigte Edelstahlwerke Voest-Alpine		Japan Steel Works
			Kawasaki Steel Corp.
Canada	Chase Nuclear		Kobe Steel
	Dominion Bridge Co.		Kubota
	Finnan Engineered Products		
	Noranda Metal Industries	Netherlands	Ameron BV
	Rio Algom (Atlas Alloys Div.)		Kawecki-Billiton Metaalindustrie
			Trent Tube
France	Creusôt-Loire		Van Mullekom
	Delattre-Levivier		Van Wijk & Boerma
	Metaux Inoxydables Ouvres		
	Vallourec	Sweden	Avesta Jernverks
			Nyby Uddeholm
Great Britain	Cabot Alloys Europe		Sandvik
	Cameron Iron Works		
	Fine Tubes	Switzerland	Zschokke Wartmann AG
	Pipework Engineering Developments	West Germany	Klockner-Werke
	RGB Pipelines		Mannesmannroehren-Werke AG
	Tioga Pipe Supply International		Schmoele, R. & G. Metallwere GmbH
Italy	Dalmine		
	Tecnitub Italiana SpA		

SOURCE: *Nuclear Engineering International* (NEI), International Buyers' Guide, 1980.

Table C-2.—Suppliers of Welding Equipment Outside the United States

Canada	Bata Engineering		Tioga Pipe Supply International
France	Polysoude		Vickers Shipbuilding Group
	Sciaky S.A.	Italy	Breda Termomeccania
			Corradi, Franco
Great Britain	Cunnington & Cooper		
	NEI Clarke Chapman Power Engineering	Japan	IHI
	Sciaky Electric Welding Machines		Kawasaki Heavy Industries
			Kobe Steel

SOURCE: NEI, International Buyers' Guide, 1980.

Table C-3.—Suppliers of Steam Generators Outside the United States

Austria	Austriatom Simmering-Graz-Paaker Voest-Alpine	Japan	IHI Kawasaki Heavy Industries Kobe Steel Mitsubishi Heavy Industries Toshiba
Canada	Babcock & Wilcox Canada Davie Shipbuilding Noranda Metal Industries	Netherlands	Neratoom Royal Scheide RSV-A
France	Creusôt-Loire Framatome Stein Industrie Sulzer	Spain	Babcock & Wilcox Espanola Equipos Nucleares
Great Britain	Babcock Power NEI Clark Chapman Power Engineering NEI International Combustion RNC Nuclear	Sweden	Uddcomb Sweden
		Switzerland	Sulzer Brothers
		West Germany	Babcock-Brown Boveri Reaktor Deutsche Babcock GHH Sterkade Klockner-Werke
Italy	Breda Termomeccanica Construzioni Meccaniche Franco Tosi NIRA		

SOURCE: NEI, International Buyers' Guide, 1980

Table C-4.—Suppliers of Pumps Outside the United States

Austria	Andritz Austriatom
Canada	Finnan Engineered Products Hayward Gordon
France	Creusôt-Loire Dresser Europe Framatome Pompes Guinard
Great Britain	GEC Reactor Equipment Haskel Hayward Tyler & Co. Holden & Brooke Weir Pumps
Italy	Fiat TTG Franco Tosi
Japan	IHI Kawasaki Heavy Industries Torishima Pump Manufacturing Co. Toshiba
Netherlands	Borg-Warner Corp. Delaval-Stork
Sweden	Karlstads Mekaniska Werkstad
Switzerland	Eschler Urania K. Rutschi, Ltd. Sulzer Brothers
West Germany	Interatom Klein, Schanzlin & Becker Orlita

SOURCE: NEI, International Buyers' Guide, 1980

Table C-5.—Suppliers of Valves Outside the United States

Canada	Canadian Worcester Controls Curran Valve Supply EPG Energy Products Group Fisher Controls Co. of Canada Velan Engineering
France	Alsthom-Atlantique Neyrpic Pont-a-Mousson Trouvay & Cauvin
Great Britain	Adams, Gebruder GEC-Elliott Control Valves Hattersley Heaton Hindle Valves Hopkinsons
Italy	Fiat TTG
Japan	Japan Steel Works Okano Valve Mfg. Co. Utsue Valve Co.
Netherlands	Borg-Warner G. Dikkers & Co.
Sweden	Karlstads Mekaniska Werkstad
Switzerland	Alfred Battig Sulzer Brothers
West Germany	Gebruder Adams ARF Armaturen-Vetrieb Deutsche Babcock Stahl-Armaturen Persta

SOURCE: NEI, International Buyers' Guide, 1980

Table C-6.—Suppliers of Containment Structures Outside the United States

Austria	RFB Veost-Alpine
Canada	Canatom Davie Shipbuilding
France	Bignier Schmid-Laurent Creusôt-Loire Neypic Spie-Batignolles S.A.
Great Britain	Babcock Power Fairey Engineering GEC Reactor Equipment Sir Robert McAlpine & Sons, Ltd.
Italy	Bosco Industrie Meccaniche Fochi
Japan	IHI Kawasaki Heavy Industries Kobe Steel Shimizu Construction Co.
Switzerland	Bureau BBR Buss Sulzer Brothers Woolley Zschokke Wartmann
West Germany	Krupp Fried Maschinenfabrik Augsburg-Nurnberg AG L. & C. Steinmuller

SOURCE: NEI, International Buyers' Guide, 1980

Table C-7.—Suppliers of Control Systems Outside the United States

Canada	Automatec Canadian General Electric Enercorp Instruments Fischer & Porter Thermo Electric
France	CGEE Alsthom Fichet-Bauche Leanord SODETEG Spie Batignolles
Great Britain	Ferranti Computer Systems Foxboro-Yoxall Honeywell Kent Process Control R.P. Automation
Italy	ELSAG Marelli, Ercole & Co. Montedel
Japan	Fuji Electric Co. Kawasaki Steel Corp. Sukegawa Electric Co. Toshiba
Sweden	ASEA Tekniska Rontgencentralen
Switzerland	Bachofen Brown Boveri & Cie High Energy & Nuclear Equipment Sulzer Borthers
West Germany	Brown Boveri & Cie Karftwerk Union Nuclear Data Siemens

SOURCE: NEI, International Buyers' Guide, 1980

Table C-8.—Essential Equipment for Coal Mining

Preparation
1. Agitators, Conditioners, Mixers
2. Crushers
3. Flotation Machines and Reagents
4. Grinders
5. Pulverizers
6. Separators
7. Washers
8. Cleaning Breakers
9. Blending Machines
10. Electrostatic Precipitators

Surface Mining Excavation
11. Draglines
12. Drills
13. Power Shovels
14. Bulldozers
15. Front-end Loaders
16. Scrapers

Transportation at Surface Mine Site
17. Coal Haulers (100 ton)

Underground Mine Excavation
18. Continuous Miners
19. Loading Machines
20. Longwall Equipment
21. Bulldozers
22. Coal Cutters
23. Heading Machines
24. Undercutters

SOURCE: C. Simeons, *Coal: Its Role in Tomorrow's Technology* (New York: Pergamon Press, 1978).

Table C-9.—West European and Japanese Suppliers of Essential Coal Mining Equipment

1. Agitators, Conditioner and Mixers

France
  Fives-Cail Babcock

England
  APV-Mitchell (Dryers), Ltd.
  GEC Mechanical Handling, Ltd.
  Johnson-Progress, Ltd.
  Joy Manufacturing Co.

Japan
  Kawasaki Heavy Industries, Ltd.

2. Crushers

England
  Aveling-Beuford, Ltd.
  British Jeffrey Diamond
  Magco, Ltd.
  Newell Dunford Eng., Ltd.
  Pegson Ltd.
  Underground Mining Machinery, Ltd.

West Germany
  Buckan-Wolf Maschinenfabrik AG
  Buhler-Miag
  Esch-Werke AG
  Hazemag GmbH & Co.
  IBAG International Baumaschinenfabrik AG
  Krupp GmbH

France
  Alsthom Atlantique
  Dragon SA Appareils
  Fives-Cail Babcock
  Joy SA
  Stephanoise de Constr

Japan
  Ishikawajima-Harima Heavy Industries
  Kawasaki Heavy Industries, Ltd.
  Kurimoto Iron Works, Ltd.

3. Flotation Machines and Reagents

England
  Machines
    Joy Manufacturing Co.
  Reagents
    Century Oils, Ltd.

West Germany
  Machines
    KHD Industrianlagen AG
    Krupp GmbH, Fried, Krupp Industriel
    Lurgi Gesellschatten

Japan
  Machines
    Kawasaki Heavy Industries, Ltd.
  Reagents
    Sanyo Chemical Industries, Ltd.

4. Grinders

England
  Beryllium Smelting Co., Ltd.
  Chapman, Ltd.
  GEC Mechanical Handling, Ltd.
  Head Wrightson & Co., Ltd.
  Helipeds, Ltd.
  Joy Manufacturing Co.
  Newell Dunford Eng., Ltd.
  Pegson, Ltd.
  Simon-Warman
  Wilkinson Process Linatex Rubber Co., Ltd.

France
  Alsthom Atlantique
  Dragon SA Appareils
  Fives-Cail Babcock
  Stein Industrie

West Germany
  Buhler-Miag
  Esch-Werke AG
  IBAG International Baumaschinenfabrik AG
  KHD Industrianlagen AG
  Krupp GmbH. Fried. Krupp Industrie
  Kulenkampff Gebruder
  O&K Orenstein & Koppel AG
  Polysius Werke

Japan
  Ishikawajima-Harima Heavy Industries
  Kawasaki Heavy Industries, Ltd.
  Kabe Steel, Ltd.
  Kurimoto Iron Works, Ltd.
  Mitsubishi Steel Mfg. Co., Ltd.

5. Pulverizers

England
  British Jeffrey Diamond

West Germany
  KHD Industrianlagen AG
  Krupp GmbH, Fried, Krupp Industrie

Japan
  Ishikawajima-Harima Heavy Industries

6. Separators

England
  Boxmag-Rapid, Ltd.
  GEC Mechanical Handling, Ltd.

France
  Fives-Cail Babcock
  Saulas & Cil
  Stein Industrie

West Germany
  Bavaria Maschinenfabrik GmbH & Co.
  KHD Industrianlagen AG
  Krupp GmbH, Fried, Krupp Industrie
  Polysius Werke

Table C-9.—West European and Japanese Suppliers of Essential Coal Mining Equipment (Continued)

Japan
  Ishikawajima-Harima Heavy Industries
  Kawasaki Heavy Industries, Ltd.
  Kobe Steel Ltd.
  Kurimoto Iron Works, Ltd.

7. Washers

England
  Aveling Barford, Ltd.
  GEC Mechanical Handling, Ltd.

France
  Alsthom Atlantique
  Dragon SA, Appareils

West Germany
  Bavaria Maschinenfabrik GmbH & Co.
  Esch-Werke AG
  IRAQ International Baumascninenfabrik AG
  KHD Industrianlagen AG

Japan
  Kurimoto Iron Works, Ltd.

8. Cleaning Breakers

England
  British Jeffrey Diamond
  Compair Construction & Mining, Ltd.
  GEC Mechanical Handling, Ltd.
  Gullick Dóbson, Ltd.
  Mining Supplies, Ltd.
  Padley & Venables
  Underground Mining Machinery, Ltd.

France
  Fives-Cail Babcock
  Stephanoise de Constr. Mecaniques, Soc.

West Germany
  Deutsche Montabert GmbH
  KHD Industrianlagen AG
  Orenstein & Koppel AG
  Westfalia Lunen

Japan
  Furukawa Rock Drill Sales Co., Ltd.
  Ishikawajima-Harima Heavy Industries

9. Blending Machines

England
  Babcock-Moxey, Ltd.
  Babcock & Wilcox, Ltd.

France
  Fives-Cail Babcock
  Realization Equipments Industriels

West Germany
  Buckau-Wolf Maschinenfabrik AG
  Demag Lauchammer Masch & Stahlbau GmbH

Japan
  Ishikawajima-Harima Heavy Industries

10. Electrostatic Precipitators

England
  Head Wrightson & Co., Ltd.

West Germany
  KHD Industrianlagen AG
  Lurgi Gesellschaften

Japan
  Ishikawajima-Harima Heavy Industries
  Kawasaki Heavy Industries, Ltd.

11. Draglines

England
  Ransomes & Rapier Ltd.
  Ruston-Bucyrus Ltd.

France
  Poclain
  Realization Equipments Industriels

West Germany
  Aumund-Forderbau GmbH
  Demag AG, ABT Bergwerksmachinen
  Demag Lauchhammer Masch & Stahlbau GmbH
  Demag Verdichtertechnik GmbH
  Krupp GmbH, Fried, Drupp Industril Una Stahlbau
  Liebherr Hydraulikbagger GmbH
  Maschinenfabrik Augsburg-Nurnberg AG
  Orenstein & Koppel AG

Japan
  Hitachi Construction Machine Co. Ltd.
  Ishikawajima-Harima Heavy Industries
  Kawasaki Heavy Industries
  Kobe Steel Ltd.

12. Drills

England
  Boart, Ltd.
  Compair Construction & Mining, Ltd.
  Eimco, Ltd.
  English Drilling Equipment Co., Ltd.
  Euro-Drill Equipment, Ltd.
  Hydraulic Drilling Equipment, Ltd.
  Mining Dev., Ltd.
  Underground Mining Machinery, Ltd.

13. Power Shovels
  See Draglines

14. Bulldozers
  See Draglines

France
  Maco-Meudon

West Germany
  Demag AG, ABT Bergwerksmachinen
  Demag Drucklufttechnik GmbH
  Demag Verdichtertechnik GmbH
  Deutsche Montabert GmbH
  Flottman-Werke GmbH
  Werth & Co.

Japan
  Furukawa Rock Drill Sales Co.
  Koken Boring Machine Company
  Mitsubishi Steel Mfg. Co., Ltd.
  Mitsui Shipbuilding & Eng. Co., Ltd.

Table C-9.—West European and Japanese Suppliers of Essential Coal Mining Equipment (Continued)

15. Front-end Loaders

England
Aveling-Barford, Ltd.
Eimco, Ltd.
Matbro, Ltd.
Mining Dev., Ltd.

France
France Loader
Realization Equipment Industriels

West Germany
Aumund-Fordererbau GmbH
Deilmann-Hanill GmbH
Eickhoff Maschinf bk-U Eisengiesserei Mb
Gutehoffnungshuttl Sterkrade AG
Orenstein & Koppel AG
Salzgitter Maschinen AG
Westfalia Lunen

Japan
Furukawa Rock Drill Sales Co.
Hitachi Construction Machine Co., Ltd.
Kawasaki Heavy Industries, Ltd.
Kobe Steel, Ltd.
Komatsu, Ltd.
Mitsuboshi Belting, Ltd.
Shinko Electric Co., Ltd.

16. Scrapers
See Draglines

17. Coal Hauler
See Draglines

18. Continuous Miner

England
Babcock & Wilcox, Ltd.
Dasco Overseas Eng., Ltd.

West Germany
Demag AG, ABT Bergwerksmachinen
Demag Verdichtertechnik GmbH

19. Loading Machine
See Front-end Loaders

20. Longwall Equipment

England
British Jeffrey Diamond
Underground Mining Machinery, Ltd.

France
Minex Mine-Expert

West Germany
Eickhoff Maschinf bk-U Eisengiesserei Mb
Westfalia Lunen

21. Bulldozer
See Draglines

22. Coal Cutters

England
Babcock & Wilcox, Ltd.
British Jeffrey Diamond
Dosco Overseas Eng., Ltd.
Mining Supplies, Ltd.
Underground Mining Machinery, Ltd.

France
Minex Mine-Expert
Stephanoise de Constr. Mecaniques

West Germany
Eickhoff Maschinf bk-U Eisengiesserei Mb
Thyssen Industrie AG
Titanit Bergbau Technik
Westfalia Lunen

23. Heading Machines

England
Eimco, Ltd.
Gullick Dobson, Ltd.
Mining Dev., Ltd.
Thymark Thyssen Group

West Germany
Becorit Grubenausbau GmbH
Salzgitter Maschinen AG
Westfalia Lunen

24. Undercutters
See Draglines

SOURCE: C. Simeons, *Coal: Its Role in Tomorrow's Technology* (New York: Pergamon Press, 1978).

Table C-10.—West European and Japanese Producers of Electric Transmission Equipment

Country	Company	Country	Company
West Germany	Siemens AEG	Sweden	ASEA (High Voltage Transformers and DC Equipment)
England	English Electric GEC	Switzerland	Brown-Boveri
		Japan	Mitsubishi
France	Thomson-Brandt Alstom-Atlantique DEL (High Voltage Circuit Breakers)	Netherlands	Philips

SOURCE: Office of Technology Assessment.

The Prospects for Energy Conservation in the U.S.S.R.

CONTENTS

LIST OF TABLES

FIGURE

The Prospects for Energy Conservation in the U.S.S.R.

Most of the attention devoted to Soviet energy in Western literature of the last several years has focused on production, especially the prospects for Soviet oil and gas and the potential role of Western technology in petroleum output. Production, however, is only half of the energy picture; the prospects for Soviet energy consumption are equally important. If the U.S.S.R. could slow its rate of growth of energy demand, the consequences of any decline in production would be correspondingly less critical. Moreover, Soviet plans to substitute among different sources of energy supply (more abundant for scarcer, nearer for farther, more efficient for less) will involve corresponding shifts in the structure of consumption. Measures aimed at controlling or modifying consumption, therefore, are not simply alternatives or complements to a "supply-side" energy strategy; the two are inseparably dependent on one another.

This chapter briefly examines the current structure of Soviet energy consumption and recent trends in Soviet energy use; describes the evolution of Soviet official policy on energy conservation and the major conservation options available to Soviet planners; reviews recent performance in achieving energy savings; and discusses the implications for Western policy of Soviet conservation strategies.

Energy conservation is not a new issue for Soviet policymakers, but an aspect of a general concern with the availability and use of all raw materials. The increasing scarcity of attractively located and high-yield natural resources is a constraint on Soviet prospects for economic growth. Nevertheless, Soviet experience in conservation has been mixed. For all the official exhortation over the years, there are only a few cases in which more than token gains have been recorded. This is not for want of opportunities. In 1975, the Soviet Union used nearly as much energy as the United States in industry, but Soviet industrial output was only three-fourths that of the United States. Similarly, in the agricultural sector in the same year, the U.S.S.R. used appreciably more energy to achieve 80 to 85 percent of America's output.[1]

Impressive Soviet conservation efforts have taken place in the energy field itself, particularly through steady improvement in the efficiency of electrical power generation and the use of cogeneration for centralized urban heat supply.[2] Unfortunately, now that these gains have been realized, further progress may be slow. The reasons, as this chapter demonstrates, stem from basic features of the Soviet command economy: the system of prices and incentives, the mechanisms for investment and technological innovation, and the distribution of power. In addition, Soviet decisionmakers are disposed to think of output performance before efficiency. This makes individual enterprises disinclined to support actions that might threaten the former in support of the latter, and planners reluctant to stake the success of their energy policy on conservation measures that they may know from previous experience are unlikely to work. In short, the future of Soviet efforts to conserve energy can be regarded as one part of a larger problem—the struggle of the Soviet system to overcome the inherited habits of Stalinist-style industrialization.

---

[1] Robert W. Campbell, "Energy Prices and Decisions on Energy Use in the U.S.S.R.," discussion paper, Department of Economics, Indiana University, February 1981, p. 1.

[2] See Leslie Dienes and Theodore Shabad, *The Soviet Energy System* (New York: John Wiley & Sons, 1979), pp. 187-194.

STRUCTURE AND TRENDS IN
SOVIET ENERGY CONSUMPTION[3]

The most striking feature of Soviet energy consumption compared to that of any major Western industrialized country is the dominance of the industrial and the comparatively small shares of the transportation and residential sectors in energy use. Table 49 shows how Soviet primary energy supply is allocated among various uses—including own use and losses in the energy industries, nonfuel uses, and various sectors of the economy—both with electric power stations shown as consumers and with consumption shown by final user (i.e., after consumption in power stations has been reallocated

among the sectors that use their output). Industry has consistently accounted for by far the largest share of Soviet domestic energy consumption over the past 20 years. This is largely due to the small stock of private automobiles and the low share of trucks in the overall transport mix, and the fact that a large proportion of Soviet buildings are heated through centralized systems using the heat cogenerated during the production of electrical power.

Another striking characteristic of Soviet energy consumption is the large share used by the energy-producing sector itself. Own-use by the gas industry, for example, is about 10 percent of total output, mostly to power pipeline compressors.[4] Much of the energy consumed by the energy industries is simply due to the circumstances under which they must operate: long transit distances to points of consumption (40 percent of all Soviet rail traffic is devoted to shipment of fuels), inhospitable environments in the producing regions, the need to work with low-grade fuels, and some energy-intensive extractive methods. For example, water-flooding has turned the Ministry of the Oil Industry into a major consumer of electricity. The industry consumed 48 billion kilowatt-hours (kWh) in 1979, and the rate is rising by 11 to 13 percent a year.[5] These factors are becoming even more important over time, as transportation distances increase and the accessibility of fuel declines. In fact,

---

[3]For a review of recent data on this subject, see Leslie Dienes and Nikos Economou, "CMEA Energy Demand in the 1980's: A Sectoral Analysis," paper presented at the 1981 Colloquium of the NATO Economics Directorate, Brussels, April 1981.

Table 49.—Soviet Energy Consumption
(million tons of oil equivalent)

	1960	1965	1970	1975	1980
Total primary energy supply (corrected for foreign trade) ...	438.9	579.5	737.4	922.0	1,086.2
Consumed in:					
Own needs and losses	56.4	67.8	79.8	115.7	119.2
Industry	132.0	169.9	209.9	253.4	306.4
Electric power stations	111.9	164.0	225.3	302.8	367.4
Household and municipal	46.3	65.3	79.5	98.0	80.5
Agriculture	21.3	29.3	42.6	43.4	54.3
Transportation ..	51.7	57.1	57.0	52.9	85.9
Construction	7.3	8.2	14.2	16.0	27.0
Nonfuel	12.0	17.9	29.1	39.8	45.5
After reallocation of electric power and cogenerated heat:					
Own needs and losses	123.4	157.5	195.7	271.8	304.6
Industry	165.8	227.5	292.9	360.0	438.9
Household and municipal	54.7	77.3	98.7	126.7	114.4
Agriculture	21.8	30.2	44.5	47.5	60.2
Transportation ..	53.1	59.9	61.2	58.6	93.5
Construction	8.0	9.1	15.3	17.6	29.1
Nonfuel	12.0	17.9	29.1	39.8	45.5

SOURCE: Robert W. Campbell.

---

[4]A. A. Makarov and L. A. Melentev, "Problems and Directions of Energy Development in the U.S.S.R.," *Ekonomika i organizatsiya promyshlennogo proizvodstva (Eko)*, No. 3, 1981, p. 28.

[5]V. D. Kudinov, "Rational Utilization of Fuel and Power Resources in the Oil Industry," *Neftyanoye khozyaystvo*, No. 9, 1980, pp. 6-9. According to Kudinov, the ministry also consumed 9.3 mtoe in liquid fuels and gas in 1979, one-quarter of which was used for gas-lift operations. Total own-use by the Ministry of the Petroleum Industry in 1979, according to another source, was 14.7 mtoe. See I. Grekhov, et al., "Use Fuel and Energy Resources Economically and Efficiently," *Neftyanik*, No. 10, 1980, p. 12, in JPRS No. 77154, Jan. 12, 1981, p. 8.

there is overall an upward trend in own-use by the energy sector.[6]

Prospects for Soviet energy conservation and substitution, therefore, turn heavily on developments in industry, and especially in the energy industries. This is in contrast to conservation prospects in the West, where major savings have recently been realized in residential and transportation uses. Analysts in both the U.S.S.R. and the West have argued that Soviet industrial consumption tends to be less flexible than that of other sectors, and potential energy savings will consequently be harder and slower to realize than were gains in the past.[7] This is not to say, however, that opportunities for most such savings have been exhausted. In any case, the shares of residential and municipal consumption, relative to industrial, are unlikely to alter much without a sharp shift in the political priorities of the Soviet leadership. There is no evidence that such a change is imminent.

The inevitable rise in the energy sector's consumption may alter the regional distribution of demand, yet the traditional concentration of demand in the European part of

---

[6]For example, in the first 4 years of the Tenth FYP, the energy required to produce and transport 1 ton of crude oil by pipeline increased by 12 and 26 percent respectively. Ibid., p. 14.

[7]Makarov and Melentev, op. cit.

the U.S.S.R. is reinforced by the tendency of industrial ministries to locate where infrastructure and labor are readily available. Further, as table 49 demonstrates, a large share of Soviet energy is consumed in the form of electricity. This is due to the nature of the technological processes employed in industry. (The transportation and residential sectors directly consume greater proportions of liquid, solid, or gaseous fuel.) Gradual technological modernization of Soviet industry and the rise of sophisticated manufacturing specialities should lead to a continuation of this trend.

But the balance among Soviet energy sources must change. Soviet planners have understood for some time that the share of oil must be cut and other fuels substituted. At the center of their effort has been a policy of investing in nuclear and hydropower for electricity generation, and of mobilizing the abundant resources of lignite and subbituminous coal from Siberia and Kazakhstan. Now, a growing emphasis on gas has emerged. Much of what is presently used as fuel oil will be converted to lighter fractions, requiring a major expansion of refinery capacity in the next decade. There have been recurring differences of opinion among Soviet leaders and planners, however, over the relative priority to accord these plans (see below and ch. 8).

ENERGY CONSUMPTION AND ECONOMIC GROWTH

The key point at issue is whether the Soviet Union can "decouple" increases in energy consumption from overall economic growth by the end of the century. There are important areas in the Soviet economy where certain energy intensities have fallen—largely accomplished through substituting gas and oil for less efficient coal; electrifying key sectors of the economy, including half of all railroad haulage; and controlling conversion losses through improvements in the heat rate and extensive use of cogeneration. But in general, once the Soviet index for gross national product (GNP) growth is

corrected (downward) to make it conform more closely to Western definitions, it is apparent that Soviet energy use is growing faster than Soviet GNP.

Analysis of Soviet energy/GNP elasticities shows a progressive deterioration since the early 1970's. Soviet economists continue to claim an elasticity of less than 1, perhaps because they use an exaggerated measure for growth rate of aggregate output, but the Soviet data in table 50 nonetheless reflect the recent rise in GNP elasticity of energy use. It must be noted, however,

Table 50.—Relationship of Economic Growth to Energy Use, 1961-78
(increments expressed in average annual percentages)

	Gross energy consumption	National income	Coefficient of elasticity
1961-65......	6.6	6.5	1.02
1966-70......	4.8	7.8	0.62
1971-75......	5.1	5.7	0.89
1976-78......	4.8	5.1	0.94

SOURCE: S. N. Yatrov, "Fuel-Energy Complexes," *Ekonomieheskaya gazeta,* No. 10, March 1980, p. 10.

that this trend appears to rest more on a decline in GNP than on growth in gross energy consumption.

The most authoritative Western estimates of energy/GNP coefficients are given in table 51. These are based on different data than the Soviet estimates, and consequently show significantly higher coefficients. Nevertheless, they reflect the same trend. Both Soviet and Western calculations, therefore, carry the same long-term implications. If one assumes an average annual Soviet growth rate of 2.5 percent and an energy/GNP elasticity of 1.0, gross energy consumption would rise from under 1,556 million tons of oil equivalent (mtoe) in 1980 to 1,615 mtoe a year by the year 2000.[8] Some forecasts for total Soviet primary energy production in that year come to very little more, leaving nothing for energy exports. This is a crude calculation, but it serves to highlight the importance of conservation in the Soviet economy.

The aggregate energy intensity of the Soviet economy can decline only if energy ef-

[8]Robert W. Campbell, "Energy in the U.S.S.R. to the Year 2000," unpublished paper prepared for the Conference on the Soviet Economy, Oct. 23-25, 1980, Airlie House, Va.

Table 51.—Relationship of Economic Growth to Energy Use, 1960-80
(increments expressed in average annual percentages)

	Gross energy consumption	GNP	Coefficient of elasticity
1965-60........	5.9	4.9	1.20
1970-65........	4.9	5.3	0.92
1975-70........	4.0	4.1	0.97
1980-75........	4.0	3.0	1.33

SOURCE: Robert W. Campbell, "Energy in the U.S.S.R. to the Year 2000," unpublished paper prepared for the Conference on the Soviet Economy, Oct. 23-25, 1980, Airlie House, Va.

ficiency continues to improve as it did before 1970; or if the overall structure of the Soviet economy evolves in the direction of sectors that are less energy- and material-intensive. The degree to which either of these conditions can be met is, at best, debatable. Several factors work against the prospect of declining energy intensity.

First, opportunities that made it possible for Soviet managers to reduce the energy input per unit of output in many processes in the 1950's and 1960's—shifts to cheaper fuel, rapid improvements in the heat rate of powerplants, etc.—dwindled in the 1970's and are vanishing in the 1980's. Extraction and transportation costs for all energy sources are climbing; the quality of coal and oil is deteriorating; and progress in lowering the heat rate in the best powerplants has nearly reached its limits. Furthermore, any structural shifts in the Soviet economy toward deemphasis of energy- and material-intensive sectors may be offset by equally strong trends in the opposite direction, i.e., toward sectors like agriculture that are highly energy-intensive. In addition, the energy-producing sector itself is growing rapidly in importance.

The prospects for improvement in either the net energy intensity or the efficiency of major conversion processes are, in short, uncertain. If present trends continue, a deterioration in the aggregate energy intensity of the Soviet economy over the next two decades is possible. It is in this context that Soviet energy conservation policy must be evaluated.

EVOLUTION OF SOVIET CONSERVATION POLICY

High-level interest in energy conservation has been evident in the U.S.S.R. since the early 1970's. Such concern is probably less attributable to specific anxiety about future Soviet oil production (Siberian supergiants were still being discovered until 1973) than to a general deterioration in the economics of energy supply caused by the rapidly growing extraction and transmission costs connected with the decline of European energy deposits and the development of Siberian resources. In 1973, the Central Committee of the Communist Party and the U.S.S.R. Council of Ministers issued a joint decree on energy conservation, "On Steps to Improve Efficiency in Using Fuel-Energy Resources in the National Economy." This decree, frequently cited as the starting point of the present conservation policy,[9] was primarily concerned with recovery of "secondary energy resources," mainly high-temperature process heat that could be profitably reused before being released and lost.

By now a number of official decrees have appeared on the subject of energy conservation, gradually broadening the scope of the policy. This policy now encompasses a wide range of investment and housekeeping measures, aimed not merely at capturing secondary heat resources but at monitoring energy use, establishing criteria of efficiency for major industries and processes, improving insulation, etc. The decrees have also created an administrative apparatus for the implementation of the conservation effort, and have attempted to increase the involvement of the party in conservation. As a result, conservation is now a prominent part of official energy policy.

Soviet decrees evince a tension between two broad conservation strategies. The first, a "high-investment strategy," is aimed at improving the efficiency of Soviet energy use through technological innovation and replacement of obsolete plant. The second, a "low-investment" or "housekeeping" approach to conservation, is aimed at saving energy through better monitoring and better production practices, but largely within the confines of existing technology and plant. For example, in the case of automotive transportation, the first strategy might aim at developing more efficient engines, perhaps through widespread adoption of diesels; the second approach would call for

better maintenance of the existing stock and for measures to curtail the thriving black market in oil and gasoline.

In principle, the investment and housekeeping strategies are complementary. The former seeks to substitute capital for energy, the latter labor for energy, including labor in the form of innovation and more stringent management. In a market economy, their precise mix is determined in principle by the relative marginal return from each, and the conservation strategy employed at any time will be a combination of investment and housekeeping, with no clear disjunction between them. In the Soviet Union, in contrast, there is a clear difference between the investment and housekeeping strategies. The first is handled through the central planning system. The second is regarded as an enforcement and monitoring problem, and is handled through so-called "public organizations," largely at the local or enterprise level. Coordination of these two strategies is a difficult process in a command economy.

THE HIGH-INVESTMENT STRATEGY

Restructuring Demand

The most urgent task of Soviet planners in the next decade is to lessen the share of oil in the overall energy balance. This involves both producing fuel substitutes and adjusting the capital stock on the demand side to accommodate them. Problems exist on both counts. Nuclear power development and electrical power construction have fallen behind schedule; and coal output has seriously lagged plan targets. The combination of these problems has caused a virtual eclipse of the ambitious program adopted at the 25th Party Congress in 1976, which relied heavily on conversion to coal-fired power-plants. Although ambitious gas targets have been adopted, it is still possible that diversification of supply will not proceed fast enough to avoid domestic oil shortages in the mid to late 1980's. These might necessitate fuel rationing or mandatory cutbacks (see below).

[9]E. I. Vertel, in *Vestnik mashinostroyeniya*, No. 3, 1980, pp. 3-7; and V. Varavka, "Gas Heats the Sky," *Ekonomicheskaya gazeta*, No. 17, April 1980, p. 16.

Altering the structure of energy demand in the Soviet Union raises some of the same problems as in any other industrialized country. Consumption patterns are the outcome of many past policies. Changing these patterns requires making adjustments to the country's basic infrastructure, a slow and expensive process, despite the fact that aging capital stock must eventually be replaced. In the Soviet Union such replacement has taken place much more slowly than in the West, and there is a large stock of inefficient and obsolete equipment. Given the present shortage of capital, alleviation of this problem in the near term is unlikely.

The existing capital stock strongly inhibits Soviet ability to shift the energy consumption mix. For example, there are about 250,000 small boiler plants in the U.S.S.R., the majority of which operate on coal. Half of the Soviets' larger nonnuclear powerplants use gas and fuel oil. Sound policy, according to Soviet fuel experts, would be to convert the small boilers to gas and the larger ones to coal.[10] But such a conversion would require a massive expansion of local gas lines and of gas storage facilities, a project that seems out of the question for a gas ministry that will be totally absorbed in the next few years in expanding gas production and bulk carriage.[11]

Coal presents other obstacles. It is unattractive to users, and its quality is rapidly declining. Enrichment facilities, treatment to remove moisture and ash, and the utilization of new boiler types that can deal with poorer grades of coal have all been slow to achieve widespread application. Coal's declining heat content also makes larger shipments necessary. The Moscow power system, for example, requires 20,000 more coal cars than formerly just to offset the poorer quality. The presence of impurities adds to downtime and maintenance costs, and short-

ens the lifetime of boilers.[12] Moreover, conversion of large powerplants to coal is expensive and time-consuming, taking powerplants out of operation for long periods. The latter is a particularly serious consideration in view of existing concern over power supplies to the European U.S.S.R., where generating capacities are strained by demanding output targets.

The result of these problems has been a tendency to convert oil-fired powerplants to gas, but even that process has been lagging.[13] In short, there is reason to believe that Soviet planners will find it difficult to restructure demand. Success will depend to a great extent on replacing an older, inefficient plant, and on planning an energy-saving investment.

Replacing Inefficient Plant

Modernizing or replacing existing stock with more energy-efficient equipment requires both capital and the active commitment of industrial planners and designers. In the U.S.S.R., technological innovation and diffusion have traditionally been hampered by a dysfunctional incentive system and problems of coordination across administrative boundaries.[14] The country has been slow in modernizing its inefficient plant, and this situation is unlikely to readily change.

An illustration of Soviet problems in this area may be found in the electrical-power sector. The Ministry of Power and Electrification is assigned an annual plan for the replacement of obsolete boiler equipment, particularly equipment operating at pressures of less than 90 atmospheres. The Power Ministry's planners nevertheless continue to include the less efficient equipment in their own specifications, because they know it has a better chance of being pro-

[10]S. N. Yatrov, "Energy Resources: Ways to Economize," *Sotsialisticheskaya industriya*, May 19, 1980, in JPRS No. 76261, Aug. 20, 1980, p. 9.

[11]S. N. Yatrov and A. Pyatkin, "Effectiveness of Utilization of Fuel and Power Resources," *Planovoye khozyaystvo*, No. 2, 1979, p. 12.

[12]N. Kovalev, N. Tikhodeyev, and I. Y. Ershov, "How To Draw Reserves," *Pravda*, Nov. 20, 1980.

[13]M. A. Styrikovich, "The Main Link," *Izvestiya*, June 1, 1980.

[14]See Paul M. Cocks, "Science Policy in the Soviet Union," in National Science Foundation, *Science Policy USA/ U.S.S.R.*, vol. 2 (Washington, D.C.: U.S. Government Printing Office, 1980).

duced in sufficient quantities to meet plan targets. No incentive exists for the enterprises that manufacture boilers to undertake difficult and time-consuming change-over operations that would inevitably affect planned output fulfillment for some time. The Ministry of Power Machine Building, which builds boilers for the Ministry of Power, claims to be indignant about the lack of progress and has raised the issue before planning authorities. Yet large numbers of small and energy-inefficient boilers continue to be produced.[15]

Because Soviet industry has problems with technological innovation in general, it of course does not necessarily follow that improved energy efficiency in particular is impossible. Energy efficiency in the Soviet iron and steel industry, for example, has long been increasing. These improvements in ferrous metallurgy can be seen in dramatic declines in the average fuel rate in Soviet open hearth furnaces over the past 40 years, declines that are likely to continue in the present decade.[16] Similar improvements have occurred in thermal power generation,[17] and a number of other energy-intensive industrial sectors—construction materials and chemicals for example—might well cut their energy use sharply with the introduction of modern equipment.

The potential for significant energy savings through the high-investment strategy should, therefore, not be lightly written off. The key to the rate and degree of success in this area lies in the Soviet economic system itself. One major inhibitor of innovation in the U.S.S.R. is the fact that ready measures of "good" innovation are lacking—a seller's market prevails in the producer-goods sector, and users must accept what they get,

regardless of whether changes actually constitute improvements. Energy efficiency is an exception. It can be based on a criterion that is readily measurable, both by users and by outside monitors. With such a criterion, Soviet industry has in the past been able to produce useful change, *once it was given the incentive to do so.* Restructuring incentives may, therefore, be the key in determining the response of Soviet industry to a high-investment conservation strategy.

Planning Mechanisms

Energy conservation cannot be built into new industrial technology simply by decree from above; it must originate within the industrial ministries themselves and be built into basic technologies and designs. Consequently, a high-investment approach to energy conservation requires more than just the cooperation of industrial ministries and research and design institutes. Enterprises must also be given the appropriate instructions and incentives to incorporate energy-saving schemes into proposals for new plant and machinery. In the absence of such cooperation and incentives, official exhortation will have little practical effect.

There is little evidence of the necessary mechanisms for progress in this area. In 1979, a deputy chief of Gosplan's energy division criticized the industrial ministries for failing to incorporate energy conservation into their planning systems. In the most important areas of fuel substitution—nuclear power, hydropower, and waste heat recovery—the necessary procedures had been "partially" implemented. But the article was especially critical of the failure to plan for greater efficiency in low-parameter use of heat resources.[18] Similarly, a high-level survey done in the same year revealed that only 2 of 67 enterprises of the Ministry of the Automobile Industry had adopted 5-year energy conservation plans, and that matters were no better in other industries.[19]

[15]*Pravda,* Oct. 28, 1980.

[16]See William J. Kelly, Hugh L. Shaffer, and Timothy P. Spengler, "Trends in the Energy Efficiency of Open-Hearth Furnaces in the Soviet Union," paper presented at the Midwest Economics Association, Louisville, Ky., April 1981.

[17]William J. Kelly, "Industrial Energy Conservation in the Soviet Union: Thermal Power Generation," paper presented at the Southern Economic Association, Washington, D.C., November 1978.

[18]A. Troitskiy, *Planovoye khozyaystvo,* No. 2, 1979, p. 24.

[19]S. Veselov, "Rational Expenditure of Fuel and Power Resources," *Planovoye khozyaystvo,* No. 2, 1979, p. 33.

Such results are hardly surprising given the past cheapness and abundance of energy resources and the relative novelty of Soviet recognition of "energy problems." No evidence has yet appeared in the West of any adaptation of industrial ministries' policy-making systems to these emerging problems. Perhaps the relative lack of official attention to this issue reflects a shift of focus away from a centralized-investment approach to energy conservation, since during the same period there has been no lack of coverage of the ministries' mechanisms for oversight and enforcement. It would appear that for now this aspect of conservation policy is still fluid.

Planning Prices[20]

Another factor inhibiting the high-investment strategy relates to the price system employed by planners in making major investment decisions. Soviet "planning prices" are intended to convey the national economic cost of various fuels in the major economic regions. Planning prices are calculated through a complex system which takes account of aggregate demand for boiler and furnace fuel in each region; capacity levels for these fuels and for certain transport links; the capital and operating costs of producing the fuels; and the capital and operating costs of transporting them.

These planning prices, however, tend to understate the real costs of producing and delivering the energy, which rose steeply in the 1970's as the centers of oil, gas, and coal production moved eastward. In addition, planning price calculations take only domestic demand into account. World market prices for energy have been rising rapidly, however, and the real opportunity cost of, for example, Soviet oil is the world market price. This is substantially higher than its planning price. The absence of a system which takes account of the true opportunity costs of exportable oil and gas makes it difficult to construct optimum production

mixes and to decide on rational fuel substitution policies. The high-investment conservation strategy is, therefore, seriously hampered because the prices used as a basis for investment decisions do not sufficiently encourage the substitution of capital for energy or of one energy source for another.

THE LOW-INVESTMENT STRATEGY

Transaction Prices

Prices play a second role in Soviet energy conservation. Beside the planning price used to make investment decisions is a separate system of transaction prices. These are the prices at which energy is actually bought and sold. Because they directly affect the consumer, transaction prices figure prominently in the low-investment or housekeeping strategy, and here too prices have proved an obstacle to energy conservation.

Soviet energy transaction prices for both industrial and residential consumers are far below actual energy costs. The fuel bills of even large enterprises can be too low even to be recorded. At the large Gorky Automobile Factory, for instance, natural gas constitutes less than 2 or 3 percent of total production costs.[21] One plan to control industrial over-consumption of gas during periods of intense cold by charging punitive rates (as much as five times above normal) for consumption above established limits had to be abandoned when planners concluded that gas prices were so low that such surcharges would have no effect if charged to an enterprise's direct production costs.[22] It is hard to imagine how a surcharge system could be applied to residential customers. Most homes are charged only a flat subscription fee for gas.[23]

Underpricing of energy was part of earlier policies designed to encourage consumption. In 1968, prices for centrally supplied heat and electricity were set low to encourage

---

[20]This section is based on Campbell, "Energy Prices . . .," op. cit.

[21]Varavka, op cit., p. 16.
[22]N. Fedorov, "Economic Flame," *Pravda*, Feb. 2, 1981.
[23]Varavka, op. cit.

users to switch to the central sources. The rates have not changed since. As a result, most power systems lose money—70 million rubles a year at the Moscow system alone. Demand for electrical power has recently been growing faster than capacity, but authorities have added to this demand by establishing a new system of preferential rates to agricultural users.[24]

Soviet planners are well aware that energy prices are too low. On January 1, 1982, transaction prices of coal, petroleum, natural gas, fuel oil, electric power, and thermal energy will be increased. According to the State Committee on Prices, the new prices will provide a better stimulus to conservation because they will make the more remote consumer and the consumer of higher quality fuels pay more. Early indications are that this price rise will be on the order of a 2.3-fold increase. If this is indeed the case, it should help to alleviate the problem—assuming that all other prices are not increased in tandem. But although a major criterion for this reform appears to have been the increasing cost of energy extraction, Western experts believe that it is still unlikely that the 1982 prices will reflect the real opportunity costs of energy.[25]

Nor is there necessarily a strong correlation between higher prices and lower energy consumption. There is some evidence that the 1967 price reform—which greatly increased the prices of oil products, natural gas, and coal—resulted in substantial energy savings.[26] However, the mechanisms which brought about that result are not well-understood and may not still be operative. Because of numerous institutional barriers in the Soviet economy, higher prices do not guarantee significantly lowered consumption.

In sum, it is far from axiomatic that higher prices will lead to large energy savings. The Soviet system is one in which the role of markets is deliberately restricted and the impact of prices therefore limited. Within the existing incentive system, factory directors who cut their energy bills are likely to be rewarded with a cut in future energy allocations. Moreover, some energy-related commodities have "no value" because no markets exist for them. Designers of petrochemical plants typically fail to provide uses for byproducts, which can amount to two-thirds of the original feedstock. The usual practice is simply to burn them—using fuel oil.[27] Finally, the numerous administrative barriers that separate producers from users may make decisionmakers remote from the costs they impose.

Measurement of Energy Consumption

The difficulties of employing a price mechanism are compounded by problems in measuring the amount of energy actually consumed. The Soviet press frequently alludes to the widespread lack of apparatus to measure all energy sources, at every stage from extraction to final use. From the oilfields of Tyumen to the gas heaters in Moscow homes, energy is produced and delivered "na glazok," as the Russians say, "by eye alone."[28] The situation is apparently less serious for electricity than for gas,[29] and less serious in the cities than in the countryside,[30] but the problem remains pervasive.

[24]Kovalev, et al., op. cit.

[25]*Ekonomika neftyanoy promyshlennvosty,* August 1980, p. 2; Campbell, "Energy Prices . . .," op. cit., pp. 30-32.

[26]William J. Kelly, "Effects of the Soviet Price Reform of 1967 on Energy Consumption," *Soviet Studies,* vol. XXX, No. 3, July 1978, pp. 394-402; Albert L. Danielson and Charles D. DeLorme, Jr., "An Alternative Analysis of the Effects of Energy Prices on Energy Consumption in the Soviet Union," *Soviet Studies,* vol. XXXI, No. 4, October 1979, pp. 581-584. Danielson and DeLorme employed different econometric specifications, but confirmed Kelly's thesis that energy prices and energy consumption in the Soviet economy are inversely related.

[27]P. V. Fedorin, "An Important Source of Savings of Fuel and Power," *Ekonomicheskaya gazeta,* No. 5, January 1981, p. 10; "It Does Not Appear in the Designs," *Pravda,* Mar. 5, 1980, in JPRS No. 75687, May 13, 1980, pp. 13-16.

[28]For comment on Tyumen, see V. P. Rosliakov, "Reserves of the Economy of Energy Resources in Western Siberian Industries," *Neftysnoye khozyaystvo,* No. 9, 1980, p. 10; on household use see Varavka, op. cit.

[29]Orudzhev, op. cit.

[30]"The Price of a Kilowatt-hour," *Pravda,* Mar. 7, 1981. Electricity charges in most collective and state farms are established according to the rated power of the electrical equipment located on the farm, regardless of actual level of

Lack of measurement apparatus causes both short- and long-term problems. In periods of intense cold, for instance, there is no way to control or even evaluate surges in demand for gas.[31] Serious long-term consequences are the impossibility of setting and enforcing rational consumption standards, and the difficulty of setting meaningful energy prices, particularly for individual or small consumers.

Minpribor, the ministry in charge of instruments, automation, and control systems, is often blamed for the lack of meters and other energy-measuring devices. Gas and electricity meters are low-cost items, and therefore unprofitable for Minpribor to manufacture. (Ironically, Minpribor was one of the first ministries to undergo "economic reform" putting it on a profit-oriented, cost-accounting basis. Its failure exemplifies the difficulties of economic reform in a command system.) But the blame does not lie with Minpribor alone. Soviet energy has been literally "too cheap to meter," and it is doubtful that the forthcoming price reforms are radical enough to change this situation.

Consumption Norms

In the absence of a realistic price system or market mechanism, Soviet planners provide industries and enterprises with energy-consumption norms or indices to set optimum levels for energy use. These are detailed specifications for the amount of energy that may be employed in any process. Consumption indices are rigid and often arbitrary, and there is ample evidence that norm-setting is a process fraught with bargaining and controversy.

Because energy consumption varies widely with the age of the plant and type of technology, each industry exhibits a considerable range between the consumption of the leading enterprises and the industry average. In 1976, for example, the leading enterprise in the production of forgings and stampings in machine-building consumed 288 kg of standard fuel for every ton of metal it used. The industry average was 342 kg/ton. Similarly, the leading iron casting enterprise consumed nearly 100 kg/ton less of standard fuel than the industry average.[32]

An optimum system would not only require separate indices for every enterprise, but for every major process within each enterprise. Even then, the system would be flawed because energy inputs are usually measured only for the enterprise as a whole, not for individual processes. In practice, little information is available to advise on the effectiveness of indices.[33]

Therefore, the most common system of norm-setting is simply to set a consumption ceiling for an enterprise as a whole. This practice entails a large measure of guesswork, and encourages bargaining by interested ministries or local regions. Ministries can assign their enterprises inflated consumption norms that allow output targets to be met without energy constraints. If an enterprise exceeds its norm, the ministry can raise it after the fact. The enterprise can then claim paper energy savings which are credited to its conservation performance.[34]

In principle, the main consumption norms are steadily lowered each year according to an official plan specifying the amount of the decline. But ministries do not always abide by these plans. One recent article charged that the Ministry of Power failed to carry out its 1979 plan for lowering the heat-rate norm in thermal powerplants, and was thereby responsible for wasting over 0.622 mtoe.[35] Moreover, some norms are exempt from the

---

Continued from p. 235.

use. This fact, combined with preferential rates for electrical power (part of a massive program of rural electrification that received high priority during the 1970's), means that there are no disincentives attached to overconsumption.

[31]Fedorov, op. cit.

[32]Yu. Sibikin, "The Efficiency of Utilization of Fuel-Energy Resources in Machine Building," *Planovoye khozyaystvo*, No. 12, 1979, p. 49.

[33]V. A. Zhmurko, "Economics of Electric Energy—A Common Concern," *Ekonomicheskaya gazeta*, No. 4, 1981, p. 9.

[34]"Fuel and Energy—Strict Accountability," *Pravda Ukrainy*, Dec. 7, 1979; S. Bogatko, "The Working Kilowatt," *Pravda*, Jan. 8, 1980; see also Zhmurko, op. cit.

[35]Dolgikh, "To Increase the Level . . .," op. cit., p. 23.

annual change. The Ministry of Power has been blamed for failing to insulate power stations and steamlines. As a result, 24 million gigacalories of thermal energy are purportedly lost each year. But the norms governing insulation have not changed since 1959.[36]

The combination of a consumption system with no meters, prices too low to encourage strenuous conservation efforts, and meaningless norms produces a vicious circle which has been very difficult to break. The deficiencies of the price system make the norms necessary, but the lack of measurement apparatus makes them easy to virtually ignore. So long as this situation persists, there is little incentive to obtain meters or to pressure Minpribor, which at present cannot be induced to produce them. The root of the problem is the fact that energy has in the past been so cheap in the true economic sense that careful monitoring was unnecessary. Now that this is no longer the case, the task of the Soviet system will be to adjust to the new circumstances. It may be expected, however, that the rigidities and dysfunctional side-effects of the command economy will make change slow, especially if the political leadership is less than fully committed to conservation.

Monitoring and Enforcement of Conservation

The Soviet leadership's commitment to housekeeping strategies might be measured by the effort put into creating effective machinery for monitoring and enforcement. In the last year, enforcement offices with modest functions have been upgraded and given greater powers; the scale of monitoring activity and enforcement has increased; and the official publicity given to the effort has grown. Officials of the monitoring agencies themselves, however, are among the most outspoken in charging that their efforts have been almost totally ineffective.

The two most visible enforcement offices, the State Power Inspectorate of the Ministry of Power and Electrification and the State Gas Inspectorate of the Ministry of the Gas Industry, are well-established bodies. Originally charged with oversight of the supply of gas and electricity in their respective ministries, their jurisdiction has been expanded to include a wide range of other industrial ministries. In theory, the State Gas Inspectorate can recommend administrative fines of up to 100 rubles, and it has the power to cut off an offender's gas supply. In practice, however, such powers are weak. The likelihood of a cutoff is very small, and even the imposition of a fine must be assessed by the apparatus of the local Soviets. The "recommendations" of the inspectorates have been widely ignored by enterprises.

The energy inspectorates are useful, however, in conducting investigations which have publicized the extent of inefficient energy use throughout Soviet industry. Such publicity in itself will do little if anything to influence the incentive structure which regulates the behavior of Soviet enterprise officials, but it may be a first step to potentially more drastic actions.

Another way of publicizing official Soviet policy is through public mobilization campaigns. Common devices in the conservation campaign include "commissions" and "staffs" located in factories to perform public inspections; "raids" and contests for energy conservation; "socialist pledges" and "personal creative plans" to save energy.

These groups lack power and are usually regarded as a symptom of low priority and inaction. This appears to be true of the conservation campaign; in fact, many enterprises are not even going through the motions. In 1978, the State Power Inspectorate found that 21 of 48 enterprises surveyed in the Ministry of Ferrous Metallurgy had made no move to apply a recent statute establishing bonuses for workers and engineers for saving energy, and about a third of the enterprises had adopted no energy-

[36]"Don't Let Heat Be Lost," *Materialno-tekhnicheskoye snabzheniye*, No. 9, 1979, pp. 42-43, in JPRS No. 75753, May 22, 1980, pp. 16-19.

Now no one will reproach us that we are devoting little attention to saving electric power.

A Soviet cartoon that appeared in *Pravda* in March 1978

saving program at all.[37] According to one account, the much-vaunted public groups are "activated only when the corresponding directives come down from above and gradually dwindle down to nothing as soon as the campaign is over."[38]

The result is widespread cheating. According to critics in the energy inspectorates, industrial enterprises have become adept at saving awe-inspiring amounts of energy—on paper. For example, in 1978 the Uzbek Ministry of Municipal Services claimed to have saved 410 tons of gasoline, 2 percent of its annual consumption. An official inspection discovered that the ministry's drivers did not log gas consumption in their trip reports, routinely exaggerated the length of their trips, and sold quantities of gasoline on the black market.[39] Such accounts have lately become common in the Soviet press.

The enforcement system is now being expanded. A recent official decree on energy conservation instructs all ministries, state agencies, and republic-level Councils of Ministers to develop conservation offices.[40] In 1980, a decree of the U.S.S.R. Council of Ministers upgraded the official rank and powers of the conservation inspectorates of the Ministries of Gas and Power.[41] The responsibilities of the Gas Inspectorate now include gas consumption throughout the economy—a change which has increased its jurisdiction from 22,000 enterprises to 150,000.[42] In addition, the Gas Inspectorate is now responsible for forwarding recommendations to the State planning apparatus on whether natural gas should be used in proposed new enterprises; passing on all proposals to install new gas-using equipment; demanding the removal from service of old and inefficient gas-burning devices; and

---

[37]Veselov, op. cit., p. 34.

[38]Iu. Kopytov, "What Does the Analysis Say?" *Sotsialisticheskaya industriya*, July 5, 1979; see also Ilinskiy, op. cit.

[39]Ilinskiy, op. cit.

[40]"To Improve the Economics of Fuels," op. cit.

[41]The official text appears in *Sobraniye postanovleniy pravitelstva SSSR*, No. 14, 1980, pp. 339-348. A statement by TASS appeared in *Sotsialisticheskaya industriya* on June 22, 1980. See also "The Kilowatt Must Be Put to Work," *Pravda*, Dec. 15, 1980; and A. S. Voytenko, "To Improve the Control of Gas Utilization in the National Economy," *Gazovaya promyshlennost*, No. 8, 1980, pp. 20-22. The change is partly a matter of nomenclature: the State Gas Inspectorate has been upgraded from the rank of *upravleniye* (administration) to *glavnoye upravleniye* (chief administration), while the State Power Inspectorate has been promoted from the Office of Power Supply and Oversight of Use to State Power Inspectorate and Sales Office—the change of order presumably reflecting a change in emphasis in the agency's official duties.

[42]Voytenko, op. cit.

screening recommendations for mass manufacture of new devices. Inspectorate representatives have also been made official members of so-called "acceptance commissions," the bodies charged with commissioning new completed buildings and plants which are about to be transferred to the user. Membership on these commissions gives the State Gas Inspectorate the power to veto the commissioning of new plants with gas-consumption systems not up to official regulations.

It will be some time before the impact of the Gas Inspectorate's new powers can be measured. On paper they resemble those technically available to similar enforcement offices, notably water-quality inspectorates, whose actual influence is known to be modest.[43] But analogous groups have made impressive claims. In 1980, the Power Inspectorate reportedly made over 60,000 plant inspections, imposed over 100 million rubles in fines, and saved 1.5 billion kWh of electricity. It is difficult to evaluate the accuracy and practical effect of these assertions, but an important inference can be drawn from the evolution of their tone over time. Power Inspectorate officials who now boast of housekeeping savings were recently writing derisively of the local energy-saving efforts, stressing instead the importance of central investment measures as the only way to meaningfully conserve energy.[44]

This change in attitude, together with the general evolution of official policy, suggests that emphasis in the conservation campaign as a whole has shifted from high- to low-investment. The stress of such a strategy would be on the small innovations that enterprises can implement without central investment—substituting stamping for cutting in the manufacture of small metal parts, for example. If a locally oriented conservation strategy has in fact been accorded priority,

and if Soviet leaders are serious about energy conservation, there should now be evidence of more prominent participation in these campaigns by the apparatus of the Communist Party. References to such involvement by the Party have hitherto been rare.[45]

OTHER CONSERVATION STRATEGIES

The preceding discussion has suggested that the U.S.S.R. lacks the capital to implement a major high-investment conservation strategy and will encounter difficulties in monitoring and enforcing low-investment measures. These do not exhaust the leadership's alternatives. It can also impose calculated fuel and power cutoffs. Such tactics would obviously be reserved for emergencies, but they would not necessarily impose unprecedented hardships on industrial or residential consumers—or seriously increase threats to economic growth. The European portion of the Soviet Union has historically experienced chronic shortages of power and fuel. Difficulties during the 1980's might, therefore, be viewed as a return to a traditional state of affairs that was interrupted by a brief period of energy abundance in the 1960's and 1970's.

One would expect Soviet authorities to be experienced in allocating shortages and cutoffs so as to preserve economic growth, and to be prepared to force unable or unwilling Soviet industry to conserve energy. There is no sign of coordinated implementation of such measures, however. The allocation of electrical power, for example, resembles a tug-of-war in which Gosplan and the industrial ministries are pitted against the Ministry of Power. The quotas assigned by the Ministry of Power to its regional power authorities are smaller than the quotas as-

[43]See Thane Gustafson, "Environmental Policy Under Brezhnev: Do the Soviets Really Mean Business?" in Donald R. Kelley (ed.), *Soviet Politics in the Brezhnev Era* (New York: Praeger Publishers, 1980), pp. 129-149.

[44]Kopytov, op. cit.

[45]One of the few references to an active role by the Party is in "The Price of a Kilowatt-hour," *Pravda*, Mar. 7, 1981. In the city of Ural'sk the local party forced the local power station to switch rapidly from fuel oil to gas, presumably using its influence to ensure allocation to the city of the necessary gas supplies.

signed by State planners to each of the consuming industries (based on the industries' own statements of their power needs). The result is power cutoffs. Those who are unable to pad their requirements so as to ensure a healthy margin of safety are those who suffer.[46]

Moreover, chronic problems in fuel supply to powerplants frequently cause the unified power grid that serves the European part of the country, the Urals, and the Transcaucasus to operate at reduced power. Shortages are not apportioned according to a system of political or economic priorities, but are spread equally among all unified grid customers in "universal brown-outs." Those users whose production depends on small electric motors, which slow or stop altogether when the power drops, are most seriously affected. Similarly, lack of sufficient capacity to cover daily peak demand forces network operators to resort to load-shedding, sometimes blacking-out entire areas. There is no sign of priority setting to determine who shall bear the costs, and indeed, there is evidence that these costs have never been systematically studied. As a result, undersupply of kilowatt-hours that cost 2 or 3 kopecks to generate cause losses of production on the order of 1 or 2 rubles.[47] These examples support the generalization that rationing and cutbacks, although increasingly frequent, are not well planned. Nor are they a particularly promising route for enforced energy savings in the economy.

[46]A. Fedosyuk, "Protect Energy," *Pravda*, Dec. 7, 1980.

[47]Styrikovich, op. cit.

OFFICIAL SOVIET CONSERVATION TARGETS

The preceding sections have described Soviet conservation strategies and the opportunities and constraints appropriate to them. The true test of any set of conservation measures is the extent of energy savings. These are particularly difficult to identify in the U.S.S.R., both because consumption data are scarce and difficult to interpret, and because of the manner in which energy savings are counted.

Energy savings in the U.S.S.R. are computed on the basis of the last year of the Five Year Plan (FYP) to which they apply. In other words, if a "savings rate" of 100 mtoe is claimed, this means that by the last year of the plan actual energy consumption was lower by 100 mtoe than it would have been at the input norms experienced at the beginning of the plan period. The results are noncumulative, and the reduced consumption rates achieved by the end of one plan become the norm for the following one.

The U.S.S.R. claims that in the Ninth FYP rates were reduced enough to save 81 mtoe and in the Tenth, 62 to 78 mtoe. The latter was considerably below the plan target which called for savings of 100 mtoe.[48] The savings target for 1985 is 100 to 106 mtoe, roughly 10 percent of 1980 domestic energy consumption.[49] Table 52 shows the way in which the savings target for 1980 was broken down in the original plan. It demonstrates how conservation is defined in the

[48]*Eko*, No. 9, 1980, p. 124 and No. 3, 1981, p. 30. *Eko* gives a 1980 figure of 62 mtoe; the 78 mtoe is from A. Lalayants, *Planovoye khozyaystvo*, No. 1, 1981, p. 35.
[49]*Pravda*, Dec. 2, 1980.

Table 52.—Tenth FYP (1976-80)
Planned Energy Savings

27 mtoe	Decline in consumption of fuel per unit of output.
24 mtoe	Increase in output of electricity from nuclear power and hydropower.
22 mtoe	Decline in consumption of electricity and heat per unit of output.
12 mtoe	Better use of secondary heat resources.
9 mtoe	Efficiency gains in consumption of light fractions.
4 mtoe	Cuts in losses from storage and transportation.

SOURCE: *Eko*, September 1980, p. 124.

U.S.S.R., and the areas in which major energy savings are anticipated. These figures include net additions to nuclear and hydropower capacity, items which would not be considered "savings" in the West.

The most noteworthy feature of the targets for both the Tenth and Eleventh FYPs is their modesty, particularly if they represent the outer bounds of official optimism. The original 1980 goal of saving 100 mtoe amounted to conserving about 10 percent of the total primary energy planned for distribution in that year. If the figures are adjusted to subtract net additions of nuclear and hydropower capacity, actual planned savings were about 7.5 percent of total primary energy. The 1985 goal is even less ambitious. Although the total primary energy available for distribution is projected to grow by over 20 percent, the total amount to be saved remains unchanged from the previous plan.

Based on past performance, however, the prospects for achieving even this are uncertain. Although Soviet energy consumption nearly doubled between 1965 and 1980, the "annual savings rate" declined from 16.1 to

5.6 percent over the same period.[50] A careful examination of Soviet predictions for the period beyond 1985 shows that experts expect this downward trend to continue, albeit at a more moderate rate, to the end of the century. One prediction is for a savings of only 49.8 mtoe by 2000,[51] a forecast predicated on the assumption that major technological advances will be achieved, i.e., that high-investment conservation strategy will be successfully implemented. Although the electric power and ferrous metallurgy sectors have continued to improve their energy efficiency, Soviet difficulties with assimilating technological innovation, and the extreme shortage of capital for investment, make the prospects for wholesale achievements across a number of industrial sectors unlikely. Without evidence of fundamental changes in investment and conservation strategies, and without basic reforms of incentive, price, and monitoring systems, there is little reason to expect more than modest energy savings in the Soviet Union over the next 10 years.

[50]Makarov and Melentev, op. cit., p. 30.
[51]Campbell, op. cit., p. 31.

PROSPECTS FOR CONSERVATION

The picture that emerges is of a level of technology and a structure of energy consumption that provide ample opportunities for energy conservation, and an economic system which impedes the implementation of promising conservation measures. Some idea of the potential for and difficulties to be encountered in conservation in the U.S.S.R. can be gleaned from an examination of several areas in which energy savings might be most easily achievable.

Perhaps the most promising target for Soviet energy savings is in boiler uses, i.e., the use of fuel for the production of electricity, steam, and hot water. Energy consumption in boiler uses is increasing rapidly. In 1970, they accounted for 54 percent of the

total primary consumption; by 1980 the share had risen to 59 percent.[52] This trend is expected to continue. One Soviet source estimates that by 2000 half of all Soviet energy consumed will be in the form of electricity.[53] Boiler uses are potentially the most flexible means of energy consumption, and they give the U.S.S.R. a measure of flexibility in its efforts to substitute coal and gas for fuel oil.

Efficiency gains in the production of electricity and heat have been the chief source of energy savings in the past two decades, but an upper limit has nearly been reached. Even the most optimistic Soviet forecasts see sav-

[52]Nekrasov and Troitskiy, op. cit.
[53]Makarov and Melentev, op. cit., p. 29.

ing no more than about 16 mtoe in this area by 1990,[54] a small fraction of the roughly 249 mtoe in conversion losses the Soviet economy will be experiencing by that time. Figure 21 shows the striking pattern of decline in efficiency gains over the last decade. The trend suggests the possibility that by 2000, efficiency in electricity and heat generation may actually be declining. Nevertheless, replacement of small furnaces and the displacement of the direct use of fuel in them should lead to continued, if small, overall gains in energy efficiency.[55] Soviet

---

[54]Troitskiy, op. cit., p. 25.

[55]There are over 280,000 small furnaces and boilers scattered throughout the country. These produce 1.5 billion of a national total of 3.2 gigacalories of heat annually. Their efficiency is low: they require an average of 200 to 220 kg of standard fuel to produce 1 gigacalorie of heat, whereas the larger boilers require only 173 to 175 kg. See Lalayants, op. cit., p. 40.

Figure 21.—Where Energy Savings Come From (1961-85)

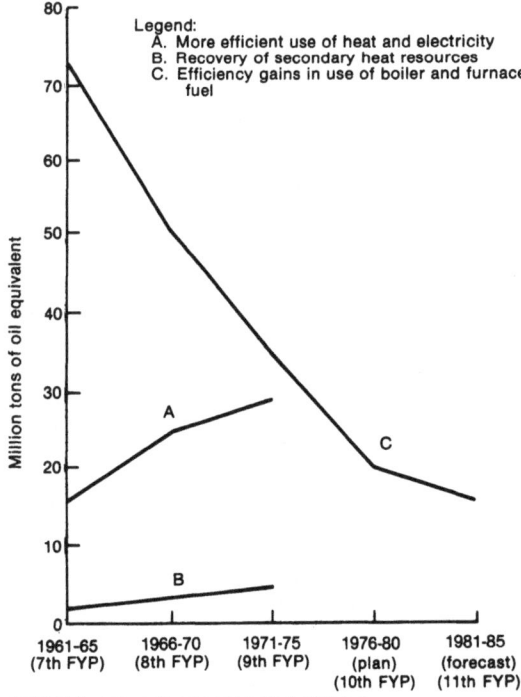

Legend:
A. More efficient use of heat and electricity
B. Recovery of secondary heat resources
C. Efficiency gains in use of boiler and furnace fuel

Million tons of oil equivalent

1961-65 (7th FYP) 1966-70 (8th FYP) 1971-75 (9th FYP) 1976-80 (plan) (10th FYP) 1981-85 (forecast) (11th FYP)

SOURCE: Vestnik mashinostroyeniya, No. 3, 1980, p. 5.

---

planners are aware that future progress will depend more on efficiency gains in the *use* of electricity and heat, rather than in their *production.*

Further insight into energy conservation prospects may be gained from analysis of individual consumption sectors. Agriculture, whose share of total energy use has grown substantially in the last two decades, is particularly interesting. A large part of the considerable agricultural investment under the present Soviet leadership has consisted of mechanization of farm and off-farm operations and food processing. Table 49 recorded the result. The amount of energy used by the agricultural sector doubled between 1965 and 1980, from 30.2 to 60.5 mtoe. The horsepower available per worker in Soviet agriculture increased from 7.7 in 1965 to 22.9 in 1979, and continues to rise rapidly.[56]

This growth in energy consumption has largely consisted of electricity, a reflection of an active rural electrification policy. Total agriculture-related use of electricity rose about fivefold between 1965 and 1979, from 21 billion to 102.3 billion kWh.[57] In 1965 most of the fuel consumed in the agricultural sector was diesel fuel and gasoline, but by 1979 electricity accounted for more than 10 percent of total agricultural energy consumption. This proportion is growing.[58]

Agriculture is an important sector for overall energy conservation policy because it may be one of the few sectors of the economy in which more or less arbitrary cuts could be made in case of emergency. Agriculture's share of total energy—just as its share of capital investment—depends on whether it continues to enjoy the high priority accorded it under the present leadership. There is at present no evidence that Soviet leaders are considering any major curtailments. Prime Minister Tikhonov's report to the 26th Party Congress called for a 50-percent increase in the horsepower available per agricultural

---

[56]*Narodnoye khozyaystvo,* 1979, p. 120.

[57]Ibid., p. 125.

[58]Campbell, op. cit.

worker during the Eleventh FYP.[59] Still, eventual successors to Premier Brezhnev may cut back the massive flow of inputs to the countryside, and it is possible that agriculture could be held considerably short of the roughly 156 to 187 mtoe it might have expected to receive by the end of the century.

Such a change in priorities would allow the Soviet economy to save some of the light-fraction petroleum products which make up a substantial portion of agricultural consumption. To the extent that agricultural energy consumption continues to shift toward electricity, the demand for light-fractions is lightened further. On the other hand, increasing electricity consumption compounds Soviet problems in shifting away from fuel-oil in electricity generation. Opportunities to ration agricultural electricity supply, therefore, will presumably be welcome.

Another promising candidate for conservation efforts is the energy producing sector itself. The importance of conservation here is magnified by the fact that own-use by energy producers is bound to increase. The refining and petrochemicals industry, for example, in 1979 consumed 7 percent of the country's total output of refined oil. As the Soviet Union increases the volume and depth of refining, this share will grow, even if efficiency is improved.[60]

Important savings could be gleaned here by reducing losses, which are currently estimated at 3 mtoe/yr in the gas industry alone. In the coal industry 6.2 to 9.3 mtoe are lost annually through faulty transportation and storage. In the oil industry in 1980, 13.5 out of a total of 47 billion cubic meters of associated gas were lost in 1979.[61] Newspaper accounts claim that 30 million tons of crude oil (i.e., 5 percent of total Soviet output) are lost annually by producing organizations, i.e., before the crude oil is shipped.[62] Further sizable losses occur in transportation, because of the poor state of repair of tank cars and wasteful methods of transferring oil and oil products from one vessel to another in transit.[63] Finally, losses of electricity in transmission networks are substantial. In 1978, these amounted to more than 9 percent of the total generated (95 billion kWh), and the share will increase as transmission distances rise.[64]

---

[59]*Pravda*, Feb. 28, 1981.

[60]G. M. Yermolov, "Conservation of Fuel and Energy Resources in the Oil Refining and Petrochemical Industry," *Khimiya i tekhnologiya topliv i masel*, No. 11, 1980, pp. 13-16, in JPRS No. L/9482, Jan. 12, 1981, p. 17.

[61]A. M. Lalayants, "Problems of the Economics of Fuel-Energy Resources in the National Economy," *Planovoye khozyaystvo*, No. 1, 1980, pp. 34-44.

[62]V. V. Kuleshov and V. M. Sokolov, "The Economics of Fuel: Possibilities and Reality," *Eko*, No. 9, 1980, pp. 114-122. It is claimed that this is an improvement. Since the 1960's, the loss rate per ton extracted has declined by one quarter. P. Kozlov, "Lose Not, But Save!" *Pravda*, June 26, 1979.

[63]Transfer losses account for a further 2 million tons of crude annually (Kuleshov and Sokolov, op. cit.). The poor condition of railroad tankcars is discussed in N. Valitov, "Containers for Fuel," *Sotsialisticheskaya industriya*, Dec. 24, 1980. The problem is substantial because 46 percent of Soviet oil is shipped by rail (Valitov, op. cit.). Doubling the ton-mileage of oil shipped by rail between 1965 and 1978 has severely overloaded the railroad network, which may account for much of the poor condition of the cars. K. B. Leikina, *The Lowering of Losses in the National Economy* (Moscow: "Nauka," 1980), p. 108.

[64]"Directions for the Economics of Fuel and Energy in the National Economy," *Planovoye khozyaystvo*, No. 2, 1979, p. 6. Seventy percent of those losses occurred in transmission lines operating at voltages of 110 kV or lower. Total line losses, including the transmission lines of the user enterprises, amounted to about 250 billion kWh (51.6 mtoe) in 1979. See V. Vladimirov and I. Tarikuliyev, "Without Losses in Transmission," *Sotsialisticheskaya industriya*, Sept. 14, 1980.

THE POTENTIAL ROLE OF THE WEST
IN SOVIET ENERGY CONSERVATION

Western energy technology might play a role in either a high- or low-investment conservation strategy. If Soviet planners implement centralized, high-investment methods, logic would dictate that it consist primarily of new plant and processes in the most energy- and capital-intensive industries—metallurgy, oil refining and petrochemicals, chemistry—and that investment be aimed at reducing own-use within the energy-producing sector itself. Promising areas for industrial conservation include the following:[65]

- *Chemicals:* improvements in the energy-efficiency of production of yellow phosphorus, chlorine, caustic soda, ethylene, acids of phosphorus and nitrogen, divinyl monomers. Investment in these processes could produce reductions of 5 to 25 percent in energy use below current levels. One example frequently mentioned in Soviet sources is process changes in the production of ammonia, which could considerably cut unit electricity consumption.
- *Computers:* institution of microprocessor and minicomputer-based process control systems in such energy-intensive processes as oil refining.
- *Metallurgy:* improvements in aluminum-refining, continuous steel-casting, combined-blast systems for blast furnaces, autogenic processes in nonferrous metallurgy.
- *Materials:* production of cement by the dry method, which consumes half as much energy as the wet method.

With the exception of computer process control, none of these require particularly advanced technologies, and if Soviet policy-makers decided to import the required capacity, they would likely find manufacturers throughout the West who could meet their needs. In fact, implementation of a high-investment conservation strategy is tantamount to industrial modernization. There is no reason to believe that the same constraints that have inhibited wholesale purchase and implementation of Western industrial technology in general—including shortages of hard currency and difficulty in absorbing and diffusing imported technology—will not continue to operate.

A serious, high-priority, low-technology approach to conservation might lead to considerable demand for Western equipment to bolster the inadequate output of Soviet industry in insulating materials, metering equipment, small boilers and furnaces, jets and burners, static condensor batteries, etc. In the past, however, Soviet ministries and foreign-trade organizations have been unwilling to expend scarce hard currency on small items of this type, partly because they are less attractive than high technology items and partly because the need for them is scattered across many separate organizations.

An important complement to conservation is the *restructuring* of energy consumption to allow substitution among primary energy sources. Here major investment is needed soon in expansion of gas pipeline capacity, particularly expansion of local feeder networks; improvement of capacity for cleaning and enriching coal; peak-coverage technology to make up for the rigidities of nuclear power in the European zone of the country; acceleration of nuclear powerplant construction and transmission lines; and development of refinery capacity. These tasks should not be postponed, yet the evidence from Soviet literature is that planners are experiencing severe delays in all of them.

---

[65]These examples are drawn from a series of articles published in *Planovoye khozyaystvo*, No. 2, 1979. Similar "shopping lists" of opportunities for energy savings through centralized investment are common in the Soviet literature. See also Iu. Sibikin, "The Efficiency of Utilization of Fuel-Energy Resources in Machine Building," *Planovoye khozyaystvo*, No. 12, 1979, pp. 48-54.

Soviet decisionmakers might resort to Western technology to eliminate the most crucial of these bottlenecks. If they do so in a manner consistent with behavior of the past two decades, Western technology will be sought to gain a degree of flexibility that compensates for the sluggishness of domestic industry. The list of sectors in which the Soviets have made the greatest use of imported technology reveals an interesting pattern. In the chemical and agrochemical industries, in the automotive and trucking industries, and in machine tools, much of the Soviet import activity has been clearly aimed at accelerating new policy initiatives. Restructuring Soviet energy demand to allow an indispensable substitution among primary sources of supply might fall into the same urgent category.

Perhaps the most important connection between conservation and technology transfer, however, lies in possible displacement effects. To the extent that conservation is tantamount to modernization, a vigorous and successful conservation program could result either in an overall reduction of the need for Western technology or in a displacement of imports toward smaller, lower technology equipment. But, since the main thrust of Soviet energy policy appears to be directed toward energy production rather than conservation, it is possible that Soviet need for large, high-technology Western items will be correspondingly greater. Production-related imports are more likely to be concentrated in a few industries and firms than are imports targeted at energy consumption.

SUMMARY AND CONCLUSIONS

The structure of Soviet energy consumption, particularly the high percentage of energy consumed by industry, presents many opportunities for conservation. This could well be accomplished both through a centralized, high-investment strategy and a local, low-investment strategy, the latter aimed at improving the efficiency of operation of equipment already in place. Conservation should be an extremely promising policy for the U.S.S.R. Effort invested in saving energy could yield a greater payoff at the margin than investment in new production, and the difference would grow as in time production costs rise.

The emphasis of the Soviet energy policy has hitherto been on production rather than conservation. As in the West, the perception of a pressing need to conserve expensive energy resources is relatively recent and serious conservation campaigns are relatively new. Stress has lately been on a local, low-investment rather than a centralized high-investment approach. Significant savings could be achieved through a low-investment strategy, but it is unlikely to produce major results very quickly because of weaknesses in the price structure and the prevailing incentives, the enforcement mechanism, the system of norms, and monitoring of measurement.

There is little reason to expect that Western nations will have significant impact on Soviet energy conservation. Short of contributing to a long-term program of extensive industrial modernization, the most that the West could provide is a variety of "low-technology" conservation equipment, on which the U.S.S.R. is unlikely to expend precious hard currency. In sum, while major opportunities for energy savings exist and indeed have brought results, rigidities in the political and economic structure could still prevent Soviet policymakers from taking full advantage of them.

CHAPTER 8

Energy and the Soviet Economy

CONTENTS

LIST OF TABLES

LIST OF FIGURES

Energy and the Soviet Economy

The energy sector importantly influences and is influenced by the nature and health of the Soviet economy as a whole. Those who formulate energy policy do so in a context which is affected by the structures and performance of the national economy. Their decisions, in turn, help to set the parameters for economic performance. This chapter explores the role of energy in the Soviet economy. It seeks to highlight the economic impacts of alternative plausible levels of Soviet energy availability, and to point out major consequences of various economic eventualities for energy production.

The chapter begins with an overview of the Soviet economy, highlighting recent growth trends. This provides a basis for examining the role of the energy sector in that economy, for identifying some of the factors that influence Soviet energy policies, and for describing the recent direction of these policies. The chapter then presents a simplified description of the Soviet economy that can be used to better understand prospects for energy and economic growth in the present decade. It culminates in the development of "best" and "worst" case scenarios for Soviet economic growth and energy trade in 1985 and 1990.

SOVIET ECONOMIC PERFORMANCE

ECONOMIC GROWTH

The rate of Soviet economic growth over the past quarter century has generally declined. This slowdown is reflected in gross national product (GNP), investment, and consumption spending growth rates. According to Western estimates, Soviet GNP grew at close to 6 percent annually in the 1950's, but growth slowed to 5.0 to 5.5 percent in the 1960's, to 3.8 percent in 1971-75, and to 2.8 percent during the Tenth Five Year Plan (FYP), (1976-80)[1] (see table 53). Investment has traditionally grown faster than consumption in the Soviet economy. In the 1950's, new fixed investment grew at an average annual rate of 10 to 12 percent, contrasted with 5 to 6 percent annual growth for consumption. Since then, the absolute and relative gap in growth rates has alternately narrowed and widened. In the period 1976-79, annual growth was roughly 4 percent for investment v. 3.2 percent for consumption.[2] The average annual growth rate of per capita consumption, a major contributor to maintaining political stability for the Soviet regime, fell from 4.6 percent per annum in the 1950's, to 3.6 percent in the 1960's, and 2.5 percent for the period 1971-79.[3]

Soviet defense spending is commonly believed to have grown roughly in line with GNP for most of the postwar period. In the past several years, however, estimated defense spending has grown at a more rapid rate than GNP. According to the Central In-

[1]Rush V. Greenslade, "The Real Gross National Product of the U.S.S.R., 1950-1975," in Joint Economic Committee, U.S. Congress, *Soviet Economy in a New Perspective* (Washington, D.C.: U.S. Government Printing Office, 1976), pp. 269-300; Central Intelligence Agency, *The Soviet Economy in 1978-79 and Prospects for 1980,* ER 80-10328 (Washington, D.C.: CIA, June 1980).

[2]Greenslade, op. cit.; and Central Intelligence Agency, *Simulations of Soviet Growth Options to 1985,* ER 79-10131 (Washington, D.C.: CIA, March 1979).

[3]See Gertrude Schroeder-Greenslade, "Consumption and Income Distribution," paper presented at the Airlie House Conference, Oct. 23-25, 1980.

Table 53.—Average Annual Rates of Growth for Soviet GNP, Factor Inputs, Factor Productivity, and Consumption Per Capita
(percent)

	1951 -55	1956 -60	1961 -65	1966 -70	1971 -75	1976 -80
GNP	6.0%	5.8%	5.0%	5.5%	3.8%	2.8%
Labor[a]	1.9	0.6	1.6	2.0	1.7	1.2
Capital..............	9.0	9.8	8.7	7.5	7.9	6.8
Land................	4.0	1.3	0.6	-0.3	0.8	—[c]
Combined factor productivity	1.4	1.8	0.9	1.5	-0.4	-0.7
Per capita consumption[b]	5.3%	4.2%	2.5%	4.7%	3.2%	1.6%[c]

[a]Man-hours.
[b]Total consumption.
[c]Refers only to 1976-79.

SOURCES: Rows 1-5 (1951-70): Rush V. Greenslade, "The Real Gross National Product of the U.S.S.R., 1950-1975," in Joint Economic Committee, U.S. Congress, *Soviet Economy in a New Perspective* (Washington, D.C.: U.S. Government Printing Office, 1976); Rows 1-5 (1971-79): Central Intelligence Agency, *The Soviet Economy in 1978-79 and Prospects for 1980,* ER-80-10328 (Washington, D.C.: June 1980); Row 6 (1951-70), Gertrude E. Schroeder and Barbara S. Severin, "Soviet Consumption and Income Policies in Perspective," in Joint Economic Committee, op. cit., pp. 620-660; Row 6 (1971-79): Gertrude Schroeder-Greenslade, "Consumption and Income Distribution," paper presented at the Airlie House Conference, Oct. 23-25, 1980, and Central Intelligence Agency, *Simulations of Soviet Growth Options to 1985,* ER 79-10131 (Washington, D.C.: March 1979). Preliminary 1980 estimates supplied by the Central Intelligence Agency.

telligence Agency (CIA), the defense share rose to a level of 12 to 14 percent of Soviet GNP at the end of the Tenth FYP period, after having stabilized at roughly 11 to 13 percent of GNP between 1965 and 1978.[4]

Although there is debate in the West regarding the relative weight of different factors in explaining the Soviet economic slowdown, identification of the basic factors is not in dispute. A country's aggregate output typically depends on the size of its labor force, its accumulated capital stock, and the combined productivity of capital and labor.[5] (Land is a third factor of production when agricultural output is included in the summary output measure.) The more rapid the

growth of capital and labor and of their combined productivity, the greater the rate of growth of output. Soviet growth rates for individual 5-year periods for each of these factors are shown in table 53.

Labor[6]

Growth in the Soviet labor force has fluctuated as a result of underlying demographic factors and changes in the labor force participation rate, i.e., the labor force as a percentage of the population of able-bodied ages. The dramatic slowdown in labor force growth in the late 1950's was caused primarily by the fall in the birthrate during World War II. The jump in the growth rate in the 1960's is attributable both to underlying demographic factors and to an increase in the labor force participation rate from 83 to 88 percent, reflecting in large part a significant increase in the number of women workers. The 1970's were characterized by a gradual decline in the labor force growth rate.

Capital

Although the overall rate of capital accumulation has slowed since the 1950's, Table 53 shows that it continues to be quite high, particularly in relation to GNP growth. Throughout this period, the capital stock has grown much faster than the labor force. This has resulted in a remarkably rapid rise in the Soviet capital-labor ratio. In industry, for example, the labor force increased about 140 percent between 1958 and 1978 (from 15 million to 36 million), while the industrial capital stock grew by 14.5 times over the same period.[7] This has led some observers to attribute part of the decline in Soviet growth

---

[4]Greenslade, op. cit.; CIA, *Simulations . . .,* op. cit.; and CIA, *The Soviet Economy ...,* op. cit.

[5]There is now, moreover, a growing 3-factor production function literature based on capital, labor, and energy.

[6]All data on Soviet labor are from Murray Feshbach and Stephen Rapawy, "Soviet Population and Manpower Trends and Policies," in Joint Economic Committee, U.S. Congress, *Soviet Economy in a New Perspective,* op. cit., pp. 113-154; and Murray Feshbach, "Population and Labor Force," paper presented at the Airlie House Conference, Oct. 23-25, 1980.

[7]Martin L. Weitzman, "Soviet Industrial Production," paper presented at the Airlie House Conference, Oct. 23-25, 1980.

rates to strong diminishing returns to capital in industry.[8]

Slowing growth of inputs has been reinforced by falling productivity. Increased productivity of the factor inputs, made possible by the introduction of new technology, and perhaps in some cases by improvements in the planning and management systems, accounted on average for 1 to 2 percentage points of the annual GNP growth rate in the 1950's and 1960's, but in the 1970's turned negative.

PROSPECTS FOR ECONOMIC GROWTH

In 1979 Soviet GNP rose only 0.7 percent, and economic growth in 1980 has been estimated by the CIA at 1.5 percent. For the period 1978-80, GNP increased by an annual average of 1.9 percent, the lowest for any 3-year period since World War II.[9] It therefore appears that the U.S.S.R. has entered a period of more fundamental constraints on economic growth. As analysts in both East and West have long anticipated, the Soviets have exhausted the potential for rapid growth based on an *extensive* strategy, i.e., the rapid accumulation of factor inputs with relatively little emphasis placed on their quality or their efficient use.

Barring a significant change in the labor force participation rate and the death rate in the 1980's, the increase in the Soviet labor force over the next decade is preordained, i.e., all its potential members have already been born. Western experts have estimated that this rate of growth will be only 0.4 to 0.5 percent annually over the next decade, about one-fourth the rate of the 1970's. This dramatic projected slowdown in the annual growth of the population of able-bodied ages is due to a number of demographic factors, including progressive aging of the population, a fall in the birth rate since the 1950's, and a recent increase in mortality rates not entirely explained by the age structure of the population.

The slow growth of the labor force is expected to continue to the end of the century. While the labor force participation rate may be influenced through economic policy, it is unlikely that it can be raised much more in the absence of coercion. At 88 percent (v. 65 percent for the United States), the rate is already the highest in the industrialized world. These aggregate labor force trends will be aggravated by the shift in the population structure towards non-Russians in Central Asia. Unless there is considerable migration of Central Asians to labor-deficit areas of the U.S.S.R., Soviet industry could face labor constraints even greater than those suggested by the aggregate labor force projections.

Capital accumulation cannot continue to grow at the high rates of the past without severely curtailing the share of output going to consumption. In any event, the productivity of such additions to the capital stock is questionable.

To counter these declines in the growth of inputs, Soviet leaders are hoping for large increases in productivity. The Eleventh FYP calls for an increase in the productivity of "socialist labor" of between 17 and 20 percent over 5 years.[10] The growth of productivity is partly determined by economic policy but, perhaps more fundamentally, it is also conditioned by the economic system. In particular, the capability of the economy to generate ever larger output levels from given "inputs" of labor and capital—in other words, to shift from extensive to *intensive* growth—is critically dependent on the nature of the prevailing decisionmaking, information, and incentive systems. These elements of the economic system will have a fundamental impact on the efficiency with which existing resources are used, and on the

---

[8]This interpretation was first developed in Martin L. Weitzman, "Soviet Postwar Growth and Capital-Labor Substitution," *American Economic Review,* vol. 60, No. 4, September 1970, pp. 676-692.

[9]CIA, *The Soviet Economy in 1978-79 . . .* , op. cit.

[10]*Ekonomicheskaya gazeta,* No. 49, December 1980.

extent to which technological progress and industrial innovation are stimulated. A significant improvement in productivity performance, therefore, would seem to presuppose important changes in the way in which the economy operates.

Soviet leaders have understandably resisted the idea that the economic system requires fundamental change, but they have accepted modifications classified as "improvements in the economic 'mechanism.'" In contrast to the limited decentralization that has occurred in other socialist countries such as Hungary, Soviet "reform" efforts since the 1950's have largely been devoted to attempting to perfect the system of central planning and to modifying organizational structures and incentive systems so as to increase the likelihood that lower management levels will operate in accordance with plan directives.[11]

The latest of these reforms, announced in a party-government decree in July, 1979, concerning the "improvement of planning and the strengthening of the influence of the economic mechanism in the promotion of production efficiency and the quality of work," called for several changes in the planning and management systems. These included emphasizing interenterprise contractual obligations, strengthening the bonus

[11]See Joseph S. Berliner, "Planning and Management," presented at the Airlie House Conference, Oct. 23-25, 1980.

system, and adopting new major success indicators for industrial management.[12] In addition, there is to be a basic reform in the wholesale price structure at the beginning of 1982. The implications of price reform for energy are discussed in chapter 7.

U.S. experts on the Soviet economy have been virtually unanimous in concluding that these changes in the "mechanism" are not fundamental, and are therefore unlikely to significantly forestall a continued slowing in Soviet economic growth.[13] Indeed, the 1979 decree has been characterized as one of a series of fairly minor reforms which began in 1965. The reform process has been likened to "being on a treadmill, for most of them amounted to reforming previous reforms that failed to work."[14] It is difficult to evaluate these predictions, however, because it is almost impossible to empirically measure the impact of changes in economic system on aggregate economic performance.

[12]*Ekonomicheskaya gazeta*, No. 32, August 1979. See also Berliner; op. cit., and Hans-Hermann Hohmann and Gertraud Seidenstecher, "Änderungen im Sowjetischen Planungssystem: Rezept gegen Wachstrumsruckschlag?" 11, Bericht Nr. 33 (Köln: Bundesinstitut für Ostwissenschaftliche und Internationale Studien, 1979).
[13]See, for example, Gertrude E. Schroeder, "The Soviet Economy on a Treadmill of 'Reforms,'" in Joint Economic Committee, U.S. Congress, *Soviet Economy in a Time of Change*, vol. 1 (Washington, D.C.: U.S. Government Printing Office, 1979), pp. 312-340; Hohmann and Seidenstecher, op. cit.; and CIA, *The Soviet Economy in 1978-79 . . .*, op. cit.
[14]Schroeder, op. cit.

ENERGY IN THE SOVIET ECONOMY

ENERGY AND ECONOMIC PERFORMANCE TO 1980

Economic Growth

Easily accessible energy played an important role in generating past high Soviet growth rates. Soviet "gross energy consumption" has increased roughly in line with Soviet GNP over the past 30 years. How-

ever, energy consumption grew more rapidly than GNP between 1950 and 1965, less rapidly in the 1965-75 period, and then again more rapidly over the past 5 years. Indeed, the elasticity of energy use with respect to GNP (the growth rate of the former divided by the growth rate of the latter) was higher between 1975 and 1980 than in any of the earlier subperiods, precisely at a time "when the government has pursued a vigorous campaign to encourage energy saving and reduce waste."[15]

Table 54 shows how the investment requirements of the energy sector compete with other sectors of the economy. The investment share of agriculture, construction, and transport-communications increased from 25 percent in 1960 to over 35 percent by the late 1970's. Most of this increase came at the expense of investment in housing and to a lesser extent, consumer goods, trade, and services.

The investment share of the nonconsumer-goods industrial branches (mainly machinery, industrial raw materials, and intermediate products) and the energy sector

---

[15]Robert W. Campbell, "Energy in the U.S.S.R. to the Year 2000," paper presented at the Airlie House Conference, Oct. 23-25, 1980. It must be noted that energy conservation only recently began receiving serious attention in the U.S.S.R. Moreover, care should be taken in interpreting such relationships in the short-term. Energy/GNP ratios are most meaningful when observed over long periods.

Table 54.—Distribution of Soviet Gross Fixed Investment by Sector in Selected Years, and of Increments to Fixed Investment for 1970-77
(percent)

	1960	1965	1970	1975	1978	1970-77 (increment)
Fuels and power	10.4%	11.6%	10.0%	9.7%	10.3%	9.1%
Agriculture	13.2	16.9	17.7	20.6	19.9	24.6
Construction	2.8	2.6	3.6	3.8	3.7	5.2
Housing	22.8	17.1	16.6	14.4	13.5	8.5
Trade and services	16.1	16.9	17.1	15.1	14.9	11.1
Transport and communications	9.3	9.6	9.4	10.8	12.5	14.6
Consumer goods	4.8	4.3	4.3	4.0	3.3	3.6
Other industry[a]	20.8	20.9	21.2	21.6	21.8	23.3
Total[b]	100.0%	100.0%	100.0%	100.0%	100.0%	100.0%

[a]Includes machinery and industrial raw materials and intermediate products.
[b]Columns may not exactly add to 100 percent due to rounding.

SOURCES: Calculations based on CIA, *Simulations. . .,* op. cit. and CIA, *The Soviet Economy. . .,* op. cit.

have remained quite stable. The energy sector's share was about 10 percent throughout this period. Between 1970 and 1977 energy's share of increments to annual total fixed investment in the Soviet economy was only 9 percent. In December 1977, however, the energy sector was declared a "leading link" in the economy. Since then it has apparently enjoyed priority status. In 1978, almost 50 percent of the *increase* in fixed investment in industry was allocated to energy. In 1979, roughly one-half of the increment in *total* fixed investment was accounted for by increased investment in energy.[16]

Energy Trade

The U.S.S.R. is a leading energy exporter and the revenues generated by its energy sales have been critical to its economy. Tables 55 and 56 highlight the important role of energy exports both in relation to output and as a source of export revenues. As shown in table 55, roughly one-fourth of Soviet production of petroleum and petroleum products is exported, with about 40 percent of these exports (in terms of quantities) going to non-Communist countries. The U.S.S.R. also imports petroleum, principally from Iraq and Libya, but it is commonly believed that a large portion of this imported oil is reexported. In any case, these imports have not amounted to much more than 5 percent of its total petroleum exports.

About 13 percent of Soviet natural gas output was exported in 1980, and this percentage has been growing rapidly in the last few years. A little under half of Soviet natural gas exports now go to the West. Soviet imports of natural gas, principally from Iran and Afghanistan, were also significant in the late 1970's, but the level of imports has fallen and their relative importance continues to fall. By 1980, when deliveries of gas from Iran had ceased, these imports were less than 5 percent of exports.

Less than 5 percent of Soviet coal output is exported. Over one-third of these exports

---

[16]*Narodnoye khozyaystvo SSSR,* 1978 and 1979.

Table 55.—Estimated Soviet Energy Production and Foreign Trade, 1980[a]

Product group	Soviet production	Soviet exports	Exports as percent of production	Percent of exports to West[b]	Soviet imports	Imports as percent of exports
Petroleum and petroleum products .	603 mmt	150 mmt	25%	39%	7 mmt	5%
Natural gas..........	435 bcm	56 bcm	13%	41%	2 bcm	4%
Coal	716 mmt	27 mmt	4%	39%	9 mmt	33%

[a]All of the foreign trade figures are estimates. Since 1976, the U.S.S.R. has not published such data for energy commodities in natural units. The estimates for coal trade, in particular, are subject to considerable margin of error.

[b]The "West" here corresponds to non-CMEA countries.
SOURCES: *SSR Tsifrax v 1980 g.* (Moscow: 1981); CIA, *International Energy Statistical Review*, Mar 31, 1981; Wharton Econometric Forecasting Service, Centrally Planned Economies Project; and OTA estimates.

Table 56.—The Importance of Soviet Energy Exports and Imports, 1972-79

	1972	1973	1974	1975	1976	1977	1978	1979
Share of energy exports as percent of ruble value of Soviet exports to:								
Socialist Countries	16.6%	17.6%	18.5%	26.0%	27.3%	29.8%	31.8%	36.0%
All other countries	19.8	21.4	33.3	39.7	44.2	42.2	41.3	50.0
Share of all energy exports to:								
Socialist Countries	65.1	57.7	53.5	60.7	58.7	57.4	60.0	55.7
All other countries	34.9	42.3	46.5	39.3	41.3	42.6	40.0	44.3
	100.0%	100.0%	100.0%	100.0%	100.0%	100.0%	100.0%	100.0%
Share of Energy Imports as percent of ruble value of all Soviet imports from:								
Socialist Countries	2.4%	2.3%	1.9%	3.0%	2.7%	2.4%	2.2%	2.4%
All other countries	4.1	5.0	5.4	4.9	4.6	5.2	5.6	5.9
Share of all energy imports from:								
Socialist Countries	64.0	59.3	54.7	52.4	52.6	57.1	60.0	56.6
All other countries..............	36.0	40.7	45.3	47.6	47.4	42.9	40.0	43.4
	100.0%	100.0%	100.0%	100.0%	100.0%	100.0%	100.0%	100.0%

SOURCE: Derived from *Vneshnyaya torgovlya*, various years.

are directed to Western markets (see table 55). Imports of coal, largely from Poland, may have amounted to over one-third of the volume of Soviet coal exports in the late 1970's.

The growing importance of energy exports in total Soviet trade during the course of the 1970's is illustrated in table 56. In 1972, fuel and electric power exports accounted for 16.6 percent of the ruble value of Soviet exports to all "socialist" countries, and 19.8 percent of the value of exports to the "capitalist" world (the industrialized West and non-Communist developing countries). By 1979, the share of these energy exports had risen to 36.0 percent for exports to the socialist countries and 50.0 percent for exports to the capitalist world. Similarly, the quantities of oil and oil products exported

Photo credit: TASS from ©SOVFOTO

First Soviet-made supertanker

grew by more than 50 percent and the quantity of natural gas exported more than sextupled (from a very low base) between 1972 and 1979.

The relative importance of energy as a source of export revenue has been further enhanced by the enormous increases in world energy prices. Indeed, it has been estimated that for the period 1970-77 alone, improvements in Soviet hard currency terms of trade permitted the U.S.S.R. to purchase $14.2 billion more in hard currency imports than otherwise would have been possible without resorting to some combination of expanded real exports, increased gold sales, or additional hard currency debt. This windfall gain amounted to 21 percent of the cumulative value of Soviet hard currency merchandise exports from 1971 through 1977.[17]

THE FORMULATION OF ENERGY POLICY

It is clear that the time of easy energy supplies is over for the U.S.S.R., and the easy answers to energy policy followed in the past two decades are unlikely to be fruitful in the future. A new strategy has become necessary, but its formulation is, and will continue to be, a complex process. There is evidence that debates have arisen over the relative priority to be accorded different

energy industries and over the best way to improve the efficiency and productivity of energy production. While decisions naturally reflect the choices of the Communist Party and its Executive Committee (Politburo) and a number of state planning and administrative organizations, Soviet leaders are influenced by a variety of ministerial, regional, and scientific connections. These groups, which compete for resources and influence, play an identifiable role in the formation of economic policy and are critical to the outcome of policy once formulated. Thus, energy decisionmaking in the Soviet Union takes place in a political context. A brief description of the process by which energy policy is set, including identification of the actors involved, is helpful in understanding the apparent outcome and consequences of these debates.

Decisionmakers

There are two important steps in energy, as in all, decisionmaking in the U.S.S.R.[18] The first is the continuous determination of *basic* policy directions by the Politburo, which then directs the Council of Ministers and other state agencies to work out the details. The second is the formal elaboration of energy policy plans by Gosplan, the State

[17]Edward A. Hewett, "The Foreign Sector in the Soviet Economy: Developments Since 1960, and Possibilities to 2000," paper presented at the Airlie House Conference, Oct. 23-25, 1980.

[18]For further information on the structure of the Soviet Government and economy and on the economic planning system, see Office of Technology Assessment, *Technology and East-West Trade* (Washington, D.C.: U.S. Government Printing Office, 1979), ch. X; and Joseph S. Berliner, *The Innovation Decision in Soviet Industry* (Cambridge, Mass.: MIT Press, 1976).

Planning Agency, in cooperation with government ministries and planning and research institutes. Ministries involved in producing and supplying energy, together with those involved in supporting functions such as the construction of necessary infrastructure, assist in the formulation of plans for the branches of the industries for which they are responsible. The ministries also have a major role in implementing the plans.

Within the general guidelines set by the Politburo, Central Committee, and the Council of Ministers, Gosplan exerts a considerable degree of influence over the allocation of priorities between the energy sector and other sectors of the economy, and over the setting of priorities among the various energy industries. Various departments of Gosplan are responsible for general planning (which must take energy supply and demand into account), for working out the balances of inputs into the energy-producing industries and balances of supply and demand for various types of energy, and for energy production. Gosplan also makes decisions regarding energy-related imports and exports, although such decisions require the participation of a number of other ministries and government agencies.

Despite the comparative centralization of the Soviet system, there is a good deal of diffusion of responsibility among a number of energy ministries. Figure 22 demonstrates the plethora of organizations involved in the discovery, production, and delivery of Soviet energy resources. There are over 60 ministries in the Soviet Government. Of these, 11 have direct responsibility for energy production and energy resource management, and another 6 provide support (e.g., construction, transportation, infrastructure).

The involvement of some 17 ministries results in considerable overlap in jurisdiction and intense competition for resources. Deciding what Western energy technology should be imported and what energy should be exported, and implementing these deci-

sions, are processes that involve complex interactions among a variety of individuals and organizations. A ministry may be responsible for producing commodities for export (such as oil) and various institutions can request Western imports (such as turnkey plants, large diameter pipe, or mining equipment), but it is Gosplan that makes the critical choices, the monetary aspects of which must be approved and executed by Gosbank, the State Bank. The Ministry of Foreign Trade carries out approved export and import plans through its various trade associations. In addition, the State Committee for Science and Technology (SCST) coordinates policy on technology imports. The decisions and actions of all of these parties are subject to approval by high Party and government organs such as the Politburo and the Council of Ministers.

Energy Policy Debates

Problems in measuring the performance of Soviet energy industries and in appropriately allocating resources recur in a fairly routine manner, as a part of energy planning and policy implementation. But at a higher level, Soviet planners have been engaged in debates over the general direction of energy policy. Disagreements over policy are seldom pursued openly, but a careful reading of the Soviet press and scientific journals reveals a variety of opinions on energy priorities among key leaders. A fundamental question here concerns which energy sector should be awarded priority in capital investments.

Energy industries usually require large-scale investments with long-term payoff periods. This makes decisions about energy-related investments particularly difficult, as increasing allocations to one sector may necessitate reductions in growth of investments in other sectors. Soviet policymakers have been faced with setting priorities among the following: investments for expanded oil and gas production in Siberia; investments designed to increase

Figure 22.—Energy Decisionmaking in the Soviet Union

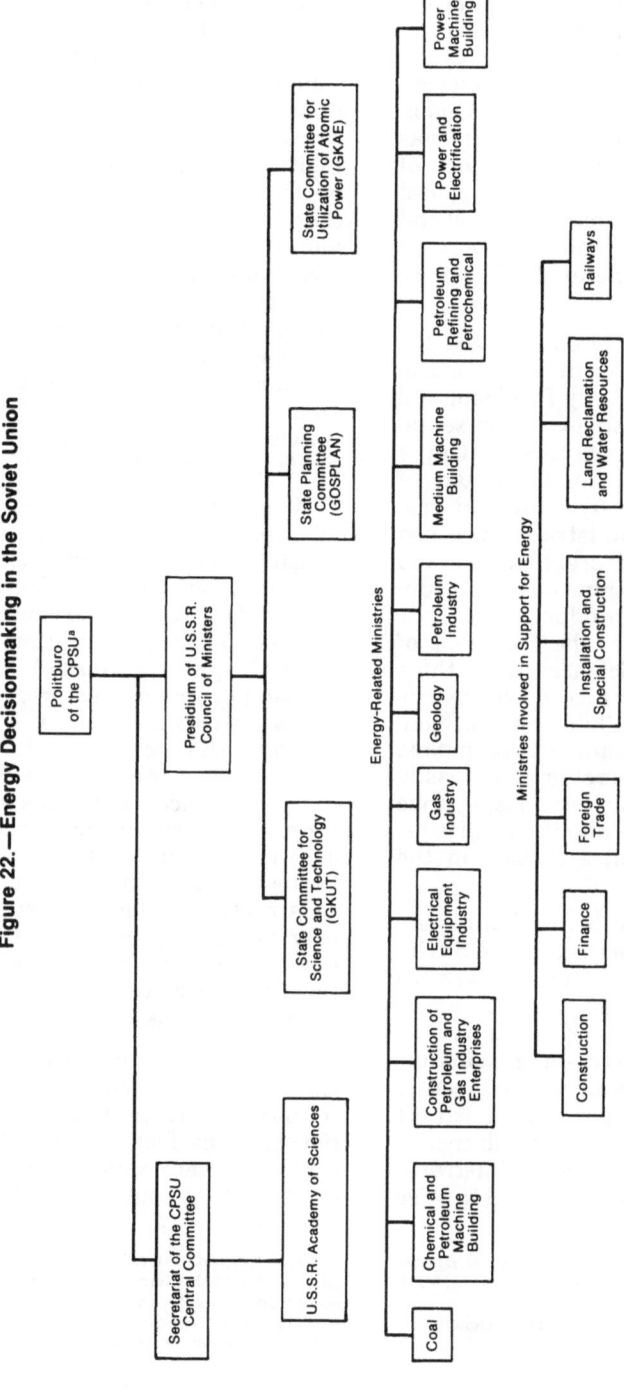

ªCPSU = Communist Party of the Soviet Union.

NOTE: The Ministries appearing here represent about one-third of total U.S.S.R. ministries.

SOURCE: Derived from Central Intelligence Agency, "Energy Decision Making in the Soviet Union," CR 80-10623, August 1980.

coal production, particularly through the development of surface mining in Siberia; investments in nuclear power stations; and commitments of resources for energy conservation, especially on a regional basis. Among the key policy debates of recent years has been controversy over the question of whether primary emphasis should be placed on the development of petroleum, i.e., oil and gas, or coal, particularly lignite.[19] This issue has been an important one in the Politburo during the last decade.

Those who have publicly emphasized the importance of oil and gas development in Western Siberia include Party President and General Secretary Brezhnev; representatives from Moscow, Western Siberia, Upper Volga, and Azerbaijan; the Chairman of Gosplan; and spokesmen from oil- and gas-related ministries and ministries concerned with automobiles, agriculture, aviation, and defense. Those who have gone on record supporting increased coal production include the late Premier Kosygin, the President of the Soviet Academy of Sciences, and others who perceive a limited future for hydrocarbon development. While Soviet controversies over energy planning are complex, normally carried out in secret, and not easily capsulized in simple dichotomies, public statements of key leaders about these issues have received widespread attention in the Soviet press.

Controversies over whether coal or oil and gas should be made the centerpiece of Soviet energy policy now appear to be resolved. In the current FYP, investment in the gas industry, mostly in West Siberia, will double, and it would seem that increased gas production is now considered the answer to meeting both growing domestic needs and export commitments.[20] But the debate itself merits examination to the extent that it illustrates the institutional conflicts which tend to arise

in Soviet energy planning. These debates can be analyzed in terms of individuals— their background, preferences, and personalities, as well as the regional or institutional interests which they legitimately represent. For example, the preference of one Politburo member, V. V. Grishin, for gas and oil might be explained in part by the fact that, as the First Secretary of the Communist Party in Moscow, he has a vested interest in assuring large and reliable supplies of motor fuel as well as heat, power, and gas for its residents. Experience with shipments of poor quality coal which caused frequent shutdowns of power-generating units in the area evidently convinced Grishin that the conversion of coal-burning plants to natural gas is necessary.[21]

It is not surprising that individuals exhibit preferences for energy policy options which promote their own regional or organizational interests. More important for long-term policy trends, however, are recurring conflicts among institutions. At the 25th Party Congress in March 1976, then Premier Kosygin championed a program for large increases in coal production. According to this proposal, during the Tenth FYP the importance of coal would increase in the total energy balance. This was to be achieved through the expansion of surface mining of lignite in the remote Kansk-Achinsk, Ekibastuz and Kuznetsk regions, and the construction of lignite-fired power stations near the mines. Extra-high-voltage powerlines would carry electricity from these stations to the European U.S.S.R., more than 2,500 km away (see ch. 5).

Both the coal advocates and the petroleum advocates had persuasive arguments to support their positions. Kosygin emphasized the fact that development of coal could facilitate savings in natural gas and oil, fuels that could be used most efficiently

[19]Leslie Dienes and Theodore Shabad, *The Soviet Energy System: Resource Use and Policies* (Washington, D.C.: V. H. Winston & Sons, 1979), p. 268.

[20]See Theodore Shabad, "Siberian Gas Field Delayed by Soviet," *New York Times*, Aug. 20, 1981.

[21]V. V. Grishin, "All Energy for the Fulfillment of the Decisions of the 25th Congress of the Communist Party of the Soviet Union, and for the Successful Completion of the Tenth Five-Year Plan," in *Selected Speeches and Articles of V. V. Grishin* (Moscow: Izd. politicheskoi literatury, 1979), p. 562.

as exports and chemical feedstocks. Coal advocates also argued that, because of the high labor productivity of surface mining, coal developed in the Kansk-Achinsk region is among the cheapest fuels available. Those who placed first priority on an oil and gas strategy asserted that, because overall coal output has not expanded rapidly and because of the low quality of much of the coal produced in Siberia, the coal industry is not a reliable energy supplier. Indeed, demand for coal from Kansk-Achinsk and Ekibastuz has consistently fallen below quota. While coal advocates expected that the European U.S.S.R. would be a great market for coal, consumers there and in Siberia have tended to prefer more reliable natural gas supplies.

Strong institutional resistance to a coal strategy evidently came from Gosplan, whose research reported unfavorably on the idea of using lignite as a major source of electricity for population and industrial centers in the European U.S.S.R. Furthermore, Gosplan has placed priority on an oil and gas strategy in its allocation and administrative functions, obstructing the construction of the long-distance powerlines. These powerlines are a critical element in the coal strategy which is oriented toward increasing supply of "coal by wire" electricity to consumers in the European U.S.S.R. Despite the fact that construction of the lines was approved by the Ministry of Power and Electrification, Gosplan delayed and reduced allocations for the project.[22] Without this crucial powerline link, the lignite strategy foundered.

Gosplan's reluctance to rapidly develop the long-distance powerlines can be explained by a number of factors. First, Gosplan experts calculated that capital investment in the transportation of natural gas was more efficient than investment in the development of coal production. This is an important point. The "coal v. gas" decision

also entails basic choices affecting the transportation sector, i.e., whether investment should be directed toward the construction of gas pipelines or additional rail capacity for coal. At a seminar held in Washington, D.C., in March 1980, a Gosplan official stated that his research institute favored postponement of Kansk-Achinsk lignite development, because capital investments could be more effectively directed toward the purchase of French gas industry equipment.[23] Gosplan's electricity cost projections were also important. It was calculated that nuclear and gas-burning power stations located in central Russia could provide cheaper electricity to consumers in that area than electricity transported from the Kansk-Achinsk and Ekibastuz mine-mouth stations.[24] In short, Gosplan's research on investment and energy costs worked against a lignite strategy and tended to favor development of the more "progressive" and efficient gas industry.

The coal and power and electrification ministries also opposed the lignite strategy, but for different reasons. Where Gosplan officials stressed investment and energy cost considerations, the ministries charged with implementing plans for coal development were concerned with the past performance of the coal industry. Surprisingly, even the Ministry of Coal has been ambivalent toward the development of lignite complexes. While it is naturally anxious to increase coal production, its officials have been slow to commit resources to the construction and equipment of new mines, evidently preferring to direct investments to older mines in areas where regional ties to the ministry are long-standing. Moreover, since the earnings of coal enterprises depend primarily on the quantity of coal shipped, the quality of

---

[22]V. M. Tuchkevich, "Speech at the Session of the General Meeting of the Academy of Sciences of the U.S.S.R.," *Vestnik an SSSR*, No. 5, May 1980, pp. 98-99.

[23]Interview with Albina Tretiakova, Demographic Division, Department of Commerce, Washington, D.C., Dec. 17, 1980, concerning R. V. Orlov's comments at the "Seminar on Energy Modelling Studies and Their Conclusions on Energy Conservation and Its Impact on the Economy," held in Washington, D.C., May 24-28, 1980.

[24]*Elektricheskiye stantsii*, No. 12 (1978), pp. 11-14, translated in "News Notes," *Soviet Geography*, Mar. 20, 1979, pp. 188-190.

the coal mined is a secondary consideration. Electricity producers are consequently vulnerable to being forced to rely on poor-quality Kansk-Achinsk and Ekibastuz coal. It is no wonder that reliable and cheap hydropower is much more popular among the electricity producers in Siberia. As the coal and power ministries each attempt to maximize their profits and performance, the result is systemic suboptimization (delays in expansion of overall coal-fired power production).

The Ministry of Power and Electrification (Minenergo) has neglected construction of lignite or coal-fired powerplants not only because hydroelectric plants are cheaper to operate, but also because of the poor quality of delivered coal. High in ash, and often certified above its actual calorific content, the coal tends to cause power equipment breakdowns and consequent loss in production time. Since Minenergo's performance is measured in terms of total output and by grams of standard fuel consumed per kilowatt-hour of electricity produced, the ministry's record is jeopardized by coal-fired power production. Although Minenergo was directed to construct coal-fired power stations in the Tenth FYP, the system of performance indicators actually embodies stong disincentives to carry out such orders. As long as the ministry maintains a good overall record in production of electricity, it is unlikely that it will be punished for failing to speed up construction of coal-fired plants.

As chapter 3 has described, efforts to increase coal production and consumption during the Tenth FYP clearly fell behind expectations. Where former Premier Kosygin had forecast a growth in coal output from 701 million metric tons (mmt) in 1975 to 790 to 800 mmt by 1980, actual output for 1980 was only 716 mmt. Stated in calorific terms, these statistics reveal an actual decrease in coal output during the plan period due to the increasing share of low calorie lignite in coal production. Furthermore, labor productivity in the coal industry has been declining since 1978. Recently published guidelines for the

Eleventh FYP now reflect diminished expectations for coal. Targets for 1985 coal production have been set at 770 to 800 mmt, equivalent to the original goals for 1980, and the coal's calorific value will continue to decline as most of this growth will come from increased production of lignite. The new FYP guidelines can therefore be interpreted as a resolution of the coal v. petroleum controversy in favor of the latter.

A second and equally important consideration is the relative priority which has been accorded oil and gas. These are handled by different ministries which compete for investment, drilling capacity, and pipeline priority. The most widespread interpretation of the current FYP—in which oil production is set to rise 7 percent and gas production 47 percent—is that the U.S.S.R. is now placing its emphasis on gas.

This view is supported by the fact that in his speech before the Party Congress on February 23, 1981, Brezhnev emphasized the importance of Siberian gas development:

> As a task of paramount economic and political importance I consider it necessary to single out the rapid expansion of output of Siberian gas.
>
> The deposits of the Western Siberian region are unique . . . . The extraction of gas and petroleum in Western Siberia and their transportation to the European part of the country are becoming a predominant link of the energy program of the 11th and even of the 12th Five-Year Plan. This is the line of the Central Committee of the Party, and I hope it will be supported by the Congress.[25]

Summary and Conclusions

Controversy among Soviet energy planners and among various energy-related institutions suggests that in order to be successful, a Soviet energy strategy needs more

[25]"Report of the Central Committee of the CPSU to the 26th Congress of the CPSU on the Immediate Tasks of the Party in the Sphere of Domestic and Foreign Policy: The Report of the General Secretary of the Central Committee of the CPSU, Comrade L.I. Brezhnev," *Pravda*, Feb. 24, 1981, p. 5.

than the formal support of the leadership. In addition to the backing of members of the Politburo and the Council of Ministers, it requires the cooperation of Gosplan, other agencies, and the several ministries directly involved in its implementation. The personalities and preferences of top leaders can be important. The decline of the coal strategy, for instance, was surely affected by the demise of a prime advocate, the late Premier Kosygin; the present emphasis on gas has been underscored by Brezhnev. The actions of many institutions and ministries, however, have also been important. This fact takes on added significance in light of the advanced age of much of the present Soviet leadership.

On the evidence of the new FYP, advocates of gas development and of nuclear power have had the most influential voice in energy planning. Current policy guidelines indicate a strong commitment to the development of these fuels. But the energy debates of the last few years suggest that competition for resources may well reappear among ministries involved in the development of oil, gas, and nuclear power—particularly when the impending change in Soviet leadership takes place. Both international and domestic developments may affect priorities placed on various types of energy development. The gas industry, because of its reliance on Western equipment imports, is likely to be more committed to pursuing a strategy of interdependence with the West than the nuclear power industry, which prides itself on the development of indigenous technology. Whatever the strategy chosen at the top, however, successful implementation will depend on the cooperation of a variety of organizations and regions.

ENERGY AND FUTURE ECONOMIC PERFORMANCE

Whatever the energy policy pursued, it will affect and be affected by Soviet economic performance in the present decade.

Any understanding of the ways in which energy availability is related to and affected by the range of Soviet economic options must carry with it a sense of the multitude of economic variables, the complexity of their interaction, and the considerable range of plausible values for many of them. A simplified and stylized way of understanding the Soviet economy is shown in figure 23. In this scheme, Soviet planners are assumed to make decisions regarding the allocation of the fixed resources at their disposal—the existing capital stock, labor force, and resource bases (e.g., timber, mineral, and energy reserves)—to produce a range of intermediate products (industrial materials and energy) which are principally valued for their use in producing other goods, and final products (capital, consumer, and defense goods). Both intermediate and final products may be earmarked for domestic use or exported, requiring decisions on the allocation of exports between other Council for Mutual Economic Assistance (CMEA) nations and the rest of the world.

Soviet planners are assumed to attempt to maximize the contribution of foreign trade to domestic growth, subject to foreign market conditions, regional balance of payments constraints, and possible "noneconomic" constraints on trade with each region. In theory, and noneconomic constraints permitting, the planners would want to expand trade with a region as long as the terms of trade (the weighted price of exports relative to that of imports) exceeded the *relative* marginal productivity of the exports to domestic growth. This is the principle behind actual "foreign trade effectiveness" indices developed by Soviet and East European economists and designed to guide decisions on the structure of foreign trade.

An example of the kind of decisions facing planners can be found in Soviet oil trade with the West. Assuming for simplicity that this trade consisted solely of the export of Soviet oil in return for oil industry technology and equipment and that the main economic goal were to maximize the amount of oil available

Figure 23.—The Soviet Economy

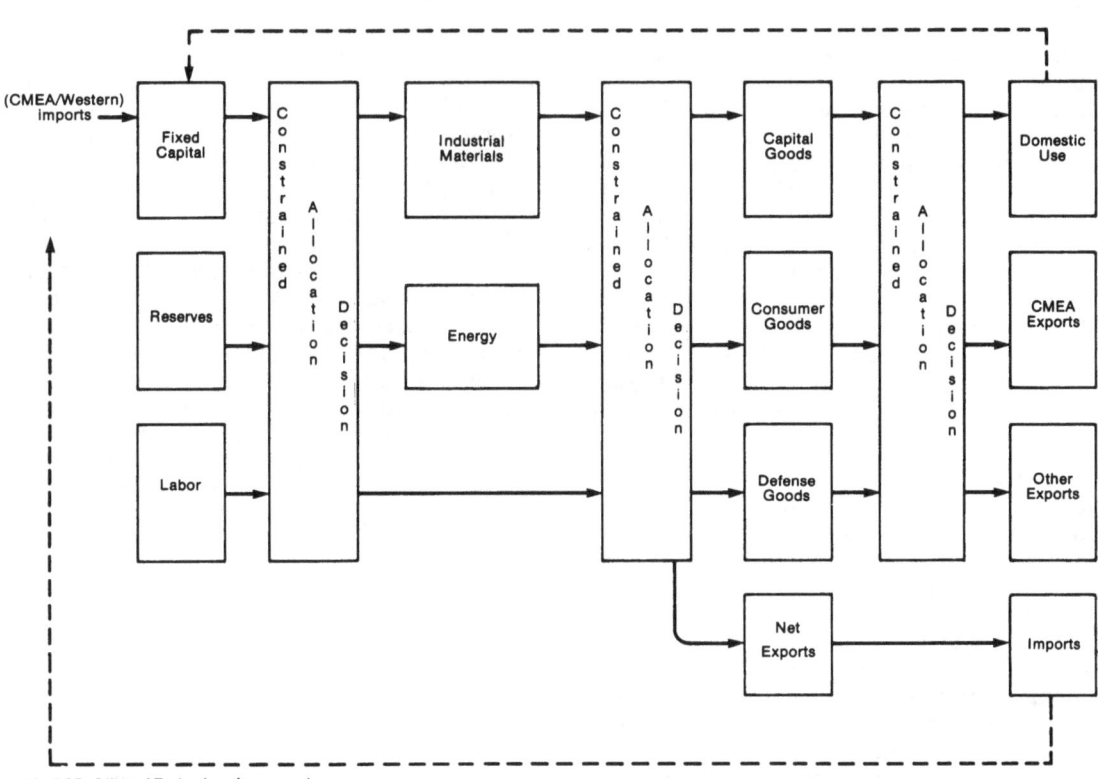

SOURCE: Office of Technology Assessment.

to domestic industry, political considerations aside, the Soviets would find it economically advantageous to expand this trade as long as their terms of trade (the export price of oil relative to the price of imported machinery and technology) were greater than the relative marginal productivity of the exportable oil in making oil available domestically. In other words, it would only make sense in this case to export oil to the West if the the proceeds could be used to buy sufficient technology to yield (on a present value basis) more oil than was exported.[26]

The actual calculations are not nearly so simple, and even in this highly stylized framework, the process of deciding whether and where oil should be exported would entail consideration of numerous tradeoffs. Obviously, Soviet planners could not make all possible calculations and comparisons given the tremendous informational requirements and the lack of a price system which efficiently generates such information. But presumably, to the extent that the planners exhibit economic rationality, these types of calculations implicitly enter into the medium and longrun planning of Soviet foreign trade. Complicating the calculus are various "noneconomic" constraints or goals. For example, the proportion of domestic oil output exported to Eastern Europe may be higher than that suggested solely on the basis of economic criteria alone.

[26]Thomas A. Wolf, "Soviet Petroleum Trade and Western Technology in a General Equilibrium Context: Some Preliminary Notes," paper presented at the Twelfth National Convention of the American Association for the Advancement of Slavic Studies, Philadelphia, Nov. 6, 1980.

The remainder of this section seeks to elucidate some of the complex relationships and the nature of the costs and benefits associated with different Soviet policy options as they concern energy. For purposes of illustration, it hypothesizes a fall in oil output at existing levels of use of capital, labor, and intermediate products in the oil industry. Faced with this disturbance, Soviet policymakers can follow one or both of two basic courses of action. They can attempt to regain the previous level of oil output, or they can attempt to "make do" with a lower level of domestic oil production.

In order to boost oil output, the planners could increase the proportion of total labor and capital available to the oil industry. Wage rates could be raised in the hope of attracting workers, but such material incentives might have to be very large to overcome the disadvantages of working in West Siberia and the East. Assuming that other money wage rates were not reduced, such a measure would increase aggregate money income in the U.S.S.R. If increased real output of consumer goods were not forthcoming, this would be inflationary and could in turn negatively affect labor productivity.

The diversion of labor and current investment from other sectors would reduce the rate of growth of output in those industries. To the extent this diversion were at the expense of investment in the machine building and heavy industry branches, the potential for future growth in *all* other sectors, including oil, would be reduced. Diversion of investment spending from the consumer goods sector would reduce the future rate of growth of real consumption, which could in turn adversely affect the rate of productivity growth throughout the economy, as dissatisfied workers work less hard and spend more time away from the job queuing for consumer goods. A decline in productivity growth would cause a further slowing in overall Soviet economic growth. In short, the diversion of resources would not only have direct adverse consequences for output in various sectors. There would also be sec-

ond, third, and higher order "multiplier" effects throughout the economy.

One way to try to reduce the adverse effects on economic growth in other sectors would be to raise the labor force participation rate. This is already very high, however, and the impact on overall economic growth of any conceivable changes would be negligible. Moreover, in order to induce additional people to enter the work force, it might be necessary to raise the output of consumer goods, at the expense of investment and future growth.

Another option would be to improve the decisionmaking, information, and incentive systems of the economy enough to raise the rate of growth of combined factor productivity. This could both raise the rate of growth of output in the oil sector itself, and stimulate higher output growth in other sectors. This approach might not involve significant economic costs, except insofar as changed indicators and norms might create considerable uncertainty among managers and workers during the transition period. Further, it would avoid the type of direct and indirect economic costs involved in any policy of resource reallocation. But such system change invokes other potential costs. The conventional wisdom among most Western observers is that any changes in planning and management systems profound enough to significantly affect productivity may well carry unacceptably high ideological and political costs for the Soviet leadership.

Finally, Soviet planners could attempt to increase and accelerate imports of Western oil equipment and technology. This strategy would presumably be based on the perception that the opportunity cost of such imports was relatively low. The calculation, however, is not so simple as might appear. Increased imports of technology for hard currency would have to be paid for with some combination of the following: increased exports of energy or industrial materials, reduced hard currency imports of other goods, greater exports of gold, and an increase in hard currency debt.

The latter two choices would carry the cost of reducing future external financial flexibility. Increased real exports of energy or industrial materials would reduce the supplies available to domestic industries, thereby slowing output growth in these sectors. A decline in hard currency imports of industrial materials or nonenergy capital goods and technology would likewise slow domestic output growth. A reduction in imports of grain or other consumer goods would reduce domestic consumption growth and indirectly adversely affect productivity growth. World market price trends for all these products, for gold, and for Western export credits, would influence the final choice. At the same time, the planners would be comparing the costs and benefits of expanded technology imports with the costs and benefits associated with other major policies, such as intersectoral reallocation of labor and capital.

In effect, each option has its economic and political opportunity costs. The economic cost of a given policy lies in the direct or indirect negative effect it has on output growth in one or more sectors, and the impact it has on future external financial flexibility. The benefit of a given policy could be measured in terms of its direct or indirect positive effect on output growth in one or more sectors. The planners' intersectoral priorities would determine the implicit weight attached to the induced change in output in each sector.

In addition to pursuing policies aimed at reviving petroleum output, the planners could seek to make a new lower level of oil production go farther, thus minimizing its negative impact on economic growth. Such an approach would involve some combination of reallocation of available energy supplies, direct energy conservation, interfuel substitution, and expansion of energy imports.

One possibility is that Soviet planners might seek to absorb any fall in oil output by cutting back on oil allocations to the capital goods and industrial material sectors.[27] This would have an especially profound effect on the overall rate of economic growth. Consumption would only be affected indirectly, through the slowdown of investment spending in that sector, but eventually production in the consumption sector would slow, and thus there might also be a decline in the rate of growth of productivity.

As output declined in certain sectors as a result of reduced energy availability (combined with unchanged energy-use coefficients), either domestic consumption of these products or exports would have to fall. In the former case, the impact on future growth would be direct; if exports are reduced, the impact would be less immediate. In that case the multiplier effect would come through an eventual fall in real imports induced by the deteriorating trade balance.

Output declines stemming from factor reallocation, energy reallocation, or other policies, could in principle be avoided through fuel conservation and substitution. But a conservation-substitution policy is not without cost. Significant retrofitting and other conversion measures would claim some new investment which otherwise would be used to expand productive capacity. On the other hand, as chapter 7 points out, much of the Soviet conservation effort is aimed at urging industry to use less energy and motivating management to reduce the materials intensity of production, which in turn indirectly reduces energy consumption.

Another way of "conserving" oil would be to export less of it. The costs of pursuing this policy are similar to those attached to increased imports of oil technology. Reduced exports of oil or other energy products to CMEA and/or the West would indirectly involve some combination of a fall in domestic output of some sectors and reduced future

[27]This is the basic policy assumption built into the econometric model of the Soviet economy used by the CIA. See CIA, *SOVSIM: A Model of the Soviet Economy*, ER 79-10001 (Washington, D.C.: CIA, February 1979); and CIA, *Simulations*, op. cit.

external financial flexibility. In the case of CMEA there are also the "political" costs of reducing oil exports.

Alternatively, the U.S.S.R. could step-up its imports from selected oil-producing less developed countries (LDCs) in return for exports of Soviet capital goods and arms. But this "soft currency solution" would not be costless to the Soviet economy. Expanded exports of Soviet capital goods would slow Soviet domestic output growth. Increased arms exports, unless from inflated inventories, might also come either at the expense of the Soviet military or at the cost of diverting investment from one or more "civilian" sectors into defense. Successful pursuit of this policy would also be predicated on the existence of sufficient demand by oil-producing LDCs for Soviet capital goods and arms. If demand is weak relative to Soviet export offers, Soviet terms of trade with this region would decline, eliminating much of the economic advantage of such trade. In other words, the relative price of LDC oil would no longer be below its relative marginal productivity to the Soviet economy.

Finally, the effect of a partial or total Western embargo of energy technology and equipment exports must be considered. An embargo policy that stopped or interrupted economically beneficial trade would mean that Soviet demand for this technology at existing prices would be frustrated. This would increase the relative attractiveness of all other policy options. It would also mean generally lower rates of growth for Soviet investment, consumption and defense than otherwise.

In sum, virtually any policy that the leadership pursues carries with it both economic costs and benefits. The task is to select that combination of policies which together yield the highest benefit-cost ratio. The remainder of this chapter attempts to suggest a plausible range of parameters within which these policies will have to be made. It seeks to shed light on the ways in which energy availability in the present decade will affect Soviet economic growth, and on the ways in which energy availability could affect Soviet hard currency trade prospects. To accomplish this, OTA has posited high and low levels of output for 1985 and 1990 in the various Soviet energy sectors and used these to generate "best" and "worst" case scenarios for Soviet economic growth and hard currency trade.

ENERGY AND THE SOVIET ECONOMY: BEST AND WORST CASE SCENARIOS

While Soviet leaders have already made and publicized their energy policy preferences for the Eleventh FYP period, their (or their successors') options for the late 1980's seem for the most part to remain open. The scenarios constructed for both 1985 and 1990 suggest some parameters and a few of the policy choices facing Soviet policymakers during the 1980's. *These are not predictions.* The intention here is simply to provide the reader with a sense, not only of the number and complexity of factors which together determine the outcomes of policy choices, but also of the sensitivities of the Soviet economy to various energy-re- lated developments. Because OTA has not relied on formal econometric modeling, all estimates are in highly aggregative terms.[28]

One basic assumption entailed in these scenarios is that Western exports of energy equipment and technology to the U.S.S.R. in coming years will have a greater effect on the Soviet energy sector after 1985 than during the Eleventh FYP. This assumption is based on the length of time usually required

---

[28]Readers wishing to consult such formal models should see CIA, *SOVSIM,* op. cit.; and Daniel L. Bond and Herbert S. Levine, "The Soviet Economy to the Year 2000: An Overview," paper presented to the Airlie House Conference, Oct. 23-25, 1980.

to consummate deals with Western firms, the lags generally encountered in utilizing Western technology and equipment, and the long lead times involved in most large energy projects. Even the West Siberian export pipeline project, discussed below and in chapter 12, is not scheduled to begin gas deliveries until the latter part of the decade. Thus, in the 1990 scenarios an attempt is made to address the question of the difference alternative "extreme" Western trade policies might have on Soviet economic growth, fuel balances, and East-West trade in the late 1980's and beyond. The extremes considered are "maximal" and "minimal" Western energy-related trade, technology, and credit assistance to the U.S.S.R. in the 1980's. It should be noted, however, that OTA does not assume that Western assistance to Soviet energy industries will have only a negligible effect before 1985. There is evidence, for example, that the U.S.S.R. is in part relying on imported pipe and possibly compressors to further expand its internal gas distribution system during the Eleventh FYP.

ALTERNATIVE SOVIET ECONOMIC GROWTH AND HARD CURRENCY TRADE SCENARIOS, 1981-85

OTA's "best" and "worst" case scenarios for the Soviet economy for the period 1980-85 are based on assumptions for Soviet economic growth, domestic energy supply and demand trends, and basic foreign trade conditions. The scenario "outcomes" are estimates of the Soviet net fuel balance after meeting domestic needs and commitments to other CMEA countries, as well as an implied maximum rate of growth for Soviet nonenergy imports from the non-CMEA region. It must be emphasized that most of the assumptions employed here are informed guesses and as such subject to question. The scenario outcomes can be visualized as order-of-magnitude indicators of the range of the plausible. But while each of the assumptions made is in itself plausible, it is far less likely

that *all* these conditions would ever be *simultaneously* either "best" or "worst." Consequently, while these cases define a reasonable universe of possible developments, the most extreme outcomes are unlikely.

As noted above, the rate of Soviet economic growth both influences and is influenced by the size as well as the composition of the Soviet energy balance. All other things being equal, the greater the supply of domestic energy supply relative to domestic energy demand, the higher the expected rate of economic growth. At the same time, the more rapidly the economy is growing, the greater will be the growth in demand for energy.

This chapter assumes that the rate of growth of Soviet GNP is basically determined by the rates of growth of the fixed capital stock, the labor force, and combined factor productivity, respectively. But changing levels of domestic energy output have an indirect influence on Soviet economic growth. The growth rate in the capital stock is influenced by current investment decisions; the size of the labor force is affected by labor market policies; and productivity growth is influenced by both economic policy and "reforms" in the system. All of these policies are affected in turn by domestic energy conditions.

Plausible growth rates for the Eleventh FYP period seem to be bracketed by "low" and "high" annual averages of 1.6 and 3.2 percent respectively. The low rate suggests a perhaps politically unacceptable growth rate for per capita consumption, well below 1.0 percent per annum, but given that estimated GNP growth for the U.S.S.R. was only 0.8 percent in 1979 and 1.4 percent in 1980, the lower bound is clearly not impossible.

The methodology used in developing these "extreme" GNP growth rates is as follows:

1. The labor force is alternatively assumed to grow at 0.4 and 0.5 percent per annum. The higher rate assumes various policy measures designed to raise the

labor force participation rate above 88 percent.

2. The growth rate for the Soviet fixed capital stock is projected on the basis of published CIA estimates of fixed capital investment and the net fixed capital stock for individual years in the late 1970's; and Soviet figures for 1980 investment and planned investment in 1981 and for the Eleventh FYP as a whole.[29] By making reasonable assumptions about the distribution of this investment over the Eleventh FYP, it can be estimated that, if plans are met, the net fixed capital stock would increase about 5.4 percent annually.

3. Combined factor productivity is alternatively assumed to decline by 0.5 percent and to rise by 1.0 percent annually. While the former prospect would be very unwelcome, it is not out of the question. As indicated in table 53, combined factor productivity in the U.S.S.R. fell at an annual average rate of 0.7 percent between 1976 and 1980. The higher growth rate assumes that the various announced measures for raising productivity in the Eleventh FYP would be enormously successful. OTA makes the conventional assumptions of 0.66 and 0.34 for the imputed shares of national product accruing to labor and capital respectively.

The Eleventh FYP projects the rate of growth of "national income utilized" to decline by about one-fifth from the average rate for 1976-1980.[30] Applying this same proportionate decline to the rate shown in table 53 above for Soviet GNP growth for 1976-80, yields a rate of 2.2 percent per annum for 1981-85, which is close to the midpoint of the "plausible" range posited here.

For each of the GNP growth rates, an estimate is made of the net energy trade balance that would result from best-worst

alternatives for domestic and foreign trade conditions with the non-CMEA world. The principal assumptions underlying both cases are listed in table 57. The worst case assumes an income elasticity of energy demand of unity (i.e., a 1-percent increase in GNP leads to a 1-percent rise in the demand for energy). This corresponds roughly to the relationship existing in the U.S.S.R. between 1965 and 1975. (For the Tenth FYP period this elasticity apparently significantly exceeded unity.) In the "best" case the energy demand elasticity is assumed to fall to 0.8. This would probably be considered highly optimistic by most experts, particularly in the near future. For example, some estimates assume that the Soviet energy elasticity will remain at about 1.00 for the next 20 years, or possibly fall to 0.9.[31]

The worst and best case assumptions for domestic output of oil, natural gas, coal, nuclear, and hydroelectric power are listed in table 57, and are based on the analyses in chapters 2 to 5 of this report. With the exceptions of gas and hydropower, these projections somewhat discount official Soviet plan targets. The worst case assumption for oil, 550 mmt, is the upper bound of the revised range estimated by the CIA.

Table 58 presents the estimated Soviet fuel balances, both aggregated and by major energy category, for 1980 and for each of the four 1985 scenarios (worst energy: high and low GNP growth; best energy: high and low GNP growth). Lacking sufficient information regarding fuel-specific conservation and interfuel substitution possibilities, OTA has refrained from disaggregating domestic consumption by energy source. In table 58, estimated domestic energy demand (calculated using the appropriate energy demand elasticities from table 57) is subtracted from the total available domestic energy supply, leaving an estimated *net* fuel export balance. To this is added an assumed level of Soviet 1985 energy imports from non-CMEA sources, leaving a *gross* fuel balance available for export outside CMEA. For sim-

[29] *Handbook of Economic Statistics*, ER 80-10452 (Washington, D.C.: CIA, October 1980); *Ekonomicheskaya gazeta*, No. 5 (1981).

[30] *Ekonomicheskaya gazeta*, No. 9, February 1981.

[31] See Campbell, op. cit., table 1.

Table 57.—Major Assumptions Underlying 1980-85 Scenarios

	Worst case	Best case
Income elasticity of energy demand	1.0	0.8
Petroleum output (mmt)	550	645
Natural gas output (bcm)	600	640
Coal output (mmt)	750	800
Nuclear power (bkwh)	170	227
Hydroelectric power (bkwh)	237	237
Average annual growth rates for real exports to non-CMEA area of (1979 share of total exports in brackets):		
Timber products (7.5%)	0%	2.5%
Platinum group metals (2.1%) ..	2.2	4.4
Raw cotton (2.0%)	0	2.5
Chemicals (6.5%)	12.3	17.7
Automobiles (1.4%)	5.0	10.0
Diamonds (1.7%)	0	5.0
Other products (23.7%)	0	5.0
Maximum permissible normalized trade balance[a]........	–0.50	–0.50
Average annual growth rates for Soviet foreign trade prices with non-CMEA area:		
Export prices	10.0%	12.5%
Import prices	10.0	7.5

[a]Merchandise trade balance divided by the value of exports, see p. 270.
SOURCE: Office of Technology Assessment.

plicity OTA assumes that these imports remain at their estimated 1979 level (roughly 7 mmt of crude oil, and 2 billion cubic meters (bcm) of natural gas from Afghanistan.)[32] This is not a forecast (indeed, gas shipments from Iran may have only temporarily ceased) but an assumption made to facilitate the computations underlying the alternative scenarios.

The figures in table 58 raise some important issues. For example, the worst energy/low growth scenario (column 3) suggests a net fuel export balance of nearly 61 million tons of oil equivalent (mtoe) (1.22 mbd), with coal consumption declining from 1980 levels, and oil consumption virtually stagnating. Thus, unless significant substitution of gas

for oil in domestic consumption occurred fairly quickly, most or all energy exports would have to be composed of natural gas shipments. An export level of 61 mtoe is imposing by 1979 or 1980 Soviet gas export standards, however. OTA has estimated that present pipeline capacity might support 27 to 29 mtoe of natural gas exports to Western Europe.[33] For the "best" case scenarios, which yield even larger net fuel balances but also larger implied natural gas deliveries, the possible pipeline capacity constraint could be even more serious.

Assessment of the foreign trade implications of these net fuel balances also involves assumptions regarding the possibilities for expanding Soviet nonenergy exports to non-CMEA countries and Soviet terms of trade with these nations. It is assumed here that all non-CMEA trade roughly reflects patterns of trade with the Soviet Union's hard crrency partners. Given the difficulty of separating hard currency from non-CMEA soft currency trade, and the very aggregative level of this analysis, it was not thought worthwhile to strive for a greater degree of accuracy. In any event, the focus here is on the net Soviet energy balance available for export to the non-CMEA region, and most of these exports are undoubtedly made for hard currency.

Western trade statistics show that Soviet energy exports accounted for an estimated 55.1 percent of the total value of Soviet exports to 17 "Industrialized West" (IW) countries in 1979.[34] Six nonenergy export product

---

[32]Campbell, op. cit.; and Jan Vanous, "Eastern European and Soviet Fuel Trade, 1970-1985," discussion paper No. 80-10 (Vancouver: Department of Economics, University of British Columbia, April 1980).

[33]1980 Soviet gas deliveries to Western Europe totaled roughly 23 bcm. The excess capacity of the Orenburg gas pipeline (i.e., after meeting annual commitments of 15.5 bcm to Eastern Europe) available for export to Western Europe is about 12 to 13 bcm. Present pipeline capacity could, therefore, support possibly 23 + 13 = 36 bcm × 0.8123 = 29 mtoe of natural gas exports. See Campbell, op. cit.; and Goldman, op. cit.

[34]These estimates are based on adjustments to unpublished data made available by the Department of Commerce. According to *Vneshnyaya torgovlya SSSR, 1979*, energy accounted for 50 percent of total 1979 exports to nonsocialist countries. A study by Jan Vanous ("Soviet and Eastern European Foreign Trade in the 1970's: A Quantitative Assessment," discussion paper 80-11 (Vancouver: Department of Economics, University of British Columbia, April 1980) suggests that 1979 Soviet energy exports

Table 58.—Fuel Balances by Category, 1985 Scenarios and 1980 Base Year
(millions tons of oil equivalent)

	1980 base year[a]	1985 (worst energy, high GNP growth)	1985 (worst energy, low GNP growth)	1985 (best energy, high GNP growth)	1985 (best energy, low GNP growth)
Hydro, Nuclear Power and "other"	107.5	138	134	150	146
Coal	319.3	300	300	320	320
Oil and products	354.1	362	364	444	446
Natural gas	249.6	368	370	396	399
Total	1,030.5[b]	1,168	1,168	1,310	1,311
Domestic and CMEA energy demand[c]		(1206)	(1116)	(1169)	(1099)
Net fuel balance	83[d]	(38)	52	141	212
1979 energy imports from non-CMEA	9[d]	9	9	9	9
Gross fuel balance (available for export to non-CMEA)	92	(29)	61	150	221

[a]From Campbell, op. cit., table 2.
[b]Net of all exports and imports of energy.
[c]The methodology for deriving the amount of energy available for domestic consumption and hard currency export relies heavily on Campbell, op. cit., and may be summarized as follows:

   [a] Assume a level of domestic energy output (oil, gas, coal, hydro and nuclear generated electricity and "other," e.g., peat, firewood, shale oil). Output for the latter is assumed (per Campbell) to decline from about 5 percent of energy demand in 1980 to 3 percent in the year 2000, at a rate of 0.5 percent every 5 years.
   [b] Subtract net losses and "internal consumption" and net "intrasector outflow," per Campbell.
   [c] Subtract assumed level of net exports to the CMEA. For simplicity, it is assumed that all electricity exported to Eastern Europe will be generated by nuclear power. (Ch. 9 estimates 16 bkWh per year in 1981-85.) Other estimates are coal (net of coal imports from Poland) —8 mmt; oil —80 mmt annually in 1981-85 to Eastern Europe, and 11 mmt annually to non-European CMEA; gas —30 bcm per year. OTA assumes that these are fixed commitments that the Soviets will honor regardless of their own internal energy situation.
   [d] Multiply the result of (a) - (c) by conversion factor to obtain amount "available for distribution."
   [e] Subtract nonfuel uses for oil and gas, assuming the following per annum growth rates:

GNP	Oil	Natural gas
3.2%	2.5%	4.0%
2.4	1.5	3.0
1.6	1.0	2.0

[d]Based on CIA, "International Energy Statistical Review," March 31, 1981; Wharton Econometric Forecasting Associates Inc., Centrally Planned Economies Project, unpublished data.

groups, listed in table 57 and together accounting for an additional 21.2 percent of Soviet exports to the IW in 1979, were analyzed and assigned individual best and worst case growth rates, in terms of real export growth. In each of these markets real export growth is determined by both supply and demand. Detailed analyses taking into account such conditions for each product group were beyond the scope of this study.

amounted to about 61 percent of total Soviet nonarms exports to non-CMEA countries. Given the rough equivalence of all these proportions, it seemed reasonable to use the more detailed IW statistics as a basis for calculating the weights of energy and selected major nonenergy exports in total Soviet nonarms exports for hard currency.

Past real export and domestic output performance was investigated for some products, however, in an attempt to generate plausible worst and best case estimates for real export growth in the 1980's. Some of these considerations are briefly set forth in appendix A.

Attempting to estimate price developments for each of these product groups is an even more speculative exercise than making real export growth projections. This is equally true for the prices of Soviet imports from non-CMEA sources. Consequently, OTA has simply assumed a uniform rate of inflation for all exportables, and a uniform rate of price increase for all importables. Indeed, assuming an unchanged relative price structure within each category of goods is more reasonable than attempting to estimate rates of inflation for separate product groups. By distinguishing clearly between exports and imports, one can still assume that Soviet terms of trade change.

Price indices developed on the basis of official Soviet trade statistics suggest that Soviet export and import prices in trade with non-Socialist countries increased at annual rates of 4.5 and 3.1 percent respectively between 1970 and 1978.[35] This implies an average annual improvement in terms of trade of about 1.3 percent. However, a possible upward bias in the export quantity index employed here may understate the rate of export price increase. It has been estimated, for instance, that between 1971 and 1977 Soviet export and import prices in "hard currency trade" (a subset of trade with non-socialist countries) increased at average annual rates of 20, 21, and 12 percent respectively. This suggests an average annual terms-of-trade improvement of about 7 percent.[36]

As a "best" case, OTA assumes that Soviet export prices in trade with the non-CMEA area increase by 12.5 percent annually, whereas import prices rise by 7.5 percent. The implied annual terms-of-trade improvement is about 4.7 percent. For the worst case it is assumed that all foreign trade prices rise at 10 percent a year, leaving Soviet terms of trade with the non-CMEA region unchanged.

Soviet import capacity is not determined solely by the growth of Soviet real exports of energy and nonenergy products and the terms of trade. Revenues from gold sales, services, and military sales to various developing countries have often accounted for more than enough hard currency to offset hard currency merchandise trade deficits. Furthermore, the U.S.S.R. has financed much of its trade deficit in recent years with Western credits.

Data on gold sales and arms shipments are notoriously poor, and erratic movements in gold prices increase the difficulty of projecting hard currency revenues from this source. Moreover, any attempt to estimate Soviet credit drawdowns would be an extremely complicated and speculative undertaking. OTA has therefore assumed that, regardless of Western credit availability and supply-demand conditions on world gold and arms markets, Soviet policymakers would avoid allowing the hard currency merchandise trade deficit to exceed, at least for any extended period of time, some reasonably conservative proportion of current Soviet exports. This analysis therefore utilizes the concept of a "normalized" trade balance, which is the merchandise trade balance divided by the value of the exports.[37]

The U.S.S.R.'s normalized balance varied between -0.11 and -0.82 in the 1970's. It reached its most negative value in 1975, due to a rapid increase in Soviet imports and a weakening of Soviet exports to the West because of recession. The U.S.S.R. was able to bring the normalized deficit to below -0.30 by 1977. Changes in the value of the normalized deficit have also typically (though not inevitably) been associated with more gradual changes in the Soviet debt service ratio, because varying degrees of the merchandise trade deficit can be financed by gold and arms sales.

For both the "worst" and "best" cases OTA has assumed that maximum permissible normalized trade deficit for 1985 is -0.5.

[35]Hewett, "The Foreign Sector," op. cit.

[36]Paul Ericson and Ronald S. Miller, "Soviet Foreign Economic Behavior: A Balance of Payments Perspective," in Joint Economic Committee, U.S. Congress, *Soviet Economy in a Time of Change,* vol. 2 (Washington, D.C.: U.S. Government Printing Office, 1979), pp. 208-243.

[37]This concept was first used by Edward A. Hewett in, "Soviet Primary Product Exports to CMEA and the West," paper presented to the Association of American Geographers Project on National Resources in the World Economy, May 1979; and has also been used by Thomas A. Wolf, "Alternative Soviet Hard Currency Scenarios: A Back of the Envelope Analysis," app. III in Phillip D. Stewart, *Soviet Energy Options and United States Interests* (Columbus, Ohio: Mershon Center, April 1980), pp. 37-56.

In other words, the hard currency deficit is allowed to equal one-half the value of Soviet exports (or imports could be 1.5 times as great as exports). The exact manner in which this deficit is financed remains unspecified. Possibly under the "best" conditions more Western credit would be available, and the U.S.S.R. would be more willing to take on debt obligations, whereas in the "worst case" scenario credit sources might dry up and the U.S.S.R. would have to more rapidly increase gold and arms sales.

Table 59 presents the different hard currency trade outcomes implied by alternative assumptions regarding: 1) GNP growth, 2) domestic energy supply and demand conditions, and 3) foreign trade conditions. Soviet hard currency import capacity has been singled out for the following reasons: 1) The Soviets view hard currency imports—whether grain, machinery and equipment, or technology—as an important stimulus to domestic productivity growth and to general economic development. 2) The U.S.S.R. cannot afford to run indefinitely a hard currency deficit above some "prudent" level, a fact which will constrain Soviet ability to import both fuels and nonenergy items, and under certain circumstances, may seriously constrain Soviet economic growth. 3) The degree to which the U.S.S.R. can increase its *real* hard currency imports will have a bearing on its foreign economic and political policies.

Most observers of the Soviet energy situation now dismiss—if indeed they ever entertained—the possibility that the U.S.S.R. itself might become a net importer of energy by 1985. This judgment is easily supported by the outcome in table 59 for the high GNP growth/worst case scenario in which the U.S.S.R. would have to import 38 mtoe (763,000 barrels per day) of energy in order to meet domestic demands *and fixed commitments to other CMEA countries.* Even if the normalized trade deficit quintuples to −0.5 by 1985, Soviet *real* imports of nonenergy products from the non-CMEA region would in this case still have to *decline* by an average of 9.2 percent a year. By 1985 real nonenergy imports would be only 56 percent as great as they were in 1979. While such a circumstance is not impossible, it would send East-West trade into sharp decline and could put enormous pressure on the Soviet Union to solve its energy problems in other ways.

Table 59.—Alternative Scenarios for 1980-85[a]

GNP growth	Worst case		Midrange[b]	Best case	
(Average annual rate)	Net fuel balance (MTOE)	Maximum growth of nonenergy imports from non-CMEA (average annual rate)	Net fuel balance (MTOE)	Net fuel balance (MTOE)	Maximum growth of nonenergy imports from non-CMEA (average annual rate)
3.2%	(38)	-9.2%		141	19.3%
		12.7%	92		15.2%
1.6%	52	3.3%		212	24.2%

[a]All growth rates are actually calculated for the period 1979-85, because base year value and export weight figures are included in the appendix expressions, and 1979 is the most recent year for which disaggregated value data on Soviet foreign trade is presently available. The actual value for the normalized trade deficit in 1979 was -0.11. The Soviet net fuel balance (as defined in table 58) for 1979 was estimated as 83 mtoe (60.8 mmt of petroleum and petroleum products, 20.7 bcm of natural gas, and roughly 10.2 mmt of coal exported to non-CMEA destinations). Adding to this estimated Soviet energy imports from non-CMEA sources of 9 mtoe gives a *gross* fuel balance (i.e., available for export) in 1979 of 92 mtoe. The total percentage growth of energy exports between 1979 and 1985 is then calculated by relating the estimated gross fuel balance for 1985 (e.g., 61 mtoe for the worst case/low growth scenario) to the base year balance for 92 mtoe.

[b]In this scenario, middle range values for GNP growth and domestic energy supply and demand conditions are assumed for *both* cases. "Worst" and "best" cases here therefore refer only to rates of real nonenergy export growth. The terms of trade are assumed to increase by the same amount in both cases.
SOURCE: Office of Technology Assessment.

While military and political solutions to the "worst case" energy situation are possible, a less drastic and perhaps more likely response would be to permit the energy constraint to slow the rate of economic growth. Maintaining some level of hard currency energy exports would lead to growing domestic energy shortages. These shortages could stimulate redoubled conservation and substitution efforts, but the near-term impact might largely be in terms of reduced economic growth. As growth slowed, energy demand would fall, and the U.S.S.R. would move toward the low growth-worst case outcome in the lower left corner of table 58.

When GNP growth slows to 1.6 percent, the ability of the U.S.S.R. to expand its real nonenergy imports from the non-CMEA region is respectable, albeit limited. A yearly 3.3-percent growth of real imports would mean a dramatic slowdown in the enormous growth of the past two decades, financed in the second half of the 1970's by windfall gains caused by exploding energy prices. (Soviet real imports from nonsocialist countries grew at annual rates of 9.3 percent in the 1960's and 12.7 percent between 1970 and 1978.[38]) Such a low GNP growth rate might be politically intolerable. Even with investment growing at only 2.5 percent a year, and with a slowdown in the growth of defense spending, annual GNP growth of 1.6 percent could easily reduce annual per capita consumption growth below 0.5 percent. This compares with an annual average growth rate of 2.5 percent in the period 1971-79.[39]

If the low extreme of the new CIA oil output estimate for 1985 (500 mmt) were used for the "worst case" analysis, the outlook for Soviet hard currency trade and economic growth would worsen. Specifically, the maximum growth rate for Soviet nonenergy hard currency imports would fall to minus 16 percent and minus 3.1 percent for the high and

low growth scenarios respectively. The corresponding net fuel balances would be (83) and 16 mtoe (1.67 million and 321,000 barrels per day respectively).

The "best case" scenarios yield much higher net fuel balances and permit annual growth of real nonenergy imports from the non-CMEA region of between 19.3 and 24.2 percent. Even in the event of rapid GNP growth, the U.S.S.R. would emerge with a net fuel balance larger than it has today. Both these instances, however, raise the question of how this much energy would be physically exported, particularly if considerable substitution of gas for oil in domestic uses were not achieved. Energy exports to Eastern Europe could be raised above planned levels, but again, if most of the increase were to be in the form of natural gas, the logistical feasibility of transporting the gas is uncertain.

It is highly unlikely that the U.S.S.R. will be faced with either the worst or best cases. Assuming that the most probable outcome falls between, OTA has also calculated the implied maximum growth of nonenergy imports from the non-CMEA region under midrange economic growth and domestic energy conditions that yield a net fuel balance of 92 mtoe (1.85 mbd). Depending on the rate of growth of nonenergy exports to the non-CMEA area, real non-energy import capacity would grow in this case by 12.7 to 15.2 percent annually, consistent with Soviet performance in the 1970's. This result is essentially due to the assumed improvement of roughly 5-percent- per-annum in Soviet terms of trade under conditions of net energy exports. Indeed, this analysis shows the very important role that the terms of trade play in determining the growth of Soviet real import capacity. For the best case scenarios, for example, a 1-percent-per-annum improvement in the terms of trade has the same impact on Soviet import capacity as an increase of 17.8 mtoe (357,000 bpd) in the Soviet net fuel balance.

---

[38]Hewett, "The Foreign Sector," op. cit.
[39]Schroeder-Greenslade, op. cit.

ALTERNATIVE SOVIET HARD CURRENCY TRADE SCENARIOS, 1990

The effects of different levels of Western assistance will be reflected much more strongly in Soviet energy output by 1990 than will be the case by 1985. An interesting issue here is the difference Western assistance might make in the Soviet Union's capacity in 1990 to import nonenergy products (grain, machinery and equipment, technology, consumer goods, intermediate industrial products) from outside the CMEA.

OTA has considered two cases: maximal Western trade, technology, and credit availability for Soviet energy projects; and minimal Western energy assistance. The former case assumes development and completion of the West Siberian gas export pipeline (see ch. 12) by 1985 or 1986, as well as other large-scale projects possibly directly involving the United States and Japan. The minimalist case assumes a virtual embargo or at best a very low level of Western energy-related technology transfer, pipe deliveries and energy credits to the Soviet Union. No attempt is made in this analysis to examine the feasibility of the maximalist case on the supply side.[40]

Rather than deal with a number of combinations of Western trade policy, Soviet energy conditions and economic growth rates, a single plausible Soviet energy situation is assumed, and a single constant rate of Soviet economic growth for the entire period is posited. The growth of nonenergy exports, the terms of trade and the maximum allowable "normalized" trade deficit are also assumed to be the same regardless of the state of Western trade policy.[41] Specifically, OTA assumes that the Soviet economy grows at an average annual rate of 2.4 percent, which is the midpoint of the extreme growth rates considered for 1980-85. An income elasticity of energy demand of 0.9 is assumed, again half way between the high and low elasticities considered for the earlier period. Nonenergy exports are assumed to increase at rates that are for the most part intermediate between the 1985 worst and best case assumptions. Soviet terms of trade are assumed to increase at 2.3 percent annually, and the normalized deficit is permitted to rise to − 0.5. Planned levels of Soviet energy exports to CMEA are retained at 1985 levels. Again, worst and best cases for energy production are based on chapters 2 to 5 in this study. These assumptions are summarized in table 60.

The discussion in the foregoing chapters leads to the conclusion that Western energy technology and equipment would have a relatively greater quantitative impact on the Soviet gas industry than on the oil industry. Translating such a judgment into numerical production levels is a highly speculative exercise. The figures in table 60 should, therefore, be seen as merely illustrative of the possible impact of Western assistance on these industries.

For oil, OTA assumes that Western assistance would make no more than a 10 percent difference in output levels. Given a policy of maximal assistance, OTA estimates that Soviet natural gas production in 1990 could be 100 bcm (or 15 percent) higher than otherwise. This assumes that the new export pipeline to Western Europe would lead to an an-

---

[40]Because of the dearth of hard data and because the focus here is on the Soviet side of the East-West relationship, the analysis proceeds not from estimates of the scale of Western exports and credit availability, but rather from aggregative but plausible assumptions regarding the impact of such assistance on 1990 Soviet output levels for different energy sources. Such issues as whether or at what terms sufficient Western financing for such ventures could be arranged are not considered.

[41]For simplicity, the normalized trade deficit is assumed not to apply to credits related to new energy projects. As contemplated, those projects would involve a very rapid buildup in credits to about 1985-86 and then equally fast repayment by about 1990.

Table 60.—Major Assumptions Underlying 1990 Scenarios

	Worst case (minimal Western energy assistance)	Common to both cases	Best case (maximal Western energy assistance)
Income elasticity of energy demand....		0.9	
Petroleum output (mmt)	500		550
Natural gas output (bcm)	665		765
Coal output (mmt) ...	850		875
Nuclear power (bkwh)	411		470
Hydroelectric power (bkwh)[a]	271		271
Average annual growth rates for real exports to non-CMEA area:			
Timber products ...		1.0%	
Platinum group Metals		3.0	
Raw cotton........		1.0	
Chemicals.........		10.0	
Automobiles......		5.0	
Diamonds		2.0	
Other products		2.0	
Maximum permissible normalized trade balance........		–0.50	
Average annual growth rates for Soviet foreign trade prices with non-CMEA area:			
Export prices......		10.0%	
Import prices......		7.5%	

[a]Based on Campbell, op. cit.

SOURCE: Office of Technology Assessment.

nual output increase of from 40 to 70 bcm, that joint U.S.-Japanese-aided projects could yield an additional 20 bcm, and other smaller scale efforts could augment Soviet gas output (but not necessarily Soviet exports) by 10 to 40 bcm.

The assumed production levels are, of course, key to the quantitative outcomes of each scenario. For oil in particular, the range

of existing 1990 production estimates is tremendous. These are from 350 to 450 mmt (CIA) to 750 mmt (The Economist Intelligence Unit). (see ch. 2). The assumed range of 500 to 550 mmt in table 60 falls roughly halfway between these two extremes.

The outcomes for the worst-best case scenarios are reported in table 61. Without Western assistance, the Soviet net fuel balance (i.e., net fuel available for export to the non-CMEA region) declines from 83 mtoe (1.67 mbd) in 1979 to a deficit of 12 mtoe by 1990. With maximal Western help, on the other hand, the net fuel balance increases by over one-third to 126 mtoe (2.53 mbd) in 1990. As with the scenarios for 1985, however, one would want to examine in some detail the technical capacity of the Soviets actually to export such large volumes of fuels, particularly natural gas.[42]

---

[42]For the best case scenario, 1990 Soviet energy demand is estimated as 1,276 mtoe. 1990 estimated Soviet "available energy" (i.e., after assumed exports to CMEA), balances for each fuel category (Campbell, op. cit., estimates of 1980 domestic energy consumption by category are in parentheses) are: hydro, nuclear, and "other"=209 mtoe (108); coal=350 mtoe (319); oil=404 mtoe (354); and gas=480 mtoe (250).

Table 61.—Alternative Scenarios for 1990 (percent)

	Worst case (minimal Western energy assistance)	Best case (maximal Western energy assistance)
1990 net fuel balance	(12) mtoe	126 mtoe
Percentage change in net fuel balance (1979-90) ...	–100%	51%
Percentage change in capacity to import non-energy products from non-CMEA area (1979-90)	14%	163%
Implied average annual growth rate for real import capacity	1.2%	10.2%

[a]This would be the average annual rate of growth of nonenergy imports from the non-CMEA region implied by the level of real nonenergy related imports that the U.S.S.R. could purchase in 1990 after repayment of energy project-related debt. The actual average annual rate of growth of such imports prior to debt repayment would be considerably smaller.

SOURCE: Office of Technology Assessment.

Given the foreign trade conditions posited above, the Soviet Union's capacity to import nonenergy products from the West grows in *real* terms between 1979 and 1990 by 14 percent in the worst case and by 163 percent in the best case. These changes translate into per annum growth rates of 1.2 and 10.2 percent respectively. The latter figure, however, must be interpreted with great care. The significant increase in energy exports implied in the best case outcome would only occur in the second half of the decade, after completion of the massive natural gas pipeline projects and after technology transfer in the coal, nuclear, and oil sectors had had an appreciable effect. Furthermore, a good portion of these incremental energy exports

would have to be used to pay off project-related debts in the 1986-90 period. Consequently, under the assumed conditions Soviet real imports of nonenergy and non-energy-project products would actually grow at a rate somewhere between 1.2 and 10.2 percent. Nevertheless, assuming that the bulk of these large Western credits had been retired by 1990, Soviet real import *capacity* at that time would have increased by 163 percent, having grown at an average annual rate of 10.2 percent since 1979. In the worst case scenario Soviet real import capacity increases at a negligible rate. In the best case, the growth in import capacity almost matches the growth rate of the 1960's.

SUMMARY AND CONCLUSIONS

Soviet economic growth has gradually slowed in recent years, and even without an energy "problem," it is likely that growth would continue to decelerate in the 1980's. The basic causes of this slowdown are falling rates of growth of the Soviet capital stock and labor force. Recently their impact has been reinforced by stagnating or declining productivity, only in part the result of adverse weather conditions. Even in the absence of a serious decline in Soviet oil output, the Soviet economy in this decade will probably not be able to attain the growth rates of the 1970's unless significant gains in productivity can be achieved.

It is generally agreed in the West that significant productivity increases are unlikely to occur in the absence of more profound changes in the Soviet planning and management systems than are presently contemplated. Superimposed on these fundamental trends and challenges to the Soviet planners is now the possibility of a plateauing or even decline in oil output. A decline would certainly cause Soviet growth to slow even more, although the magnitude of such a slowdown is not at all obvious. The impact of falling domestic energy supplies will depend on a

system of complex relationships in the economy, and on Soviet priorities and actual policies regarding the composition of the future energy balance and foreign trade patterns. The formulation of Soviet energy policy takes place in a political context and involves a number of different interests and actors. It now appears that this policy broadly favors gas and nuclear development, partly at the expense of the oil and coal sectors.

The foregoing discussion has attempted to suggest the major direct and indirect economic ramifications of basic Soviet economic policy options. There is nothing to keep Soviet planners from pursuing some or all of these policies simultaneously. They probably will pursue most of them in some measure. But every policy carries with it costs and benefits to the Soviet economy. In an effort to give some rough order-of-magnitude sense of how the Soviet economy might be affected by the energy situation, OTA has developed several alternative scenarios for Soviet energy and aggregate economic conditions in the 1980's. The "worst" and "best" case scenarios are meant to bracket plausible outcomes for Soviet economic growth, energy balances, and growth in hard currency import capacity. These scenarios should not be viewed as predictions. They are simply attempts to set forth plausible ranges for the parameters under which Soviet economic policymakers will have to operate over the next decade. Each scenario is based on a long list of simplifying assumptions.

The scenario outcomes for the period 1981-85 suggest that if most or all of the "worst" case assumptions materialize, Soviet economic growth could slow considerably during the Eleventh FYP. Annual rates of GNP growth would probably be much lower than the 2.8-percent-per-annum average recorded for 1976-80, and could result in small and perhaps politically unacceptable increases in real per capita consumption for the Soviet population. Under such conditions the ability of the U.S.S.R. to increase its real nonenergy imports from the West would also be seriously impaired. This would negatively affect the overall growth prospects for East-West trade and would in turn place further strains on the Soviet economy.

Under a series of "best" case conditions, the Soviet Union would be able to continue to grow at a rate approximating overall Soviet performance for the Tenth FYP. At the same time, its net fuel balance available for export to the West would increase over 1979-80 levels. This, combined with continued improvements in Soviet terms of trade under "best" conditions, would permit the U.S.S.R. to expand real hard currency imports at historic rates and possibly to divert more energy than presently contemplated to Eastern Europe.

Actual conditions will likely fall somewhere between these extremes. If the U.S.S.R. encountered economic growth, energy, and foreign trade conditions midrange between those assumed for the worst and best cases, the Soviet Union might be able to maintain energy exports to the West at about 1979-80 levels and continue to increase its real hard currency imports at rates established over the past 15 to 20 years. This

would make possible annual per capita consumption growth well above 1 percent.

OTA assumed that Western assistance in the development of Soviet energy resources would have its greatest quantitative impact after 1985. The 1990 scenarios therefore consider the possibility of minimal Western energy assistance (the "worst" case) and maximal Western cooperation (the "best" case). Such help, in the form of exports of energy-related equipment, materials and technology, and extensive export credits, would be forthcoming principally in the 1981-85 period. The credits are assumed to be more or less fully repaid by 1990. Most of the assumptions regarding energy and foreign trade conditions are essentially "midrange" estimates.

In the worst case scenario, Soviet fuel exports would disappear by 1990. Soviet capacity for hard currency imports (in real terms) would grow at a little more than 1 percent per annum in the 1980's, contrasted to annual growth of over 12 percent in the 1970's. With massive Western assistance in energy development, on the other hand, once these project debts were repaid, Soviet import capacity would more than double. This would mean that the Soviet *capacity* to purchase real imports from the non-CMEA area would have increased at an effective annual rate of over 10 percent a year. Real non-energy-related imports in the interim would not have grown so rapidly because most of the increase in the net fuel balance would occur only after 1985, and debt repayment would eat into energy export revenues up to 1990. In both cases, real GNP was assumed to grow at a "midrange" value of 2.4 percent, a rate that would be compatible with real per capita consumption growth in excess of 1 percent a year.

Sizable increases in the Soviet fuel balance available for export to the West raise the question of whether all of the implied balance could really be exported. The issue arises because the big gains in domestic energy production are likely to come in natural gas. In most cases sizable oil exports could only be maintained if very significant substitution of gas (not coal) for oil were possible. If gas is to replace rather than augment oil exports, a much-expanded natural gas pipeline network, perhaps even beyond the scale of ongoing and contemplated projects, is needed. More precise judgments about these constraints, however, would only be possible after examining much more closely the degree to and rate at which gas can really be substituted for oil in domestic consumption.

Assuming that the worst-best case scenarios for 1990 are at all close to the range of plausible outcomes, they suggest that the simultaneous maintenance of a politically feasible rate of economic growth in the U.S.S.R., the further expansion of real energy exports to Eastern Europe after 1985, and a reasonably high rate of growth of East-West trade (in real terms), will hinge importantly on whether or not the West plays a significant part in developing Soviet energy sources, and particularly gas, in the 1980's.

Appendix A. – Export and Domestic Output Performance for Selected Nonenergy Products

TIMBER

Although subject to considerable cyclical fluctuations, Soviet real exports to the West of timber products (mainly sawn lumber to Western Europe, sawlogs to Japan, and pulpwood) tended to stagnate in the 1970's. Real exports of sawlogs increased only about 4 percent in the course of the 1970's and declined between 1975 and 1979. Lumber exports declined by 20 percent between 1970 and 1975 and then increased to slightly more than their 1970 levels by the end of the decade. Timber export prices tended to grow at 10 to 12 percent per annum in the first half of the decade and at about 10 percent annually after 1975. According to Soviet statistics the volume of timber-cutting in the U.S.S.R. actually declined between 1975 and 1979.[1] The Eleventh Five-Year Plan (FYP), however, calls for a 17- to 19-percent increase in output in the wood products sector over the next 5 years.[2] On the basis of this information, OTA assumed a plausible range of zero to 2.5 percent annual growth in Soviet real timber product exports to the non-Council for Mutual Economic Assistance (CMEA) region. The upper limit is based on an optimistic Western assessment of Soviet timber production by 1990.[3]

PLATINUM

Soviet platinum group exports, primarily directed to Japan, West Germany, and the United States, have fluctuated considerably over the past decade, presumably because of rapidly changing demand conditions and the possible tendency to utilize platinum exports as a residual financing mechanism, much the way gold appears to be used.[4] Estimated Soviet production of these metals increased at an average annual rate of 4.1 percent between 1970 and 1979, although growth slowed to 2.2 percent per annum after 1975.[5] OTA

has taken this latter rate as a basis for the lower bound of real export growth up to 1985, with the upper bound assumed to be twice this rate (4.4 percent).

COTTON

Soviet production of cotton grew considerably in the 1970's, with output rising at an average 2.7 percent annually from 1970 to 1975 and 3.8 percent between 1975 and 1979.[6] From a fairly low base, Soviet exports of raw cotton to the West (principally Western Europe and Japan) rose by an average 30 percent a year in the first half of the decade, slowing to about 2.5 percent per annum after 1975.[7] In his 26th Party Congress speech, Soviet Prime Minister Tikhonov indicated a goal of 9.2 million to 9.3 million tons average annual cotton production between 1981 and 1985, which suggests essentially no growth over 1979 output.[8] OTA's range for export growth was therefore set at an annual rate of zero to 2.5 percent, the worst case figure assuming a constant ratio of hard currency exports to output, the latter assuming that the export growth rate of the late 1970's could be maintained.

CHEMICALS

About three-quarters of Soviet chemical exports to the West in 1979 were accounted for by "radioactive chemical elements" shipped primarily to France and West Germany. No data are available on these exports in real terms, but real growth has obviously been dramatic, as the total value of shipments to the IW rose from only $60 million in 1975 to $922 million in 1979.[9] OTA has fairly arbitrarily assumed that these exports would increase by another $250 million to $500 million by 1985, in 1979 prices.

The Organization for Economic Cooperation and Development (OECD) has estimated that annual CMEA deliveries to the West of chemicals on the basis of recent compensation agreements will total some $1.0 billion to $1.5 billion in the

[1]*Narodnoye khozyaystvo SSSR, 1979.*

[2]*Ekonomicheskaya gazeta*, No. 49, December 1980.

[3]Brenton M. Barr, "Domestic and International Implications of Regional Change in the Soviet Timber and Wood-Processing Industries," Association of American Geographers Project on Soviet Natural Resources in the World Economy, June 1978, manuscript.

[4]See Ronald G. Oechsler and Hedija H. Kravalis, "Complementarity in U.S. Import Needs and Soviet Export Capabilities: Platinum," unpublished study, Feb. 15, 1979; and Thomas A. Wolf, "Soviet Primary Product Exports to the West: An Empirical Analysis of Market Power and Price Sensitivity," study prepared for the Office of External Research, U.S. Department of State, September 1980.

[5]Calculated from CIA, *Handbook of Economic Statistics*, op. cit.

[6]*Narodnoye khozyaystvo SSSR, 1979.*

[7]*Vneshnyaya torgovlya SSSR*, various issues.

[8]*Ekonomicheskaya gazeta*, No. 10, March 1981.

[9]Hedija H. Kravalis, et al., "Soviet Exports to the Industrialized West: Performance and Prospects," in U.S. Congress Joint Economic Committee, *Soviet Economy in a Time of Change*, vol. 2 (Washington, D.C.: U.S. Government Printing Office, 1979), pp. 414-461; and U.S. Department of Commerce data.

early mid-1980's, presumably in constant (i.e., 1979 or 1980) prices. Roughly 90 percent of these deliveries are to come from the Soviet Union. OTA has, therefore, taken $1.0 billion to $1.5 billion as a plausible range for the increase by 1985 in Soviet chemical exports to the West in 1979 prices. (As OECD points out, however, whether *all* of these compensation deliveries will augment rather than replace current deliveries is not known.) Combining the very rough estimates for enriched uranium and compensation deliveries yields a range for average annual growth of real Soviet chemical exports to 1985 of 12.3 and 17.7 percent.

AUTOMOBILES

Soviet exports of automobiles to the IW countries increased from 8,000 units in 1970 to 60,000 in 1975 and 110,000 by 1979.[10] The average annual growth rate for 1975-79 was about 16 per-

cent. Assuming a continued strong Soviet push in this area, OTA set a 5- to 10-percent-per-annum plausible growth rate range for automobile exports in the 1980-85 period.

DIAMONDS

Soviet diamond exports are in many respects more of a mystery than platinum group metals. It has been estimated, however, that real diamond exports fluctuated relatively little and without trend between 1971 and 1977.[11] Rather arbitrarily, real diamond exports are estimated to increase at an average annual rate of zero to 5 percent to 1985.

All other Soviet exports to the West, (no one of which, at the SITC 4- to 5-digit level of aggregation, exceeded more than one percent of Soviet exports to the IW in 1979), are also arbitrarily assumed to grow within the range of 0 to 5 percent to 1985.

---

[10] *Vneshnyaya torgovlya SSSR,* various issues.

[11] Ericson and Miller, op. cit.

CHAPTER 9

East European
Energy Options

CONTENTS

LIST OF TABLES

LIST OF FIGURES

East European Energy Options

Eastern Europe is now struggling to adjust to an energy-expensive world. During the present decade, the six Council for Mutual Economic Assistance (CMEA or CMEA-6) countries examined here—Bulgaria, Czechoslovakia, East Germany, Hungary, Poland, and Romania—will attempt to resolve their energy problems while simultaneously increasing living standards at a politically acceptable pace, and without falling more deeply into debt with Western banks. All this must be done without the degree of reliance on Soviet subsidies in the form of cheap energy which has played such an important role in East European energy supplies in years past. The outcome will have important implications not only for political stability and economic development in Eastern Europe, but also for Soviet-East European relations.

The most important international dimension of Eastern Europe's energy problem lies in its relations with the U.S.S.R. Soviet oil subsidies to Eastern Europe at present are enormous, their value in 1980 amounting to half of all Eastern Europe's exports to the West or to half the value of all Soviet imports from developed countries.[1] As the U.S.S.R. has faced the prospect of increasing constraints on its own energy supplies, its willingness to supply Eastern Europe with cheap energy has diminished. But it is by no means clear that the Soviets could quickly reduce these subsidies without precipitating a degree of economic crisis and political unrest in some East European countries.

The Soviet Union's strategy with regard to East European energy combines plans to stabilize the level of its energy exports (i.e., by cutting the increments of energy, espe-

cially oil, to be supplied) with assistance to the CMEA countries in their efforts to develop their *own* energy resources and to use energy more efficiently. At the basis of this policy seems to be the assumption that tight energy supplies in the U.S.S.R. preclude increments to shipments comparable to those of the 1960's and 1970's.

The purpose of this chapter is to illuminate the degree to which this Soviet strategy is feasible, given the problems and opportunities in domestic energy production and consumption which will confront the nations of Eastern Europe in the next decade. The chapter briefly reviews energy trends in Eastern Europe over the last 20 years. It then analyzes the energy problem from both the supply and demand sides, identifying constraints and opportunities and evaluating the political and economic implications for East European energy imports—particularly imports from the Soviet Union. The discussion addresses the important issue of whether Eastern Europe will be able to cover its energy needs with Soviet imports supplementing domestic production, or whether large energy deficits which must be met with other imports may occur. The latter situation could augur heightened domestic political and economic difficulties in Eastern Europe and might necessitate alterations in Soviet policy.

The East European countries make up a fairly well-defined energy system, characterized by relatively few options for expanded domestic energy production and well-established historical trends of energy usage. Therefore, it is possible to make reasonable guesses about future trends. In the analysis that follows, East European plan targets for energy production are evaluated and contrasted to best and worst case projections for energy production, demand, economic growth, net energy imports,

[1]Figures for exports, imports and debt are based on 1979 data in CIA, *Handbook of Economic Statistics,* ER 80-10452, October 1980; and on 1980 estimates provided to OTA by Edward A. Hewett.

and hard currency debt. These cases, constructed by OTA, are judgments of what appear to be most and least optimistic alternative futures, and are included in order to delineate a range of possible alternatives. **They are not predictions, but informed guesses about likely possibilities.**

INTRODUCTION

The key goals of the Soviet energy strategy for CMEA are outlined in the "Long-Term Target Program for Cooperation in the Areas of Energy, Fuels, and Raw Materials." They are: 1) the development of natural resources to their fullest in every member country through expanded exploration and quick development of newly discovered deposits; 2) strong promotion of nuclear power, particularly through intra-CMEA cooperation; 3) promotion of the development of energy-saving technologies and the application of energy-saving processes; and 4) changes in the structure of output designed to reduce the share of energy-intensive products in gross national product (GNP).[2]

The emphasis in this program, one evidently supported by East European planners, has been on supply-side remedies to the energy problem. But while there has been little discussion of conservation, and few concrete measures have been designed to promote it, at least one Soviet expert has recently raised serious doubts concerning the viability of a supply-side approach in view of recent difficulties in expanding East European production of coal and other energy sources. He suggests that conservation measures deserve serious consideration since the only other alternative—energy imports from third countries—is simply too expensive.[3]

The energy-saving approach recognizes that it is cheaper to conserve than to increase production, but the problem of how genuine conservation can be achieved remains.[4] One way is to change the structure of GNP to reduce the share of energy-intensive industries and product. This approach has adherents, but it gives rise to other difficulties, since energy-intensive sectors (chemicals, fuels, metallurgy, and construction materials) are so important in CMEA economies. The development of energy-saving technology is a promising long-term solution, but the quickest short-term option—reform of the economic system—is understandably downplayed. An enhanced role for meaningful prices, and for profits, combined with a workable set of bankruptcy and enterprise reorganization laws, would probably help to reduce energy wastage by industries. The political costs of such a strategy could be quite high, however, and such reforms have not received widespread support.

Understanding East European domestic political and economic considerations, as well as those that govern relations with U.S.S.R., is important for comprehending the policy choices which East European leaders have made, and those which they are likely to make in the years ahead. The approaches they take to the energy problem will combine three elements: 1) increased domestic energy supplies; 2) reduced energy demand; and 3) increased imports. Each country calculates the costs and benefits of these strategies differently, depending on its energy situation and on perceptions of the political consequences of one or another path.

[2]P. Bagudin, "The Long-Term Target Program for Cooperation in the Area of Energy, Fuels, and Materials, and its Realization," *Vneshnyaya torgovlya*, October 1980, pp. 13-18.

[3]Vladimir M. Gzovskiy, "The Economics of Energy Resources in the CMEA Countries," *Voprosy ekonomiki*, December 1980, pp. 96-103.

[4]Miklos Szocs, "Program for the Next Five Years," *Figyelo*, Dec. 24, 1980, pp. 1, 4.

ENERGY SUPPLY AND DEMAND IN EASTERN EUROPE, 1960-79: MAJOR THEMES

This section reviews major trends in the development of East European energy production and consumption over the past 20 years. These themes will form the context for the choices facing planners, and will become the basis for OTA's own projections for Eastern Europe's energy future in the coming decade.

RESERVES

East European energy reserves are small and dwindling. The size of Eastern Europe's oil reserves is only about 3 percent of that of the estimated proved oil reserves of the U.S.S.R. Gas reserves are even smaller. Most of this petroleum is concentrated in Romania, which has about 89 percent of all Eastern Europe's proven oil reserves. In 1976, however, Romanian oil production peaked and now appears to be in long-term decline. If reserves are exploited at late 1970's production rates, they will be exhausted in a little more than 10 years. Romania also holds Eastern Europe's largest gas reserves, about 40 percent of the total. Here too declining production trends are clear.

Forty percent of Eastern Europe's coal reserves, and almost all of its hard coal, are located in Poland. These reserves have been the sole source of net energy exports from Eastern Europe, but recent events in Poland put continued exports in considerable doubt. The other major deposits, located in Czechoslovakia and East Germany, are coals with low calorific value that have served as the backbone of those countries' primary energy and electricity production.

ENERGY IMPORTS

East European net energy imports have been rising rapidly. In every year since 1961, Eastern Europe as a whole has been a net energy importer, with imports rising to 23 percent of consumption in 1978. The rate of increase, too, has been growing. One indicator of this has been the recent rise in the marginal import to consumption ratio, which records the proportion of the increment to consumption that is covered by net imports. In recent years, two-thirds of the increase in energy consumption has been covered by increases in energy imports. The increasing import/consumption ratio represents a significant policy problem in that it creates an added strain in export requirements necessary to pay for the additional energy. Almost all of these imports are of oil and gas and the percentage of the latter is rising.

Figure 24 shows the growing importance of net energy imports in Eastern Europe. Some perspective may be gained by comparing the East European energy situation with that of Western Europe.[5] In 1978 the countries of the European Economic Community (EEC) produced 81 million barrels per day of oil equivalent (mbdoe) or 438.7 million tons of oil equivalent (mtoe), importing 54 percent of all their energy. During the same year, consumption in Eastern Europe was about 9.5 mbdoe (473.1 mtoe) and production 6.24 mbdoe (310.7 mtoe). Thus, it was necessary to import only 23 percent of energy consumed. But the EEC's far higher import dependence is declining over time, and between 1974 and 1978 net imports to the EEC actually fell as North Sea oil production began. The opposite is true in Eastern Europe where import dependence is growing and where there is no prospect of a North Sea.

THE ENERGY BALANCE

Figure 25 illustrates the strikingly dominant position of coal in East European energy consumption. In 1979, 78 percent of the energy produced in the area was coal;

[5]International Energy Agency, Organization for Economic Cooperation and Development (OECD), *Energy Balances of OECD Countries 1974/1978* (Paris: OECD, 1980).

Figure 24.—Consumption, Production, and Net Imports of Energy for All of Eastern Europe, 1960-79

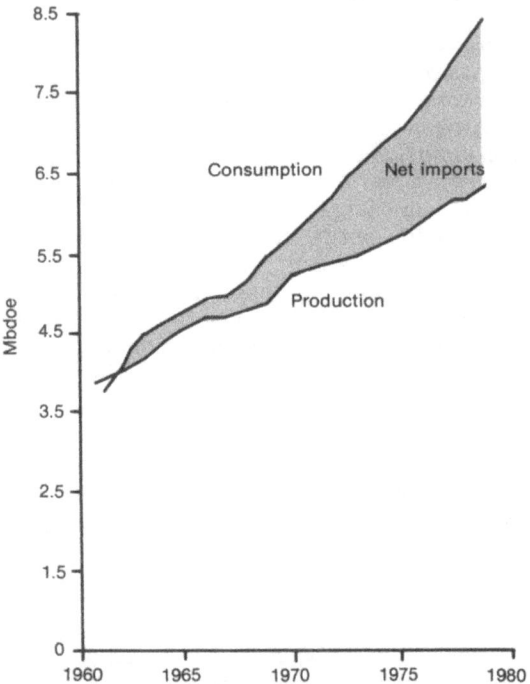

SOURCE: Data are from CIA, "Energy Supplies in Eastern Europe: A Statistical Compilation," ER 76-10624, December 1979, and CIA, *Handbook of Economic Statistics*, ER 80-10452, October 1980.

Figure 25.—East European Energy Consumption by Energy Source, 1960-78

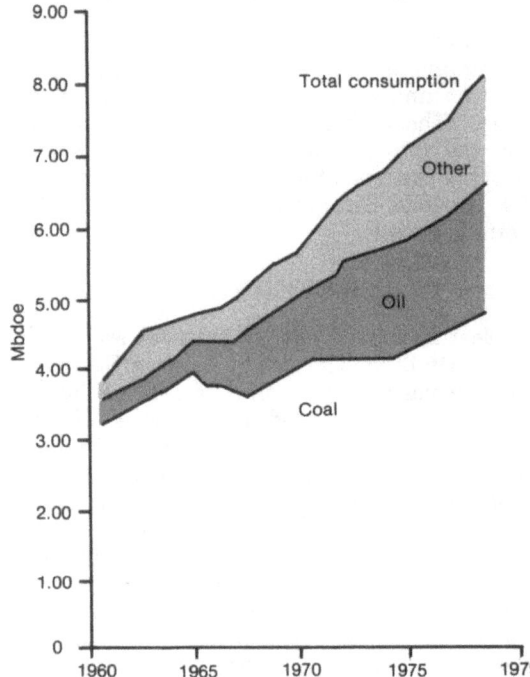

SOURCE: Data are from CIA, "Energy Supplies in Eastern Europe: A Statistical Compilation," ER 76-10624, December 1979, and CIA, Handbook of Economic Statistics, ER 80-10452, October 1980.

natural gas constituted 14 percent, and nuclear and hydropower together 3 percent. Poland, the mainstay of this production, provided 2.7 mbdoe (134.4 mtoe) in 1979—42 percent of all energy output for the region. The second largest energy producer, Romania, contributed 17 percent in the same year, but Romanian energy production has been falling since 1976. Figure 26 illustrates the crucial importance of Polish coal. Poland was primarily responsible for increases in energy production during the 1970's. Without those increases the small output gains made by other countries would have been completely canceled by Romania's decline.

Coal also plays a dominant role in East European energy consumption. Among EEC nations in 1978, coal accounted for 22 per-

cent of total energy consumption; in Eastern Europe in 1979, 57 percent of energy was consumed in the form of coal. In contrast, oil and gas—which made up 74 percent of EEC energy consumption—provided 40 percent of the total in Eastern Europe. Consumption of petroleum *has* increased over the last two decades in Eastern Europe, but world price rises have slowed that process. This gives CMEA one advantage relative to the rest of the world. As many nations attempt to switch back to coal, Eastern Europe can merely slow its transition to oil.

THE ROLE OF THE U.S.S.R.

The Soviet Union is overwhelmingly important as an energy supplier to Eastern Europe. Table 62 shows estimated Soviet

Figure 26.—Energy Production in Eastern Europe, Total, and by Country, 1960-79

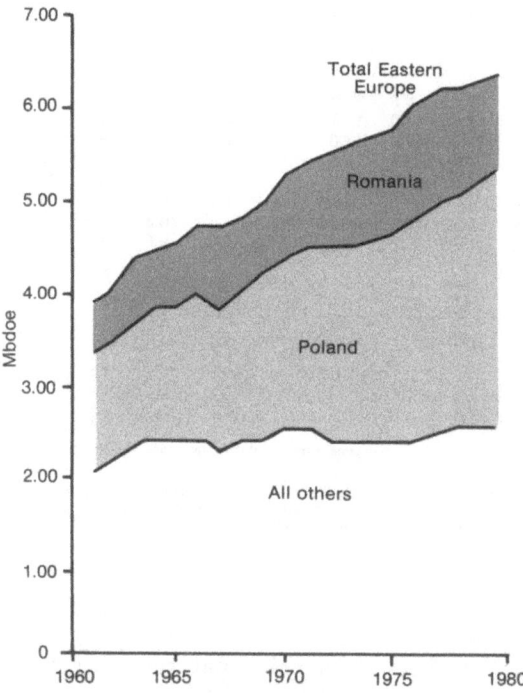

SOURCE: Data are from CIA, "Energy Supplies in Eastern Europe: A Statistical Compilation," ER 76-10624, December 1979, and CIA, Handbook of Economic Statistics, ER 80-10452, October 1980.

crude oil and product exports to CMEA during the 1970's. The Soviet Union's oil exports to all nine CMEA countries have generally been about 55 percent of its exports to the world. In 1979, for example, the Soviet Union shipped to CMEA 87.1 million tons (1.74 mbd) of crude oil and oil products, more than half of all its exports (158.1 million tons or 3.16 mbd). This percentage has remained relatively constant, although the rate of growth of Soviet oil exports to CMEA has slowed. This reflects a reduction in total Soviet oil exports over the past 5 years.

Even more important than the huge quantities of Soviet oil shipped to Eastern Europe is its relatively low price. This has constituted a substantial subsidy. The average price per ton charged by the Soviet Union for crude oil shipped to Eastern Europe in 1980 was about half of the world price.[6] This price is calculated according to a method, adopted in 1975, called the "Five Year Moving Average." This system uses world oil market price averages over the previous 5 years as a basis for annual price negotiations in intra-CMEA oil trade. The result is a considerable lag in CMEA prices for oil. The advantage to Eastern Europe has been enormous. While world oil market prices rose almost thirteen-fold between 1972 and 1980, prices paid by East European countries for Soviet oil rose only about 4.5 times. In 1980, Soviet exports to Western nations brought an estimated average price of $230/ton, while export prices to CMEA countries were $105/ton. Assuming a subsidy amounting to the difference between the two prices, the 1980 subsidy was 81.1 million tons × $125 = $10.1 billion. This was one-half the value of all Eastern Europe's exports to the West. If East European nations were forced to pay the full world market value for this oil, they would have had to increase their dollar exports by 50 percent, or double their annual hard currency borrowing. Such loans would be very difficult, if not impossible, to obtain.[7] Conversely, by selling this oil on the world market, the U.S.S.R. could increase its hard currency imports 50 percent. Obviously, however, the subsidy is so large that if the Soviets tried to eliminate it quickly, the result would be chaos for Eastern Europe.

It must be noted, however, that Eastern Europe actually imports slightly more energy from the Soviet Union than the total of its net energy imports. The reason is that while Eastern Europe has been a net importer of energy from the Soviet Union, it has also exported energy, mostly coal, to

---

[6] Jan Vanous, "Eastern European and Soviet Fuel Trade, 1970-75," in *Eastern Europe Assessment*, U.S. Congress, Joint Economic Committee (Washington, D.C.: U.S. Government Printing Office, 1981).

[7] Ibid. In fact, the oil subsidy to Eastern Europe is probably even higher if one allows for the fact that the dollar/ruble exchange rate is overvalued. The total subsidy to CMEA is, moreover, higher than the oil subsidy alone since other primary products are subject to similar kinds of subsidies.

Table 62.—Soviet Exports to CMEA of Crude Oil and Oil Products, 1970-80

	1970 C	1970 T	1975 C	1975 T	1976 C	1976 T	1977[a] C	1977[a] T	1978[a] C	1978[a] T	1979[b] C	1979[b] T	1980[b] C	1980[b] T
Exports to:														
Bulgaria	4.8	7.1	9.9	11.6	10.0	11.9	10.8	12.9	11.3	13.4	13.0	14.1	13.0	14.0
Czechoslovakia	9.4	10.5	15.5	16.0	16.3	17.2	17.0	17.0	17.7	17.7	18.3	18.3	19.2	19.2
East Germany	9.2	9.3	15.1	15.0	16.0	16.8	17.0	17.0	17.8	17.8	18.5	18.5	19.0	19.0
Hungary	4.0	4.8	6.9	7.5	7.7	8.4	7.7	9.1	8.5	10.2	8.6	11.0	9.5	12.0
Poland	7.0	8.6	10.9	13.3	11.7	14.1	12.8	14.7	13.4	15.5	12.9	14.0	13.1	15.9
Romania	—	—	—	—	—	—	—	—	—	—	0.4	0.4	1.0	1.0
CMEA-6	34.4	40.3	58.2	63.4	61.7	68.4	65.3	70.7	68.6	74.6	71.7	76.3	73.8	81.1
Cuba	4.3	6.0	5.8	8.1	6.0	8.8	6.2	9.2	6.4	9.6	6.7	9.6	7.0	10.0
Vietnam	—	0.4	—	0.4	—	0.4	—	0.5	—	0.5	—	0.6	—	0.6
Mongolia	—	0.3	—	0.4	—	0.4	—	0.5	—	0.5	—	0.6	—	0.6
CMEA-9	38.7	47.0	64.0	72.3	67.7	78.0	71.5	80.9	75.0	85.2	78.4	87.1	80.8	92.3
Entire world	66.8	95.8	93.1	130.4	110.8	148.5	NA	152.5	NA	165.6	NA	158.1	NA	NA

C = Crude, T = Crude plus products.
NA = not available.

[a]These are estimates necessitated by the fact that the Soviet Union stopped reporting quantity data on its energy exports in 1977; but they are probably fairly reliable indicators of actual shipments.

[b]These are estimates, but somewhat less reliable than the 1977-78 figures. They should be taken only as indicators of general magnitudes; in some cases the actual number could be easily 1 ton larger or smaller.

SOURCES: The data through 1976 are from Soviet foreign trade yearbooks (*Vneshnyaya torgovlya SSSR*). Figures concerning the proportion of crude and products for Cuba, Vietnam, and Mongolia are estimated. Data beginning in 1977 are estimates based on CMEA, *Vneshayaya torgovlya* (Statistical Yearbook of the Member-Countries of the Council for Mutual Economic Assistance) (Moscow: "Statistika," 1979 and 1980), and *The Journal of Commerce.*

nonsocialist countries. In effect, Eastern Europe as a whole is reexporting in the form of coal some of the energy it imports from the Soviet Union in the form of oil.

Table 63 shows estimated Soviet natural gas shipments to CMEA in the last decade. The Soviet Union only began to develop its natural gas export capabilities in the 1970's, with the completion of the Orenburg (or Soyuz) gas pipeline, the result of a 3 billion transferable ruble joint development project involving the U.S.S.R. and the CMEA-6. Each of the East European countries provided some combination of equipment, labor, and hard currency to buy Western equipment for construction of the 2,750-km pipeline, 22 compressor stations, gas treatment plant, and gas condensation unit at Orenburg. The pipeline began operating at full capacity (transporting 15.5 billion cubic meters or bcm of gas per year to Eastern Europe) during 1980.

It is likely that most future increments in Soviet energy shipments to Eastern Europe will be in the form of natural gas. The im-

plications of this trend are important because, in contrast to oil, the Soviet Union has not been subsidizing natural gas prices paid by Eastern Europe. In fact, it appears that while the Soviet Union has sold Eastern Europe oil at one-half the world market prices, it has sold gas at about the world price level.

Table 64 provides a rough comparison of average 1976 prices of Soviet oil and gas exports to both Eastern and Western Europe. This table shows that Soviet gas prices to Western Europe on a per calorie basis were about one-half oil prices, 35 rubles/ton of oil equivalent of gas compared to 68.6 rubles/ton of oil. This is consistent with world practice: gas prices generally are lower on a calorific basis than are oil prices, reflecting the higher transport costs and the fact that gas is an imperfect substitute for oil. In contrast, Soviet gas and oil exports to Eastern Europe in that year cost about the same per calorie—39.5 rubles/ton of gas v. 38.1 rubles/ton of oil. Moreover, the gas price was actually higher than the average price to West European countries.

Table 63.—Soviet Natural Gas Exports, 1970-80

	1970	1975	1976	1977[a]	1978[a]	1979[b]	1980[b]
Bulgaria	—	1.19	2.23	2.90	3.00	3.40	5.80
Czechoslovakia	1.30	3.69	4.29	5.20	5.30	7.30	8.10
East Germany	—	3.30	3.37	3.55	3.62	4.33	5.70
Hungary	—	—	—	1.00	1.03	2.50	3.83
Poland	1.00	2.51	2.55	2.77	2.76	3.99	5.56
Romania	—	—	—	—	—	0.75	1.49
Total, CMEA-6	2.30	10.69	12.44	15.42	15.71	22.22	30.48
Total, all countries	3.30	19.33	25.78	31.23	NA	NA	55.00

NA = not available.

[a]These are estimates necessitated by the fact that the Soviet Union stopped reporting quantity data on its energy exports in 1977; but they are probably fairly reliable indicators of actual shipments.

[b]These are estimates also, but somewhat less reliable than the 1977-78 figures. They should be taken only as indicators of general magnitudes; the actual figures could differ from these.

Table 64.—Comparison of Unit Values for Soviet Gas and Oil Exports, 1976

	1976 Gas		Oil	Rubles per:	1979 Gas		Oil
	Bcm	Toe[d]	Ton	Bcm	Toe	Ton
Exports to:							
West Germany	22.7	27.8	65.5	NA	NA	NA
Finland	47.3	57.8	80.9	NA	NA	NA
France	25.6	31.3	66.4	NA	NA	NA
Italy	14.0	17.1	65.0	NA	NA	NA
Austria	33.5	40.9	65.4	NA	NA	NA
Average, West	28.6	35.0	68.6	NA	NA	NA
Bulgaria	33.4	40.8	37.5	45.9[a]	56.1	64.4
Czechoslovakia	34.5	42.2	34.1	45.9	56.1	56.1
East Germany	27.7	33.9	32.0	50.5	61.7	55.3
Hungary	33.9	41.1	44.7	50.6	61.9	76.0
Poland	32.0	39.1	42.0	49.5	60.5	71.6
Romania	—	—	—	51.1	63.0	—
Average, CMEA	32.3	39.5	38.1	48.9	59.9	64.7

NA = not available.
Bcm = Billions of cubic meters.
Toe = Tons of oil equivalent (1,000 cubic meters of gas = 0.818 toe).
[a]Assumed equal to the Czech unit value.
SOURCE: Vneshnyaya torgovlya SSSR.

If, as it appears, Soviet gas sold in Eastern Europe is not being subsidized,[8] one possible explanation lies in the fact that large gas shipments only began after the 1974 world oil increases, and the Soviets must have been able to force through a full market price for gas. This would have been possible because even at that price, Soviet gas remains attractive for Eastern Europe. In any case, CMEA countries can buy additional increments of energy only at the world market price, and the world price of gas is still much lower than that of oil on a calorific basis. Transport costs also make the closest supplier the cheapest.

[8]Several additional factors temper these gas prices, however. First, the official exchange rate understates the rate of subsidy. Even at roughly equivalent gas prices at official exchange rates, Eastern Europe is receiving some implicit subsidy in being able to pay for the gas with manufactured goods sold to the Soviets at what are, in effect, inflated prices. Secondly, gas prices are complicated because of compensation deals. For example, the Orenburg pipeline agreement involved machinery and equipment, labor, and money capital in exchange for gas. It is possible that those East European investments of labor, money, and goods were overvalued relative to the final prices of gas.

Photo credit: Oil and Gas Journal

Orenburg gas pipeline

Imports of Soviet gas have, therefore, grown rapidly. In 1977 these amounted to about 15.42 bcm or 18 percent of the 70.7 million tons of the Soviet crude oil and products imported. By 1980 the Soviets were shipping 30.48 bcm of gas to Eastern Europe. This equals 24.93 mtoe, or 31 percent of the 81.1 million tons of crude and products shipped to Eastern Europe that year. Assuming that oil shipments from the U.S.S.R. will not increase above 1980 levels, the critical issues for Eastern Europe are how rapidly the oil subsidy will be reduced, and how quickly imports of natural gas will increase.

These major themes—dwindling East European energy reserves, rising imports, the continuing importance of coal, and the overwhelming importance of the Soviet Union as an energy supplier—delineate the policy context of future East European energy planning. The next two sections, on supply and demand prospects for the 1980's, investigate critical opportunities for and constraints on future East European energy strategies.

EAST EUROPEAN ENERGY SUPPLIES IN THE 1980'S

A centerpiece of the East European energy strategy is the commitment to increased coal and nuclear development to cover increases in demand over the next decade. Almost all of the additional coal output will be used in conventional powerplants, condensing stations, and cogeneration units. Coal and expanded nuclear power together will thus cover incremental electricity demands. At the same time, petroleum will be freed for use as chemical industry feedstocks, for automobile fuel, and other uses where solid fuels are inappropriate. In Romania, where state policies already prohibit the commissioning of any new heat or power stations operating on oil and gas, the proportion of electricity generated from these fuels is projected to drop from the current level of 65 to 40 percent by 1990.[9] Other East European nations, lacking Romania's domestic petroleum reserves, have been attempting to cover all of their incremental energy needs with coal and nuclear power. For example, Czechoslovakia employed this strategy during its 1976-80 plan period, but did not succeed by the amount originally targeted.[10]

Coal is likely to remain the predominant source of domestically produced energy in Eastern Europe, at least in the near future, and official plans for the next 10 years feature expanded output of coal, particularly lignite. **Eastern Europe's success in domestic energy output over the next decade will rest mainly on its ability to increase coal output.**

A second major source of energy will be nuclear power. Despite the currently tiny fraction of East European domestic energy production (1.25 percent) accounted for by

[9]"WPC Host Country Romania Swaps Tools and Technology for Crude," *Oil and Gas Journal,* vol. 77, No. 35, Dec. 27, 1979, p. 76ff.

[10]"The National Economy . . .," 1980, p. 11. See also Radio Free Europe (RFE), Czechoslovak SR No. 4, Jan. 31, 1979, pp. 1-2.

nuclear power, future prospects for the industry are promising. It is the focus of one of the major, and apparently rather successful, cooperative efforts within CMEA. (See below and ch. 4.) Finally, prospects for the Romanian petroleum industry are poor. The sections that follow explore the constraints and options for each of these energy sectors, as a foundation for the projection of future net energy import needs.

THE COAL INDUSTRY

Table 65 summarizes the plans of various East European countries for coal output in the 1980's. Since the quality of information varies considerably for different countries, these data should be viewed only as general approximations. All six countries are planning a major expansion of coal output, mostly lignites. Taken together, the plans suggest a growth rate of lignite and brown coal production of 5.4 percent per annum during 1980-85, and a growth rate of 3.8 percent for hard coal production during the same period. Lower growth rates are projected for the second half of the decade, but these are less certain, as a number of the countries have not yet announced formal plans.

The projected growth rates for coal are ambitious. During the last 5 years (1975-80) lignite and brown coal output grew 2.9 percent and hard coal grew 1.9 percent per year.

1980 production was slightly below that of 1979, reflecting a drop in Polish coal production, and continued stagnation in Hungarian, Czech, and East German output. **The plans for the 1980's thus call for a doubling of the growth rates of the past 10 years, and a substantial improvement over 1979-80 performance.** While this is not impossible, there is normally a considerable lag between the time that the decision is made to increase output, and the actual completion of the mine capacity necessary to implement this decision. Investment processes to increase lignite production have, to some extent, been set in motion in all six East European countries. When these investments will begin to bear fruit, and what type of coal will be produced, remain to be seen.

There are at least three potentially serious obstacles to the fulfillment of coal output targets. The first applies mainly to open-pit mining of lignite and is a technology constraint: East European machinery in this area is apparently of low quality, but planners are reluctant, or unable, to approve large imports of Western equipment because of severe foreign currency constraints. Second, there may be labor problems in underground mining of both brown and hard coals. Finally, there may be an environmental constraint, again associated primarily with lignites. The following sections briefly discuss each of these problems, describe intra-

Table 65.—East European Plans for Coal Output in 1985 and 1990

| | \(1980 estimate\) | | | | 1985 \(plan\) | | | | 1990 \(plan\) | | | |
| | | | Total | | | | Total | | | | Total | |
	HC	BC + LIG	Mtnat	Mbdoe	HC	BC + LIG	Mtnat	Mbdoe	HC	BC + LIG	Mtnat	Mbdoe
Bulgaria	0	31	31	0.12	0	37	37	0.15	0	45	45	0.18
Czechoslovakia	28	95	123	0.91	28	104	132	0.96	28	109	138	0.99
East Germany	0	250	250	1.05	0	275	275	1.16	0	300	300	1.26
Hungary	3	23	26	0.14	3	32	35	0.18	3	34	37	0.19
Poland	193	37	230	2.30	235	85	320	3.20	260	115	375	3.75
Romania	8	32	40	0.20	13	74	87	0.44	15	82	97	0.49
Total	232	467	700	4.72	279	607	887	6.09	306	685	992	6.86

HC = hard coal; BC = brown coal; LIG = lignites; Mtnat = million tons in natural units; mbdoe = million barrels per day of oil equivalent. Hard coal contains at least 5.7 million kilocalories per ton.
All tonnages rounded off to the nearest ton. Thus some "0's" simply indicate a figure of less than 0.5 tons, and some totals will not equal the sum of the components of the columns due to rounding.

SOURCE: Office of Technology Assessment.

CMEA cooperation in coal, and evaluate the feasibility of plans to increase coal output.

Technology

The Polish and Czech cases, about which there is a good deal of information, illustrate the types of measures which East European nations are taking to increase lignite production. Hoping to set aside as much hard coal as possible for export, Polish planners are pushing lignite for domestic consumption.

Much of Poland's planned expansion of low-calorie coal and electric power centers on the Belchatow Power Plant and the Szczercow lignite mine. The Belchatow station is designed to have twelve 360-MW generators by 1985 (completion of which may be delayed by such problems as water incursion). The entire station will run on lignites with calorific contents ranging from 1.600 to 1.900 kilocalories/kg.[11] At full capacity, the station should produce 26.5 billion kWh of electricity per year, the equivalent of 23 percent of all the electricity produced in Poland in 1979. A second station (Belchatow II), with a total generating capacity of 2,880 MW, is to come online in the period 1985-90. This complex is the major source of Polish incremental demand for lignites in the 1980's. Between 1980 and 1985, the complex will use 33 million tons of lignite, or 87 percent of the total increase in Poland's planned lignite production.[12]

While it is difficult to judge the progress of the Polish project, there are indications of problems similar to those which have developed in Czechoslovakia in attempts there to expand lignite production. First, open-pit mining equipment of East European design is evidently inferior in quality to comparable Western equipment; it breaks down frequently, causing delays in removing overburden, mining the coal, and moving it by conveyor belt. The Czechs, who have reported repeated problems with excavators and conveyors, were finally forced to import conveyors from the West. A second problem is that East European suppliers of mining equipment are not meeting their orders on time. It may well be that the sudden increase in demand for mining equipment is pushing the suppliers beyond their capabilities. At the same time, severe hard currency shortages preclude imports of high-quality Western mining equipment. These common and recurring problems suggest that East European plans to expand lignite production may be overly ambitious.[13]

Labor

A second potentially important impediment to the fulfillment of East European plans for expanded coal output lies in the unwillingness of the labor force to work long hours in underground mines. The labor constraint is a key to Polish coal production, but it is also significant in Hungary, where new brown coal mines run 24 hours a day, 6 days a week; and in Romania, where miners in the Jiu Valley have shown reluctance to work for low wages.

Indeed, Poland's plans for hard coal—about 42 million tons, constituting almost all

[11]Bagudin, op. cit., p. 15, reports these calorific values for the coal and calls it "lignite." Bartosevich calls it "brown" coal of approximately 2,000 kilocalories/kg. See Abignev Bartosevich, "The Significance of Cooperation in the Long-Term Development of Energy in Peoples' Republic of Poland," *Ekonomicheskoye sotrudnichestvo stran-chlenov SEV*, February 1980, pp. 37-42. The calculations assume that the lower figure is correct.

[12]These calculations assume a 70-percent load factor in the station and 40-percent efficiency. Each 1,000 kWh equals 0.0875 toe, and at 40-percent efficiency, it will require 0.219 toe to produce that 1,000 kWh. The lignites feeding Belchatow run from 1,600 to 1,900 calories. Assuming the average is 1,750, they convert at 0.175 toe. Thus 1,000 kWh, which require 0.219 toe, will require 0.219/0.175 = 1.25, or 0.00125 tons of lignite per kWh. 26.5 billion kWh multiplied by 0.00125 yield 33.125 million tons of lignite to produce the 26.5 billion kWh.

[13]Hungary may be the one CMEA country that has tried to directly deal with the technology problem by buying the proper equipment in the West, and then using it effectively. In the early 1970's it completed construction of the Visonta Thorez Open-Pit mine which was then responsible for a 20-percent increase in the output of surface-mined coal. It was equipped with what was called "world level technology." Judging from the pictures accompanying the story, this consisted of Western equipment. *Nepszabadsag*, Dec. 7, 1980, p. 1. The Hungarians are now finishing construction of two highly mechanized brown coal mines, Markushegy and Nagyegyhaz, which will increase brown coal production by 9.6 million tons by the mid-1980's, an increase of 40 percent. Again, it would appear that they are relying heavily on Western equipment.

of Eastern Europe's planned increment to 1985—seem unattainable. Goals for lignite production increases in all the countries of Eastern Europe amount to 140 million tons, but in calorific value this equals only 30 mtoe. If Polish targets for hard coal are not met, two-fifths of Eastern Europe's projected coal increment will be eliminated.

Nevertheless, it is difficult to imagine how such growth in Polish coal output can be attained. Increased production in the 1970's was made possible by the implementation of a four brigade system of working the mines, supplemented with overtime and, more recently, Sunday work. This effort was part of a frantic search for hard currency exportables brought on by Poland's severe debt problems. One source of Polish labor unrest in the summer of 1980 was the rapid pace of mining activity, and one concession won was reversion to a 5-day workweek. Polish hard coal output for 1980 fell to 193 million tons, 14 million tons below plan. The original 1981 plan of 188 million tons was abandoned in midyear after output during the period January-June amounted to only 81.3 million tons. Now planners feel that they can achieve approximately 170 tons for all of 1981 only if workers will again agree to work 6 days a week; otherwise, an output of 160 million tons—41 million tons below the 1979 peak of 201 million tons—appears likely.[14]

As of this writing, no one can say how the Polish situation might be resolved, and in particular, how the resolution will affect coal production. Certainly the old targets of 235 million tons in 1985 and 260 million tons in 1990 seem unattainable, having been constructed on the now impossible assumption that Polish coal mines could be worked 7 days a week. On the other hand, the low coal outputs in 1980-81 are surely below what is possible with given capital stock and a normal 5-day workweek.

Environmental Costs

Environmental costs form another potentially important political and economic constraint on lignite production. Two major problems are the loss of land to other uses as open-pit mines are developed, and the effects on air and water of mining and coal burning.

The land costs have not been quantified, but they are potentially important. In both East Germany and Czechoslovakia, countries which are heavily engaged in open-pit mining, cities and rivers have been moved to obtain access to coal deposits.[15] Aside from the enormous costs in capital and labor, the dislocation of people and the disruption of the countryside may create popular dissatisfaction which cannot be discounted.

Even more serious, however, are the air and water pollution which result from heavy reliance on coal. Environmental damage caused by coal mining dates back to the 1950's in Eastern Europe. At that time heavy fallout of particulate matter and emissions from chemical plants in some areas in Czechoslovakia reduced morning light by 50 percent, killed 37,000 spruce trees, and cut 90 percent of the ultraviolet rays.[16] These impacts evidently presented enough of a political problem that the Dubcek government responded by halting work in the North Bohemian brown coal basin in 1968. Today official statements recognize the environmental problems associated with open-pit mining.[17] While there is insufficient evidence of how deep feelings run, it seems possible that major attempts to increase coal output will do enough damage and displace enough people to turn this into a political issue.

---

[14]See RFE, Polish SR No. 10, May 9, 1979, pp. 1-4; and RFE, Polish SR No. 1, Jan. 26, 1981, p. 13; RFE, Polish SR No. 12, July 3, 1981, p. 14; and *Petroleum Economist*, August 1981, p. 359.

[15]Leslie Dienes, "Energy Prospects for Eastern Europe," *Energy Policy*, June 1976, p. 126; V. Belianov, "Working Successes of the Miners," *Ekonomicheskaya gazeta*, Feb. 9, 1978, p. 21.

[16]J. G. Polach, "The Development of Energy in East Europe," in Joint Economic Committee, Subcommittee on Foreign Economic Policy, U.S. Congress, *Economic Developments in Countries of Eastern Europe* (Washington, D.C.: U.S. Government Printing Office, 1970), p. 360.

[17]RFE, Czechoslovak SR No. 4, Jan. 31, 1979, p. 2.

CMEA Cooperation in Coal

There are currently no joint CMEA coal mining efforts, but some cooperative activity is aimed at developing coal-fired plants using low-calorific coals, and at cogeneration from conventional plants. For example, Bulgaria and the U.S.S.R. are reconstructing power stations to burn low-calorie coals without prior preparation. East Germany, the U.S.S.R., Poland, and Hungary are working on a joint project in East Germany involving the construction of mines and power stations for low-calorie coal. At various research institutes, cooperative work is being conducted in cogeneration, high-productivity steam boilers, and centralized steam production. However, none of the cooperative CMEA projects hold real promise of increasing coal supplies in the next decade. Such efforts may have some effect in efficient utilization of coal, but even in that area the joint R&D effort is modest. Prospects for East European coal in the next decade will depend to a great extent on the initiatives taken by individual countries.

The Feasibility of Meeting Coal Output Targets

The success of aggregate East European plans for the coal industry rests on developments in Poland and Romania. These two countries account for all of the planned increment to hard coal production in the 1980's, and 65 percent of the planned 140-million-ton increase in brown coal and lignites. This equals approximately 80 percent of planned increases when these figures are converted to tons of oil equivalent. The key problem in evaluating these targets is the uncertainty associated with the Polish crisis. At the very least, it seems impossible for Poland to reintroduce 7-day workweeks for miners.

Mindful of these uncertainties, OTA has constructed "best" and "worst" case estimates for coal output. Table 66 summarizes these best and worst case projections, and compares them to plan targets.

Table 66.—Projected and Planned East European Coal Production (million barrels per day of oil equivalent)

	1980	1985	1990
Actual	4.72		
Plan		6.09	6.86
Best case projection		5.46	5.98
Worst case projection		4.99	5.44

SOURCE: Office of Technology Assessment.

For Poland, the best case assumes that output levels in 1985 will reach 210 million and in 1990 220 million tons (as opposed to the planned targets of 235 million and 260 million tons). These best case figures are still above what the Minister of Mining has implied are possible; they assume a settlement of labor troubles, some new investment, and productivity improvements. OTA's worst and increasingly probable case assumes that although output will fall below long-term trends in the early 1980's, it will recover in the latter half of the decade, reaching 190 million tons in 1985 and 200 in 1990. Current events in Poland are a reminder that things could be worse in 1985 than even this worst case. But the year 1985 here is merely representative of the mid-1980's, and it is unlikely that the current level of chaos in the Polish economy can or will be sustained for that long. This is, therefore, not the worst imaginable case (which is no coal output), but rather the worst case within the range of likely coal outputs.

Romanian plan targets appear to be no more realistic. Indeed, since 1977 these have been consistently underfulfilled. In 1980, for example, Romania planned to produce 54 million tons of of coal, but probably attained an output of no more than 40 million tons. The situation has been serious enough that the army has been called in to assist the miners.[18] While it is difficult to estimate

[18]RFE, Romanian SR No. 18, Dec. 10, 1980, p. 15. It seems certain that they will not meet the 1979 plan, but the figure of 40 is a guess which assumes that they underfulfill in 1980 by the same amount they underfulfilled in 1975.

future output, OTA's projection of a reasonable best case is 60 million tons in 1985, 20 million tons above actual performance in 1980. Since there are no Romanian plan figures now available for 1990, projections are necessarily speculative. But it would appear unlikely that total output could rise above 80 million tons for that year. A worst case projection would put 1985 output at 40 million tons, and 1990 output at 60 million tons.

The other countries of Eastern Europe are less important in coal production, but if their planned targets are taken as a best case, and somewhat lower levels as a more realistic case, it appears unlikely that Eastern Europe will be able to attain regional production goals. Even under the best case conditions, therefore, the combined coal output for Eastern Europe as a region is likely to grow only half as much between 1980 and 1985 as planned (0.74 mbdoe or 36.8 mtoe v. 1.37 mbdoe or 68.2 mtoe).

THE NUCLEAR POWER INDUSTRY

Nuclear power is the only other feasible source of substantial increases in domestic energy production in Eastern Europe, although at the moment it contributes only a small portion of the electric power produced in the region. Eastern Europe's nuclear power program has been behind schedule for some time. As late as 1976, forecasts of 10,000 MW of installed nuclear capacity by 1980 were common.[19] But in 1979 installed nuclear capacity in the six countries was 3,100 MW, or about 3.4 percent of total generating capacity. This amounted to about 4.7 percent of all electricity generated in Eastern Europe during that year.[20] In 1980 installed capacity was increased to 4,440

MW by the addition of new reactors in Bulgaria, Czechoslovakia, and Hungary.

Current plans for nuclear power production contrast sharply with past performance. The six individual East European country plans, added together, call for a fourfold capacity increase between 1979 and 1985, and a tripling of that between 1985 and 1990 (see table 65). These plans depend heavily on an extremely complex cooperative CMEA effort involving specialization and cooperation in the production of, and trade in, equipment for nuclear powerplants. This program aims at raising nuclear capacity to 37,000 MW by 1990. With the exception of Romania, which is developing its own industry using Canadian "Candu" reactors, East European nuclear power is being built with Soviet technology—Soviet-designed VVER-440 reactors, produced in Czechoslovakia and the U.S.S.R. These 440-MW pressurized water reactors are apparently both reliable and economical. Another series of Soviet reactors, the 1,000-MW VVER-1000, is now under development and in the mid-1980's East European countries plan to commission stations using the new design. (see ch. 4 for a detailed description of the Soviet nuclear power industry). Plans call primarily for VVER-440 reactors to be introduced through 1985, and for VVER-1000s to be added thereafter. One exception is Bulgaria, which hopes to have a VVER-1000 in operation by 1985. Romania hopes to produce its own 660-MW "Candu" reactors and to have six of these operating by 1990.[21]

Two joint enterprises have been established to assist in this effort. One, *Interatominstrument*, oversees the manufacture of high-technology equipment for nuclear powerplants; another, *Interatomenergo*, handles shipment of equipment, parts, materials, and apparatus for nuclear powerplants. It has already been decided that two 4,000-MW nuclear power stations will be constructed in the Ukraine with Polish, Hungarian, Czech, and Romanian participation. Repayment to

---

[19]For 1976 forecasts, see *Figyelo*, Sept. 11, 1976, p. 9; M. Virius and J. Balek, "Cooperation Between CMEA Countries in Securing Supplies of Fuels and Energy," *Czechoslovak Economic Digest*, No. 3, May 1976, p. 39.

[20]These calculations simply assume a 0.7 load factor for nuclear powerplants, and then divide the electricity generated at that load factor by all electricity generated that year.

[21]RFE, Romanian SR No. 2, February 1981, pp. 12-13.

Eastern Europe will be in the form of electricity, shipped via 750-kV powerlines. The first of these powerlines, between Albertirsa near Budapest in Hungary and the Ukraine, was completed in 1978.

The first Ukrainian 4,000-MW station is scheduled to be completed in 1984-85. Czechoslovakia, Poland, and Hungary will pay half of the estimated total cost of 1.5 billion transferable rubles (TR). Poland's contribution of 400 million TR will be made in goods and services; the Czechs will cover 240 million TR through supplying equipment and machine tools; and the Hungarians will supply 110 million TR. Half of the capacity of the new station will be dedicated to shipments to each of the East European nations in proportion to their contributions to its construction. A second 4,000-MW station, Konstantinovka, is projected for the latter part of the decade, but the details remain unclear. In addition, draft agreements on cooperative development of atomic cogeneration units and atomic boilers for producing steam for industry are in preparation. CMEA agreements on specialization and cooperation in manufacturing generally lack substance, but it appears that those relating to the nuclear power industry may be exceptions. The most plausible explanation for the exceptional efforts being made in the area of cooperative nuclear programs is that Soviet leaders recognize the importance of East European development and are determined to retain control of the technology.

One great attraction of nuclear power is that it relieves pressure on the coal industry, which otherwise would be called on to provide the fuel necessary to generate equivalent amounts of electricity. But while the amount of coal "displaced" by an extensive nuclear power program could be significant, nuclear development will not much diminish the importance of Eastern Europe's coal industry.

Assuming that it replaces a conventional thermal powerplant working at 40-percent efficiency, a new 440-MW reactor operating at 70-percent capacity can displace 2.94

million tons of lignite per year.[22] Nuclear powerplants are now usually built with at least four VVER-440s (i.e., with installed capacities of 1,760 MW). Commissioning such a plant would thus obviate the mining of nearly 12 million tons of lignite per year. This amount of lignite equals 2.3 mtoe or 0.047 mbdoe. The best case increase in coal output between now and 1985 was 0.74 mbdoe. Thus, it would take almost sixteen 1,760-MW nuclear powerplants (each with four 440-MW reactors) to match that increment. This is clearly impossible. Even in the best case, therefore, nuclear power's contribution to increased supplies of domestic energy in Eastern Europe in the 1980's will be much smaller than that of coal.

The potential obstacles to fulfillment of nuclear targets differ sharply from those likely to be encountered with coal production. As in the U.S.S.R., there are no East European counterparts to the Western antinuclear groups. Nuclear power is officially considered much cleaner than coal, and therefore a very attractive energy source. Press reports include little discussion of safety issues, and individuals who may worry about the safety of nuclear power have no easy way to express their concern. Barring an accident near a major population center (a prospect not to be dismissed since current plans call for cogeneration using nuclear reactors situated in heavily populated areas), it is unlikely that safety and environmental concerns will impede the development of nuclear power. Labor problems are also unlikely to be a major factor, since the nuclear industry is not as labor-intensive as the coal industry.

---

[22]A new 440-MW reactor operating 70 percent of the time can produce 2.6981 billion kWh of electricity per year (24 hours per day × 365 days × .70 × 440 MW = 2.6981 billion kWh). Modern fossil-fired plants can produce 1,000 kW of electricity using approximately 0.218 toe of energy equivalent inputs. 1,000 kW equals approximately 0.0875 toe of energy, and therefore energy out divided by energy in is 0.0875/0.218 = 0.4 efficiency of energy conversion. A 440-MW lignite-fired plant working at that efficiency level will require 2.6981 billion kWh × 0.000218 mtoe = 0.588 mtoe of energy inputs per year. Assuming the lignite inputs average 0.2 mtoe per ton, that requires 0.58/0.2 = 2.941 tons of lignite.

The two potentially significant constraints on the development of nuclear power in Eastern Europe are technology and capital costs. While there appear to be few problems with the VVER-440 reactors, the introduction of new VVER-1000s and their attending support equipment is tantamount to an experimental program and delays are not unlikely. The capital costs of nuclear power may also impede the realization of plans for nuclear power development in the 1980's. Nuclear powerplants are huge, highly visible investment projects. If, as seems likely, East European GNP growth rates are low during this decade, the need to maintain living standards might result in a slowdown in nuclear power development, and possibly a heightened interest in conservation.

The Feasibility of Meeting Installed Nuclear Capacity Targets

East European plans for 1985 nuclear power capacity appear to be fairly realistic (see table 67). These foresee a capacity of 12,000 MW in 1985, 2,000 MW above mid-1970's forecasts for 1980. This installed capacity would support electricity production amounting to 0.320 mbdoe (15.9 mtoe). Most of the planned increment for the 1980-85 period is associated with the introduction of additional VVER-440 reactors, with which the East European and Soviet power industries now have considerable experience. Therefore, 12,000-MW installed capacity in 1985 can be viewed as a reasonable best case. A worst case could assume that Bulgaria's VVER-1000 is not operational until after 1985, that the Czechs only manage to achieve the low end of their plan (2,200 MW of capacity operating in 1985), that the Poles can get none of the capacity at Zarnowiec operational in 1985,[23] and that Romania's first reactor does not make its scheduled commissioning in 1985. It is entirely possible that all of these delays could coincide. Under these worst case conditions, 9,250-MW capacity might be in place by that year, which would produce 56.66 billion kWh or 0.247 mbdoe (12.3 mtoe) of electricity.

It is difficult to determine best and worst cases for nuclear power for 1990 since in-

[23]See RFE, background report No. 11-Eastern Europe, Jan. 20, 1981.

Table 67.—Nuclear Power Capacity, and Production of Electricity From Nuclear Powerplants, 1979 (actual), and Planned for 1985 and 1990

	Actual 1979			Planned					
	Capacity (mkW)	Production[a] (bkWh)	(mbdoe)	1985 Capacity (mkW)	Production[a] (bkWh)	(mbdoe)	1990 Capacity (mkW)	Production[a] (bkWh)	(mbdoe)
Bulgaria	0.88	5.4	0.024	2.76	16.9	0.074	3.76	23.1	0.101
Czechoslovakia	0.44	2.7	0.012	2.42	14.9	0.065	10.52	67.5	0.295
East Germany	1.76	10.8	0.047	(3.52)[b]	(21.6)[b]	(0.094)[b]	(3.52+)[d]	(21.6+)[d]	(0.094+)[d]
Hungary	0	0	0	1.76	10.8	0.047	(2.76)[e]	(16.9)[e]	(0.074)[e]
Poland	0	0	0	0.88	5.4	0.024	4.90	30.1	0.132
Romania.	0	0	0	0.66	4.1	0.018	3.96	23.9	0.104
Eastern Europe, total	3.08	18.9	0.083	12.0	73.7	0.320	37.00[c]	226.9[c]	0.991[c]

[a]These are estimates assuming a 0.7 load factor.
[b]This is the German Democratic Republic's 1980 plan quote. OTA had no information for 1985 or 1990, and assumed that the GDR plans to fulfill this early version of the 1980 plan by 1985, and that more plants are planned in 1990.
[c]This is an independent estimate of total nuclear capacity by 1990. The elements in the column add up to 29.42 mkWh. Some of the remaining unexplained portion must be in the East German plans; and whatever is left apparently represents upward plan revision(s) that have not been published.
[d]Assumes that the German Democratic Republic plans more capacity by 1990 than the plan (guessed) for 1985.
SOURCE: Office of Technology Assessment.

formation on the plans themselves is incomplete. Several sources indicate that all CMEA countries except the U.S.S.R. plan to have a total nuclear capacity by 1990 of 37,000 MW[24] (see table 57). The available fragmentary 1990 official plan data shows that over the 1986-90 period Czechoslovakia apparently expects to bring on line eight VVER-1000 reactors; Poland, at a much earlier stage in its nuclear program and therefore less experienced, is planning to add four; and Bulgaria's target calls for the construction of an additional two. By 1990 Romania plans to have six Candu reactors in operation, at least three of which will be produced by the Romanians themselves under Canadian license. Romanian expectations for solving both the problems of construction and manufacturing of nuclear reactors in the next decade are especially ambitious. The country's first 660-MW reactor is not scheduled for commission until 1985.

Based on past experience, OTA believes that goals for nuclear power development which rest on the timely installation of all of the planned reactors are too optimistic to serve as a best case projection. A realistic best case would assume that some, but not all, of the VVER-1000 reactors are in operation by the end of the decade, and that total capacity reaches about 30,000 MW. This assumes that Poland and Romania each fall 2,000 MW short of their 1980 targets, and that Czechoslovakia falls 3,000 MW below its plans. The scenario is still optimistic in that it assumes no slippage in Bulgarian and Hungarian plans. This best case projection for 1990 is 7,000 MW below the estimates referred to above by East European sources. If there are considerable investment constraints, or if there are significant difficulties in introducing the large VVER-1000 reactors, a worst case for 1990 would be total nuclear capacity of 20,000 MW (see table 68).

Even under best case conditions, nuclear

power generation is likely to supply only a small increment to the growth in domestic energy production through the end of the decade. In the best case, production will reach 184 billion kWh in 1990 (0.804 mbdoe or 40 mtoe), accounting for a little over 1 percent of the yearly growth rate in energy production to 1990. If, instead, the worst case obtains, then the 20,000-MW capacity (0.536 mbdoe or 26.7 mtoe) would add 0.6 percent per annum to the growth in energy production over the same period.

It is interesting to compare feasible nuclear and coal production increments to get a sense of the contributions each is likely to make to Eastern Europe's energy supply in the next decade. Total energy production in Eastern Europe in 1979 was 6.737 mbdoe (336 mtoe). Production in 1980 was certainly no higher, due to leveling coal output. Under the best case conditions, nuclear power will add 0.201 mbdoe (10 mtoe), 3 percent of 1979 production, by the year 1985. In the best case coal could contribute more than four times that amount of energy over the same period, 0.86 mbdoe (43 mtoe). The worst case for coal is nearly equivalent to the best case increment to energy supplies from nuclear. Under the best of circumstances, by 1990 nuclear could provide a 0.684-mbdoe (34.1-

Table 68.—Planned and Projected Nuclear Capacity to 1990 (million barrels per day of oil equivalent)

	1979	1985	1990	1990 (Coal)
Actual	0.119 (out of 6.37 total energy production)			
Plan		0.320	0.991[a]	(6.86)
Best case projection		0.320	0.803	(5.98)
Worst case projection.........		0.247	0.536	(5.44)

Best case increment provided by nuclear (1979-90) = 0.684
Worst case increment provided by nuclear (1979-90) = 0.417
Best case increment provided by coal (1979-90) = 1.36
Worst case increment provided by coal (1979-90) = .62

[a]Estimate using 37,000 MW, and a load factor of 0.7.

SOURCE: Office of Technology Assessment.

---

[24]Iu. Savenko and M. Samkov, "Cooperation of the Member-Countries of CMEA in the Development of Electric Energy," *Ekonomicheskoye sotrudnichestvo stran-chlenov SEV*, February 1980, p. 52.

mtoe) increment to 1980 total energy production; under the worst case nuclear power in 1990 would provide a 0.417-mbdoe (20.8-mtoe) increment. Coal, on the other hand, would provide under best conditions 1.36 mbdoe (67.7 mtoe), and under worst conditions an increment of 0.62 mbdoe (30.9 mtoe), by 1990. In sum, even in the best of circumstances, throughout the 1980's nuclear power will provide no more of an increment to domestically produced energy in Eastern Europe than coal.

SOVIET ELECTRICITY EXPORTS TO EASTERN EUROPE

Soviet electricity exports to Eastern Europe have never been large. In 1979 these amounted to 12.6 billion kWh, about 3 percent of Eastern Europe's own electricity production that year. One important constraint on exports of electricity has been the lack of high-voltage transmission lines to link up with the East European system. The previously mentioned 750-kV line running from the Soviet Ukraine to the Hungarian electric power grid near Budapest was the first step in breaking this bottleneck. The Soviets built the portion of the line within their borders, while the Hungarians built the rest with Polish and Czech assistance. At full capacity of 6.4 billion kWh the line will increase Soviet electricity export capacity about 50 percent above 1978 levels.[25] This line accounts for all of the increment to Soviet electric energy export capacity to Eastern Europe in 1978 and 1979.

The transmission network will grow further. When the two jointly built Ukrainian nuclear power stations are operating at full capacity in the latter half of the decade, 20 billion to 22 billion kWh of electricity will be shipped to Eastern Europe through the completed Hungarian line, and two similar lines to Poland and Romania. This will triple present levels of Soviet exports of electricity to Eastern Europe.

In the absence of projections for total electricity production in Eastern Europe in the 1980's, it is difficult to say how large the contribution of imported Soviet electricity will be. An additional 20 billion to 22 billion kWh of electricity would amount to 0.096 mbdoe (4.8 mtoe), or less than 25 percent of the worst case increment to domestic energy supplies coming from nuclear power in the next 10 years. Clearly, the exports of electricity from the Soviet Union will not make a contribution as large as is likely to come from coal and nuclear power development.

OTHER DOMESTIC SOURCES OF ENERGY

Romania is the only significant East European producer of petroleum, supplying most of Eastern Europe's oil and two-thirds of its gas in 1979. However, Romanian output of both oil and gas has been declining since 1977 and it is generally agreed that this trend will continue. Romanian plans call for maintenance of oil production at the 1979 level (0.25 mbd or 12.4 mtoe) and a reduction in natural gas production to 0.51 mbdoe or 25.4 mtoe (down from 0.61 mbdoe or 30.4 mtoe in 1979). Hungary, which produces 0.04 mbd (1.99 mmt) of oil and 0.11 mbdoe (5.48 mtoe) of gas expects to maintain, but not increase, those levels. Poland's natural gas production plans are unknown, but it seems unlikely that output will increase significantly. Therefore, it is realistic to assume that East European hydrocarbon production, which totaled 1.22 mbdoe (60.8 mtoe) in 1979, will probably fall to about 1.10 mbdoe (54.8 mtoe) for the 1980's.

The only remaining East European domestic energy source is hydroelectric power, the contribution of which is small in comparison to coal, nuclear, or even oil and gas. Since plans for hydropower either have not been developed or are not available, it is very difficult to make meaningful predictions. A few hydropower development projects can

[25]Leslie Dienes and Theodore Shabad, *The Soviet Energy System: Resource Use and Policies* (Washington, D.C.: V. H. Winston & Sons, 1979), pp. 239-40.

be identified, however. Two of these—in Bulgaria and Hungary—are aimed at the development of pumped storage capacity to handle peakloads and have evidently received preliminary approval in CMEA.[26] A number of East European nations are also discussing the development of minihydro stations (less than 100 MW) to supply small rural areas. Finally, there are traditional hydroelectric stations under construction in several countries. Overall, however, it does not appear that hydroelectric power can make a major contribution to added domestically produced energy supplies.

CONCLUSIONS

OTA's best and worst case projections of East European energy supply in the 1980's are summarized in table 69, which also shows actual and official energy production

---

[26]Bagudin, op. cit., pp. 39-40.

in 1979, and present plans where these were available.

Between 1970 and 1979 total East European energy production grew at an annual rate of 2.4 percent.[27] Projections of total energy production in Eastern Europe for 1985 are at best 7.07 mbdoe (352.1 mtoe) and at worst 6.43 mbdoe (320.2 mtoe). The best case figure represents annual growth in energy production of 1.8 percent, while the worst case represents virtual stagnation in production over the 1980-85 period. The 1990 projections range from a best case of 7.88 mbdoe (392.4 mtoe) production, which implies a 2-percent annual growth rate over the decade, to a worst case of 6.89 (343.1 mtoe) mbdoe which implies a growth rate of 0.7 percent.

---

[27]This is a simple compound growth rate, computed from 1980 CIA data.

Table 69.—East European Energy: 1979-90, Plans and Projected Actual Supplies
(million barrels per day of oil equivalent)

		1979 Coal	1979 Oil	1979 Gas	1979 Electric	1979 Total	1985 Coal	1985 Oil	1985 Gas	1985 Electric[a]	1985 Total	1990 Coal	1990 Oil	1990 Gas	1990 Electric[d]	1990 Total
Bulgaria	Plan[b]	0.11	0.01	0	0.04	0.16	0.15	NA	NA	0.09	NA	0.18	NA	NA	0.12	NA
	Projected (b)[c]	—	—	—	—	—	0.15	0.01	0	0.09	0.25	0.18	0.01	0	0.12	0.31
	Projected (w)[c]	—	—	—	—	—	0.12	0.01	0	0.06	0.19	0.15	0.01	0	0.09	0.25
Czechoslovakia	Plan[b]	0.93	0	0.01	0.03	0.97	0.96	NA	NA	0.08	NA	0.99	bd	bd	0.33	NA
	Projected (b)	—	—	—	—	—	0.96	0	0.01	0.08	1.05	0.99	0	0.01	0.23	1.23
	Projected (w)	—	—	—	—	—	0.93	0	0.01	0.08	1.02	0.96	0	0.01	0.12	1.09
East Germany	Plan[b]	1.06	0	0.05	0.05	1.17	1.16	NA	NA	NA	NA	1.26	NA	NA	NA	NA
	Projected (b)	—	—	—	—	—	1.15	0	0.05	0.09	1.29	1.26	0	0.05	0.09	1.40
	Projected (w)	—	—	—	—	—	1.06	0	0.05	0.09	1.20	1.15	0	0.05	0.09	1.29
Hungary	Plan[b]	0.14	0.04	0.11	0	0.29	0.18	0.04	0.10	0.05	NA	0.19	0.04	0.10	0.07	0.40
	Projected (b)	—	—	—	—	—	0.18	0.04	0.10	0.05	0.37	0.19	0.04	0.10	0.07	0.30
	Projected (w)	—	—	—	—	—	0.16	0.04	0.10	0.05	0.35	0.18	0.04	0.10	0.05	0.37
Poland	Plan[b]	2.56	0.01	0.12	0.01	2.70	3.20	NA	NA	0.03	NA	3.75	NA	NA	0.14	NA
	Projected (b)	—	—	—	—	—	2.82	0.01	0.12	0.03	3.28	3.06	0.01	0.12	0.09	3.28
	Projected (w)	—	—	—	—	—	2.42	0.01	0.12	0.01	2.56	2.60	0.01	0.12	0.03	2.76
Romania	Plan[b]	0.16	0.26	0.61	0.05	2.08	0.44	0.25[c]	0.51	0.07	1.27	0.49	NA	NA	0.15	NA
	Projected (b)	—	—	—	—	—	0.30	0.25	0.51	0.07	1.13	0.40	0.25	0.51	0.10	1.26
	Projected (w)	—	—	—	—	—	0.20	0.25	0.51	0.05	1.01	0.30	0.25	0.51	0.09	1.15
Total	Plan[b]	4.96	0.32	0.90	0.19	6.37	6.09	NA	NA	NA	NA	6.86	NA	NA	NA	NA
	Projected (b)	—	—	—	—	—	5.56	0.31	0.79	0.41	7.07	6.08	0.31	0.79	0.70	7.88
	Projected (w)	—	—	—	—	—	4.89	0.31	0.79	0.34	6.43	5.34	0.31	0.79	0.45	6.89

NA = not available.
[a]This includes electricity produced in 1979, plus the planned increment from nuclear powerplants.
[b]The 1979 figures are actual production.
[c]Projected (b) is best projected case; Projected (w) is worst projected case.
[d]This includes electricity produced in 1979, and the increment planned from nuclear power through 1990.
SOURCE: Office of Technology Assessment.

The results of this analysis suggest that Eastern Europe's plans for domestic energy supplies in the 1980's are overly optimistic. The plans assume that growth in energy production can be maintained at levels attained in the last decade—an assumption which seems unjustified. While there are no data available on total planned energy output in Eastern Europe in 1985, it is reasonable to assume that planners' expectations for energy production resemble OTA's best case projections for all fuel sources except coal, where their plans show much higher produc-

tion levels. The implied plan for 1985 is 7.6 mbdoe (378.4 mtoe) total energy production, an annual growth of 3 percent over 1979; for 1990 the plan is 8.66 mbdoe (431.2 mtoe) or 2.8 percent per annum growth rate over the 1979-90 period. These growth rates are higher than those actually achieved in the last decade, and probably unattainable. It would be more reasonable to assume that, at best, East European energy production will grow at 1.8 percent, and at worst at less than 1 percent yearly.

EAST EUROPEAN ENERGY DEMAND IN THE 1980'S

In the last 15 years, energy consumption in Eastern Europe has grown at approximately 4 percent per year, while production has been growing at about 2.5 percent per year. The difference has been met with Soviet oil. This situation cannot continue. As previously noted, the U.S.S.R. has announced that henceforth it intends to maintain its oil exports to Eastern Europe at 1980 levels. Thus, even maintaining the status quo would involve East European countries' increasing imports of OPEC oil at world market prices. As the analysis in the previous section has shown, even under the best of circumstances the growth rate of East European energy production in the 1980's will be no more than 2 percent, and under worst case conditions the growth rate might fall below 1 percent. These statistics underline the importance of efforts to moderate growth in energy consumption in the years ahead.

A critical determinant of the growth of energy demand in any country is the production of goods and services, as measured by GNP. One indicator of the relationship between GNP and energy is energy/GNP elasticity, i.e., the percentage change in the consumption of energy divided by the percentage change in GNP. Table 70 summarizes estimates of historical trends in energy/GNP elasticity for Eastern Europe. A coefficient

Table 70.—Energy GNP Elasticities in Eastern Europe: Past Trends and Future Projections[a]

	1965-78	1974-78	1981-90 (projection)
Bulgaria	1.79	1.50	1.65
Czechoslovakia	1.06	2.34	1.00
East Germany	.76	.79	.75
Hungary	1.20	1.34	1.00
Poland	1.01	1.29	1.00
Romania	1.34	1.10	1.21
Total - Eastern Europe[b]	1.09	1.28	1.00

[a]None of these elasticities is significantly different from the elasticity for the entire 1965-78 period.

[b]Weighted average from a regression for all of Eastern Europe.
SOURCE: All the elasticities are from the equation:

$$\log (C_1{}^e) = a_1 + b_1 \log (DUM) + c_1 \log (GNP) + d_1 \log (DUM) \cdot \log (GNP)$$

where $C_1{}^e$ = consumption of energy in the ith country

GNP_1 = GNP of the ith country

DUM = a dummy variable, = 1 through 1973, and "e" for 1974-78.

For a discussion of the econometric technique used here, see Edward A. Hewett, "Alternative Econometric Approaches for Studying the Link Between Economic Systems and Economic Outcomes," *Journal of Comparative Economics*, 4 (1980), pp. 274-290.

(an indicator of energy/GNP elasticity) of "1" for a given period means that when GNP has grown by 1 percent, average energy consumption has grown by a like amount.

Several important conclusions can be drawn from table 70. First and most important, the energy/GNP relationship has been above "1" for each covered period for every Eastern European nation except East Germany. For the six countries together, between 1965 and 1978 a 1-percent increase in GNP was associated with a 1.12-percent increase in energy consumption.

Secondly, the figures for the brief 1974-78 interval show no statistically significant differences from the elasticity for the entire period, although clearly the trend was for elasticities to *increase* during this period. There is no evidence that frequent public statements on energy conservation in Eastern Europe have had any tangible effect. On the other hand, one should not place a great deal of weight on the 1974-78 elasticities. The sample period is quite short, and the fact that there is no statistically significant difference between these estimates and those for 1965-78 indicates high variability in energy/GNP elasticities. Hence these 1974-78 coefficients are averages of a broad range of numbers.

Third, there is evidence that levels of economic development influence the energy/GNP elasticity. In countries such as Bulgaria and Romania where industrialization is occurring at a fairly rapid pace, energy/GNP elasticity is high. East Germany and Czechoslovakia, on the other hand, have lower energy/GNP coefficients. This suggests that the elasticities may fall over time as development proceeds.

East European leaders have become increasingly concerned with energy/GNP elasticity, and their concern has been enhanced both by recognition of the costs involved in expanding domestic energy production and awareness of the U.S.S.R.'s intentions not to increase its energy exports. Not surprisingly, therefore, discussions of conservation as the least expensive policy for avoiding a full-fledged energy crisis are becoming increasingly frequent,[28] and a variety of energy conservation measures are now being contemplated or employed. The most important of these include the following:

1. reducing the proportion of energy-intensive products in total output;

2. development and installation of energy-saving machinery in energy-intensive sectors, including efforts to recapture heat lost in power stations;
3. introduction of various administrative regulations designed to control household and industrial energy demand; and
4. increasing prices to cut energy use throughout the economy, including in the household sector.

These efforts appear to have had little effect so far on the energy/GNP ratio. Slowdowns in the growth of energy consumption in Eastern Europe in the late 1970's were due to slowdowns in production, not to conservation. Now, as concern about energy supply increases, energy conservation is being taken more seriously. Soviet leaders have also apparently urged their East European counterparts to adopt conservation measures, particularly those aimed at restructuring GNP and modernizing capital stock in energy-intensive industries.

These policies may have some effect, but it is doubtful whether they will seriously reduce energy/GNP elasticity. Excess use of energy in Eastern Europe is a manifestation of a general pattern of excess use of all industrial materials. The situation is similar to that in the U.S.S.R.: traditionally, economic institutions have emphasized high output growth rates over economizing on inputs to the production process. In most cases, East European leaders are attempting to deal with the energy crisis by using the same administrative techniques they have relied on in the past. In 1979 and 1980, for example, East European countries introduced special plan targets designed to reduce energy use by industry, with penalties for noncompliance. However, this consumption target is only one of a variety of targets which enterprise managers must consider. Often managers give preference to the fulfillment of production tasks rather than to rationalization measures. For example, a survey in East Germany, a country which has been fairly successful in controlling energy demand, showed that one-third of all enter-

---

[28]Gzovskiy, op. cit.

prises simply ignored their energy-economizing targets.[29]

Enterprises will probably avoid energy conservation until they are forced to do otherwise. Administrative measures have limited impacts since enterprise managers know a number of ways to get around them. The only effective method of introducing conservation may be to weave energy conservation into a package of comprehensive economic reform. This is presently being attempted in Hungary. If the Hungarian experiment proves successful, it may suggest a promising approach for other countries.[30]

Overall, it appears unlikely that the energy/GNP coefficient will fall sharply in the absence of economic reforms. As the industrialization drives in Romania and Bulgaria slow, coefficients for those countries could fall slightly, but Hungary is the only country in which the energy/GNP elasticity may fall significantly. Assuming no other major economic reforms, it is reasonable to expect that during the next decade the energy/GNP coefficient for all of Eastern Europe will fall from 1.09 to 1.00, if Hungary is able to achieve relatively higher levels of conservation (see table 70).

In order to make energy demand projections based on energy/GNP elasticity, it is necessary to consider likely developments in economic growth. Analyzing the prospects for East European GNP is complicated by two problems. First there are no reliable GNP projections for Eastern Europe. Second, it is possible that tight energy supplies may constrain GNP in the 1980's, reversing the situation of the last decade. This would occur if the supply projections developed in the last section result in import and balance of payments problems which force East

European countries to curtail imports and GNP. With these caveats in mind, OTA has attempted projections of likely rates of economic growth in the next 10 years without considering a feedback from energy constraints.

During the 1970's, GNP growth rates in Eastern Europe declined from about 5 percent to less than half that figure. Although the reasons for this are multifarious and complex, the energy crisis exacerbated already existing problems by putting tremendous pressure on Eastern Europe's balance of payments. OTA here assumes as a best case that Eastern Europe will be able somehow to maintain the levels of growth achieved for the latter half of the 1970's (2.9 percent per year) in the next decade. A worst case would involve growth at half of this rate—i.e., rates of GNP for 1981-90 of 1.5 percent. These GNP projections, combined with the energy/GNP elasticity projections above, yield the energy demand projections in table 71. The final section of this analysis combines these demand projections with the supply projections developed above. The result is a projection of Eastern Europe's energy import needs in the next decade.

Table 71.—Energy Demand Projections for Eastern Europe, 1985 and 1990 (million barrels per day of oil equivalent)

	1979 energy consumption	1985 energy consumption		1990 energy consumption	
		High	Low	High	Low
Bulgaria	0.65	0.80	0.72	0.93	0.79
Czechoslovakia	1.59	1.77	1.68	2.08	1.75
East Germany	1.74	1.97	1.85	2.18	1.96
Hungary	0.60	0.71	0.65	0.81	0.70
Poland	2.38	2.86	2.62	3.33	2.83
Romania	1.39	2.00	1.68	2.68	1.96
Eastern Europe, Total	8.35	10.10	9.20	12.01	9.99

SOURCE: The 1979 figures are an estimate based on 1978 figures, the energy/GNP elasticities in table 70, and 1979 GNP growth rates reported by CIA for each East European country. The 1985 and 1990 projections are derived by applying the high and low GNP growth rates assumed for 1981-90, and the assumed energy demand elasticities to estimated energy consumption in 1979. For example, the figure for the high case for Bulgaria in 1985 is derived as 2.1 percent × 1.65 = 3.465, which is the estimated per annum growth rate of energy consumption. $1.03456^6 × 0.65 = 0.80$ mbdoe.

---

[29]Szocs, op. cit.

[30]Edward A. Hewett, "The Hungarian Economy: Lessons of the 1970's and Prospects for the 1980's," in Joint Economic Committee, U.S. Congress, *East European Economic Assessment* (Washington, D.C.: U.S. Government Printing Office, 1981).

EAST EUROPEAN ENERGY IMPORTS IN THE 1980'S: IMPLICATIONS FOR TRADE WITH THE SOVIET UNION AND THE REST OF THE WORLD

IMPORT NEEDS: BEST, WORST, AND MIDDLE CASES

Combining projections of domestic production and energy demand can yield a reasonable estimate of energy import needs for the next decade. Table 72 summarizes these estimates, and outlines best, worst, and middle cases for net imports in the 1980's. The worst cases for both 1985 and 1990 assume that energy consumption continues to grow at a high rate, while simultaneously domestic energy production follows the worst case in table 69. The best case is based on the lowest projections for growth in energy consumption and the highest for production. One middle case is also identified. This assumes that while energy consumption grows at a high rate, the six East European nations are successful in attaining the best case energy production projections.

Table 72 posits a best case in which Eastern Europe maintains a stable net import/energy consumption ratio throughout the 1980's. This outcome is not impossible, although it is based on the assumption that everything goes well for Eastern Europe; i.e., demand grows slowly and production increases at best case rates. Under these conditions, which are certainly less optimistic on the production side than those outlined in the official plans, net imports for the six Eastern European countries together will actually fall from about 24 percent of all energy consumed in 1979 to about 21 percent in 1990.

The worst case is dramatically different. **Under worst case conditions even Poland becomes an importer of energy and by 1990 Eastern Europe as a whole will import 43 percent of all the energy it uses.** For all countries except Poland, the net import/consumption ratio is even higher by 1990—52 percent. Like the best case, this is a possible outcome. It is conceivable that energy production will grow slowly while energy conservation programs fail to have significant impacts.

The middle case, which is probably the most likely, assumes favorable develop-

Table 72.—Projected Energy Consumption and Production in Eastern Europe, 1985 and 1990
(million barrels per day of oil equivalent)

	1979 actual			1985 projected							1990 Projected						
	Con-sump-tion	Pro-duc-tion	Net Imp.	Consump-tion		Produc-tion		Net Imp.ᵃ			Consump-tion		Produc-tion		Net Imp.ᵃ		
				High	Low	Worst	Best	Worst	Mid	Best	High	Low	Worst	Best	Worst	Mid	Best
Bulgaria..........	0.65	0.16	0.49	0.79	0.72	0.19	0.25	0.60	0.54	0.47	0.93	0.79	0.25	0.31	0.68	0.62	0.48
Czechoslovakia	1.59	0.97	0.62	1.77	1.68	1.02	1.05	0.75	0.72	0.63	2.08	1.75	1.09	1.23	0.99	0.85	0.52
East Germany	1.74	1.17	0.57	1.97	1.85	1.20	1.29	0.77	0.68	0.56	2.18	1.96	1.29	1.40	0.89	0.78	0.56
Hungary	0.60	0.29	0.31	0.71	0.65	0.35	0.37	0.36	0.34	0.28	0.81	0.70	0.37	0.40	0.44	0.41	0.30
Poland	2.38	2.70	-0.32ᵇ	2.86	2.62	2.56	2.98	0.30	-0.12	-0.36ᵇ	3.33	2.83	2.76	3.28	0.57	0.05	-0.45
Romania	1.39	1.08	0.31	2.00	1.68	1.01	1.13	0.99	0.87	0.58	2.68	1.96	1.15	1.26	1.53	1.42	0.70
East Europe, Total..........	8.35	6.37	1.98	10.10	9.20	6.43	7.07	3.77	3.03	2.16	12.01	9.99	6.89	7.88	5.12	4.13	2.11

ᵃThe worst case is when consumption is "high" and production is "worst." Thus for Bulgaria in 1975, the worst case is 0.79 –0.19 = 0.60 mbdoe. The "mid" (middle) case is when consumption is "high" and production is "best." The best case is when consumption is "low" and production is "best."
ᵇIndicates net energy exports.

SOURCE: Office of Technology Assessment. Actual 1979 consumption data are estimated by multiplying the 1978 consumption data by actual 1979 GNP growth rates, and those by the elasticities in table 70.

ments in energy production, combined with high-GNP growth rates. This case assumes that for political reasons East European planners will place high priority on attempting to maintain relatively high economic growth rates, hence acceptable growth rates for personal consumption.

SOVIET ENERGY EXPORTS TO EASTERN EUROPE

Thus far, this chapter has approached the question of East European energy in the 1980's from the perspective of the East European planners themselves. But Eastern Europe's energy future rests heavily on the actions of the Soviet Union. The amount of energy that the Soviet Union is willing to export to Eastern Europe for transferable rubles will to a large extent determine the amounts which Eastern Europe will be forced to buy on world markets—at world prices, for hard currency.

In order to make precise projections here, one would have to know the intentions and capabilities of the Soviet Union regarding energy exports to Eastern Europe. All that is known about Soviet intentions is embodied in several statements by the late Premier Kosygin, indicating that energy exports to all CMEA countries will be about 20 percent more in the 1981-85 period than they were in the 1976-80 period, and that crude oil exports to CMEA during the first half of the decade will total 400 million tons.[31] If these statements are accurate, the Soviet Union will export about 117.98 mtoe (about 2.36 mbdoe) of energy annually between 1981-85 to Eastern Europe. This is a substantial *cut in increments* to energy exports as compared to the situation in the last decade. Table 73 summarizes plans for Soviet energy exports to CMEA.

Table 74 combines estimates and projections of Eastern Europe's net energy imports in 1979, 1985, and 1990, with esti-

---

[31]A. N. Kosygin, "Speech of the Head of the Delegation of the Union of Soviet Socialist Republics, Comrade A. N. Kosygin," *Ekonomicheskoye sotrudnichestvo stran-chlenov SEV,* April 1980, p. 30.

Table 73.—Soviet Energy Exports to CMEA, Actual in 1976-80, and Planned for 1981-85

	1976-80 (estimates)[c]		1981-85 plan	
	Natural[a]	Mtoe	Natural[a]	Mtoe
Natural gas..........	97.8	80.1	152.4[d]	124.8[d]
Crude oil............	370.4	370.4	400.0[e]	400.0[e]
Oil products.........	55.0	55.8	64.4[i]	65.3[i]
Electricity	55.6	19.5	80.0[h]	28.09
Coal and coke[b]......	41.0	28.1	41.09	28.19
Total..............	—	553.9	—	646.2[f]
Eastern Europe ...	—	501.2[j]	—	589.9[k]

[a]Natural units of measure: bcm for natural gas; metric tons for crude oil, oil products, coal and coke; and bkWh for electricity.
[b]These figures are net of East Europe's coal exports to the Soviet Union.
[c]Each of these figures include estimates for 1980, as well as earlier years in some cases. The hydrocarbon exports are from tables 1 and 2. Electricity exports are given in table 11. For coal and coke, the assumption is that 1976-80 exports remained at the 1976 level since value data suggest that this trade is quite stable.
[d]Assumes 1980 levels of Soviet gas exports through 1985 since Orenburg was at full capacity in 1980.
[e]Statement by Premier Kosygin, 1979.
[f]Kosygin in 1980 stated that energy exports to Eastern Europe from the Soviet Union will rise 20 percent in 1981-85 over 1976-80, and he gives the figures for 1976-80. Those figures, which equal 538.5 mtoe, multiplied by 0.12 yield the 646.2 mtoe for 1981-85. Note that Kosygin's preliminary figures were apparently low.
[g]Assumes 1976-80 delivery levels will be maintained.
[h]Assumes 1979-80 levels of electricity exports, and that Khmel'nitska nuclear powerplant begins full shipments in 1984 (optimistic).
[i]This is a residual obtained by subtracting all other elements from the total derived from Kosygin statement.
[j]This subtracts the 52.7 mtoe of oil and products shipped to Cuba, Vietnam, and Mongolia.
[k]Assumes that 56.3 mtoe of oil and products will be shipped to Vietnam, Cuba, and Mongolia during 1981-85, which is 5 x the 1980 level.
SOURCE: Office of Technology Assessment.

mates and projections of Soviet energy exports to Eastern Europe in those years. The table shows that in 1979 Eastern Europe as a whole was a slight net exporter of energy to the world outside the Soviet Union. This was due to Poland's imports of energy from the Soviet Union and simultaneous exports of coal to the rest of the world. The other East European countries were able to cover most of their needs with Soviet energy, except for Romania, which had significant imports from outside CMEA.

If the *best case* obtains, Eastern Europe will be able overall, and in each individual case, to cover virtually all of its net energy needs with imports from the Soviet Union. Overall, Eastern Europe would remain a slight net exporter of energy to the rest of the world. If the *worst case* should occur, by 1985 Eastern Europe will switch from being a small net energy exporter to a net importer of 1.41 mbdoe (70.2 mtoe); by 1990 imports would reach 2.74 mbdoe (137 mtoe). Even in

Table 74.—East European Projected Net Energy Imports, 1985 and 1990: Total, From the Soviet Union, and From the Rest of the World (million barrels per day of oil equivalent)

	1979 actual			1985 projected							1990 projected						
	Total	From U.S.S.R.	Net from ROW^a	Total Worst	Mid	Best	From U.S.S.R.	Net from ROW^a Worst	Mid	Best	Total Worst	Mid	Best	From U.S.S.R.^b	Net from ROW^a Worst	Mid	Best
Bulgaria..........	0.49	0.45	0.05	0.60	0.54	0.47	0.49	0.13	0.05	-0.02	0.68	0.62	0.48	0.49	0.19	0.13	-0.01
Czechoslovakia...	0.62	0.53	0.09	0.75	0.72	0.63	0.57	0.18	0.15	0.06	0.99	0.85	0.52	0.57	0.42	0.28	-0.05
East Germany	0.57	0.51	0.06	0.77	0.68	0.56	0.55	0.22	0.13	0.01	0.89	0.78	0.56	0.55	0.34	0.23	0.01
Hungary	0.31	0.30	0.01	0.36	0.34	0.28	0.36	0	-0.02	-0.08	0.44	0.41	0.30	0.36	0.08	0.05	-0.06
Poland...........	-0.32	0.22	-0.54	0.30	-0.12	-0.36	0.30	0	-0.42	-0.66	0.57	0.05	-0.45	0.30	0.27	-0.25	-0.75
Romania	0.31	0.05	0.26	0.99	0.87	0.58	0.09	0.90	0.78	0.49	1.53	1.42	0.70	0.09	1.44	1.33	0.61
Eastern Europe, total	1.98	2.06	-0.08	3.77	3.03	2.16	2.36	1.41	0.67	0.20	5.10	4.13	2.11	2.36	2.74	1.77	0.25

^aNet imports from the rest of the world, derived by subtracting the imports from the U.S.S.R. from 1979 actual in the case of 1979, and for the projections, by subtracting imports from the U.S.S.R. from the worst, medium, and best cases. Thus the worst case for East German net imports from the rest of the world in 1985 equals their worst case net imports from all sources (0.77) minus net imports from the U.S.S.R. (0.55) = 0.22.
^bThere are no public commitments for Soviet energy shipments to Eastern Europe after 1985. This assumes the same commitments made for the 1981-85 period.

SOURCE: Office of Technology Assessment.

the worst case, Hungary and Poland would be able to cover their energy needs with Soviet imports. However, Bulgaria, Czechoslovakia, East Germany, and Romania would be forced onto world markets to purchase substantial amounts of energy. Energy imports would surely be in the form of oil or gas, the most easily transported fuels. The *middle case* projects net energy imports to Eastern Europe from the rest of the world at 0.67 mbdoe (33.4 mtoe) in 1985, and 1.77 mbdoe (88.1 mtoe) in 1990. In the middle case, Poland remains a net energy exporter, and Hungary exports a small amount (0.02 mbdoe or 1.0 mtoe) to the rest of the world. Romania accounts for two-thirds of all net imports in the middle case.[32]

It is possible, although unlikely, that Polish energy production will fall below OTA's worst case projections. Should that occur, it will probably be accompanied by a

general economic slowdown which will also cause a reduction in Polish energy demand below the low case. Because of this connection between energy supplies and energy demands, OTA's forecasts of energy balances in Poland (and Eastern Europe) are less sensitive to unforeseen events than are either the production or demand forecasts by themselves. Therefore, even in light of recent events in Poland, OTA regards these balances as a realistic view of the range of possible outcomes in 1985 and 1990.

HARD CURRENCY REQUIREMENTS

In order to evaluate the feasibility of any of these outcomes, it is important to ascertain whether the foreign exchange burden implied by a projection can actually be handled by the East European nations, given their export capacities and their abilities to absorb new debt.

Table 75 shows Eastern Europe's hard currency debt in 1979, and projections of the hard currency requirements for 1985 and 1990 based on the energy import levels

---

[32]The only other comparable estimates have been done by Jan Vanous with results strikingly similar to OTA's projections. Vanous estimates that in 1985 Eastern Europe will be buying from the Middle Eastern suppliers at worst 1.23 mbdoe (OTA's projection is 1.41), and as a medium projection 0.88 mbdoe (0.67 in the OTA projection). See Vanous, op. cit.

Table 75.—Projected Hard Currency Burden on Eastern Europe of Various Projected Net Imports of Energy, 1985 and 1990 (millions of dollars)

	Net Hard currency debt 1979	Hard currency exports (dev. countries)	Net oil imports at $30/barrel					
				1985			1990	
			Best	Medium	Worst	Best	Medium	Worst
Bulgaria...............	3.73	1.29	-0.22	0.55	1.42	-0.11	1.42	2.08
Czechoslovakia........	3.07	2.85	0.66	1.64	1.97	-0.55	3.07	4.60
East Germany	8.44	4.10	0.11	1.42	2.41	0.11	2.52	3.72
Hungary	7.32	2.64	-0.88	-0.22	0	-0.66	0.55	0.87
Poland.................	20.00	5.04	-7.28	-4.60	0	-8.21	-2.74	2.96
Romania...............	6.70	3.45	5.37	8.54	9.86	6.68	14.56	15.77
Eastern Europe, total	49.23	19.37	-2.24	7.34	15.66	-2.74	19.38	30.00

SOURCE: Office of Technology Assessment.

outlined above at 1980 oil prices ($30/barrel). These data provide a very conservative estimate of the hard currency burden implied in each of the three cases, since the world market price of oil may rise faster than the value of Eastern Europe's exports in the next 10 years.

The *best case* is a possible scenario for Eastern Europe as a whole and for each country. It is, however, rather implausible, since it assumes that per capita consumption growth rates will stagnate. The *worst case* would impose extreme difficulties, both for Eastern Europe overall, and for each individual country with the possible exception of Hungary. If the worst case actually occurred, Romania would be spending three times its 1979 dollar exports for energy imports. Bulgaria, Czechoslovakia, and East Germany would all be forced to significantly increase their debts. For all of these countries, the 1990 worst case is even more unattractive. Hungary appears to be the only country which might be able to surmount the worst case with no critical difficulty.

The worst case should therefore not be viewed as a feasible outcome. The nations of Eastern Europe have neither the export reserves nor the borrowing capacity to handle such hard currency problems. If the conditions underlying this case actually begin to develop, a number of factors are likely to intervene and prevent its fulfillment. In the

short run, growth rates would fall sharply if hard currency constraints hold back imports of energy and other inputs. Stagnating production, which would accompany such a situation, would create political tensions over declining living standards, and perhaps even rekindle discussion of significant economic reforms. Under such conditions, the Soviet Union would surely participate in all decisions, and might choose to alleviate part of the crisis by increasing energy exports (particularly natural gas). Soviet preoccupation with a politically stable Eastern Europe would probably stimulate the U.S.S.R.'s assistance if worst case conditions developed. The other eventuality which might redirect a worst case scenario would be the introduction of significant economic reforms aimed at reducing energy demand, and increasing production of manufactured goods which could be exported for hard currency. This would not be an easy road, nor one that the East European nations are likely to freely choose. In a worst case situation, however, there might be no alternative.

The *middle,* and probably most likely, case is closer to the worst outcome than to the best. It suggests that in 1985 an amount equal to 38 percent of Eastern Europe's 1979 hard currency export proceeds will be required for the purchase of oil, and that all of the amount of 1979 export sales will be necessary to cover oil imports in the year

1990. In this case, Hungary would be under no apparent pressure and Poland would be much better off than it is likely to be in the worst case. The pressure would be greatest on Romania, which would spend more than double its hard currency export proceeds on oil imports, and on Czechoslovakia, which would be spending more than half. The pressure would be comparatively strong on Bulgaria and East Germany as well, but neither of these nations would face such strong hard currency constraints as Romania and Czechoslovakia.

If, as is likely, hard currency burdens in the middle case are actually greater than is indicated by table 74 because of rapidly rising oil prices, then Bulgaria, Czechoslovakia and East Germany would all face added pressure. If these countries are forced to spend more than half their hard currency exports on oil, they will be less able to import the machinery and industrial materials necessary to expand output. **If the medium case actually transpires, there will be pressure on the Soviet Union to increase energy exports. In the absence of such assistance from the Soviet Union, pressure for economic reform within Eastern Europe, as well as growing difficulties in the East European-Soviet energy relationship, will likely result.**

This analysis suggests that most East European nations can make it through the 1980's without major crisis, if their domestic energy production develops according to the best case, and if the Soviet Union will continue to provide them heavily subsidized energy shipments at the quantities promised in the early 1980's. Should an energy crisis in the Soviet Union cause a cutback in Soviet energy exports, or should Eastern Europe's energy production stagnate along worst case lines, there will be serious difficulties.

SUMMARY AND CONCLUSIONS

A review of Eastern Europe's energy options in the 1980's suggests the following conclusions: First, there is a wide disparity in the energy situations of various East European nations. Eastern Europe's natural resources are concentrated largely in Poland and Romania. Some nations such as Romania appear quite likely to encounter difficulties associated with requirements for additional imports of energy in the decade ahead, as domestic supplies are depleted; others such as Hungary may be capable of withstanding even worst case developments. Thus, while this chapter has treated Eastern Europe as a region, there are good reasons to watch the developments in individual nations. For example, a continuing and severe Polish crisis might strain domestic energy production for the region as a whole.

Second, and even more important, is the crucial position of the Soviet Union as an energy supplier to Eastern Europe. East European economic development has been significantly assisted by the subsidization of its oil imports from the Soviet Union. If this subsidy were abruptly removed, the negative impacts would be serious. While it is unlikely that the Soviet Union will opt to quickly end the subsidy, the transition from oil to gas exports in itself embodies a decisive change, since the U.S.S.R. is selling its gas to CMEA at world market prices.

An energy crisis in the Soviet Union would seriously impact the nations of Eastern Europe. If, for some reason, the U.S.S.R. decisively reduced its energy exports to the CMEA-6, these nations would be faced with a difficult set of choices. Hard currency constraints would preclude massive purchases of oil on the international market, but demand-reduction measures might be politically problematic.

While it is true that Eastern Europe as a region is much less dependent on imported energy than is Western Europe, there are a variety of additional constraints which bound the energy options available to

CMEA planners. With limited prospects for increased energy production, and with relatively energy-intensive economies, Soviet energy exports occupy a critical position in the energy situations of these countries.

Regardless of whether best, worst, or middle cases actually transpire, East European energy plans and strategies will be significantly affected by those of the U.S.S.R.

CHAPTER 10

The Soviet Bloc and World Energy Markets

CONTENTS

LIST OF TABLES

The Soviet Bloc and World Energy Markets

One important theme in the debate over the Soviet Union's energy future has been the potential impact on the West of a decline in Soviet oil production. The prospect of such a decline has been greeted with the apprehension that it could cause the Council for Mutual Economic Assistance (CMEA) as a whole, or even the U.S.S.R. itself, to become a net oil importer. Many have argued that, by increasing demand, net oil imports by the countries of the CMEA could initiate additional competition on world markets and further push up the price of oil. OTA's analysis indicates that this is improbable. A more likely eventuality is that CMEA's net exports will decline. This would have repercussions for the countries of both the Western alliance and the Eastern bloc. Such an outcome would certainly place strains on the economies of the U.S.S.R. and Eastern Europe, strains which would have both domestic and foreign policy consequences.

This chapter addresses the question of the likelihood and implications for both the East and West of the CMEA's changing its position as a net energy exporter. Informed discussion of the probability and consequences of the Soviet bloc's importing or exporting less oil is hampered by a number of complicating factors, foremost among them the enormous range between plausible best and worst case oil production scenarios extending 5 and 10 years into the future. But oil production is not the only important variable. Oil is obviously important to Soviet and East European energy balances, but it is only part of a far larger energy picture. Future prospects—for energy self-sufficiency or dependence—will be determined by total energy production and consumption in all energy sectors. Thus, the continued ability of the U.S.S.R. to fill most of the energy needs of Eastern Europe on favorable terms and to earn large amounts of hard currency by exporting energy to the West will rest on a complex array of factors. These include the volume and mix of total CMEA energy production and consumption (the latter strongly correlated in the past with economic growth rates and also dependent on the success of conservation programs); and perhaps most important, on the degree to which other fuels—i.e., gas—can be substituted for oil.

Given the range of outcomes possible for each of these variables, attempting to make firm predictions on this subject is futile. OTA has instead chosen to devise and analyze a scenario which will illuminate likely prospects for the present decade. This scenario is constructed from the foregoing material. Chapters 2 through 5 of this study culminate in sector-by-sector projections of reasonable levels of Soviet energy production for 1985 and 1990; chapter 8 employs these projections to construct plausible best and worst case energy production, consumption, export, and hard currency import scenarios for the U.S.S.R.; and chapter 9 consists of a similar exercise for six East European countries. The present chapter combines these separate analyses into one that focuses on the CMEA as a whole.

Although previous chapters have presented both best and worst case scenarios, here only one "midrange," outcome is considered. OTA's decision to employ a midrange scenario in the analysis is based on the expectation that, while extreme developments are of course possible, the most probable outcome will lie between them. For either extreme possibility to materialize, a large number of parameters must simultaneously exhibit either "best" or "worst" characteristics. This is improbable if for no other reason than that political events are likely to intervene to prevent extreme consequences. If, for instance, the most optimistic energy production targets were fulfilled, planners might reallocate investment away from the energy sector. On the other hand, if a number of worst case developments occurred simultaneously, the U.S.S.R. could well be forced to any of a number of drastic actions—e.g., military adventurism, economic reform, or massive Western imports. Neither of these extremes illuminates the more likely intermediate outcome. A far more informative discussion can, therefore, result from consideration of a medium case.

CMEA ENERGY TRADE: A MIDRANGE SCENARIO FOR 1985

From the perspective of the Soviet bloc, the future of CMEA participation on world energy markets will be determined not simply by production in each energy sector, but by a number of other factors as well. These apply to the countries of Eastern Europe as well as to the U.S.S.R. and include overall levels of economic growth (which affect rates of growth of energy consumption), the degree of substitution among fuels, and levels of debt and hard currency requirements. If, for instance, the worst possible conditions prevail in the U.S.S.R.—i.e., energy production in all sectors falls far short of targets; and oil is replaced with gas to only a limited extent—the Soviet Union may itself experience an oil deficit. Beyond a certain point, however, hard currency constraints will almost certainly preclude Soviet purchases of oil on the world market. Instead, there may be no alternative but to cut back on economic growth and energy consumption.

On the other hand, if the Soviet economy continues to grow comparatively slowly (about 1.6 percent), and domestic demand for oil can be kept down, the U.S.S.R. might be able to maintain oil exports even in the face of declining growth in oil production. Moreover, to the extent that gas can replace oil as an export to the West, the criticality of oil in Soviet hard currency exports will decline, and the key question for the U.S.S.R. will become not whether it can maintain its oil exports, but whether it can continue to earn hard currency as a net *energy* exporter.

Chapters 8 and 9 show the enormous range of outcomes in the Soviet and East European energy balances which are possible from different combinations of assumptions regarding economic growth, the growth in energy demand, and domestic energy production. These scenario outcomes are summarized, in terms of net hard currency energy balances, in table 76. In each case it is assumed that Soviet energy exports to Eastern Europe (the CMEA-6) remain as planned for 1981-85 (in other words, about 118 million tons of oil equivalency (mtoe) annually, or about 18 percent above the average annual level in 1976-80). Soviet exports

Table 76.—CMEA-Seven Net Hard Currency Energy Exports, 1979 and 1985

	1979	1985		
	(Estimated)	Worst case	Mid-range	Best case
Net hard currency energy exports (mtoe)				
U.S.S.R........	83	(38)	92	212
CMEA-6.......	4	(70)	(33)	10
CMEA-7.......	87	(108)	59	222
Change in net hard currency energy exports (mtoe)				
U.S.S.R..............		(121)	9	129
CMEA-6..............		(74)	(37)	6
CMEA-7..............		(195)	(28)	135

SOURCE: Chs. 8 and 9.

to other CMEA countries (notably Cuba) are assumed to reach about 11 mtoe annually, a rate slightly higher than in 1976-80. The total of assumed Soviet energy exports to CMEA countries (129 mtoe), as well as the 1979 level of Soviet energy imports from outside CMEA (9 mtoe), is netted out of the figures shown in table 76.

As the table indicates, CMEA was a net exporter of energy in 1979, the last year for which reliable estimates are available, of roughly 87 mtoe. By 1985, under OTA's worst and best case scenarios, the net hard currency energy balance for CMEA could range from a deficit of 108 mtoe to a surplus of 222 mtoe. As noted earlier, a much more likely outcome would fall between these two extremes and the analysis which follows concentrates on the midrange scenario.

In this case, Soviet gross national product (GNP) is assumed to increase at 2.4 percent annually, and midrange estimates are used for the income elasticity of Soviet energy demand and for Soviet energy production (see ch. 8, tables 53-55). The midrange assumptions for Eastern Europe include GNP growth comparable to that achieved in the late 1970's (about 2.9 percent annually), combined with a lower income elasticity of energy demand and favorable developments in domestic energy production (see ch. 9, tables 70-72).

Given these midrange assumptions, the Soviet Union in 1985 would be in a position to export a slightly greater amount of energy (net) than in 1979, about 92 mtoe. Eastern Europe, on the other hand, would change from being a net exporter of energy outside the CMEA of 4 mtoe in 1979, to a net importer of energy for hard currency of 33 mtoe by 1985. Overall, CMEA would remain a net energy exporter (59 mtoe), but it would be offering 28 mtoe less to the world market in 1985 than it was in 1979.

The impacts of this midrange situation for the West and for the Soviet bloc itself are equally important. The relevant question for

Western nations is twofold. First, what are the implications of this outcome for world oil markets; and second, what are its implications for the volume and composition of the U.S.S.R.'s energy exports to the West? The issues faced by Eastern nations have to do with their hard currency situations and with the implications of Soviet energy export decisions for the economies of Eastern Europe.

CMEA IMPACT ON WORLD OIL MARKETS

From the point of view of the West, any assessment of the likely impact of the CMEA on world energy markets must take into account the worldwide availability of petroleum during the 1980's. OTA has elsewhere provided a basis for estimates of world oil production.[1] Table 77, which is based on this work, shows that between 1980 and 1990 world oil production, excluding that produced by the U.S.S.R. and other centrally planned economies, is unlikely to rise significantly.

Such predictions are complicated by the fact that a number of oil-producing countries do not produce at full capacity. If the capability of these "swing" nations is considered, the capacity for oil production is increased[2] by as much as 500 million metric tons (mmt). The rather conservative estimates[3] of excess capacity in table 77 show that world oil production could be significantly increased in this decade if Iraq, Kuwait, Libya, Iran, and Saudi Arabia (which alone accounts for over half of the excess capacity shown in table 77) so wished.

A variety of economic and political factors—the price and demand for oil, and the

---

[1] See Office of Technology Assessment, *World Petroleum Availability, 1980-2000*, October 1980.

[2] See Department of Energy, *International Energy Evaluation Systems*, VI, Sept. 1, 1978.

[3] See Congressional Budget Office, *The World Oil Market in the 1980's, Implications for the United States*, May 1980. The estimates for excess capacity are 300 mmt higher for both 1985 and 1990 than those in this chapter.

Table 77.—World Oil Supply—Noncentrally Planned Economies, 1985 and 1990

(million metric tons to nearest 25 mmt)

	1980	1985	1990
OPEC medium production[a]	1,350	1,600	1,650
Non-OPEC LDC's[b]	} 1,625	375-450	375-500
Developed countries[c].......		650-775	550-750
World production[c]	2,975	2,625-2,825	2,575-2,900
Excess capability of OPEC swing countries over production[4]	775	550	500
World capacity[c] ..	3,750	3,175-3,375	3,075-3,400
Capacity above 1980 levels	775	200-400	100-425

SOURCES: ¹1980 estimate—*Monthly Energy Review* May 1981; DOE; 1988 and 1990 projections are mean figures in *World Petroleum Availability 1980-2000*, OTA October 1980, with 1990 figures obtained through interpolation.
²1980 estimates—*Monthly Energy Review*, May 1981, DOE; 1985 and 1990 projections are from *World Petroleum Availabilty 1980-2000*, OTA October 1980.
³Excluding centrally planned economies.

international and domestic political situations of the "swing" countries—will affect decisions to use excess production capacity. Barring intensified political instability in the Middle East, the pressure of growing world demand would likely result in an increase in the capacity utilization level of the OPEC "swing" producers. This could mean that in 1985 and 1990, there would be an additional 200 to 400 mmt and 100 to 425 mmt respectively of oil available in the world market from noncentrally planned economies. This would more than compensate for even the worst case Soviet production declines.

But while these additional supplies are possible, it would be a mistake to count on them. It is by no means clear that demand is the most important stimulus for increased capacity. The most important limits are political, and, as recent events in Iran have shown, cannot be forecasted. Even if this were not the case, experience following OPEC oil price increases in 1979-80 has

shown the limitations of demand forecasting models which rely on historical price elasticities. At present, oil demand has slackened and further production cutbacks have been announced.

To the extent that these uncertainties allow reasonable projections, however, it is clear that the outcome described in OTA's midrange scenario would likely have only a negligible impact on the supply-demand balance in world energy markets. A decline of net CMEA exports of 28 mtoe would equal only about one percent of estimated petroleum production capacity in the non-Communist world in 1985, as reflected in table 77.

THE VOLUME AND COMPOSITION OF ENERGY EXPORTS TO THE WEST

Table 78 shows that under midrange assumptions the Soviet Union could entirely cover Eastern Europe's incremental energy needs if it chose to do so, and still have some energy left to export for hard currency. The issue, however, is not just one of aggregate energy balances. It is also important for the CMEA countries to ensure that energy is supplied in volume and form appropriate to meet local demand. Thus, an important consideration for both energy producers and consumers is **the composition of incremental 1985 supplies.**

In 1979, the Soviet Union exported an estimated 83 mtoe to countries outside Eastern Europe. Of this, 60 mtoe (more than 70 percent) was oil and oil products; 16 mtoe was gas; and 7 mtoe was coal. But while in 1979 oil exports clearly dominated CMEA net energy exports, by 1985 the situation may change. Although it is difficult to anticipate the precise contribution of each energy sector to Soviet incremental energy production or exports, it is clear that even under best case conditions oil production in the U.S.S.R. is unlikely to increase rapidly enough to carry the primary weight of incremental energy exports. Likewise, OTA ex-

**Table 78.—Possible Composition of Soviet
Net Energy Exports**

	1979 (Estimated)	1985 Midrange Oil = 50 percent	Oil = 40 percent	Oil = 30 percent
	(million tons of oil equivalent)			
Net export	83	92	92	92
Oil	(60)	(46)	(37)	(28)
Coal	(7)	(7)	(7)	(7)
Required gas and electricity exports	16	39	48	57
Estimated present capacity for gas exports to West		(29)	(29)	(29)
Required increase in gas export MTOE capacity if no electricity exports		10	19	28
(bcm equivalent) ..	12	23	34	

SOURCE: Office of Technology Assessment.

pects that by 1985, coal production will at best rise little above 1980 levels. Gas production, however, is projected to increase substantially. The only remaining energy sector which is growing rapidly is electricity produced from nuclear and hydropower. This suggests that gas, in conjunction with electric power, must supply the preponderance of additional energy available both for export and for internal substitution.

Chapter 2 demonstrates clearly that the U.S.S.R. can produce as much gas as it needs, **provided it can be moved and utilized.** This raises two issues, the feasibility of replacing oil with gas in hard currency-exports, and the prospects for internal substitution of gas for oil.

The countries of Western Europe have made it clear that they are willing, indeed eager, to import substantially greater quan-

tities of Soviet gas. Table 78 shows the level of Soviet gas exports to the West in 1985 necessary to maintain net energy exports of 92 mtoe. This table assumes that coal exports are maintained at estimated 1979 levels and that oil exports fall, alternatively, to 50, 40, and 30 percent of total net energy exports.

These calculations raise important questions concerning the logistics of such sales. At present, there is limited pipeline capacity in place to support additional gas exports. In 1980, the excess capacity of the Orenburg pipeline (after meeting annual commitments to Eastern Europe of 15.5 bcm) was 12 to 13 bcm. During that year an additional 23 bcm of gas were exported to the West. Present pipeline capacity could therefore support 23 + 13 bcm = 36 bcm of natural gas exports to the West. This is equivalent to 29 mtoe/yr. When this figure is subtracted from indicated required gas exports (see table 78), it appears that by 1985 additional gas export pipeline of from 12 to 34 bcm might have to be constructed—if oil exports do decline within the indicated range and to the extent that the export shortfall is not made up by sales of electricity.

According to the U.S. Defense Intelligence Agency, two new lines are already under construction, and altogether six to seven are contemplated during the present Five Year Plan (FYP) period, including the controversial pipeline which will carry West Siberian gas to Western Europe. Four of these pipelines should be available for supporting growth in domestic gas consumption.[4] The West Siberian export pipeline, discussed in chapter 12, is scheduled to support from 40 to 70 bcm of additional gas exports to Western Europe, but whether it will be completed by 1985 remains an open question.

The second important issue is the ability of the Soviet Union to substitute other types of energy for oil. Current Soviet plans reflect

[4]Statement of Major General Richard X. Larkin, Deputy Director, Defense Intelligence Agency, before the Joint Economic Committee, Subcommittee on International Trade, Finance, and Security Economics, Sept. 3, 1981.

a high level of optimism in this area. The targets for rapidly increasing gas production in the next 5 years imply a good measure of domestic substitution. The capability of the CMEA to maintain net oil exports—or avoid the need to import more oil—will depend on its success in substituting gas and non-oil-fired electric power for domestic oil consumption. In other words, the greater the success of energy policies promoting substitution, the more oil will be available for export in 1985. Some idea of the sensitivity of the CMEA export position to substitution can be gained from considering the consequences of the U.S.S.R.'s achieving a rather limited level of substitution.

OTA has amassed very little hard data on substitution of gas for oil, but it seems reasonable to assume that while complete substitution is unlikely, 20 percent may be attainable. For purposes of illustration, OTA has assumed that gas is substituted for 20 percent of Soviet oil consumption. This level of substitution is equivalent to 6.8 percent of total U.S.S.R. energy consumption—87 mtoe or 1.75 mbdoe. Complete substitution of oil would imply a major effort—displacing 439 mtoe (almost 9 mbd of oil). But since the U.S.S.R. uses oil extensively to generate electricity, and since ECE data for 1980 show that both the U.S.S.R. and Eastern Europe depended on oil as a source of energy to a much lesser extent than did many Western nations,[5] there would seem to be fair potential for substitution on the order of 20 percent.

Under these circumstances, roughly 40 mmt less oil would be available for export. In this case, the Soviet Union could probably meet the projected incremental East European energy import needs (about 33 mtoe) with oil exports. However, the U.S.S.R. would have only about 10 mtoe of oil above 1980 levels available for export to countries outside the CMEA. In other words, the U.S.S.R. could actually have as much as 70

mtoe of oil for export **if a substantial degree of domestic substitution were possible.** Assuming continued exports to Eastern Europe at 1980 levels and low levels of substitution, domestic oil demand would preclude an expansion of energy exports in the form of oil.

In sum, the U.S.S.R.'s great gas potential could allow it to compensate on world energy markets for stagnating or even declining oil production. For this to occur, gas will have to replace oil to a certain extent in domestic consumption, but more importantly it will have to become much more prominent as a hard currency export. Since the countries of Western Europe are already eager to import more Soviet gas, and since this gas is widely regarded as replacing rather than supplementing current Soviet oil deliveries, such an outcome need not present problems for the West. It is contingent, however, on the successful and timely completion of sufficient pipeline capacity to transport the gas.

THE HARD CURRENCY POSITION OF THE CMEA

Under the midrange conditions of moderate GNP growth, energy production and consumption posited here, it does not appear that the U.S.S.R. itself will face a hard-currency crisis by 1985. Indeed, under the midrange scenario, the analysis in chapter 8 shows that the Soviets would be in a position, in terms of the aggregate energy balance, to possibly increase the amount of energy they export for hard currency at roughly 1979-80 levels and, given favorable terms of trade developments, continue to expand hard currency imports at a respectable rate.

The U.S.S.R. cannot be considered in isolation from Eastern Europe, however. The energy position of the entire CMEA-7 will set the parameters for the Soviet leadership. The situation facing the bloc as a whole is rather less sanguine. The midrange case shows a drop in net energy exports for hard-currency of 28 mtoe. **Where this burden falls**

[5]United Nations, Economic Commission for Europe, Economic Bulletin for Europe, June 1981, p. 162. In 1980, Soviet dependence on liquid fuel was 38 percent of total energy consumption. East European dependence was 25 percent.

will be determined by Soviet policymakers. The "energy squeeze" could conceivably be borne by the U.S.S.R. itself in an effort to ameliorate Eastern Europe's economic problems; it could be shared; or the U.S.S.R. could leave the CMEA-6 to purchase energy—most likely in the form of oil—on world markets.

If the Soviets were to make up the entire 1985 shortfall of Eastern Europe (33 mtoe), hard currency pressures on these countries would be reduced. They would not be eliminated because presumably Eastern Europe would have to divert increasing amounts of relatively high-quality exportables away from the West and towards the Soviet market, as payment for stepped-up Soviet energy deliveries. But such a policy would also reduce Soviet energy deliveries to the world market by one-third, and as chapter 8 suggests, would seriously erode Soviet hard currency import growth.

On the other hand, if the net East European hard currency energy balance deteriorates through the purchase of 37 mtoe (see table 76), it will be extremely difficult for most of these nations to pay for imports. Romania will be particularly hard pressed. As chapter 9 points out, Romania alone may be responsible for one-third of all energy imported by Eastern Europe in 1985. Romanian energy imports, moreover, are expected to triple between 1979 and 1985. Changes in Poland's energy situation could also affect the overall position of the group—Poland is the only East European country with a chance of remaining an energy exporter through the decade.

Assuming for purposes of illustration that incremental East European net hard currency energy requirements reached 37 mtoe in 1985, and that they were met entirely with imports of oil from the world market (priced at $36/barrel), hard currency requirements for the region would increase by almost $10 billion annually. Because one or two of these countries are likely to remain net energy exporters, an even greater burden would actually fall on the others, particularly

Romania. Romania would be forced to use from one-half to three-quarters of its export earnings to pay for oil imports—a situation which is neither feasible nor likely. (Use of 25 percent of export earnings to finance oil imports is considered a reasonable hard currency "breakpoint.")

There are, of course, a number of developments which might ease the hard currency constraints on Eastern Europe. Poland could improve its hard currency position if oil consumption could be held at 1980 levels, and if coal and electricity were used to meet additional energy needs. Even more beneficial from the perspective of the CMEA as a group would be measures taken by Romania, the nation most dependent on oil, to meet all of its incremental energy needs by importing gas instead. This would considerably improve Romania's hard currency situation, since gas is currently priced at half the cost of oil per Btu. The overall situation of the CMEA-6 could, furthermore, be ameliorated by conservation and improvements in energy efficiency. Even in the absence of such measures it is unlikely that Eastern Europe will be able to rapidly increase purchases of energy (particularly oil) from outside the CMEA. If energy demand should increase in line with high consumption scenarios, it is far likelier that economic growth will slow and energy demand consequently fall. Hard currency constraints thus reduce the probability of a sudden increase in oil purchases on world markets by Eastern Europe.

SOVIET ENERGY AND EASTERN EUROPE

This exercise has shown that **under midrange conditions for GNP growth, energy production, and substitution, the CMEA as a group is not likely to become a net energy importer by 1985.** Increases in aggregate Soviet energy production will overall offset rising Eastern European energy requirements. Soviet leaders are thus faced with a tradeoff between supplying cheap energy to the Eastern alliance and potential hard currency earnings through energy exports to the West.

As chapter 9 has pointed out, the critical linkage between Soviet energy exports and East European energy supplies cannot be overemphasized, and since prospects for expanded energy production in East Europe are dim, the U.S.S.R. is certain to continue as an important supplier. Thus, while oil sales to Western Europe are obviously attractive to the U.S.S.R., it is fully cognizant of the risks to itself should Eastern Europe be faced with economic chaos. When East European countries suffered shortfalls in oil imports from Iran and Iraq in 1980, the Soviet Union expanded its own exports to its allies—at the expense of hard-currency-earning sales to Western Europe. (It must be noted, however, that given the rising world market price of oil, the U.S.S.R. can maintain its hard currency earnings while exporting less oil.)

But the extent to which the U.S.S.R. will be willing to continue this assistance remains to be seen. In late summer 1981, Romania requested increased Soviet deliveries of both oil and gas. The U.S.S.R. had already offered to export additional gas to Romania—in return for Romanian participation in gas pipeline, nuclear power, and iron ore mining projects. As of this writing, it is not known whether Romania has accepted these terms or whether the U.S.S.R.'s willingness to supply additional energy will extend to oil. It might be expected that the U.S.S.R. will encourage its allies to import incremental energy supplies wherever possible in the form of gas.

CMEA ENERGY TRADE IN 1990

World oil production in 1990 is likely to be only slightly higher than that for 1985—reaching a maximum of 2,900 mmt (compared to 2,825 mmt for 1985). If the excess capacity of the swing producers is taken into consideration, world production capacity for 1990 could reach 3,400 mmt (v. 3,375 for 1985) (see table 77.)

This production differential is enormous. When the range of scenarios constructed for Soviet energy trade in chapter 8 are taken into account, the range of 1990 possibilities widens even further, significantly beyond those postulated for 1985. Chapter 9 shows too that a similarly wide range of possibilities exists for the CMEA as a whole. Under worst case conditions, the Soviet Union by 1990 could become a net hard currency energy importer (ch. 8: table 60), and Eastern Europe would have *incremental* net hard currency energy import requirements well in excess of contemplated Soviet energy exports in 1981-85 (ch. 9: table 73). On the other hand, if optimistic assumptions are used as the basis for calculation, the U.S.S.R. would be in a position to expand its hard currency energy exports over 1979-80 levels, and Eastern Europe would remain a net hard currency exporter of energy.

This tremendous range of possibilities makes the construction of a midrange scenario for 1990 an extremely tenuous exercise—and one of little utility. What can be said with some degree of certainty, however, is that the same constraints operating on energy trade outcomes for 1985 will be relevant in 1990. Regardless of whether the U.S.S.R. is able to reach its energy production targets, levels of Soviet economic growth, the degree to which gas and electricity are substituted for oil, and the ability of Eastern Europe to hold down oil imports will all influence CMEA incremental oil import needs. The message here is that a variety of factors, amenable at least in part to policy direction, could significantly ameliorate or aggravate the CMEA's oil import/export situation.

CONCLUSIONS

Regardless of whether the Soviet Union is able to meet its own energy production targets, a variety of additional factors will significantly influence its ability to maintain its status as an energy exporting nation. Those factors include the degree to which economic growth proceeds at a moderate or low (rather than a higher) level, and the ability of Eastern Europe (particularly Poland and Romania) to hold down demand for imported oil, but the most crucial are the ability of the U.S.S.R. to substitute gas for oil in domestic consumption and the rate of construction of new pipelines for gas exports to the West. **The ability of the CMEA to develop an energy policy which results in the better case conditions for substitution, demand, and economic growth will be as important as its ability to meet production targets in determining the degree to which CMEA's net hard currency energy balance will deteriorate.**

The formulation of such policy will confront the U.S.S.R. with difficult choices involving tradeoffs which will inevitably be most difficult if worst case conditions develop. The most obvious example here is the trade-off of hard currency earned through oil exports to the West against supplying subsidized energy to Eastern Europe. There are also costs involved in decisions over gas exports, where the primary problem is not production, but rather transportation of the gas both to Eastern and Western Europe and within the U.S.S.R.. To the extent that the Soviet bloc is able to increase its domestic use of gas, nuclear power, and other energy sources, it frees oil for export to the West. The development of gas and other energy sources, however, requires considerable investment and economic adjustment. While expansion of gas production and consumption is an attractive option, it is not a costless alternative.

Should it become necessary for Eastern Europe to increase its purchases of oil on the world market, these nations will be faced with decisions about the reliability of supply similar to those that must be made by policymakers in the industrial West. One approach, consistent with past patterns of energy imports to the CMEA, would be to strengthen special relationships with a few key Middle East oil producers like Iran and Iraq, perhaps through an expansion of barter trade. The difficulty here is that this policy would increase CMEA vulnerability to interruptions in supply by one of these key suppliers. Indeed, the Iranian revolution has already demonstrated precisely such vulnerability. Thus, while the "special relationship" option may appear just as attractive to CMEA leaders as it has to certain Western policymakers intent upon building bilateral ties with producer countries, it offers no easy solution. Even military occupation of an oil producing nation would not necessarily eliminate such supply uncertainties—the ongoing costs of a military solution are clearly high, albeit difficult to measure precisely, and oil supplies could be highly vulnerable to sabotage.

In the final analysis, oil and energy import problems must be viewed as critical threads in the fabric of CMEA economic viability. If the U.S.S.R. and Eastern Europe together find it increasingly difficult to produce energy needed for both internal consumption and export earnings, it will be more difficult to sustain a growing economy. While constrained energy supplies are commonly assumed to lead directly to increased purchases in the international market, OTA's analysis makes it clear that the domestic economic impacts of such problems are extremely important. In fact, **if the worst conditions materialized and the CMEA faced an oil deficit, hard currency constraints would almost certainly preclude large purchases on world markets and, therefore, the most likely immediate impact would be to reduce levels of economic growth and domestic consumption.**

In short, a growing CMEA energy crisis would signify difficult and long-term economic and social adjustment—as has been the case in the West. Energy must be viewed as one of a number of critical policy factors which could either severely constrain or greatly enhance the economic and political viability of the Soviet bloc. **Shortfalls or surpluses in Soviet oil are probably more significant from the perspective of the domestic economic adjustments that they will engender within CMEA than in their implications for the nature of CMEA participation in world energy markets.**

The significance of this analysis for U.S. policymakers, of course, rests on the question of the maximum possible oil import needs of the CMEA relative to projected world oil production in the decade ahead. Assuming that the most likely future for the CMEA lies somewhere between the extremes sketched in chapters 8 and 9, **through 1985, at least, it appears that if moderate conditions of production, substitution and economic growth prevail, the CMEA as a bloc will not become a net energy importer.** The U.S.S.R. could meet all incremental East European oil needs by reducing its energy exports to the West—if it chose to do so—although this would have a significant impact on Soviet hard currency import capacity. Regardless of which policy the Soviets pursue, the decline in net CMEA energy balances available for hard currency export by 1985 would probably have a far less significant effect on world energy markets (amounting to roughly 1 percent of expected non-Communist oil production capacity) than on the economies of the Soviet Union and its East European allies.

CHAPTER 11

Japanese-Soviet Energy Relations

CONTENTS

Japanese-Soviet Energy Relations

Japan's postwar energy-related trade with the Soviet Union has been limited. Although Japanese leaders are committed to cooperating in Sovet energy development in East Siberia, Japan depends on the U.S.S.R. for only a miniscule part of its energy supply. Similarly, Japan is the West's largest supplier of energy-related technology and equipment to the Soviet Union, yet these exports constitute a relatively small part of Japan's total world exports. Both of these facts reflect a situation in which Japan's political relations with the U.S.S.R. have tempered, but not precluded, energy interaction between the two countries.

A variety of factors—political economic, and energy-related—provide a mix of incentives and disincentives for Japanese energy interaction with the U.S.S.R. On the political side, Japan's orientation has been clearly toward the West. It is a member of the International Energy Agency (IEA), the Organization for Economic Cooperation and Development (OECD), and CoCom,* and its foreign policy has been anchored on the U.S.-Japan Security Treaty. Despite a number of persisting disputes between Japan and the Soviet Union, however, Japanese leaders consider joint energy development projects to be an important signal to the Soviet Union that they are committed to peaceful coexistence in Asia. From energy and economic perspectives, there is clear complementarity between Japan's energy import requirements and Soviet plans for energy and economic development. Japan is understandably anxious to diversify its sources of imported energy so as to reduce dependence on Middle East oil, and its

leaders have for years looked to the Soviet Union as a potential—and nearby—energy supplier. The Soviet Union also provides a significant market for Japanese energy equipment and technology exports. **The balance of these factors favors a positive, albeit cautious, Japanese approach to energy relations with the Soviet Union.**

A systematic look at the way in which Japanese leaders evaluate the potential risks and benefits of trade and energy cooperation with the U.S.S.R. is essential for an evaluation of past trends and future prospects for **Japanese-Soviet energy relations. The purpose of this chapter is to explore from the Japanese perspective the dimensions and dynamics of Japan's energy and trade relationship with the Soviet Union. The focus underscores Japan's importance—for both the Soviet Union and for the United States.**[1] **Japan is the single most important market for Soviet timber and coal, and a potential market for gas produced in Eastern Siberia. Thus, Japanese policy is a critical factor in Soviet economic calculations in Asia. But Japan is also the strongest non-Communist economy in Asia, and Japanese cooperation is important for the success of American foreign policies, globally and toward the region.** The chapter outlines the nature of Japan's trade and energy relations with the U.S.S.R.; explores the domestic organizational and international political context of Japanese policymaking; examines three primary examples of Soviet-Japanese joint energy development; and assesses likely future developments in energy relations between the two nations.

---

*CoCom is the informal multilateral export control organization which includes NATO countries (minus Iceland, plus Japan).

[1]Allen S. Whiting, "The Japan Connection," Siberian Development and East Asia: Threat or Promise? (Stanford, 1981).

JAPAN'S ENERGY AND TECHNOLOGY TRADE RELATIONS WITH CMEA DURING THE POSTWAR PERIOD

JAPANESE ENERGY IMPORTS FROM THE SOVIET UNION

Japan is highly dependent on imported energy and other resources. It must purchase over 90 percent of the energy it consumes (see table 79). A large portion of these imports consists of oil, for Japan is more dependent on oil for its total energy requirements than any other Western nation examined here (see table 80 and ch. 12, tables 86-89). Oil accounts for more than 78 percent of the nation's total primary energy supply, and virtually all of it is purchased abroad. Japan has for years sought to relieve its dependence on the Persian Gulf, which supplies 75 percent of its imported crude oil, by decreasing the share of oil in its energy balance and by increasing imports from non-OPEC nations.

Despite this extreme energy dependence, however, Japanese energy imports from the U.S.S.R. have been small, both in value terms and as a percentage of total energy supplies. During the last 5 years, the value of all Japan's imports (including energy and other commodities) from all communist nations has annually averaged less than 5 percent of its total imports, and the relative importance of energy-related imports has actually fallen. The U.S.S.R. is the only Council for Mutual Economic Assistance (CMEA) nation that exports energy—oil and coal—in significant quantities to Japan, but the dollar value of this trade has been consistently low. In recent years the total value of all Soviet energy exports to Japan has not exceeded $300 million annually. This amounts to less than 1 percent of all Japan's imports.

Table 81 illustrates Japan's very limited dependence on Soviet energy, which comprises a miniscule part of total Japanese energy imports and available primary energy. In recent years energy imported from the

Table 79.—International Comparison of Dependence on Imported Energy—1979
(million tons of oil equivalent)

	Japan	West Germany	France	Italy	United States
Total energy requirements[a] 1979	327.5	283.3	184.9	132.5	1,747.0
Energy imports as percent of total energy requirements	94%	67%	96%	109%[b]	25.2%
Oil imports as percent of total energy requirements	78%	53%	75%	91%	24%
Gas imports as percent of total energy requirements	4%	12%	9%	10%	2%
Coal imports as percent of total energy requirements	12%	2%	11%	7%	0.1%

Conversion factor: 1,000 million tons coal equivalent (MTCE) = .6859 MTOE.

[a]Total energy requirements by commodity - observed consumption data is used wherever available for coal and natural gas, due to the limited availability of inventory data; otherwise requirements are computed by the following formula: domestic primary production + imports – exports – international bunkers – inventory changes. Total energy requirements are computed only if inclusive of all commodities (oil, gas, coal, primary electric power, and net electricity imports). "Other electricity" includes net electricity imports. Graphs of total energy requirements do not account for inventory changes if production and import data are separated.

[b]Italy re-exports imported energy.

SOURCE: Business Information Display, *World Energy Industry*, Volume 1, First Quarter, 1980.

Soviet Union has not exceeded 1 percent of Japan's total primary energy requirements. Even Soviet coal, the most important of these imports, represented only 5 percent of all hard coal imports in 1979.[2]

---

[2]The 1979 total import figure is taken from *World Energy Industry*, while the U.S.S.R. import figure is taken from MITI data. See table 81.

During 1979 imports from the U.S.S.R. probably represented less than 3 percent of total Japanese coal imports. For the first 9 months of 1979 imports of Soviet hard coal totaled 1.7 million tons, while Japan's total coal imports from all sources for that year amounted to 60 million tons. See *Soren Too Boekikai Chosa Geppo* (hereafter Chosa Geppo) (Monthly Report of the Soviet—East European Trade Association), November 1980, p. 3; for total coal import data for 1979, see Japan Economic Journal, *Industrial Review of Japan*, 1980, p. 60.

Table 80.—Japanese Energy Balance—1979
(million tons of oil equivalent)

	Oil	Gas	Coal	Nuclear	Hydro and imported electricity	LNG
Total energy requirements:						
100.0 percent............	74.0%	0.6%	15.5%	1.3%	1.5%	7.1%
327.5 mtoe.............	242.3	2.0	50.7	4.4	4.9	23.2
Energy Imports:						
—as percent of total energy requirements	78.4%	—	12.1%	—	—	3.5%
308.3 mtoe.............	256.7	—	39.9	—	—	12.0
Energy exports:						
11.2 mtoe						

SOURCE: Business Information Display, *op. cit.*

Table 81.—Japanese Energy Dependence—1979
(million tons of oil equivalent)

	Oil/oil products	Gas	Hard coal	Nuclear	Imported electricity	LNG	Total energy
Requirements....................	242.3	2.0	50.7	4.4	4.9	23.2	327.5
Imports from world..............	256.7	—	39.9	—	—	12.0	308.3
Imports from U.S.S.R.............	0.7	—	2.0	—	—	—	2.7
Imports from U.S.S.R. as percent of total imports..............	0.3%	—	5.0%	—	—	—	0.9%
Imports from U.S.S.R. as percent of requirements...................	0.3%	—	9.9%	—	—	—	0.8%

Conversion factors: 1 kiloliter = 6.289 barrels; 1 barrel = 0.1358 thousand metric tons oil equivalent; 1,000 mtce = 0.6859 mtoe.

SOURCE: For imports from world: Business Information Display, *World Energy Industry*, Vol. 1, No. 3, First Quarter, 1980, p. 116. For imports from U.S.S.R.: Ministry of International Trade and Industry (Japan), *Energi Tokei Nenpo* (Yearbook of Coal, Petroleum and Coke Statistics), (Tokyo: Tsusho Sangyo Chosa Kai, 1979), pp. 30, 39, 82, 176ff.

From the Japanese perspective, then, Soviet energy has been relatively unimportant. From the Soviet perspective, however, Japan is a very important customer, purchasing virtually all of the lignite and more than half of the hard coal that the Soviet Union has sold to the industrialized West. In 1979, about a quarter of the U.S.S.R.'s total petroleum product exports to Japan, West Germany, France, Italy, the United Kingdom, and United States were purchased by the Japanese. **In short, Japan, is presently more important to the U.S.S.R. as a customer than the U.S.S.R. is to it as an energy supplier.**

This situation may change in the years ahead, but probably only in limited ways. Current Japanese official energy forecasts suggest that, theoretically at least, Soviet energy might play a role in meeting projected needs. Japan's official long-term energy forecast, first drawn up at the end of 1979 and now under revision, shows a dramatic reduction in the use of oil over the next decade. Recent revisions for 1990 call for the share of oil in the energy supply to fall from 74 percent to 47 to 48 percent,[3] and

[3]"Sekiyu Izondo, Hanbun Ikani" (Oil Dependence to be Reduced to Less Than Half), *Nihon Keizai Shimbun*, Apr. 13, 1981, p. 1.

for the proportion of oil-fired electricity generation to be reduced by nearly half— from 46 percent to about 24 percent. These ambitious plans assume rapid increases in consumption of coal, nuclear power, and liq- uefied natural gas (LNG); rapid development of new energy sources; and continuing suc- cess in energy conservation (see fig. 27). The plans will require imports of all types of coal to rise rapidly from about 60 million tons

Figure 27.—Japan's Provisional Long-Term Energy Supply and Demand Outlook

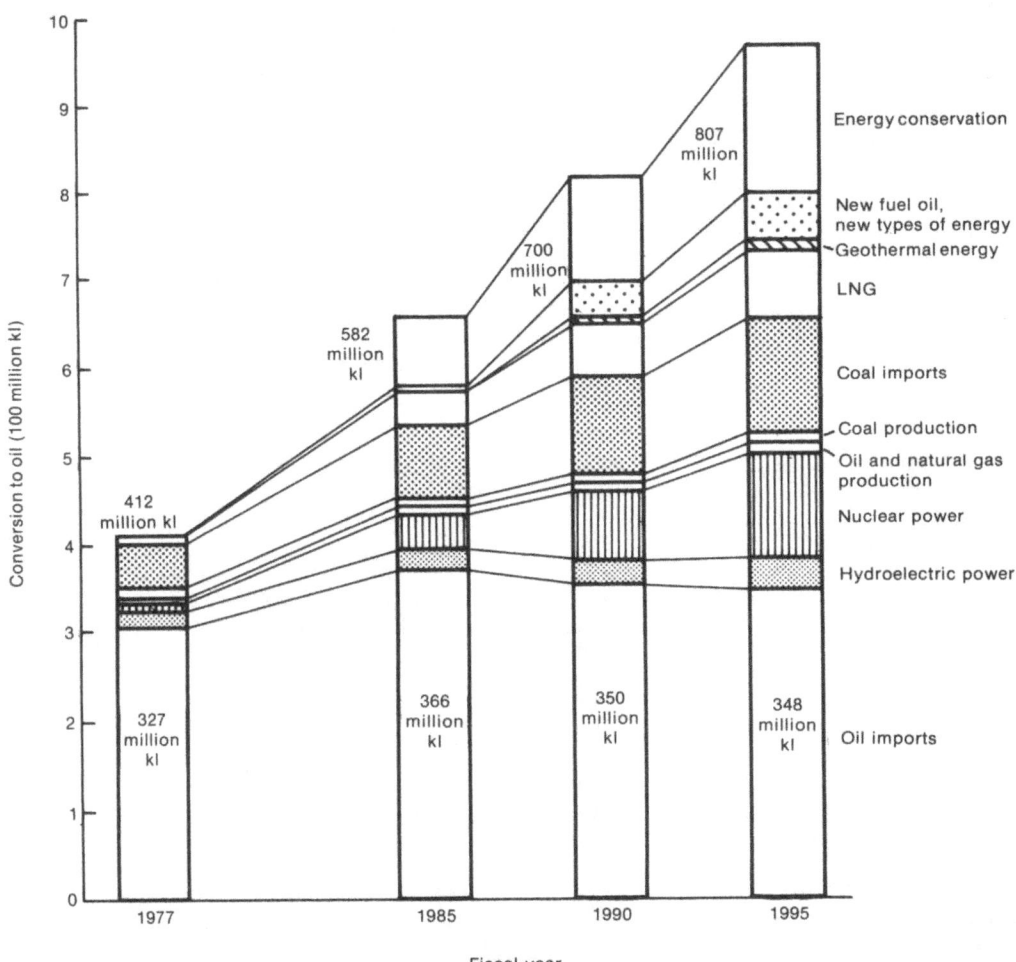

Note: The projections shown in this diagram are now under revision. According to the long-term oil supply plan published in May 1981, Japan's oil imports during 1985 will total 308 million kl in 1985. This figure includes imports of crude oil and refined products, excluding LNG. The new oil supply plan reflects a reduction in oil imports to Japan from a level of 6.3 - 6.9 million barrels per day for 1985 set at the time of the Tokyo summit in 1979 to a level of about 5.7 million barrels a day for the year 1985. See Tsusho Sangyosho (MITI), *Showa 56-60 Nendo Sekiyu Kyokyu Keikaku* (Oil Supply Plan for 1981-1985), May 27, 1981.

SOURCE: "News from MITI," NR-213 (79-28), Tokyo, Sept. 29, 1979, p. 9.

(40 million tons of oil equivalent (mtoe)) in 1979 to 143 million tons (98 mtoe) by 1990. Steam coal imports are expected to soar from less than 1 million tons per year in the late 1970's to more than 50 million tons (34 mtoe) by the end of the decade, as the cement and steel industries reduce their oil consumption by substituting coal, and as more coal is used to generate electricity.

Soviet coal could contribute to this planned energy transformation, but not on any massive scale. Japan is now participating in a joint effort to develop Siberian coal in South Yakutia (see below). But the projected 4 million to 6.5 million tons (2.7 to 4.4 mtoe) of coal for export to Japan which the project is expected to produce by the mid-1980's will still constitute only a small portion of Japan's anticipated 1985 coal imports of more than 100 million tons (68.59 mtoe).[4] In sum, Japan's urgent need to diversify its energy imports, both by geographic source and by type of energy, means that the Soviet Union will continue to be seriously considered as a potential energy supplier. At the same time, however, energy imports from the Soviet Union will not rapidly increase as a share of total supplies.

[4]*Nihon Keizai Shimbun*, Dec. 25, 1980. The article cites a revised delivery schedule for South Yakutian and Kuznetsk coal: 1 million tons annually 1979-82; 4.5 million tons, 1982-85; and 6.5 million tons, 1985-99. *Nihon Keizai Shimbun*, Feb. 9, 1981, reported that it is unlikely that supplies will reach 3.2 million tons by 1983.

JAPANESE ENERGY TECHNOLOGY TRADE WITH CMEA

The second dimension of Japan's commercial relationship with the Soviet Union and Eastern Europe has been in exports of Japanese manufactured equipment and plants to CMEA. Japan has been an important contributor to both Soviet and East European economic development, and has traditionally maintained a positive overall balance of trade with CMEA. The relative volume of Japanese exports to the U.S.S.R., however, has not been large. Between 1975 and 1979, these accounted on average for less than 3 percent of Japan's total yearly exports ($2.4 billion in 1979). As a rule, the value of Japan's exports to the U.S.S.R. has been two to four times as great as those to Eastern Europe. Thus, even if all CMEA nations are included, exports to the Soviet bloc represented less than 4 percent of Japan's total exports during most of the 1970's.[5] Similarly, in 1979 Soviet goods made up less than 2 percent of all Japanese imports. In fact, from the Soviet perspective overall trade with Japan has diminished in importance during the last decade, falling from a high of 12.7 percent of all Soviet trade with industrial nations in 1975 to 8.8 percent in 1980 (see table 82).

[5]"1979 Nen no Nisso Boeki" (Japan-U.S.S.R. Trade in 1979), *Chosa Geppo*, April 1980.

Table 82.—Soviet-Japanese Trade—1975-80
(millions of rubles)

	1975	1976	1977	1978	1979[a]	1980
A. Soviet imports from Japan	1,253.5	1,372.1	1,444.4	1,583.7	1,653.5	1,772.6
B. Soviet exports to Japan	668.9	748.4	853.4	736.1	944.4	950.2
C. Total Japanese-Soviet trade turnover	1,922.4	2,210.5	2,297.8	2,319.8	2,597.5	2,722.8
D. Total trade turnover between U.S.S.R. and industrial nations[b]	15,843.9	18,658.1	18,741.6	19,679.9	25,753.8	31,583.1
C/D	12.8%	11.4%	12.3%	11.8%	10.0%	8.6%

[a]Revised figures.
[b]"Industrial nations" is a standard Soviet trade classification which includes Western Europe, North America, Japan, and Australia.
SOURCE: *Moscow Narodny Bank Press Bulletin*, 1975-1980, *inter alia*; *Soviet Foreign Trade*, 1975-1980, *inter alia*.

In certain industrial sectors trade with the U.S.S.R. is disproportionately important. The bulk of Japan's exports to the U.S.S.R. and Eastern Europe (86.7 percent in 1979) has been in plants and equipment for heavy industry. In 1979, almost half of these were in iron and steel.[6] This trade is concentrated in areas which complement Japan's own efforts to restructure its domestic industries. Heavy and petrochemical industries have for years dominated Japan's industrial structure, but current government plans foresee a diminished importance for these sectors. Exports are viewed as a way to help declining industries. Moreover, declining plant exports in 1980, due mainly to contract cancellations by the Chinese, have led Japanese plant exporters to hope for a compensating growth in CMEA markets.[7] Before the Soviet invasion of Afghanistan, Japanese plant exports to the U.S.S.R. and Eastern Europe were growing briskly, earning Japan about 10 percent of the world plant export market.[8]

Japan has made an important contribution to energy-related technology trade with the Soviet Union and Eastern Europe (see ch. 6). In 1979, Japanese worldwide energy equipment and technology exports were valued at more than 6 billion. Exports to the Soviet Union accounted for 15.9 percent and those to Eastern Europe 2.6 percent of this total. (Exports to all communist nations, including the People's Republic of China, totaled almost 30 percent of all Japan's energy-technology related exports during that year.)[9] **Between 1975 and 1979, Japan alone supplied almost 30 percent of all Western energy-related exports to the U.S.S.R. About 45 percent of Japan's total exports to the U.S.S.R. during 1979 were of energy-related equipment.**

Such trade has been heavily concentrated in a few areas—pipes, tubes, pumps, and light vessels. Japan has not been a major manufacturer of seismic equipment for oil exploration, but Japanese companies have been important suppliers of pipe and other petroleum production equipment. Japanese exports of "tubes, pipes, and fittings" accounted for 34 to 53 percent of all trade in these commodities between the United States, Germany, France, Italy, United Kingdom, and Japan and CMEA between 1975 and 1979.[10] **Between 1975 and 1979, Japan ranked first among Western nations in the dollar value of energy equipment and technology trade with the U.S.S.R.**

Japan is undoubtedly a major exporter of energy-related equipment and technology to CMEA; opinion as to the significance of these exports differs, however. Many Japanese businessmen believe that American technology in these areas is superior to their own; U.S. manufacturers suggest that there are many items which can be produced in Japan as well as anywhere in the world. Japanese drill pipe, for example, incorporates the latest technology, including inertial welding of the tool joints at the end of the pipe, and it is widely recognized as at least comparable to that produced by American firms.[11] Japanese firms such as Mitsui, Sumitomo, Nippon Kokan, Kawasaki, and others have supplied quality pipelines for transporting oil and gas, as well as pipeline valves, pipeline booster stations, and pipe-laying equipment. Mitsubishi has built quality offshore semisubmersible rigs such as the Hakuryu II (White Dragon), used in exploration around Sakhalin Island. Japanese firms are capable of producing almost all of the major pieces of equipment needed

---

[6]Kin Murakami, "TaiSo Boeki no Genjo to Hatten" (Current Status and Future Prospects for Japan-U.S.S.R. Trade), *Nikiren Geppo* (Japan Machinery Industry Federation Monthly), November 1980, p. 4.

[7]Interview with President of New Japan Steel, *Nihon Keizai Shimbun*, Feb. 2, 1981, p. 3.

[8]"Plant Exports to the Soviet Union and East European Countries," *Digest of Japanese Industry and Technology*, No. 144, 1980, p. 41.

[9]Data from the Japan Tariff Association, *Japan Exports and Imports: Commodity by Country, 1979* (Tokyo: 1980). CCFTS categories were chosen to correspond with those developed by OTA for U.S. Department of Commerce data. Compiled by Stephen Sternheimer for OTA.

[10]U.N. SITC data, compiled by Stephen Sternheimer for OTA.

[11]See ch. 6, on Western Energy Equipment and Technology Exports to the U.S.S.R.

The Japanese-built Hakuryu oil rig, off Sakhalin Island

for coal mining. In almost every major category of energy technology, therefore, there are Japanese companies competing with those from the U.S. and Western Europe. In some cases, Japanese manufacturers are able to produce items at low cost, making them attractive suppliers for CMEA energy development projects.[12]

_____
[12]Japanese steel firms have been supplying seamless pipe to the United States because American firms do not produce

In sum, past patterns of trade between Japan and the U.S.S.R. in fuels and energy-related equipment and technology show that—despite the potential for mutually beneficial exchange of Japanese equipment and know-how for Soviet energy and raw materials—the interaction between these two nations has been limited. Japan's present reliance on Soviet energy is very small. However, in certain sectors, including energy-related technology and equipment, Japan has been a major Soviet supplier. Except in specialized areas, Japanese energy equipment is on a par with equipment produced by other Western nations.

Japan's consistently positive trade balance and its low level of energy imports from the Soviet Union indicate a cautious approach to energy and trade relations. This brief outline of past patterns of interaction shows that while there are strong underlying incentives for Japanese participation in Siberian energy and economic development, there has been no rapid development of such ties. Japan has avoided dependence on the U.S.S.R. for energy, although its exports of energy-related technology have increased.

_____
enough pipe to fill domestic demand. See "Japanese Makers of Seamless Pipe Swamped With Foreign Orders," *Asian Wall Street Journal*, Dec. 8, 1980.

JAPANESE POLICY TOWARD ENERGY TRADE WITH THE CMEA: THE INSTITUTIONAL AND INTERNATIONAL POLITICAL CONTEXT

Japan's postwar policies concerning energy cooperation with the Soviet Union have developed both informally and officially. Semigovernmental and private organizations, as well as government agencies, have played important roles in exploring potential trade and joint energy projects, and in carrying out agreements. These organizations —which include the large trading companies (*sogo shosha*), companies manufacturing various types of machinery and equipment, Japanese utilities and other potential consumers of energy, the Federation of Economic Organizations (*Keidanren*), the Ministry of International Trade and Industry (MITI), the Foreign Ministry, and the Export-Import Bank (Ex-Im Bank) of Japan—all participate in development and implementation of joint Japanese-Soviet energy projects. The persistence of an identifiable group of institutions responsible for these policies has ensured a degree of policy continuity. This section briefly identifies the central actors in this institutional setting, and then examines the international political context of Japanese energy trade with the U.S.S.R.

THE DOMESTIC INSTITUTIONAL SETTING FOR POLICY

Private Organizations

The Keidanren has played a leading role among the private organizations, companies, and institutions involved in negotiating and participating in joint Japanese-Soviet energy projects. Since the 1960's Keidanren leaders have taken a strong personal interest in prospects for Siberian development, and businessmen from Keidanren, as well as other economic organizations such as the Japan Chamber of Commerce, have participated in these negotiations with the Soviet Union.

The Keidanren's Japan-Soviet Economic Committee (*Nisso Keizai Iinkai*), which is made up of more than 100 Japanese businessmen, works to coordinate opinions among interested Japanese firms. The committee includes a number of subcommittees, each of which has primary responsibility for a particular type of project area (gas, coal, oil), and is made up of corporate executives from these firms. At times one individual or firm may exert decisive influence. An individual from Tokyo Gas, for example, has been the leading figure in negotiations over Siberian gas development.

Preparatory negotiations over potential energy development projects normally span a number of years. During this time a series of meetings are held to discuss the project's possibilities and to specify the nature of participation on both the Japanese and Soviet sides. A Soviet organization, the U.S.S.R.-Japan Business Cooperation Committee, parallel's Keidanren's committee, and is headed by the Soviet First Deputy Minister of Trade. The Keidanren and Soviet committees hold discussions; a protocol agreement is signed; and finally a "general agreement" specifies the overall commitment of both sides. The latter agreement outlines the financing, cost estimates, and plans for equipment purchases, and carries the commitment of both the governments.

In addition to their participation as members of Keidanren, a number of private trading, manufacturing, and energy companies play important roles at various stages of the development of joint energy projects. The sogo shosha have handled the bulk of trade between Japan and the Soviet Union since 1956. In 1980 the primary trading companies dealing with the U.S.S.R., in rank order, were Mitsubishi, Mitsui, C. Itoh, Nissho Iwai, Sumitomo, and Marubeni. The first three of these were responsible for about a third of all Soviet trade.[13] These companies all have Moscow offices. Since these firms are associated with other related corporations in company groups *(keiretsu)* they are in a good position to bring affiliated companies into projects as they develop. Many of the trading companies have specialized departments, comprised of Soviet area specialists, who deal in trade with CMEA. All of these factors make the trading companies important participants in the development of joint Japanese-Soviet energy projects, both during preliminary negotiations, and in later discussions of supply contracts.

Another secondary actor on the private side is the *Soren Too Boekikai*, the Association for Trade with the Soviet Union and Eastern Europe. This group specializes in economic and trade research, publishing monthly journals and assembling trade data. When requested by member firms, the Association undertakes special studies. It also arranges for visits of Soviet delegations, assists Soviet participation in Japanese trade fairs, and helps Japanese businessmen interested in trading with U.S.S.R. and Eastern Europe.

Consortiums may be formed to organize participation of Japanese firms in joint ven-

---

[13]*Chosa Geppo* April 1978, p. 212. See also Alexander K. Young, *The Sogo Shosha: Japan's Multinational Trading Companies* (Boulder, Colo.: Westview Press, 1979), p. 8.

tures. The Sakhalin Oil Development Corp. (SODECO), for example, is composed of firms that represent a variety of Japanese industries—manufacturers of equipment, energy corporations, banks, and trading companies engaged in joint Japanese-Soviet oil and gas development offshore Sakhalin. Such consortia spread the financial risk among a group of firms and facilitate coordination among them. The businessmen involved in joint energy projects with the Soviet Union are generally also in close touch with Japanese Government officials. In some cases, such as the unfruitful negotiations that took place over joint oil development in Tyumen, businessmen have preferred to move more positively toward cooperation than have government officials.[14] But despite a natural difference in the perspectives of government officials (particularly those in the Foreign Ministry) who have broad policy concerns, and Japanese businessmen interested in expanding trade, the two sides are normally in fairly close agreement.

Government Organizations

The Ministry of Foreign Affairs is the formal coordinator of Japan's overseas policies. However, in general, the ministry has been less involved in the development of specific joint Japanese-Soviet energy projects than other government agencies. Since 1956, when Japan and the Soviet Union officially resumed diplomatic relations, overall trade agreements have been reviewed and revised every 5 years. These 5-year trade agreements do not generally spell out the precise details of joint energy development projects. When a proposed project becomes a matter of political debate—and this has generally not been the case—the Ministry of Foreign Affairs can play a decisive role in the negotiations.

The government's trade and financial agencies (MITI, the Ministry of Finance,

and the Ex-Im Bank) all routinely play more important roles in the development of energy projects. MITI, through its International Trade Policy Bureau, supports Japanese firms with export insurance, and tax and credit incentives. Since MITI implements the foreign exchange and trade control laws, including Export Control Division oversight of items restricted by CoCom, it plays an important role in the development of trade and exchange with the Communist nations. Through its support for the Japan National Oil Corp. and other public energy corporations, MITI has helped to provide financing for overseas energy development.

Government financial institutions are also directly involved in negotiations over energy projects with the U.S.S.R. The Ministry of Finance is authorized to provide policy guidance to financial institutions making overseas loans and investments; it normally plays an important indirect role through its budgetary oversight of the Ex-Im Bank. Through its loans, the Ex-Im Bank supplies the major share of government funding for large-scale development projects in the Soviet Union. These loans can be made to Japanese importers and exporters and to foreign governments for financing imports of Japanese plants and equipment.

Officials of the Ex-Im Bank are usually consulted by Japanese firms at a number of stages in project negotiations. Through its assessments of risk and projections of credit needs and appropriations availability, the Ex-Im Bank determines which Soviet projects should be supported, and eventually the Bank signs a loan agreement with the U.S.S.R.'s Bank of Foreign Trade. This establishes bank-to-bank credits used by the Soviet bank to reimburse Japanese firms.[15] As early project stages are completed, progress is reviewed and financing arrangements

[14]Gerald L. Curtis, "The Tyumen Oil Development Project and Japanese Foreign Policy Decision-Making," in *The Foreign Policy of Modern Japan,* Robert Scalapino (ed.) (Berkeley, Calif.: University of California Press, 1977), p. 160.

[15]For a description of Japanese financing of joint Japanese-Soviet coal development in Siberia, see *Coking Coal Manual,* 1980, p. 289. According to experts, the bank provided $4 million in loans for Siberian development projects as of 1979. See Allen Whiting, op. cit., p. 137, and Henry Scott Stokes, "Japan Facing Complex Policy Issues About Sanctions on Soviet Union and Iran," *New York Times,* Jan. 9, 1980.

renegotiated. Under routine circumstances, the financial arrangements form the framework for development and review of joint projects.

Japanese commercial banks are legally prohibited from lending more than 25 percent of their capital funds to any one recipient; therefore, firms interested in participating in large projects often turn to the Ex-Im Bank for assistance. While project loans generally involve a combination of government credits and monies from commercial banks, the Ex-Im Bank's commercial assessment of the project is important in determining loan rates.[16] The Bank does not usually make public the exact proportion of the loan it provides, or the differential between the loan rate charged by it and that charged by private banks. Since its purpose is to stimulate Japanese exports, Ex-Im Bank financing is concentrated in those parts of a project that involve purchases of Japanese-manufactured plants and equipment, rather than in those involving purchases of Soviet-made goods.

The institutional and financial support provided by the Japanese Government through the bank and other agencies has been a distinguishing feature of the joint development projects in which Japanese firms have participated.[17] Even when Japanese Government and business officials are favorably disposed toward a project, financial considerations can delay or significantly modify it. The example of the joint Japan-U.S.S.R. oil and gas project offshore Sakhalin illustrates the central role of financial institutions.

In the first stages of the Sakhalin negotiations in the early 1970's, Keidanren's Japan-Soviet Economic Committee held discussions with Soviet representatives. Once the two sides reached a preliminary understanding, Keidanren leaders undertook extensive consultations with various Japanese Government agencies. The Ministry of Foreign Affairs apparently was not extensively involved, but the Ex-Im Bank made an important contribution to these discussions through its project risk assessments. MITI officials helped to develop the consensus among Japanese parties that provided the working basis for a new round of more detailed negotiations between the Keidanren committee and the Soviets.

After both sides agreed on a protocol, discussions took place with the U.S.S.R.'s Bank of Foreign Trade. A Japanese consortium was formed to organize corporate participation in the project. This consortium, the aforementioned SODECO, signed the basic contract with the Soviet Ministry of Foreign Trade in January 1975. A "general agreement" outlined the financial participation of both sides, and set various project targets.

The formation of SODECO and the arrangements for financial backing from the Ex-Im Bank and other institutions were crucial to the progress of this project. The Japanese initially advanced risk money of $100 million for drilling at Sakhalin. Much of that capital came from the Japanese corporate shareholders in SODECO, as well as from the Japan National Oil Corp. The Ex-Im Bank's risk assessment and its financing signaled the commitment of the Japanese Government.

The Political Context of Japanese-Soviet Energy Relations

Political Relations With the U.S.S.R.— Japan's lack of indigenous energy resources and history of export-led economic growth both suggest strong incentives for cooperation in Soviet energy development. However, the historical and political context of Japanese-Soviet relations is marked by a variety of complicated and persisting dis-

---

[16]In 1976, the OECD instituted guidelines for lending rates charged by member nations for large-scale projects such as those established to promote Soviet energy development. Before that time, interest rates set by the Ex-Im Bank and other financial institutions could be the key determinants of whether Japanese firms or their competitors won associated export contracts. Interest rate charges continue to be a focus of negotiations over Soviet energy projects.

[17]Terutomao Ozawa, *Japanese Multinationalism* (Princeton, N.J.: Princeton University Press, 1980), p. 33.

putes and tradeoffs. These are important factors that Japanese policymakers weigh in their negotiations with Soviet leaders. While the political factors are usually viewed as constraints on interaction between these two nations, incentives for cautious Japanese interdependence with the U.S.S.R. can also be identified.

It is often noted that Japanese public opinion surveys demonstrate acute public dislike and distrust of the Soviet Union. This is generally assumed to indicate fundamental opposition to expanded Japanese-Soviet relations. The implications of such surveys are, however, far from clear. A recent survey of Japanese elite views on security issues indicates that Japan's policymaking leadership holds no clearly distinguishable or coherent view of the "Soviet threat."[18] Most of the respondents considered a Soviet military attack unlikely, supported only a modest Japanese defense build-up, and perceived the "Soviet threat" as primarily psychological and political rather than military. Like numerous other polls on the subject, this survey provides no conclusive indications about what policy Japanese leaders are likely to initiate, but it does indicate that the Soviet Union is not perceived by them in black-and-white terms.

A second factor commonly viewed as an obstacle to increased cooperation is the Northern Islands issue (see fig. 28). Persisting disputes over territorial claims to four northern islands were reflected in the failure of Japan and the Soviet Union to conclude a peace treaty following World War II. The Northern Islands issue is a recurrent theme in the Japanese media. The problem has been recently exacerbated by a Soviet military build-up on these islands between 1978 and 1980.

The emotional significance of the issue, however, has not prevented the establishment of diplomatic relations between the two nations or the development of a number of joint Siberian projects.[19] In other words, it is not clear that there is any direct policy link between the Northern Islands and Japanese-Soviet interaction in Siberia. Publicly, the domestic salience of the issue precludes any Japanese public official from conceding the improbability of the return of the islands to Japan. Privately, however, many admit that there is no precedent for the U.S.S.R. returning territory it has occupied for so long.

The China Factor.—The "China Factor" is shorthand for another set of policy issues which are often assumed to inhibit Japan's interactions with the U.S.S.R. It has been suggested in the West that China and the Soviet Union pose an either/or choice for Japanese economic involvement, for reasons of both politics and competitive economics. According to this argument, Japan's historic ties to, cultural compatibility with, and nearness to China mean that priority is placed on Japanese-Chinese relations.

Japanese leaders nevertheless strongly disagree with the idea that they can or should choose between their two Asian neighbors. A prime concern is that the Soviet Union not be provoked to take aggressive action in Asia. Japan's basic allegiance is clearly to the United States, but Japanese leaders worry that Moscow may perceive Tokyo as cooperating (tacitly or explicitly) with Washington and Peking in an anti-Soviet alliance. As a result, Japanese leaders attempt to signal the U.S.S.R. as to Japan's peaceful intent, without alienating China.[20]

---

[18]International Communications Agency, Office of Research, Research Memorandum, "Japanese Elite Views on Key Security Issues," Nov. 25, 1980.

[19]Richard L. Edmonds, "Siberian Resource Development and the Japanese Economy: The Japanese Perspective," paper prepared for Association of American Geographers Project on Soviet Natural Resources in the World Economy, August 1979.

It is interesting to note that a parallel dispute between Japan and China over the Senkaku Islands—located between Taiwan and the Ryukyus—has not inhibited joint oil development between these nations. Administered by the United States and then returned to Japan at the time of the return of Okinawa, these islands have received less media attention but the dispute over them is no less sensitive. This is another case of an unresolved conflict which has not prevented a wide range of interactions—including Japanese participation in offshore oil development in other areas near China.

[20]Whiting, op. cit.

Figure 28.—Soviet East Siberia and Japan

*Possible route

SOURCE: Office of Technology Assessment.

To this end, Japanese leaders both in and out of government depict cooperation in Siberian energy development as necessary to Japan's political interest—not for abstract reasons such as winning good will, but in order to reduce perceptions of hostility. Sakhalin offshore oil and gas development is a prime example. Sakhalin was contested by Japan and the Soviet Union for nearly a half century, finally falling under Soviet control in the final stages of World War II after the U.S.S.R. hastily renounced a neutrality treaty and entered the war against Japan. If the island represents a symbolically sensitive piece of lost territory, the oil and gas resources there are also strategic commodities important for both Japan and the U.S.S.R. Japanese participation in the development of these resources symbolizes commitment to peaceful cooperation in the Asian region. Japan's Sakhalin "signal" has continued— despite the Soviet invasion of Afghanistan. Furthermore, China has not protested joint energy development there. Indeed, the same Japanese firm exploring for oil offshore Sakhalin won the first foreign contract to explore in the Bohai Gulf near China.

Nor does it appear that such signals involve unusually great economic risks, at least compared to those incurred in participation in energy and development projects in other nations. Following China's readjustment of national planning priorities in 1980-81, Japanese contracts valued at $1.5 billion were canceled as projects were postponed. While China has indicated its willingness to renegotiate the bulk of these contracts, the incident cautions against overly optimistic expectations about the China market. No similar renegotiation or reversal has affected Japanese-Soviet interaction since the abortive Tyumen pipeline proposal of 1974-75, which ended before any contracts were signed. Nor is there any concrete evidence that firms which participate in Chinese economic development are denied access to the Soviet market, or vice versa. Indeed, a number of Japanese firms have figured prominently (and simultaneously) in both Soviet and Chinese development proj-

ects, supported by export credits from the Japanese Ex-Im Bank.

Policy Stance. — Contrary to popular stereotypes, it appears that Japan's attitude toward energy cooperation with the U.S.S.R. is based on a careful assessment of both political and economic tradeoffs. Politically, Japan hopes to avoid strong association with either China or the U.S.S.R. Economically, Japan needs diversified sources of energy and prospers through expanded plant, equipment, and technology exports. The Japanese organizations and institutions involved in formulating policy toward energy cooperation with the U.S.S.R. naturally weigh the potential risks and benefits of various projects from different perspectives —financial, political, energy, and trade. **However, there appears to be widespread agreement on the broad outlines of Japanese policy regarding Japanese-Soviet interaction. This is best described as cautiously optimistic.** Disagreements within Japan's policymaking leadership inevitably arise over the details of specific projects, and recur when international political or economic conditions change—but the general orientation of Japanese policy has been fairly consistent. For Japanese policymakers who are officially committed to diversifying Japan's energy supplies—both by types of fuels and geographic sources—both the Soviet Union and China are viewed as potential alternatives. During the 1980's Japan hopes to import a modest amount of oil from China, some gas from the Soviet Union, and considerable coal from both.

The Soviet Union is thus viewed as a potential supplier of additional energy—in limited increments. So long as the U.S.S.R. **remains merely one among many more-or-less equal suppliers, the Japanese believe that the political leverage likely to accrue through a threat of a cutoff will be minimal, if not nonexistent.** A further and commonly held extension of this view is that to the extent **that the Soviet Union relies on Japan for capital, technology, and equipment to develop its energy resources, the likelihood** of political manipulation or pressure is reduced and international tensions obviated.

Japan's willingness to cooperate with the Soviet Union in developing the latter's energy resources is in no sense an unbounded commitment. Informally and unofficially, Japanese leaders often cite 20 percent as the maximum safe level of reliance on Soviet imports for any commodity, energy included. Japanese are quick to point out that this falls well below the dependence of a number of West European nations in some fuels, and that actual Japanese energy imports from the Soviet Union are likely to fall far short of this level in the next few years. Another indication of the bounds to Japanese cooperation is the fact that, despite Soviet proposals, Japan has not entered into comprehensive trade agreements lasting longer than 5 years.[21] Additionally, Japanese leaders have been reluctant to move ahead in some cases of Siberian development without American approval, and even participation. This reluctance is illustrated by the case of the now dormant proposed gas development project in Yakutia (see below), in which Japanese firms under the leadership of Tokyo Gas strongly requested American participation.

While Japan's positive attitude toward energy and trade interaction with the U.S.S.R. has been cautious, at the same time there is little sympathy among Japanese leaders for a policy of strong controls over trade and technology exports to the Soviet Union. Nor do they support the idea of attempting to employ trade as a political lever in order to promote long-term Western security interests. Despite the continuing concern, particularly of Foreign Ministry officials, that Japan not take a position that isolates it from the United States and Western Europe, expanded controls on both equipment and technology trade are viewed as intrinsically unattractive options.

This position is in part based on the apparently widespread view within the Japanese bureaucracy that where there is trade, there is bound to be some technology leakage. Japan participates in CoCom and Japanese leaders believe that some export controls are feasible and necessary. But MITI officials in particular contend that controls on technology transfer are both difficult to construct and to implement. Such controls, they say, are best applied to limiting the sale of spare parts and manufacturing know-how, and then only to technology that is easily identifiable and separate from products.

In sum, the political context of Japanese-Soviet energy relations includes complicated and persisting issues such as the Northern Islands dispute and the "China factor." These lie behind Japan's policy of cautious interaction with the U.S.S.R. The economic and cultural complementarity of Japan and China make it unlikely that Japan's relationship with the Soviet Union will be promoted to a position of equal importance with its relationship with China. However, from the Japanese perspective, it is important that, in principle at least, Japan offer similar opportunities for economic cooperation to both nations. Japan's "omnidirectional diplomacy" thus implies involvement with *both* China and the U.S.S.R. in energy development and trade.

---

[21]Soren Too Boekikai, *Showa 53 nendo Nisso Keizai semmonka kaigi hokokusho* (Report of the 1978 Conference of Japanese and Soviet Economic Experts) (Tokyo: Soren Too Boekikai, 1979), p. 40.

JAPANESE PARTICIPATION IN SIBERIAN ENERGY DEVELOPMENT

Japanese-Soviet energy development projects have been beset by repeated problems and delays. Joint development of oil in Tyumen never even began, despite the interest of Japanese firms. Oil and gas development offshore Sakhalin is progressing very slowly; 5 years after the signing of the initial agreement, the exploration stage has still not been completed. Technical problems elsewhere have caused coal shipments from the Soviet Union to fall below anticipated levels.

The following sections describe three key Japanese-Soviet energy projects—Yakutian natural gas and coal development schemes and the Sakhalin offshore oil and gas project—analyzing the nature and prospective results of Japanese participation. These are the most significant examples of Japanese-Soviet joint energy cooperation to date, and illustrate the type of interactions likely to be feasible in years ahead.

YAKUTIAN NATURAL GAS

The Yakutian gas development project is a trilateral effort involving Japan, the Soviet Union, and the United States. Since the early negotiations, Japanese firms and organizations have taken the lead. Japanese participation is organized in a consortium of 21 firms called Siberia Natural Gas. The consortium includes major trading companies, utilities, banks, steelmakers, and other plant exporters. Hiroshi Anzai, President of Tokyo Gas and Chairman of Keidanren's Japan-U.S.S.R. Joint Economic Committee's subcommittee on gas, is Chairman of the consortium, and has been the leading figure in the development of the project. El Paso Natural Gas, Occidental Petroleum, and Bechtel Inc. were involved on the American side.

At the time of the preliminary negotiations and signing of the first contracts in 1975, the extent of Yakutian gas reserves had not been determined.[22] Under the terms of the original agreement, the beginning of actual production would await initial exploration, during which an anticipated 1 trillion cubic meters of gas reserves would be verified. This stage is now 90 percent completed but production has yet to commence. When and if it does, one-third of the gas will be retained by the U.S.S.R. and the other two-thirds divided equally between Japan and the United States at prevailing market prices—roughly 7.5 million tons of LNG annually to each country over a period of 25 years.[23]

Initial estimates of development costs for Yakutian gas stood at $3.4 billion. A loan of $50 million—half from Japan's Ex-Im Bank and half from the Bank of America—was provided for purchases of exploration equipment. The bulk of the Japanese financing came directly from the Ex-Im Bank, with only about 20 percent supplied by private companies.[24] U.S. law (i.e., Jackson-Vanik Amendment), however, precluded U.S. Export-Import Bank loans to the Soviet Union.

The project developed slowly, due to delays caused by cold weather and export licensing problems with the shipment of a U.S. computer. By the time of the Soviet invasion of Afghanistan, proved reserves were still about 10-percent short of the target; after the invasion, a tripartite meeting scheduled for the spring of 1980 was canceled. It has now been tabled. At this meeting a second general agreement to cover the next stage of the project was to have been developed. Those involved believe that it will take at least another year to conclude a second-stage agreement, and that even under such best-case conditions, actual commercial production of gas will not begin until 1987. Moreover, informal Japanese estimates suggest that the costs of the project may climb to $7 billion to $8 billion, double the initial figure.

In addition to the costs for exploration and production, huge invesments will be required to complete the necessary infrastructure. Gas produced in Yakutia was to be shipped by a new pipeline to the Soviet port of Olga on the Pacific Coast, where liquefac-

[22]U.S. Senate, *U.S. Trade and Investment in the Soviet Union and Eastern Europe,* staff report for the Subcommittee on Multinational Corporations, Committee on Foreign Relations (Washington, D.C.: U.S. Government Printing Office, 1974), p. 19.

[23]Soren Too Boekikai, *Handbook of the U.S.S.R.* (Tokyo: Soren Too Boekikai, 1978), pp. 444-46.

[24]Raymond S. Mathieson, *Japan's Role in Soviet Economic Growth* (New York: Praeger, 1979), p. 112.

tion plants and additional port facilities will be constructed. Original plans called for the United States and Japan to share the costs. The planned pipeline would be longer than the 1,700-mile Orenburg pipeline and would run over extremely cold and mountainous terrain. Japanese experts view this venture as being at least on a par with the efforts which will be required to construct the gas export pipeline across West Siberia to Western Europe. The cost of the equipment (pipeline and liquefaction facilities) needed for the next stages of the project will be very large, and any previous consensus that might have existed regarding who should pay for facilities to be built within the Soviet Union has broken down.

For the Japanese, Yakutian gas offers a market for energy equipment and technology and a source of LNG. To date, Japanese firms have supplied drill pipe and bits, gas detectors and masks. The Japanese firm IHI Heavy Industries has apparently offered to sell 36-MW compressors for the pipeline, and there will certainly be opportunities for sales of a variety of other Japanese equipment if the gas is actually developed.

But even assuming that all goes according to plan, it is not likely that Yakutia will render Japan greatly dependent on Soviet natural gas. The planned 7.5 million tons of LNG, to be supplied to Japan by the year 1990, will amount to less than 15 percent of Japan's projected total LNG imports for that year (45 million tons), or a little more than 1 percent of Japan's total primary energy supply. Soviet gas from Sakhalin (see below) might add 3.5 million tons per year, however. The two projects together, therefore, *could* provide one-quarter of Japan's gas imports by 1990 (see table 83).

These considerations belong to the future, however. With delay in proving reserves, mounting cost estimates, and delays due to international political tensions, the ultimate fate of the tripartite Yakutian gas project remains uncertain. Participants continue to maintain that it is economically and technically feasible, but there is little likelihood

Table 83.—Projected Japanese Energy Imports From the U.S.S.R. During the 1980's
(million tons of oil equivalent)

		Coal	Oil	Gas (liquefied natural gas -LNG)
A.	1979 Imports from the U.S.S.R.	2.0	0.683	—
B.	Projected incremental imports from U.S.S.R. + 1979 imports	6.4[a]	2.051[b]	2.4—1985[c] 5.1—1990
C.	Projected *total* Japanese imports from all sources, 1985	69.2	312	19.8
D.	Projected *total* Japanese imports from all sources, 1990	98.0	294	30.8
B/C		9.2%	0.65%	12.1%[d]
B/D		6.5%	0.69%	24.3%[d]

Conversions: 1000 metric tons coal × 0.6859 = 1,000 metric tons oil equivalent.
1000 barrels/day × 0.1358 = 1,000 metric tons oil equivalent.
1 kiloliter = 6.28 U.S. barrels.

[a]Assumes 6.5 million tons (4.4 mtoe) additional coal imports from South Yakutia.
[b]Assumes 10,000 barrels/day (1.6 mtoe) additional oil from Sakhalin.
[c]Assumes 3.5 million tons from Sakhalin (2.4 mtoe) + 7.5 million tons (5.1 mtoe) from South Yakutia. In all likelihood, the Yakutia gas will *not* come on stream until 1990 or later.
LNG imports from *all* sources will provide 7.2 percent of Japan's primary energy supply in 1985, and 9.0 percent in 1990 according to official Japanese Government forecast.

SOURCES: Ministry of International Trade (Japan) *Energi Tokei Nenpo* (Tokyo: 1suko Sangyo Chosa Kai, 1979) and Japanese government long-term energy forecast.

that all three parties will quickly move ahead in the current international environment.

SOUTH YAKUTIAN COAL

In contrast to the situation with Yakutian gas, where adequate recoverable reserves have not yet been established, a sufficient amount of coking coal, much of it located in thick seams, is known to exist in the area, and feasibility studies by the Soviets and on-site inspection by Japanese experts have led to the joint development of the Siberian Yakutian coalfields. Although South Yakutian production is still very small,[25] generally high-quality coal has been mined here for a number of years.

---

[25]*1980 Coking Coal Manual*, op. cit., pp. 290-92.

An initial agreement, signed in 1974, provided $450 million in Japanese credits for Yakutia, $60 million of which were to be spent within the U.S.S.R. for onsite costs connected with the labor force. The rest of the credits were set up to facilitate Soviet purchases of plant and machinery. The loan period extended from 1975 through 1982, with repayments in the form of coking coal to begin in 1983. Japan is to receive 85 million tons of medium-quality coking coal by the end of the century. In addition, the general agreement provides for the import of Soviet coal from the Kuznetsk basin at a rate of 1 million tons annually from 1979 to 1999.

The South Yakutian coal project encountered no major negotiation problems, but technical obstacles appear to have further delayed delivery schedules of both Kuznetsk and Yakutian coal. In December 1980, a revised timetable projected exports of 4.5 million tons to Japan between 1982 and 1985; and exports of 6.5 million tons between 1985 and 1999.[26] In the opinion of Japanese experts, schedules may be set back by as much as 2 years, due to difficulties associated with the use of equipment in such harsh climates.[27] In addition, inadequacies in the "coal chain" on the Soviet side—insufficient numbers of coal tankers and poorly developed transportation facilities—may raise further difficulties. A 400-km section of railroad connecting the mine site to the Trans-Siberian Railroad was completed in 1978. The city of Neryungri and nearby regions of Chulman are expected to experience a population influx, and the progress of regional infrastructure development will greatly influence the delivery schedules of coal produced in the mines.

Japan has sold the U.S.S.R. a variety of coal mining equipment for use in South Yakutia. This has included coal rotors, draglines, coal-washing and sorting equipment,

and earth-moving and excavation equipment. The Komatsu Co. has sold 190 bulldozers for the project, worth about $40 million. Other Japanese companies have supplied transport vehicles, electric locomotives, a crusher station, coal-washing equipment, and a coal terminal. The latter went into operation in 1978.[28] (Since U.S. firms are able to produce larger capacity trucks, draglines, and excavators than are available elsewhere, some American equipment has also been used here.[29]) Japanese firms expect to have expanded opportunities for sales as the project continues.

Even if produced on schedule, Yakutian coal is unlikely to provide a major portion of Japan's future coal imports. If Japan's imports reach the projected 100 million tons by 1985, 4 million to 6.5 million tons per year of South Yakutian coal would make up less than 1 percent of 1985 total Japanese primary energy supplies. In 1979, Japan imported about 2.0 million tons of Soviet coal (about 30 percent below the level contracted), all of it from the Kuznetsk basin. Even combined with this, Yakutian coal will still represent well under 10 percent of Japan's total imports of coal in 1990—a much lower percentage than supplies from the United States, Canada, or Australia (see table 83).

Yakutian coal development has progressed more rapidly than the tripartite gas effort, but deliveries from this area are not scheduled to start before 1983 and even then may not proceed before the latter part of the decade. Nevertheless, the Yakutian project is one of the centerpieces of joint Japanese-Soviet energy development. At the end of 1980, after holding up new loans for Siberian development in the wake of the invasion of Afghanistan, the Japanese Government approved a loan of $42.3 million for this project.[30] This, as much as anything, demonstrates its importance to Japan.

---

[26]*Nihon Keizai Shimbun,* Dec. 25, 1980.

[27]Information in English on South Yakutian coal reserves is scanty. The Central Intelligence Agency (CIA) expects that the Neryungri mine, which will supply Japan, will produce 13 million tons of coal after 1985. CIA, *U.S.S.R.: Coal Industry Problems and Prospects,* ER 80-10154, March 1980, p. 15.

[28]Mathieson, op. cit., pp. 76-78; *Soviet Geography,* February 1979, pp. 125-126.

[29]Ibid, p. 77.

[30]*Asian Wall Street Journal,* Dec. 29, 1980.

SAKHALIN OFFSHORE OIL AND GAS

Sakhalin Island is located between Japan and the continental Soviet Union (see fig. 28), about 50 miles from the northernmost Japanese island of Hokkaido; at points on its northwest coast, the island is even closer to the continental Soviet Union. Japan's participation in Sakhalin onshore oil development began before World War II, when the southern half of the island was Japanese territory. Offshore development dates from the 1960's, when Japanese oil refiners, as well as Gulf Oil Corp. (which had been supplying independent Japanese firms with oil from other regions) became interested in the project. Inspections began at the site in the early 1970's.

As noted above, Japanese participation here is organized through the consortium SODECO. SODECO is comprised of 18 corporate shareholders, the largest of which is a public corporation—the Japan National Oil Corp. (JNOC). JNOC holds more than 40 percent of the equity, as well as stock in a number of other shareholding firms. Japanese oil and trading companies also have shares in the project, and Gulf Oil holds about 5.7 percent of the total equity.[31]

In 1975, SODECO and the Soviet Union signed a basic agreement. The contract provided some $100 million to $150 million in Japanese credits, to be used for exploration equipment, including excavators, drilling rigs, drill casing, and computers. In return, for 10 years Japan is to receive 50 percent of any crude oil or gas produced offshore at a discounted price.[32]

The Soviet Union is the project operator. Day-to-day operation is supervised by a secretariat which has offices on Sakhalin Island. Onsite work teams are composed of technicians from a variety of different companies and nationalities. On the Japanese side, technical experts from the various participating companies are periodically "detailed" to the project, allowing the consortium to draw on a wide range of skills. Working in teams with Western technicians, Soviets gain "hands on" experience in operating the equipment.

The U.S.S.R. is contributing money as well as labor to the project. Soviet expenses have run about $100 million, paid in rubles to cover the costs of labor and construction for the 1980-82 exploration period. To date, SODECO and other Japanese sources such as the Ex-Im Bank have probably provided as much as $170 million.

Western technology has played an important role at Sakhalin. In 1976, SODECO leased a French geophysical vessel and computer equipment, and a variety of Japanese-manufactured rigs have been used. The semisubmersible White Dragon II, built by Mitsui, as well as the Okha jack-up rig, built by the same firm to a design patented by Armco, have been used for offshore test drilling. In July, 1979, the marine department of C. Itoh trading company sold a Mitsui-Livingston Class III jack-up rig for use at Sakhalin. This rig was especially designed for very cold conditions.[33]

One of the project's most important technological requirements will be for ice-penetrating rigs. Because of the thickness of ice around Sakhalin, Western technology developed for the Alaskan slope cannot be used without modification. In instances where specialized equipment is needed, American companies will probably be given market opportunities. However, the general pattern to date has been for Japanese firms to do the basic hull construction, finally assembling the rigs with equipment from a variety of companies.

The last phase of test drilling has now begun, with exploration concentrated in two fields, where 13 test wells have been drilled, seven of which have proved promising. Three more test wells will be sunk in 1982 to

---

[31]*Japan Petroleum and Energy Handbook* (Tokyo: Japan Petroleum Consultants, 1978), pp. T 118-19.

[32]Soren Too Boekikai, *Handbook of the U.S.S.R.* (Tokyo: Soren Too Boekikai, 1978), p. 446.

[33]*C. Itoh News*, No. 48, September-October 1979, pp. 5-6.

complete the exploration phase. If all goes well thereafter, plant construction and the installation of equipment will begin. This is scheduled to be completed by 1986, when a third stage will feature production and shipment of LNG.[34] The final stages of the project will involve the most costly outlays. Costs of building an LNG plant, extending the pipeline system on the island, tankers, and receiving facilities could add $5 billion to investment requirements.

An oil discovery has already been made in the Sea of Okhotsk, northeast of Sakhalin,[35] and as exploration has progressed, prospects for gas production have appeared more and more promising. Test wells sunk in the same area have confirmed gas reserves adequate to produce 5 bcm annually and oil deposits producing 6,600 bcm (0.328 mtoe/yr).[36]

In recent months, the major point of discussion among project participants has been how offshore gas will be transported to Japan. The initial Soviet proposal was to build a north-south pipeline on Sakhalin, with an underwater connecting link to Hokkaido. The U.S.S.R. favored the pipeline because it would be cheap and technically feasible to build, given the shallow waters. The Japanese have opposed this plan for security reasons, i.e., the pipeline might tie the northern part of Japan too strongly to one Soviet source of energy, and because it entails piping gas to the rural island of Hokkaido, where demand is low and where domestic coal producton is significant.

At the beginning of 1981, agreement was reached on a different option—construction of LNG facilities on Sakhalin.[37] Japanese LNG tankers will transport the LNG directly to areas on the island of Honshu where demand is strong. As mentioned above, this plan will involve huge capital outlays for

building an extended pipeline to the LNG facilities, the construction of the liquefaction plants, and the related harbor and loading facilities. While many details must still be worked out and the financing arranged, the agreement in principle indicates serious commitment by both sides to the continuance and development of the project.

Problems in U.S. participation have also delayed Sakhalin development. The export license for drilling equipment to be supplied by an American division of Armco was temporarily held up by the U.S. Department of Commerce.[38] This led to concern that American equipment could not arrive in time for the very short Sakhalin drilling season. After clarifications were sought by both the Japanese firm assembling the rig for SODECO and the U.S. firm, the export license was reinstated. The decision was taken quickly enough so that drilling proceeded on schedule.

Other problems have been largely technical, and generally related to the cold climate and difficult terrain. The ice around Sakhalin is so thick that drilling can only be carried out during a few months of the year, and special equipment is needed. Test rigs have been hauled out and then transported back to Japan when the drilling season ends. Storms and difficult weather conditions have periodically damaged equipment. This does have one bright side. Experience with drilling offshore Sakhalin will be invaluable to the U.S.S.R. as it exploits its potential for offshore exploration in other very cold regions.

The Japanese do not expect Sakhalin to ever provide them with large quantities of oil. But the hoped-for 3 million to 3.5 million tons of LNG per year would be a significant contribution to Japan's LNG imports which are expected to reach 29 million tons by 1985, and 45 million tons by 1990. Sakhalin gas alone could represent more than 10 percent of Japanese LNG imports in 1985, but it is far from certain that deliveries will begin

[34]Information on Sakhalin project stages from SODECO officials in Tokyo, Japan, March, 1981, and from *Nihon Keizai Shimbun,* June 9, 1981.

[35]Mathieson, op. cit., p. 99.

[36]*Daily Industry News,* Dec. 8, 1980; *Nihon Keizai Shimbun,* Nov. 11, 1980 (Converted at 1 kiloliter = 6.2898 barrels); and June 9, 1981, for information on test wells.

[37]"Nisso LNGka de Goi" (Japan and U.S.S.R. Agree on LNG Facilities), *Nihon Keizai Shimbun,* Feb. 2, 1981.

[38]Based on interviews with officials from National Supply, a division of Armco in March 1981.

so soon. If LNG makes up the expected 7.2 percent of Japan's total primary energy supply in 1985, the contribution of Sakhalin gas to Japan's overall energy supply will be minimal—less than 1 percent of total primary energy in 1990 (see table 83). If Sakhalin gas came onstream at 3.5 million tons per year, and if Yakutian natural gas were available at the projected 7 million tons per year, Japan could receive almost 24 percent of its LNG imports from the Soviet Union in 1990. This is optimistic, however, considering the delays that have developed at Yakutia and the fact that the decision was only recently made to set up an LNG facility on Sakhalin, the financing for which must still be worked out. It seems more probable that Sakhalin development will progress more quickly than Yakutian. Prospects for Soviet natural gas are fairly certain. Therefore, a respectable contribution to Japan's energy needs may be anticipated.

In the final analysis, the real significance of the Sakhalin project is as a test case for joint Japanese-Soviet development. With the Yakutian gas project stalled, and Siberian coal development proceeding slowly, Sakhalin remains the brightest spot in Soviet-Japanese energy cooperation for the next decade. For the Japanese, it offers a potential for diversification of energy supplies as well as a market for equipment and technology. For the Soviets, it offers a chance to develop exploration expertise and perhaps production capability in offshore oil and gas. Should Sakhalin become a significant source of energy to Japan, the proximity of the island to Japanese territory would certainly heighten chances for interchange between Soviet and Japanese industrial and technical personnel.

OTHER PROSPECTS FOR JAPANESE PARTICIPATION IN SOVIET ENERGY DEVELOPMENT

A few other areas of Japanese energy technology assistance to the U.S.S.R. are also worth mentioning. The Japanese Science and Technology Agency has signed an agreement with the U.S.S.R. State Committee on Science and Technology to promote scientific and technical exchanges between the two nations. These exchanges do not appear so far to have directly aided energy development. Between 1968 and 1978, the U.S.S.R. sent more than 100 missions to Japan under the auspices of the agreement, but only about 12 percent of these were even peripherally related to energy.[39] During the 1970's these energy-related missions visited power plants, and factories producing generators and steel pipe. One recent Soviet delegation has studied high-voltage transmission technology, necessary to bring power generated in Siberia to the European U.S.S.R. Missions from East European nations have focused primarily on the study of Japanese energy conservation techniques.

Another potential area for energy technology transfer between Japan and the U.S.S.R. is in nuclear power. The Soviet Union has approached Japan several times with requests for cooperation in this area, but although Japan and the Soviet Union signed a Cooperative Agreement on the Peaceful Use of Nuclear Energy in 1978, there have been few results. Initially the Soviets hoped to obtain a pressurized water reactor which Mitsubishi manufactures under license from Westinghouse, but these negotiations never proceeded far. Exports of nuclear reactors are controlled by CoCom, and Japan announced in 1978 that it was in principle willing to fabricate a nuclear reactor for the U.S.S.R., but only if construction was to Soviet design.

Japan has a large, well-integrated and technologically sophisticated nuclear industry. There is natural interest in exporting its equipment and technology for peaceful purposes. However, exports to other Asian neighbors are more likely than to the Soviet Union. As of 1981, the only Japanese sale to the U.S.S.R. in the area of nuclear power pro-

---

[39]Soren Too Boekikai, *Soren keizai to taigi kagaku gijutsu kyoryoku* (Tokyo: Soren Too Boekiaki, 1978), pp. 94-97.

duction has been a 15,000-ton press for manufacturing heads for atomic vessels. This utilized a type of manufacturing process already well-established in the U.S.S.R.

The U.S.S.R. has offered to provide uranium enrichment services for Japan: Difficulties with the U.S. over consignment commitments led in 1976 to the Chairman of Keidanren's energy policy committee broaching the possibility of using either French or Soviet enrichment services.[40] The U.S.S.R. offered in 1977 to enrich and return Japanese-supplied uranium. Japan, with longterm enrichment contracts with the United States and France, would consider purchasing enriched uranium from the U.S.S.R. on a commercial basis, but was not interested in providing the feedstock. Nothing more came of this deal.

In a few instances, Japan has actually imported energy technology from the Soviet Union. Japanese steel companies have purchased Soviet technology for the treatment of coking coal and for top furnace gas turbines, which have reportedly resulted in significant energy conservation. Most of these transactions, however, occurred in the

late 1960's and early 1970's. Today perhaps the only area in which Japanese energy experts are studying Soviet techniques is in high voltage electricity transmission.

This overview of Japanese participation in Soviet energy development reveals cautious participation on the part of Japanese firms. Joint Japanese-Soviet energy projects have evolved slowly and unevenly. Japanese firms, like their counterparts in the United States and Western Europe, produce equipment and possess technology which can assist Soviet energy development. As Japanese firms develop their technological expertise in electronics and other areas, there may be even greater demand for their products in the U.S.S.R. In militarily sensitive areas like nuclear power, however, the Japanese have been reluctant to deal with the Soviet Union. Security issues aside, Japanese businessmen have been inclined to participate in technology trade, both because of the prospect of expanded worldwide energy supplies and because of the potential market for Japanese exports, although this predisposition has been tempered to some extent by the changing shape of international politics following the Soviet invasion of Afghanistan. The final section of this chapter briefly reviews these recent developments in Japan's trade and energy relations with the Soviet Union.

---

[40]Soichi Matsune, "Genshiryoku hatsuden ni okeru kakunenryo saikoru no kakuritsu ni tsuite," *Keidanren Geppo*, August 1976, pp. 516-19.

RECENT DEVELOPMENTS IN JAPANESE-SOVIET RELATIONS: PROSPECTS FOR THE FUTURE

The worsening of U.S.-Soviet relations following the invasion of Afghanistan, and the second oil crisis triggered by the suspension of oil production in Iran, have provided a new context for Japan's interactions with the U.S.S.R. Beginning in early January, 1980, when President Carter ordered sanctions, Japan's policy toward the U.S.S.R. has been under review and reconsideration. American policy, as it gradually evolved, included expanded restrictions on the export of products (such as grain and phosphoric acid used in the manufacture of fertilizer)

and of high technology, particularly computers. The U.S. Department of Commerce was to act on all applications to export industrial technology for manufacturing oil and gas production and exploration equipment with a presumption of denial.[41]

Japan, like the West European nations, actually increased its exports to the U.S.S.R. during the period of the sanctions, although

---

[41]"U.S.-U.S.S.R. Trade After Afghanistan," *Business America* (Washington, D.C.: U.S. Department of Commerce Publication, Apr. 7, 1980), p. 12.

probably not as much as would otherwise have been the case. During 1980, Japanese exports to the U.S.S.R. rose almost 25 percent over the 1979 levels. The Japanese response to the U.S.-initiated sanctions was thus similar to that of Western Europe—lukewarm (see ch. 12). Officially, Japanese policy prohibited the extension of new government credits for the U.S.S.R., and suspended high-level diplomatic exchanges. This effectively froze all financing through the Japanese Ex-Im Bank. At the private level, however, trade continued unabated.

During most of 1980, those joint energy projects already underway were continued, with no new official funding. Japan's official policy thus amounted to a kind of holding pattern—maintaining prior commitments, but studiously avoiding their extension or the initiation of any new ones. This policy was pursued despite the fact that the Soviet Union both warned Japan about possible Soviet retaliation if it participated in such sanctions,[42] and invited it to join in the pipeline project designed to carry Soviet gas from West Siberia to Western Europe.[43]

By the end of 1980, Japanese businessmen were publicly criticizing the sanctions, arguing that government policy disadvantaged Japanese firms vis-a-vis their competitors in Western Europe but had no real effect on Soviet foreign policy.[44] Industrial leaders in Japan claimed that the sanctions accounted for the loss of 14 plant export contracts worth some $4 billion to $5 billion. They pointed to examples of trade lost to other Western nations which failed to participate in the sanctions. For example, the planned export of an electrical steel sheet plant by the U.S. firm Armco International and Nippon Steel Corp. fell through after more than 3 years of preliminary negotiations. The contract was awarded to the French firm Creusôt Loire.[45] In other cases vacillations in U.S. policy—e.g., denying license applications for Sakhalin drilling equipment and then reinstating them,[46] as well as the decision to grant export licenses to the American firm Caterpillar for pipelaying equipment (an item which Japanese firms were also interested in selling to the Soviet Union) were carefully noted in the Japanese press.

By the end of 1980, there were indications that the Japanese government was ready to relax the measures it had imposed. It made a significant step in approving new loans, which were reported to carry a 7.25-percent interest rate repayable over a 5-year period, through its Ex-Im Bank for the continuation and expansion of two Siberian development projects. The loans included $42.3 million for coal development in South Yakutia, and $96.3 million for a Siberian forest resource development project. In return, the Soviet Union committed itself to increased exports of coking coal to Japan.[47]

In April 1981, the Japanese Government resumed official trade talks with the Soviet Union. These had been suspended, although the previous bilatral trade agreement had expired. Under the new trade accord, which runs until 1985, Japan will import about 90 Soviet commodities including coal and oil; in return the U.S.S.R. will import items in 70 different categories from Japan. Another signal of a thaw in the trade freeze was the announcement in late January 1981, that agreement had been reached over the construction of the Sakhalin LNG facility.[48] Finally, additional new loans of $949 million for two Siberian development projects were approved in June 1981, evidence of the loosening of sanctions following the U.S. decision to end the grain embargo. The bulk of this money is for forestry, with about $40 million earmarked for coal development.[49] The loans will allow the U.S.S.R. to purchase

[42]*Asian Wall Street Weekly,* May 12, 1980.
[43]"Soren Daikei Shakkan o Yosei" (U.S.S.R. Asks for Large-Scale Loans), *Yomiuri Shimbun,* Sept. 9, 1980.
[44]*Asian Wall Street Journal,* Nov. 10, 1980; "Japan Fears Sanctions Breakdown," *Washington Post,* Sept. 23, 1980.

[45]See *Japan Economic Journal,* Sept. 2, 1980, p. 1.
[46]*Soviet Business and Trade,* May 21, 1980.
[47]*Asian Wall Street Journal,* Dec. 29, 1980.
[48]"Nisso, LNGka de Goi" (Japan and the Soviet Union Agree on LNG), *Nihon Keizai,* Jan. 28, 1981.
[49]*New York Times,* June 11, 1981.

equipment and services from Japanese firms.

The U.S.S.R. has also sought Japanese participation in the West Siberian gas pipeline project. At present, there are no firm indications of the role Japan might play, but the Soviet Union has been calling for Japanese financing amounting to as much as $3 billion. Japan would not receive gas, but its prospects for sales of large diameter pipe and other related commodities and equipment are good. The Soviet Union has apparently approached two Japanese firms—Hitachi and Marubeni—about the possibility of buying at least 10 gas boosters, each worth more than $1 million, and Nippon Electric Co. was reported to be considering bidding on contracts for the central pipeline control system.[50] The Japanese firm Komatsu is negotiating for a sale of pipelaying equipment worth $1.5 million to the U.S.S.R.[51] In late May 1981, press reports indicated that Japan's four largest steel firms had reached agreement with the Soviet Union to supply 750,000 metric tons of large diameter pipe over the next year. It was further reported that the Japanese Ex-Im Bank would extend $500 million in credit for the sale. Evidently the pipe is to be supplied on a regular commercial basis, without any clear specification that it will be used for the West Siberian project.

All of these developments reflect significant controversy within Japan over the economic sanctions, and a general reconsideration of policies toward the Soviet Union during the last year and a half. Throughout most of that period official Japanese foreign policy statements showed a chill in relations with the Soviet Union. The 1980 Foreign Ministry Blue Book, for example, stressed the need for close alliance with the free world in a period of growing international tensions.[52] Additionally, February 7 was designated as "Northern Islands Day."[53] While there were apparently a variety of domestic political reasons for the decision to institute the new commemorative day, the choice, as well as the rising salience of defense issues, reflected growing concern with East-West tension.

By mid-1981, however, signs were that Japanese leaders were moving back to their cautious but positive approach to energy and trade with the Soviet Union. Soon after his appointment, Foreign Minister Sonoda announced plans to "review" policies toward the Soviet Union with an eye toward renewing Japan's "omnidirectional diplomacy."[54] The events of the 18 months following the Soviet invasion of Afghanistan illustrate the fact that **international political tensions can act as an effective brake on Japan-Soviet energy and trade relations. Even in the presence of such conditions, however, Japanese leaders tend to favor continuing cooperative energy development. For Japan's** Government and business leaders, questions of trade with and technology transfer to the Soviet Union are just as much energy and economic as they are political issues.

---

[50]"Russia Said to Offer One Billion Dollar Contract to Japanese Firms," *Asian Wall Street Journal,* Apr. 27, 1981. "Niden, Taiso Shodan Kaiso" (NEC Begins Talks With the Soviet Union), *Nihon Keizai,* Feb. 4, 1981.

[51]*Japan Times,* May 20, 1981. There were $152 million in credits reportedly extended by the Japanese Government to cover Soviet purchases of pipelayers and bulldozers from Komatsu Co. for a pipeline from Urengoy, south of Yamburg in West Siberia. See *New York Times,* Aug. 20, 1981.

[52]Statements by Cabinet Secretary Miyazawa, "Caution Necessary About the Soviet Union," *Yomiuri Shimbun,* Sept. 12, 1980.

[53]"Hoppo Ryodo Yoron Moriage" (Whipping Up Public Opinion About the Northern Islands), *Nihon Keizai Shimbun* Jan. 6, 1981); "Kono Koe Todoke Hoppo Ryodo" (Northern Islands Campaign), *Nihon Keizai Shimbun,* Jan. 23, 1981.

[54]"Taiso Seizai Rosen o Shussen" (Toward a Revision of Sanctions on Trade With the Soviet Union), *Nihon Keizai Shimbun,* May 21, 1981), p. 2.

SUMMARY AND CONCLUSIONS

This overview of past patterns of interaction between Japan and the Soviet Union, as well as recent developments during the period of economic sanctions following the invasion of Afghanistan, leads to the conclusion that Japan will probably continue to pursue a positive approach to energy relations with the Soviet Union. Available data indicate that it is **unlikely that Japan will become very dependent on the Soviet Union for energy in the next decade—even if all the projects currently underway come to fruition.** Table 83 shows projections for Japanese imports of Soviet energy as a percentage of the nation's total energy imports, and total primary energy supply. If Japan imports 6.5 million tons of coal from South Yakutia, and if other Soviet fuels such as gas from Sakhalin are available as planned, the Soviet Union will still supply only about 3 percent of Japan's total energy imports, and about 2.2 percent of the nation's total primary energy. The only sector in which Soviet energy would be of more than marginal significance is gas, where it could account for nearly one-quarter of Japanese imports by 1990. Even this, however, is a relatively small portion of total Japanese energy requirements. Thus, in contrast to some West European countries, Japan does not risk any significant degree of "energy dependence" on the U.S.S.R.

While it is unlikely that Japan will become very dependent on Soviet energy in the years ahead, it is certain that Japan will remain a very important supplier of energy-related equipment to the Soviet Union. Japan's unique geographical proximity to the East Siberian energy development projects ensures a continuing Japanese role in Siberian energy development. Over the last 5 years,

Japan has ranked first among all Western nations as an exporter of such equipment to the U.S.S.R. A few industrial sectors play a dominant role in this trade—and Japanese business leaders from those sectors have traditionally taken the lead in trade negotiations with the U.S.S.R. **In the last analysis, Japan may be much more important as a supplier of energy-related equipment to the Soviet Union than the U.S.S.R. is to the continued dynamism of Japanese trade worldwide.** Japan's energy-related exports to the U.S.S.R. make up only a tiny portion of total Japanese exports to all nations worldwide, but Japan is the largest supplier in dollar value of energy equipment to the U.S.S.R. Any policies aimed at affecting the volume or nature of Western energy-related exports to the U.S.S.R. must necessarily take into consideration the role of Japan.

When the variety of economic, energy, and political factors influencing Japanese-Soviet energy relations are weighed, the result is a general Japanese orientation that favors expanded energy and trade interaction. The potential gains—in increased exports, diversified energy supplies, and political signals to Moscow that Japan is committed to peaceful coexistence in Asia—appear normally to outweigh the persisting political disputes and the technical and financial constraints on joint energy development efforts with the Soviet Union. Only under extraordinary circumstances would Japanese leaders support proposals for a policy of embargo or leverage against the U.S.S.R. From the Japanese perspective, the benefits of expanded but limited energy and trade relations with the Soviet Union clearly counterbalance the potential risks.

CHAPTER 12

West European-Soviet Energy Relations

CONTENTS

LIST OF TABLES

LIST OF FIGURES

West European-Soviet Energy Relations

In Western Europe, growing energy interdependence with the Council for Mutual Economic Assistance (CMEA)—the Communist world equivalent of the Common Market—is a fact of life. It is generally viewed as a natural and desirable extension of historic trade patterns, which is not disadvantageous so long as it is kept within prudent bounds. Evaluations of what constitutes a "reasonable" or "dangerous" level of East-West interdependence hinge on a variety of factors, including the availability of alternative export markets and energy supply sources. Controversy over West European-CMEA energy relations has grown among observers—primarily on this side of the Atlantic—who weigh these factors differently.

This chapter explores the dimensions of West European energy relations with CMEA, and particularly with the Soviet Union. It focuses on four Western nations—the Federal Republic of Germany (FRG or West Germany), France, Italy, and the United Kingdom (U.K.), examining first past trends in both energy-related equipment exports to, and energy commodity imports from, the U.S.S.R. The chapter then turns to the proposed West Siberian gas export pipeline, the controversial project which will significantly increase Soviet gas exports to much of Western Europe. "Yamburg," the popular name for this project, which will transport gas initially from the Urengoy field in West Siberia, is used here to illuminate the differing perspectives of these Western nations on increasing East-West trade and energy interdependence, including the costs and benefits associated with such interdependence.

Photo credit: Oil and Gas Journal

56-inch gas pipeline being laid in West Siberia

INTRODUCTION

Trade between Western Europe and CMEA constitutes a relatively small, but growing, part of Western Europe's worldwide trade. Between 1972 and 1978, the percentage of European Economic Community (EEC) exports that went to CMEA grew from 2.9 to 3.7 percent. Similarly, commodities from the CMEA constituted 2.9 percent of EEC imports in 1972. These increased to 3.2 percent in 1978.[1] Energy and energy-related equipment and technology have made up an important portion of this trade.

Western Europe has been importing energy from CMEA for decades, but during the course of the last 10 years these imports—particularly those from the U.S.S.R.—have grown. Exports of Soviet oil to Western Europe rose from a level of 33.8 million metric tons (mmt) or 0.678 million barrels per day (mbd) in 1971 to 54.8 mmt (1.1 mbd) in 1979.[2] While exports of coal from the Soviet Union to Western Europe fell during the same period, gas exports grew exponentially from 0.005 to 0.06 billion cubic meters (bcm) per day.[3] In 1980, Soviet gas exports to the region were estimated at 24.5 bcm.[4] This is the equivalent of 20 million tons of oil equivalent per year (mtoe/yr) or 0.40 million barrels per day of oil equivalent (mbdoe). In short, energy trade between the CMEA and Western Europe has been one of the most dynamic sectors of East-West trade during the last decade.

These trends are even more striking when the energy situation of Western Europe is compared with that of Eastern Europe. Western Europe's dependence on imported energy is more than twice as high (54 percent) as that of Eastern Europe (23 percent) (see ch. 9). Despite the fact that Western Europe's *overall* dependence on imported energy has declined since the first oil crisis in 1973-74, imports of energy from the Soviet Union have been increasing.

Total Western exports to CMEA have also increased over the last decade. Historically, Western Europe's exports to CMEA, unlike those of the United States, have been heavily concentrated in industrial goods. Energy-related equipment and technology have been an important part of this trade. As was shown in chapter 6, the U.S.S.R.'s largest Western supplier of energy-related technology and equipment has been Japan, but West Germany is a close second. Between 1975 and 1979, Japan captured nearly 29 percent of this market; West Germany, 28 percent; Italy, 15.7 percent; France, 13 percent; the United States, 8 percent; and the United Kingdom, 0.2 percent.

Moreover, as table 1 demonstrates, these energy-related items constituted over one-third of Italy's exports to the U.S.S.R., about one-quarter of France and West Germany's and 10 percent of Britain's. Thus, while overall trade between Western Europe and CMEA remains a relatively small part of Western Europe's worldwide exports, energy equipment and technology make up a significant part of those exports which do go to the U.S.S.R..

There is disagreement about the significance of West European energy and industrial trade with the Soviet Union. On the one hand, proponents of interdependence

---

[1]International Monetary Fund, *Direction of Trade Yearbook*, 1979, pp. 60-61.

[2]Central Intelligence Agency, *International Energy Statistical Review*, ER IESR 81-003, Mar. 31, 1981, p. 25. (Western Europe here is defined as France, Italy, Finland, the Netherlands, Italy, Sweden, and West Germany.)

[3]Coal exports to EEC from the U.S.S.R. fell from 448,000 metric tons in 1970 to 3,700 in 1975, and to 2,800 in 1979. Data for 1970 and 1975 is from Soviet trade statistics. After 1976, however, Soviet data records coal exports in ruble value rather than in volumes. Data for 1979 come from Business Information Display, *World Energy Industry*, vol. I, No. 3, first quarter, 1980, p. 195. It should also be noted that coal imports to EEC from Poland rose substantially until recent months. In 1979 EEC imported 15 mmt of hard coal from Poland (out of total hard coal imports of 81 mmt). During the same year, however, 18 mmt of hard coal were exported by EEC. Ibid., p. 313.

Data on gas exports from Ibid., p. 26. Here, Western Europe consists of Austria, Finland, France, Italy, and West Germany.

[4]Jonathan P. Stern, *Soviet Natural Gas Development to 1990* (Washington, D.C.: Heath, 1980), p. 99.

Table 84.—Western Trade With the CMEA—1979 (million U.S. dollars)

	United States	Japan	France	West Germany	Italy	United Kingdom
A. Energy-related exports to U.S.S.R. (see table 1)	237	1,097	474	906	408	90
B. Total exports to U.S.S.R.	3,607	2,442	2,005	3,619	1,217	889
A/B	6.5%	44.9%	23.6%	25.0%	33.5%	10.1%
C. Total exports to CMEA-6 + U.S.S.R.	5,672	3,243	4,028	11,270	2,633	2,059
D. Total exports to world	181,801	102,802	97,981	174,092	72,123	90,810
C/D	3.1%	3.1%	4.1%	6.4%	3.6%	2.2%

SOURCE: *Ch. 6; and OECD, Statistics of Foreign Trade.*

argue that growing trade and energy relations between the East and West should be encouraged. Adherents of this position in all major political parties in Western Europe advance both economic and political reasons for such interdependence being ultimately beneficial to both sides. Economically, it contributes to expanded energy supplies and produces new trade opportunities. The countries of Western Europe need imported energy and, perhaps even more importantly, wish to export, particularly steel pipe. Politically, such trade is expected to lock the U.S.S.R. into long-term economic relationships which will give it a stake in maintaining the political status quo and increase the chances of its moderating its policies—toward Berlin, for instance.

On the other side, critics of interdependence fear it will lead to heightened reliance of Western countries on the Soviet Union for both energy and export markets, rendering these nations more susceptible to pressures from the East. The inevitable result as economic ties are strengthened becomes the "Finlandization" of Europe—i.e., the moderation of West European policy to placate Soviet demands.

These opposing views provide the foundation for dramatically different policy proposals, one aimed at promoting expanded interaction and the other at controlling it. A prime example of such policy disputes has been the controversy over the proposed 5,000-km natural gas pipeline, which would bring 40 to 70 bcm of gas (32.9 to 57.3 mtoe or 0.66 to 1.15 mbdoe) annually from Soviet West Siberia to Western Europe. This project is the largest and most recent in a series

of gas deals between the U.S.S.R. and Western Europe. It could more than double the present level of imports (24.5 bcm). For Western Europe, the proposed pipeline offers prospects for significantly expanded gas supplies, as well as for exports of energy-related equipment and technology. At the same time, however, the deal would mean that Western Europe's dependence on Soviet energy would increse.

Assessing either the economic benefit to Western Europe or the potential degree of dependency which this project raises is hampered by both practical and conceptual problems. The most fundamental of these is that much of the readily available energy trade data have not been standardized to allow analysis of CMEA-West Europe flows. Since 1976, for instance, the Soviet Union has recorded only the ruble value—not the volumes—of energy exports.

Policy debate about interdependence is also confounded by definitional issues. "Western Europe" is sometimes used as a shorthand for any of a variety of multilateral organizations: the European Economic Community (EEC), the Organization for Economic Cooperation and Development (OECD), the International Energy Agency (IEA), or the Economic Commission for Europe (ECE—part of the United Nations). Each of this confusing array of organizations has a slightly different list of members. In 1979, for example, Soviet oil accounted for 7.2 percent of all oil and oil product imports by the European OECD, and of 6.7 percent of such imports by the EEC. During the same year, Soviet oil provided 0.2 percent of all oil imported by the total OECD membership (which includes the United States, Japan, and their nations).[5] Similarly, the magnitude of Communist world coal exports changes, depending on whether one includes exports from Poland. If West German im-

ports of Polish hard coal are added to those from the Soviet Union, its dependence on CMEA coal in 1979 rises from 2.4 to almost 30 percent of total coal imports.

There are also different ways to calculate "energy dependence." It can be seen from table 85 and figure 29 that, except in the case of Italy, levels of dependence are higher

[5]OECD, *Quarterly Oil Statistics 1979*, No. 1, 1980, Paris, pp. 248 and 188; *World Energy Industry*, pp. 195 and 260.

Table 85.—Western Energy Dependence, 1979
(million tons of oil equivalent)

	Oil and oil products	Gas	Hard coal	Nuclear	Geothermal, hydro and imported electricity	Total energy
Federal Republic of Germany						
A. Total energy requirements[a]	145.4	49.6	83.3	3.4	1.6	283.3
B. Total energy imports from world	150.8	33.5	6.0	—	—	190.3
C. Total imports, from U.S.S.R.	9.3	8.0	0.1	—	—	17.4
D. Imports from U.S.S.R. percent of total imports	6.2%	23.9%	1.7%	—	—	9.1%
E. Imports from USSR as percent of total energy requirements	6.4%	16.1%	.1%	—	—	6.1%
France						
A. Total energy requirements	117.8	22.7	34.2	3.2	7.0	184.9
B. Total energy imports from world	139.1	16.1	20.2	—	1.4	176.8
C. Total imports from U.S.S.R.	6.5	1.6	0.5	—	—	8.6
D. Imports from U.S.S.R. as percent of total imports	4.7%	9.9%	2.5%	—	—	4.9%
E. Imports from U.S.S.R. as percent of total energy requirements	5.5%	7.4%	1.5%	—	—	4.7%
Italy[b]						
A. Total energy requirements	93.4	24.1	10.6	0.2	4.2	132.5
B. Total energy imports from world	120.9	13.2	9.1	—	0.6	143.8
C. Total imports from U.S.S.R.	6.8	5.7	0.6	—	—	13.1
D. Imports from U.S.S.R. as percent of total imports	5.6%	43.2%	6.6%	—	—	9.1%
E. Imports from U.S.S.R. as percent of total energy requirements	7.2%	23.7%	5.7%	—	—	9.9%
United Kingdom						
A. Total energy requirements	90.3	43.2	87.2	2.8	0.5	224.0
B. Total energy imports from world	70.4	8.2	3.0	—	—	81.6
C. Total imports from U.S.S.R.	2.9	—	—	—	—	2.9
D. Imports from U.S.S.R. as percent of total imports	4.1%	—	—	—	—	3.6%
E. Imports from U.S.S.R. as percent of total energy requirements	3.2%	—	—	—	—	1.3%

[a]Total energy requirements is similar but not identical to apparent consumption—observed consumption data is used where available for coal and natural gas. Otherwise requirements are computed by the following formula: domestic primary production + imports - exports - international bunkers - inventory changes. Total energy requirements are computed only if inclusive of all commodities (oil, gas, coal, primary electric power, and net electricity imports). "Other Electricity" includes net electricity imports. Graphs of total energy requirements do not account for inventory changes if production and import data are separated.
[b]Italy reexports imported energy.
SOURCE: Business Information Display, *World Energy Industry*, Vol. 2, No. 3, First Quarter, 1980; *Petroleum Economist, Natural Gas Across Frontiers*, December 1980; *Petroleum Intelligence Weekly*, July 21, 1980.

Figure 29.—West European Energy Imports From U.S.S.R. and World—1979
(million tons of oil equivalent)

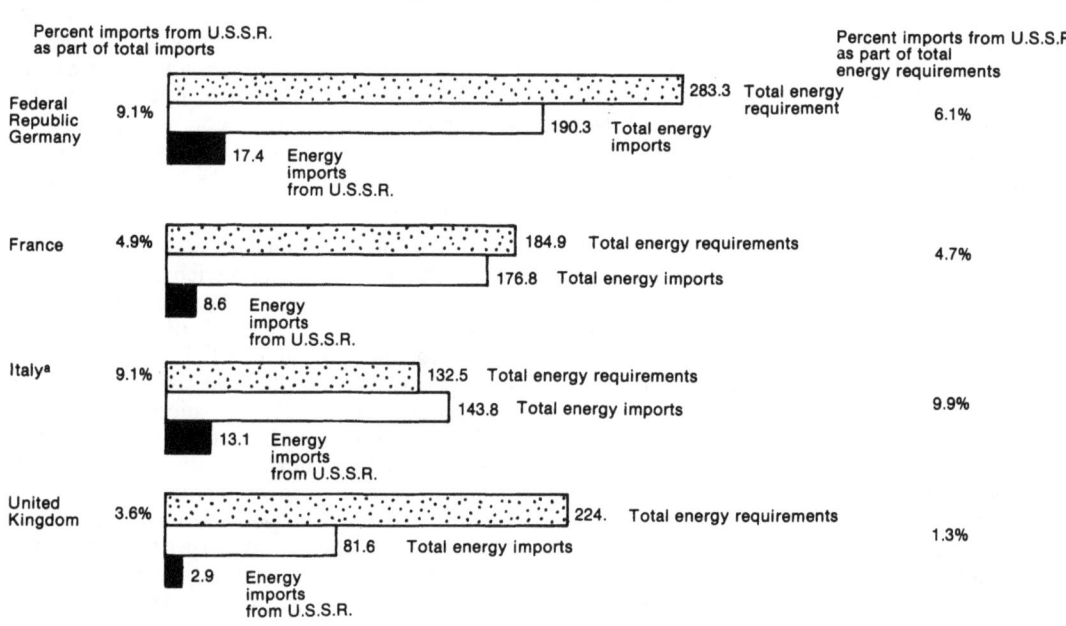

^aItaly reexports imported energy.
SOURCE: Table 85.

if one looks at Soviet energy *imports* as a percentage of all energy imports to each nation (fig. 29) than if one looks at energy imports as a percentage of a nation's total energy *requirements* (table 85). Italy purchased 43 percent of its imported gas from the U.S.S.R. in 1979, but the Soviet Union provided less than 10 percent of the total energy that Italy apparently consumed. The data thus provide no unambiguous indicators of levels of risk or dependence. Individuals make risk assessments based on their perceptions of vulnerability and their judgments about energy supply alternatives—factors which cannot be precisely measured.

WEST GERMANY

Over the past decade, West Germany has become a major Western supplier of energy equipment to the U.S.S.R. Meanwhile, the Soviet Union has become a major source of natural gas for West Germany, and for political and economic reasons, Bonn is officially committed to greater energy interdependence with CMEA. The following sections outline past and present patterns in West German energy and trade relations with the U.S.S.R..

ENERGY COOPERATION BETWEEN WEST GERMANY AND THE U.S.S.R.

There are three important explanations for West Germany's interest in energy cooperations with the U.S.S.R. These relate to the energy, economic, and political realms. First, and perhaps most important, West Germany imports 67 percent of its energy, and is anxious to diversify sources of energy supply (see table 86). In 1979, 51 percent of the FRG's total energy requirements came from oil; 29 percent from coal; 17 percent from gas; and less than 2 percent from nuclear and other energy sources. Even though West Germany produced nearly all of the coal and more than a third of the natural gas it consumed in 1979, it remained dependent on imports for 96 percent of its oil and 62 percent of its natural gas.[6]

Official government energy forecasts show West German oil imports remaining level and thus declining in importance within the overall energy balance during the next decade. These plans require a doubling in nuclear power production between 1979 and 1984, and a quadrupling over the decade.[7] Although politicians from all three parties have expressed support for nuclear power, opposition from the left wing of the Social Democratic Party (SPD), segments of the Free Democratic Party (FDP), and various environmental groups has slowed construction of additional nuclear power plants. Oil will also be replaced with natural gas and coal, augmented by intensified conservation measures.

Soviet energy—particularly natural gas— offers an alternative for West German

energy planners anxious to reduce dependence on Middle East oil. Currently West Germany imports the bulk of its natural gas from the Netherlands; the Soviet Union is the second-largest supplier, and Norway the third. As figure 29 shows, energy imports from the Soviet Union made up 9 percent of West Germany's total energy imports in 1979 and 6.1 percent of its total energy requirements.

The second reason for West German interest in energy relations with CMEA is that the FRG depends on foreign trade for 30 percent of its gross national product (GNP). West Germany is the U.S.S.R.'s most important Western supplier of machinery and equipment, especially chemical equipment, and the largest supplier of high technology.[8] West German exports to the U.S.S.R., like those of Japan, are concentrated in a few key industries. Between 1973 and 1979, half of its large-diameter pipe exports went to the U.S.S.R.; indeed, pipe exports were the largest single export item in West German-Soviet trade during the period.[9] Moreover, specific firms are strongly involved in East-West trade. The giant steel firm Mannesmann exports 60 percent of its total produc-

---

[6]Jochen Bethkenhagen, "Energy Policy of the Federal Republic of Germany," paper presented at the NATO Colloquium, 1980, p. 11.

[7]"Germany," in International Energy Agency (IEA), *Energy Policies and Programmes of IEA Countries* (Paris: OECD, 1980), pp. 115-121.

[8]John Young, "Quantification of Western Exports of High Technology Products to Communist Nations" (Washington, D.C.: U.S. Department of Commerce, 1977).

[9]Deutsches Institut Für Wirtschaftsforschung, *Wochenbericht*, No. 15 (Berlin, 1980). Exports of large-diameter pipe made up 12 percent of all West German exports to the U.S.S.R. during that period.

Table 86.—Federal Republic of Germany—Energy Balance, 1979

	Oil	Gas	Coal	Nuclear	Geothermal hydro and imported electricity
Total energy requirements:					
	51.3%	17.5%	29.4%	1.2%	.6%
283.3 MTOE	145.4	49.6	83.3	3.4	1.6
Energy imports:					
—as percent of total energy requirements	53.2%	11.8%	2.1%	—	—
190.3 MTOE	150.8	33.5	6.0		
Energy exports:					
17.2 MTOE					

SOURCE: Business Information Display, *op. cit.*

tion of large-diameter pipe to the U.S.S.R.; Salzgitter, the other large West German steel producer, sells 40 percent of steel exports to CMEA nations. Thus, while East-West trade represents only a small portion of FRG's total worldwide trade, in energy-related equipment, trade with the Soviet Union is disproportionately significant. Some experts estimate that 92,000 West German jobs are dependent on trade with the Soviet Union, and 220,000 on trade with CMEA.[10]

The third dimension of West German-CMEA energy interdependence is political. The ruling SPD-FDP Party coalition officially supports East-West trade as an incentive for detente and as a basis for long-term political interdependence which reduces the likelihood of conflict between East and West.[11] The West German Government has, moreover, specifically stated that it is in the interest of the West to assist the U.S.S.R. in developing its natural resources.[12] The 1978 Soviet-West German economic cooperation agreement contains a section committing both sides to cooperate in joint energy development, and in May 1980, official spokesmen from the two nations agreed to intensify such efforts. Support for this position, among businessmen and others, is widespread.

FRG-SOVIET ENERGY TRADE, 1970-80

FRG Equipment and Technology Exports

The FRG has been exporting energy-related equipment to the Soviet Union for more than 30 years, very often as part of compensation deals in which the exports are paid for in other goods or commodities. In recent years, a pattern of growing equipment exports linked to increasing imports of

Soviet energy has evolved. The greater part of FRG exports has been in pipe, compressor stations, and pumps for pipelines. Additionally, West German firms have supplied equipment for petrochemical plants in the CMEA region. Added together, these categories of energy-related exports make up about 25 percent of all West German exports to CMEA (see table 84).

West German energy equipment trade with the U.S.S.R. first came to prominence in 1963, when three large West German steel companies signed a contract to supply the U.S.S.R. with 163,000 tons of 40-inch steel pipe, valued at $28 million. In November 1962, a North Atlantic Treaty Organization (NATO) resolution, passed at U.S. initiative, embargoed all sales of large-diameter pipe to the Soviet Union. The intention of the United States was to impede the completion of the Friendship oil pipeline from the U.S.S.R. to Eastern Europe. While the governments of some other West European nations refused to cooperate, the FRG compelled its firms to cancel their contracts. This understandably provoked considerable outcry from the West German business community. One company, Phoenix-Rheinrohr, was forced to close one of its plants, and others suffered substantial financial losses and were forced to cut back capacity.

Since the United States did not have as great political leverage over its other allies as it did over Bonn at the time, it was unable to prevent Great Britain, Italy, or Japan (which was not a member of NATO) from selling similar pipe. Ultimately, the U.S.S.R. found alternative suppliers and completion of the pipeline was delayed by only 1 year. Moreover, many contend today that the embargo forced the U.S.S.R. to develop its own pipemaking capability. This episode still rankles. West German leaders still recount it and regard it as misguided.[13] It remains an important event in the shaping of West German opinion on the utility of economic sanc-

[10]Deutsches Institut Für Wirtschaftsforschung, *Wochenbericht*, No. 13 (Berlin, 1981).
[11]Statement by Chancellor Helmut Schmidt, Feb. 28, 1980, quoted in Bundesminister für Wirtschaft, *Der Deutsche Osthandel 1980* (Bonn: June 1980), p. 62.
[12]Ibid., p. 19.

[13]Angela E. Stent, *From Embargo to Ostpolitik: The Political Economy of West German-Soviet Relations, 1955-1980* (New York: Cambridge University Press, 1981), ch. 5.

tions and the ineffectiveness of such sanctions in preventing the acquisition of technical capabilities in the U.S.S.R. Since then, pipeline equipment has played an increasingly important role in West German-Soviet trade.

In 1969, 3 years after the NATO embargo was lifted, the U.S.S.R. concluded a $25 million contract with the West German firm Thyssen for joint construction of pipe-making factories in the U.S.S.R. and the FRG. The pipes were used to transport gas from the Soviet Union to Eastern Europe. This contract marked the beginning of intensified West German exports of energy-related equipment to the U.S.S.R.

The first major exchange of West German equipment for Soviet energy was arranged in February 1970. West German banks, steel, and gas firms were all involved in the deal, which provided for the sale of 1.2 million tons of pipe, worth $400 million, as well as increased imports of natural gas. A consortium of 17 West German banks under the leadership of the Deutsche Bank supplied the credits at an undisclosed, but reportedly very favorable, rate of interest. By charging high prices for the pipe they sold, steel firms were evidently able to help to compensate the banks for losses suffered in interest rate charges. This deal assisted the ailing West German steel industry, which, as noted above, is highly dependent on foreign trade. The exchange was viewed as a success by both sides; it was followed by similar contracts for West German pipe arranged in 1972 and 1974.

Another important gas project was the "triangular" West European-Soviet-Iran contract signed in 1975. Here, West European equipment was sold to Iran, Iranian gas to the Soviet Union, and Soviet gas to Western Europe. In 1977, the deal was enlarged to include construction of the IGAT-II, a 1,440-km pipeline designed to carry gas from fields in southern Iran to the town of Astara on the Soviet border (see fig. 30). France, West Germany, and Austria were scheduled to receive a combined total of

11 bcm of this gas in exchange for supplies of equipment and cash.[14] The project has now been abandoned.

West German firms also took a leading role in the Orenburg pipeline project. The 2,750-km Orenburg or "Soyuz" pipeline is a joint project involving the U.S.S.R. and the CMEA-6, which will supply the bulk of additional Soviet gas exports to Eastern Europe over the next decade. After only 2½ years of construction, the pipeline was completed on schedule in 1978, and heralded as a prime example of CMEA cooperation. West German firms such as Mannesmann supplied pipe and other ancillary equipment, including drill pipe, casing, and equipment for 22 compressor stations. Japanese, French, American, and Italian companies were also involved.

The possibility of exporting West German equipment for Soviet nuclear power stations has been under discussion for some years. One project, initially discussed by the U.S.S.R. and the FRG in 1975, eventually fell through. The German Kraftwerkunion was to construct a 1,200-MW nuclear power-plant at Kaliningrad in the U.S.S.R. at a cost of $600 million. One major point of contention was the West German stipulation that some of the power produced by the plant should supply West Berlin. West Berlin's energy situation is precarious—the city is not connected to the electrical grids of either East or West Germany. The project failed, not only because the Soviets asked a very high price for the electricity, but also because East Germany objected to transmitting electricity to West Berlin.[15] Some West German firms have supplied equipment for Soviet powerplants, and while discussions of prospects for new projects surface periodically, FRG Government spokesmen say that there is little likelihood of West German nuclear plant exports to the U.S.S.R. in the immediate future.

[14]Stern, op. cit., p. 79.
[15]Walter B. Smith, "Securing Energy for West Germany and West Berlin," paper presented at Department of State Executive Seminar on National and International Affairs, 1978-79, pp. 13-15.

Figure 30.—Gas Pipeline Feeding Western Europe

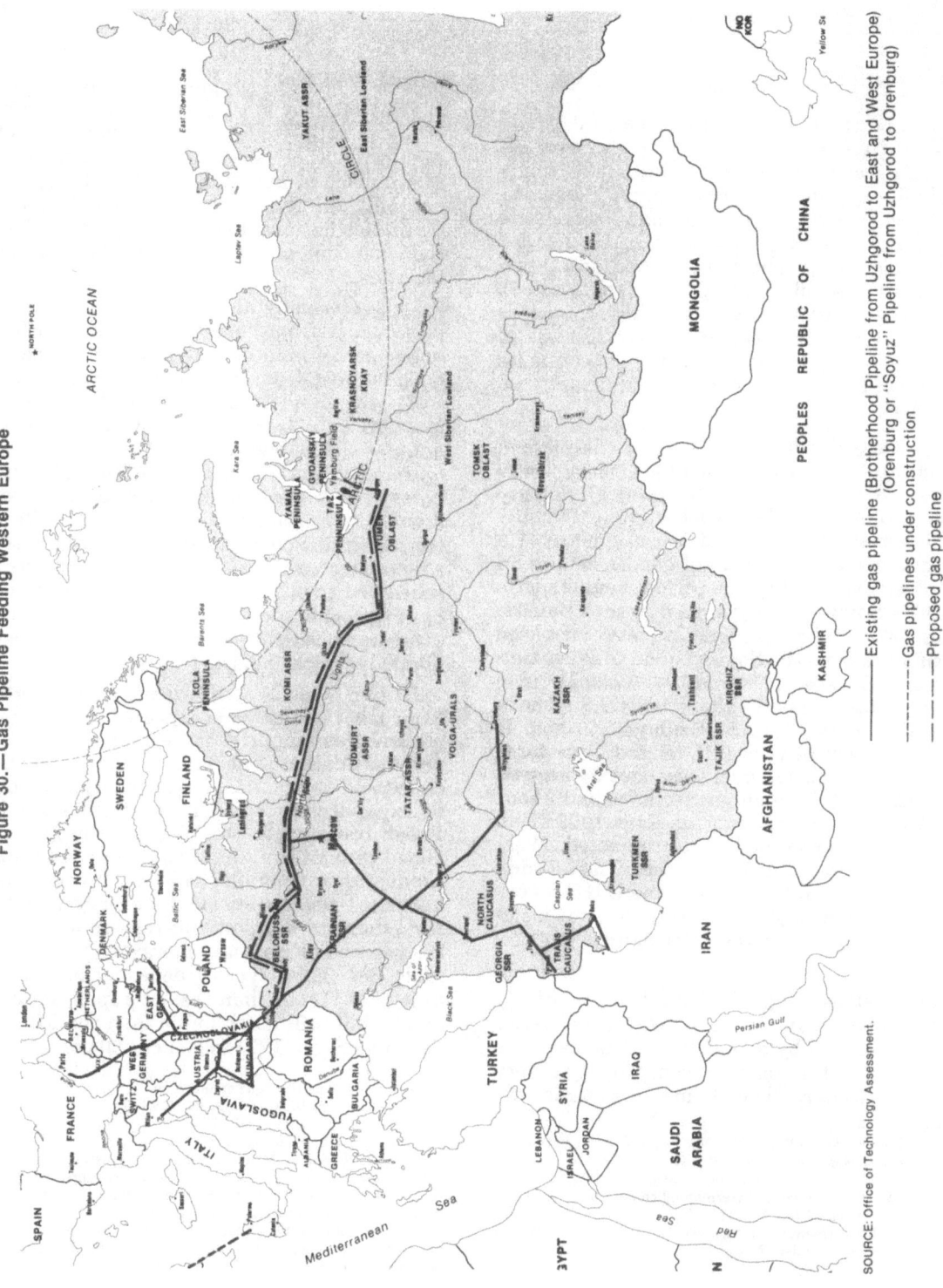

——— Existing gas pipeline (Brotherhood Pipeline from Uzhgorod to East and West Europe)
      (Orenburg or "Soyuz" Pipeline from Uzhgorod to Orenburg)

- - - - - Gas pipelines under construction

— — — Proposed gas pipeline

SOURCE: Office of Technology Assessment.

FRG Energy Imports

Imports of Soviet oil and natural gas to West Germany have grown steadily over the last decade, and have become increasingly significant in the FRG's energy balance. Soviet exports of crude oil and products to the FRG rose from 6.6 mmt in 1972 to 9.3 mmt in 1979,[16] about 0.187 mbd or 6.2 percent of total oil consumption[17] (see table 85). However, the West German Government is cognizant of the declining rate of growth of Soviet oil production and is not expecting its level of Soviet oil imports to significantly increase.

Clearly the most promising area for potential Soviet hydrocarbon imports is natural gas. When the first gas-pipe contract was signed in 1970, the Soviet trading company Soyuzneftexport agreed to supply German Ruhrgas with 51.5 bcm of gas (equal to 41.8 mtoe) over a 20-year period beginning in 1973. Soviet gas began to flow into Bavaria in 1973. Subsequent deals have increased supplies from the Soviet Union to West Germany. Between 1974 and 1979 West German natural gas imports from the U.S.S.R. have grown from 2.1 to 9.8 bcm/yr (9.8 bcm is about 8.0 mtoe/yr or 0.16 mbdoe).[18] As table 85 shows, this represented about 24 percent of all West German gas imports and about 16 percent of gas consumed in 1979. Had IGAT II come on line as projected, West Germany might have been receiving an additional 5.7 bcm of gas from the U.S.S.R. Imports of Soviet gas have thus become a significant part of West Germany's gas imports.

Soviet gas supplies have been relatively dependable. In 1979, however, deliveries were reduced by as much as 25 percent due to technical problems accompanying a very severe winter. Those living in Bavaria, the region primarily supplied by Soviet gas, have apparently not been adversely affected by these supply cuts due to substitution of gas from other parts of West Germany.[19] The impact of any future cuts will depend not only on the region involved, but the nature of the consumer (industry or residential), the time of year and available storage facilities.

West Germany also imports CMEA coal, particularly from Poland. In 1979, the FRG imported 2.4 mmt of hard coal (1.65 mtoe or 0.03 mbdoe) from Poland and 0.210 mmt from the Soviet Union. Together these amounted to 30 percent of all coal imports, or about 2.1 percent of West German coal requirements for the year.[20] The FRG also imports substantial amounts of brown coal (lignite) from Eastern Europe, and the West German and Polish governments are jointly supporting a coal gasification project scheduled to produce nearly 1 bcm/yr of gas by 1983. In addition, 55 percent of the uranium used in West German nuclear plants is enriched in the U.S.S.R.[21]

In sum, while West Germany's oil imports from the U.S.S.R. are expected to decline, natural gas imports are expected to rise, perhaps doubling in volume by the end of the decade.[22] There appears to be little likelihood of expanded imports of coal from Poland; Polish coal exports to Western Europe declined dramatically in 1980 due to faltering production.[23] Overall, the FRG now imports about 6.1 percent of all the energy it uses from the Soviet Union; this represents about 9.1 percent of its total energy imports (see table 85). This level of dependence is significantly higher than that of Japan, for example, but still much lower than the FRG's dependence on OPEC oil (about 60 percent).

[16]Marshall Goldman, *The Enigma of Soviet Petroleum* (Boston: George Allen & Unwin, 1980), p. 64.

[17]The 1979 imports from the total CMEA region to the FRG made up 8 percent of German oil consumption. See *Der Deutsche Osthandel 1980*, p. 19.

[18]U.S. Department of Energy, *World Natural Gas 1978*, and Petroleum Economist, *Natural Gas Across Frontiers*, December 1980.

[19]Interview with Ruhrgas spokesman.

[20]*World Energy Industry*, op. cit., p. 78.

[21]*Der Deutsche Osthandel*, 1980.

[22]Stern, op. cit., p. 105.

[23]*New York Times*, May 21, 1981, p. 1, Business Section. "Western governments and banks alike are concerned that in recent months the Soviet Union has been receiving a growing proportion of Poland's faltering coal production, always considered the country's best collateral. While the West used to get two out of every 3 tons of coal shipped out of Poland, it now receives only about half, bankers say."

FRANCE

Like the Federal Republic of Germany, France has been interested in promoting greater interdependence with the Soviet bloc. Since the Presidency of General de Gaulle, French leaders have sought to maintain a foreign policy stance distinct from that of the United States. In addition, pursuit of detente with the Soviet Union has long been a major policy goal.

French energy-related interaction with the Soviet Union is, however, tempered by two factors. First, while France's dependence on imported energy is much higher than that of the FRG, French energy planners have in the past committed themselves to a very ambitious nuclear power program. The fate of this program has now been called into question as the new Mitterrand government has placed a moratorium on new nuclear reactor construction, but if targets were met, nuclear power, which made up only a miniscule portion of total energy requirements in the 1970's, would provide 30 percent of total energy by the year 1990. This has now been scaled down to 21 percent.[24] (See tables 86 and 87 for a comparison of French and West German energy balances in 1979.) Even this would minimize—theoretically at least—the need to seek alternative foreign energy suppliers. Secondly, French political relations with the U.S.S.R. have not been as sensitive or complicated as those of West Germany. There is no French Berlin; nor are Soviet troops stationed on French borders.

ENERGY COOPERATION BETWEEN FRANCE AND THE U.S.S.R.

French energy relations with CMEA are, like those of West Germany, informed by energy policy, more general economic, and political concerns. France is seeking to change its energy balance so as to reduce its dependence on imported oil. In 1979, 63.8

percent of France's total energy requirements were provided by oil (crude and products); 12.3 percent by gas; 18.6 percent by coal; and about 5.3 percent by nuclear, hydropower, and other sources. At the time of the first oil shock, only Italy among the EEC nations had a higher oil dependence. Since that time French energy policy has operated with the assumption that such dependence creates national security risks. Hence its emphasis on nuclear power.

France has a tradition of strong state intervention in the economy, which has provided a supportive context for strong guidance of energy industries.[25] There are virtual state monopolies in electricity supply, coal mining, and gas sales, and extensive regulation of the oil sector. Official government statements recently reflected a strong resolution to develop nuclear power: "France has no viable alternative to nuclear power other than economic recession and dependence."[26] The vigorous nuclear program is supported by the huge Eurodif uranium enrichment plant, which started production in 1979. It was envisaged that by 1985 49 reactors would provide 50 percent of French electricity.[27] (In August 1980, there were 19 French reactors in operation with a capacity of 10,000 MW.) The French Communist Party, and the Communist-dominated trade union federation CGT, have supported this direction. Socialists and environmental groups, however, have opposed nuclear power in the past and the Mitterrand government now appears to be formulating a policy which compensates for reduced growth in nuclear power with coal and massive investment in conservation and alternative energy sources.

France's attempts to reduce dependence on oil have been accompanied by stress on

---

[24]Ministère de l'Industrie, *The Energy Policy of France* (Paris: 1980).

[25]See Robert J. Lieber, "Energy Policies of the Fifth French Republic: Autonomy Versus Constraint," in C. Andrews and S. Hoffman (eds.), *The Fifth French Republic* (New York: State University of New York Press, 1980).

[26]Ministère de l'Industrie, op. cit., p. 7.

[27]Nicholas Wade, "France's All-Out Nuclear Program Takes Shape," *Science*, Aug. 22, 1980.

Table 87.—France's Energy Balance, 1979

	Oil	Gas	Coal	Nuclear	Geothermal hydro and imported electricity
Total energy requirements:	63.9%	12.3%	18.6%	1.7%	3.8%
184.9 mtoe.........................	117.8	22.7	34.2	3.2	7.0
Energy imports:					
—as percent of total energy requirements	75.5%	8.7%	10.9%	—	0.7%
176.8 mtoe.........................	139.1	16.1	20.2	—	1.4
Energy exports: 18.0 mtoe					

SOURCE: Business Information Display, op. cit.

diversifying geographical sources of oil supply. In 1979, Saudi Arabia and Iran supplied more than 40 percent of France's crude oil imports. As table 85 shows, France relied on Soviet oil and products for less than 5 percent of all imports in this category during 1979 and for some 5.5 percent of oil and products consumed in that year. Soviet gas amounted to 9.9 percent of total gas imports and 7.4 percent of gas consumption. While gas from the U.S.S.R. could contribute to a reduction in OPEC oil imports, spokesmen in the previous French Government have expressed concern that such imports not rise too quickly.[28] (More than 60 percent of French gas supplies come from the Netherlands.) Soviet coal constitutes an even smaller share of French coal consumption. While, overall, France is very dependent on imported energy, supplies from the Soviet Union make up less than 5 percent of its total fuel imports and total energy consumption (see table 85).

French leaders are perhaps more interested in energy cooperation with the U.S.S.R. because of the export possibilities it raises. East-West trade makes up only a very small share of French trade worldwide, but certain industrial sectors such as steel have important stakes. France was the third largest supplier of energy-related equipment and technology to the U.S.S.R. during the 1975-79 period, and this trade has become increasingly important (see table 84).

[28]Interview with French Foreign Ministry officials.

The political incentives for energy cooperation with the U.S.S.R. relates to French desire to strengthen detente. Both business and government leaders in France have viewed trade with the U.S.S.R. as normal and desirable. This attitude was reflected in French reluctance to participate in the post-Afghanistan economic sanctions initiated by the United States. In 1980, the French steel firm Creusôt-Loire won a Soviet contract for a steel sheet manufacturing plant, a contract originally awarded to Japanese and American firms and canceled when participation of the latter was prohibited by the United States on political grounds. France thus appears unwilling to use trade as a political lever in East-West relations, and indeed, the French Foreign Ministry has regarded efforts to assist CMEA—particularly the U.S.S.R.—as necessary and mutually beneficial. By helping the Soviet Union to develop its resources, they say, the Soviet Union will be less likely to extend its influence—military and otherwise—into the Persian Gulf.

FRANCO-SOVIET ENERGY TRADE, 1970-80

French Energy Equipment and Technology Exports

France has never been as important a supplier of energy equipment and technology to the U.S.S.R. as has West Germany, but French exports to the Soviet Union have been significant in certain areas. French steel firms, especially Creusôt-Loire and Vallourec, have shipped large amounts of pipe and plants for pipe manufacturing, and French companies have also sold petrochemical equipment and plants to the U.S.S.R. One French firm recently signed a contract with the U.S.S.R. to produce offshore oil exploration equipment. In many cases the U.S.S.R. has paid for these in shipments of oil or gas.

At the same time that French firms have attempted to enlarge their sales to the Soviet Union, the French government has sought economic ties. The first Franco-Soviet Economic Cooperation Agreement, signed in 1971 and later extended, identified several areas for cooperation, including power generation and the development of new energy technologies,[29] and commissions have been established to discuss mutual research in such areas. These commissions and numerous associated working groups provide a forum where French and Soviet specialists share technical information. Since both industry experts and government officials participate, information about potential contracts and projects is disseminated. French participants have repeatedly demonstrated their reluctance to discuss certain types of energy projects—such as exchange of technical information about nuclear powerplants, or Soviet proposals for joining the European and Soviet electricity grids—at these meetings. There is, however, shared interest in fusion technology, and a French delegation has recently visited a Soviet fast breeder reactor.

In sum, like the West Germans, the French take a positive view of sales of

energy-related equipment. At the governmental level, mechanisms for cooperation in energy technology development have been established and energy trade with the U.S.S.R. is supported by official policies as well as by the informal efforts of businessmen.

French Energy Imports From the U.S.S.R.

Soviet energy has played a modest role in the French energy balance. During the period following the Iranian revolution, the Soviets raised oil prices to France. These increases presented particular problems for companies like French BP, which depends on Soviet crude oil for as much as 25 percent of its supplies. In 1980, the Soviets cut back oil deliveries to France by 15 percent; French Government leaders were particularly concerned about these reductions since they included some 300,000 tons of oil contracted under a compensation agreement.[30] It appears unlikely that Soviet oil exports to France in the next decade will claim a higher proportion of total French oil imports than in 1979.

France has imported gas from the Soviet Union since 1976, although until 1980 Soviet gas was traded for Dutch gas originally destined for Italy. Since then, Soviet gas has flowed directly to France via a pipeline which crosses into Eastern Europe at the Czech border. In 1979, France imported about 1.9 bcm of Soviet gas (1.54 mtoe/yr or 0.03 mbdoe), almost 10 percent of total gas imports and 7.4 percent of the gas consumed during that year. Thus, French reliance on Soviet gas has not been as great as that of West Germany or Italy (see table 85).

---

[29]Axel Krause, "France Slows Down on Soviet Gas Deal," *International Herald Tribune*, Jan. 22, 1981.

[30]*Petroleum Intelligence Weekly*, Jan. 21, 1980.

ITALY

Italy has been importing energy from CMEA since the late 1950's, and Italian policies toward energy relations with the

U.S.S.R. and Eastern Europe resemble those of West Germany and France. Italy's political relationship with the U.S.S.R. is less sensitive than that of West Germany. However, Italy's position is distinguished by the existence of a Communist party, the PCI, with 30 percent of the national vote. Despite its differences with Moscow, the importance of PCI influences Italian-Soviet political and economic relations.

ENERGY COOPERATION BETWEEN ITALY AND THE U.S.S.R.

Italy's dependence on imported energy in general and on oil as a share of its total energy requirements is higher than that of any other Western nation considered in this report. In 1979, oil made up more than 70 percent of total energy consumed, and Italy is well aware of the need to reduce its dependence on oil and diversify its energy suppliers. In late 1979 the Italian Government adopted measures, including higher gasoline prices, to restrain energy demand, but progress in implementing these plans was slow.[31] Official energy forecasts project rising levels of oil imports during the present decade.

The main reason for this projected trend is that nuclear power development has fallen behind plans. Forecasts for nuclear power, which in 1979 contributed less than 1 percent to total Italian energy requirements, have been scaled down dramatically to less than one-half the levels projected a few years ago. If all goes according to this revised plan, nuclear power will contribute about 7 percent of total energy consumed in 1990. Reaching this goal will require 12 new nuclear powerplants, only one of which was under construction in 1980.

A dominant trend in Italy's energy balance has been the rising importance of natural gas. Since the early 1970's, Italian natural gas imports have grown steeply, while domestic production has remained stable.[32] As table 85 indicates, in 1979 Italy imported over 43 percent of its natural gas from the

U.S.S.R., which provided 23.7 percent of Italy's gas requirements in that year. Italian dependence on Soviet gas is currently the highest of any of the Western industrial nations OTA has studied, although Italy plans to lessen this dependence through a trans-Mediterranean pipeline, currently under construction, which will carry Algerian gas.

Italy is facing continuing economic problems, particularly in export competitiveness, and its leaders tend to take a positive view of trade with the Communist world. Italy has been an important supplier of pipe to the U.S.S.R. for many years, and continued exports are important for Italy's steel industry, which, like steel industries in other West European nations, has been experiencing a recession. Italian corporations, with government encouragement, hope to increase cooperation with the Soviet Union in hydrocarbon exploitation and production. In recent years, Italy has had a negative balance of payments with the U.S.S.R. and would like to remedy that situation.[33]

Italy also has a political interest in the promotion of trade with the Soviet Union. While Italian officials are less likely to argue that Western trade with the U.S.S.R. can act to moderate Soviet political ambitions, they view cooperation in energy development as mutually beneficial and overall trade as part of their traditional relationship with the U.S.S.R.

ITALIAN-SOVIET ENERGY TRADE, 1970-80

Italian Energy Equipment and Technology Exports

The U.S.S.R. and Italy signed their first postwar bilateral trade agreement in 1948, and since the early 1960's trade in energy-related equipment has been a major component of Italian exports to the Soviet Union.

---

[31]See IEA, op. cit., p. 139
[32]"Italy," in OECD, *Economic Survey*, March 1980, p. 47.

[33]In 1979 total imports from the CMEA amounted to 5 percent of all imports; exports were valued 3.6 percent of all exports for the year. Department of State, Bureau of Intelligence and Research, "Trade of NATO Countries With Communist Countries, 1976-79," report No. 31-AR, Dec. 1, 1980.

Italian firms have exported large-diameter pipe, refinery and telecommunications equipment, gas turbines, electricity generators, and compressor stations.

Some Italian firms do as considerable a business with the U.S.S.R. as do German companies such as Mannesmann. Finsider, a subsidiary of state-owned IRI, has been selling large-diameter pipe to the U.S.S.R. since 1962, when Italy defied the U.S.-initiated NATO pipe embargo. Under current contracts, Finsider will sell 2.5 million tons of large-diameter pipe and 5,000 tons of steel pipe and special pipe to the U.S.S.R. over the next 5 years.[34] In all, some 25 to 30 percent of the firm's annual production in pipe and other steel products goes to the Soviet Union. Finsider has concluded 5-year agreements that provide for partial payment by the U.S.S.R. in coal, iron ore, and scrap metal. The company issues promissory notes which are subsequently discounted and repurchased, an unusual practice in East-West trade.

There is considerable cooperation between Italy and the Soviet Union in energy development. Finsider is currently considering a proposal to assist in the construction of a coal slurry pipeline from the Kansk-Achinsk basin in Siberia to the Western U.S.S.R. Italian firms such as Nuovo Pignone, a subsidiary of ENI, have been important suppliers of equipment for gas pipelines—compressor and booster stations. ENI has also discussed prospects for cooperation with the U.S.S.R. in offshore oil development.

As with West Germany, there have been discussions for years about joint nuclear power development involving Italy and various CMEA members. In the Italian case, agreement in principle has been reached to build a nuclear power station—in the U.S.S.R. or Czechoslovakia.[35] Since more than 4,000 workers are now unemployed due

to setbacks in Italy's domestic nuclear program, there is strong interest among the Italian corporations involved in such a project. If constructed, the joint Soviet-Italian powerplant would supply some electricity to Italy. Reports also indicate that a consortium (Ansaldi Nucleari), in which both the Italian firm IRI and General Electric are participants, will build two nuclear powerplants in Romania, each with a capacity of 700 MW. Since Romania plans to build more than 10 power stations in the next decade, this contract may provide the basis for additional Italian participation in CMEA nuclear power development.[36] Italy has thus come closer than any other European nation to joint development of nuclear power with CMEA.

In order to encourage Italian exports, the Italian Government, like the French and British but unlike the West German, subsidizes credits to Communist nations.[37] Until 1972, this system of cheap credits evidently worked fairly smoothly, but since that time the consensus on interest has broken down and many Italian firms have criticized the policy of granting low-interest rates to the U.S.S.R. In 1980, previously arranged credits having expired, the Soviet invasion of Afghanistan led to controversy over the renewal of these cheap credits. The Italian Government, heeding U.S. wishes, stopped all negotiation with the U.S.S.R. on the subject, but Socialist and Communist deputies in the Italian parliament sharply criticized Italy's support of the United States. In the end, Italy announced that a new credit agreement would not be extended, but that loans would be granted on a case-by-case basis. This move was unprecedented for Italy, given its past commitment to East-West trade. There may, however, have been domestic economic reasons for halting the credits.

---

[34] *Soviet Business and Trade,* June 15, 1980, p. 4.

[35] Interview with Dimitri Zhimerin, Vice President of Soviet State Committee for Science and Technology, reprinted in *Il Fiorino,* July 29, 1980.

[36] Rupert Cornwell, "Romania to Buy Reactors From Italian-U.S. Group," *Financial Times,* Mar. 3, 1981.

[37] In 1966, for example, Italy extended a $367 million credit with a 14-year maturation period for the construction of a Fiat plant in the U.S.S.R. See Glen Alden Smith, *Soviet Foreign Trade* (New York: Praeger, 1973), p. 166.

Finally, a number of bilateral forums have been established for discussions of energy cooperation with the U.S.S.R. involving government officials, businessmen, and technicians. There have been a number of private joint symposia on Soviet energy development. At one such meeting in Moscow in 1979, a major topic was proposed nuclear cooperation; at another, held in Italy in 1980, new developments in energy technology—including nuclear, geothermal, biomass, and pipeline technologies—were discussed.[38] Government-to-government energy meetings, the most recent of which was held in March 1981, also provide opportunities for exploring potential energy development projects.

Italian Energy Imports From the U.S.S.R.

In 1979, more than 9 percent of all Italy's energy imports and nearly 10 percent of all the energy it used came from the Soviet Union. Since 1958, Italy has been importing Soviet oil under the terms of a series of 5-year agreements. In 1979, these imports amounted to 6.7 mmt (0.135 mbd) of oil and oil products,[39] about 5.6 percent of Italy's oil imports and 7.2 percent of its total oil requirements for that year. During 1976-81, the Italian company ENI alone imported 4 mmt of oil annually (about 80,000 bd) from the Soviet Union. In 1981, however, for the first time, the supply agreement was not renewed. At the same time all Western importers of Soviet oil—with the exception of Finland—were the subjects of delivery cuts, which some attribute to Soviet failure to achieve production targets. Italian experts believe that a new supply agreement will soon be signed and that ENI will receive Soviet oil at about the same level as in the past. Since official energy plans project growing oil imports, the share of Soviet oil is likely to decline as part of total Italian oil imports.

Italy has imported gas from the U.S.S.R. since 1974. In 1979, 7 bcm (5.7 mtoe/yr or 0.11 mbdoe)—over 40 percent of Italy's gas imports and nearly 24 percent of its total gas requirements came from the U.S.S.R. Most of the Soviet gas is imported by SNAM, a state-owned company which is part of the ENI group, and distributed throughout Italy. Soviet gas is evidently competitively priced.

SNAM has arranged a variety of contracts with Libya, the Netherlands, the U.S.S.R. and Algeria, to cover Italy's projected rise in gas requirements over the decade. Beginning in the fall of 1981, Italy will receive Algerian gas via a new submarine pipeline to Sicily. Arrangements have been made for as much as 15.9 bcm (12.9 mtoe) to be supplied to Italy through this pipeline in the event of an emergency.

Italy's official energy plan foresees a rise in gas requirements from about 30 bcm (24.1 mtoe) in 1979 (see table 88) to about 40 bcm in 1985. While domestic production is expected to remain stable at about 12 bcm, imports will rise sharply to about 28 bcm. Informed estimates show that in 1985 Libya will supply about 3, the U.S.S.R. 7, the Netherlands 6, and Algeria 12 bcm.[40] If this does in fact occur, the Soviet Union will be providing about one quarter of Italy's total gas imports in 1985—considerably less than the current percentage.

Until 1981 Italy received stable and dependable gas supplies from the U.S.S.R. These imports came during the early part of the 1970's, primarily from the Ukraine, and later from Urengoy in West Siberia. It was reported that in 1981 Soviet supplies of gas to Italy fell 30 percent—a fact which some attribute to "technical difficulties," and others to rising Soviet exports to Eastern Europe.[41] If such supply shortfalls persist, the U.S.S.R. is unlikely to be able to provide

---

[38]See *Staffetta Quotidinia Petrolifia*, July 2, 1980.

[39]*Petroleum Intelligence Weekly*, July 21, 1980, compiled by Petro Studies.

[40]These import projections, and much of the information in this section, can be found in U.S. Department of State, Outgoing Telegram, Rome, Mar. 28, 1980, "The Italian Petroleum and Natural Gas Industry, 1978-79."

[41]Interview with officials from Italian Ministry of Foreign Trade.

Table 88.—Italy's Energy Balance, 1979

	Oil	Gas	Coal	Nuclear	Geothermal hydro and imported electricity
Total energy requirements:	70.9%	18.3%	8.0%	0.15%	3.2%
132.5 mtoe.........................	93.4	24.1	10.6	0.2	4.2
Energy imports:					
—as percent of total energy requirements	91.7%	10.0%	6.9%	—	0.46%
143.8 mtoe.........................	120.9	13.2	9.1	—	0.6
Energy exports: 22.4 mtoe					

SOURCE: Business Information Display, *op. cit.*

the projected amounts of gas mentioned above.

Italy also imports CMEA coal. In 1979, nearly 7 percent of its hard coal imports and almost 6 percent of its hard coal requirements came from the U.S.S.R. (see table 85). Coal imports from the U.S.S.R. have fallen steadily over the past decade—from more than 2,000 metric tons in 1970 to only 925 tons in 1979.[42] Poland has remained a more important source of Italy's coal, providing 26 percent of its coal imports in 1979. (Italy imports the bulk of its coal from the United States; deliveries from America made up 32 percent of Italian coal imports in 1979.) Since late 1980, coal imports from the Soviet Union have fallen still further, as have those from Poland.

---

[42]A 1970 figure from Soviet trade data; 1979 figure from *World Energy Industry*, op. cit., p. 108.

Italy, like West Germany and France, has used Soviet uranium enrichment services. However, due to delays in the Italian nuclear program, there will be excess capacity in Eurodif, the French enrichment facility in which Italy participates. This means that Italy could easily depend on Eurodif if necessary.

In sum, Italy is more reliant on CMEA for imports of natural gas than the other West European nations examined here. In addition, Soviet oil and coal have played significant roles in its energy imports. Altogether, the U.S.S.R. fulfills nearly 10 percent of Italy's total energy requirements. This comparatively high level of dependence, together with past patterns of Italian exports of pipe and other energy-related commodities, indicate an overall relationship of energy and trade interdependence with the Soviet bloc stronger than that of France and on a par with that of West Germany.

THE UNITED KINGDOM

The United Kingdom has also been favorably disposed toward East-West energy cooperation, but a variety of factors differentiate the British approach from those of the other countries discussed here. Most important among these is the fact that the United Kingdom is more nearly self-sufficient in energy than any of these other na-

tions. Britain's enviable position derives from North Sea oil and gas, discovered in 1969. In 1979, the United Kingdom imported 81.6 mtoe of energy commodities and exported more than 51 mtoe. Its total energy requirements amounted to 224 mtoe (see table 89). Therefore, Britain imported energy to meet only about 13 percent of all its

Table 89.—United Kingdom Energy Balance, 1979

	Oil	Gas	Coal	Nuclear	Geothermal hydro and imported electricity
Total energy requirements:	40.3%	19.3%	39.0%	1.3%	.2%
224 mtoe.........................	90.3	43.2	87.2	2.8	.5
Energy imports:					
—as percent of total energy requirements	31.4%	3.6%	1.3%	—	—
81.6 mtoe.........................	70.4	8.2	3.0		
Total exports: 54.9 mtoe					

SOURCE: Business Information Display, op. cit.

energy requirements. It can afford to be quite distanced from the U.S.S.R. in its role as energy supplier.

A political factor also distinguishes the current British approach. The United Kingdom refused to follow the 1962 NATO pipe embargo, and in the past it has generally pursued a course of separating trade with the Soviet bloc from politics. However, the Thatcher government has been the strongest supporter of U.S. initiatives aimed at using trade sanctions as a lever against the Soviet Union. Dissenters have questioned the efficacy of such an approach, but Britain's political stance has been considerably more distanced than that of the continental Europeans to East-West trade and energy cooperation issues.[43] Britain's weaker involvement with the Soviet bloc is illustrated in the trade data. In 1979, only about 1 percent of all British exports went to CMEA, and energy-related trade represented only about 12 percent of all British exports to the Soviet Union (see table 84).

ENERGY COOPERATION BETWEEN BRITAIN AND THE U.S.S.R.

As noted above, Britain's very limited energy relationship with the U.S.S.R. is primarily determined by its own fortunate

energy situation. The United Kingdom has substantial reserves of oil and gas and enough coal to last 300 years at current rates of extraction. Table 89 shows Britain's 1979 energy balance. Energy use by the United Kingdom is more balanced among a variety of fuel sources than in most other Western industrialized nations, and its dependence on oil is low—about 40 percent of its energy requirements. Coal occupies a position of about equal importance to oil, with gas third.

Britain's long-term national energy policy is based on the development of a balance among four fuels (oil, gas, coal, and nuclear), and the further reduction of energy imports through increased domestic production. By 1985, the government hopes to achieve a net surplus in oil. In the year 2000—when North Sea oil and gas production will have peaked—the plan calls for an equal balance among various types of energy. National energy planning in Britain is facilitated by the fact that many energy industries are owned or guided by the government.

---

[43]For a dissenting opinion, see House of Commons, Fifth Report for the Foreign Affairs Committee, Session 1979-80,

Afghanistan: The Soviet Invasion and Its Consequences for British Policy (London: Her Majesty's Stationery Office, 1980), p. xxxi. The report warns:" . . . Despite large Soviet reserves of oil, technical difficulties with extraction could produce circumstances in which a Western embargo of oil technology might lead the Soviet Union or its allies to action in the Gulf to acquire oil on terms which would be detrimental to Western interests."

BRITISH-SOVIET ENERGY TRADE, 1970-80

British Energy Equipment and Technology Exports

Only a very small portion of total British exports go to CMEA, and the United Kingdom has not been as important an exporter of energy-related equipment to the U.S.S.R. as have the other Western nations studied here. Between 1975 and 1979, the United Kingdom ranked a distant sixth among Western nations in such sales.

British experts are skeptical about claims that Western technology is critical for Soviet energy development, but they do believe that Western exports can nevertheless make a significant contribution in speeding or easing this development. In the past, British companies such as John Brown Engineering and Rolls Royce have sold gas turbine compressor units and engines for use in the Orenburg pipeline. The most promising area for energy-related exports in the future is undoubtedly in oil and gas exploration and production equipment. Britain has considerable experience in offshore oil and gas development in the North Sea, and such technology could contribute to the development of the Soviet Baltic and Barents seas.

In 1976, British Petroleum signed a technological cooperation agreement with the Soviet Union which covered certain energy-related areas—modernization of refineries, secondary and tertiary oil recovery technologies, and offshore exploration. But despite the fact that the British government has extended credits to facilitate this agreement, little concrete action has yet been taken on the Soviet side, and there exists no elaborate system of working groups (as is the case in France) to manage the details of specific projects.

The British Government provides subsidized credits to Communist nations through its Export Credit Guarantee Department. In recent years, the Soviet Union has not taken full advantage of this cheap credit. The long-term U.K.-Soviet trade agreement expired in early 1980, but as was the case with Italy, the Afghanistan invasion led to no new agreement being reached, and since then credits have been supplied on a case-by-case basis.[44]

British Energy Imports From the U.S.S.R.

The United Kingdom imports no Soviet gas and in 1979 imports of Soviet oil and oil products amounted to 2.9 mtoe (two-thirds of it in the form of crude oil), about 4 percent of the nation's oil imports and 3 percent of its oil requirements (see table 85). During the same year the United Kingdom exported more than 49.7 mtoe of oil and oil products. Soviet oil imports are clearly not critical to Britain's energy balance. Nor could it be argued that coal imports (from Eastern Europe) are important, since domestic coal production is so large.

In sum, Britain's relatively high level of energy self-sufficiency gives it less incentive for involvement in energy trade with the U.S.S.R. than other West European countries. In addition, as past patterns of trade in energy and energy-related equipment with CMEA show, the United Kingdom is less involved with CMEA in this area than any of the other four West European nations reviewed here. Should these trends change and a policy of expanded trade with the Soviet bloc be instituted, however, Britain's experience with North Sea oil and gas could put it and its corporations in a good position to assist Soviet offshore petroleum development.

SUMMARY

The previous sections have briefly examined the energy-related trade relations between West Germany, France, Italy, the United Kingdom, and the Soviet Union. This survey has shown that, although East-West

---

[44]Testimony of Christopher Mallaby from the Foreign and Commonwealth Office in the House of Commons, op. cit., p. 10.

trade makes up only a small portion of the overall trade of these nations, energy-related exports in 1979 constituted about one-third of Italian, approximately one-quarter of West German and French, and about 10 percent of British exports to the U.S.S.R. (This may be compared with 45 percent for Japan and 7 percent for the United States.) In absolute amounts, this translates into nearly $1 billion worth of energy-related exports in 1979 for West Germany, and nearly one-half billion each for France and Italy. Particularly in West Germany, much of this trade is concentrated in the steel industry, where a significant number of jobs depend on it. Important industrial sectors in Western Europe, therefore, have strong interest in trade with the U.S.S.R.

These nations also import energy from the Soviet Union. The most important Soviet energy export to Western Europe is gas. At present, 43 percent of Italy's and about 20 percent of West Germany's imported gas comes from the U.S.S.R. The corresponding figures for gas consumption show that Italy imports from the U.S.S.R. nearly 24 percent and West Germany about 16 percent of the gas they require. These figures may be interpreted in several different ways. In no country examined here, for instance, does Soviet energy constitute more than about 9 percent of total energy imports or 10 percent of total energy consumption. Once again, it is clear that while overall "dependence" on the U.S.S.R. is low, the importance of certain forms of Soviet energy—i.e., gas—may be disproportionately prominent in the imports of some West European countries.

FUTURE PROSPECTS FOR ENERGY INTERDEPENDENCE: THE WEST SIBERIAN GAS PIPELINE PROJECT

OTA's examination of past patterns of energy cooperation and trade between Western Europe and CMEA reveals increasing, but at present still limited, levels of interdependence. Except in specific sectors of energy or equipment trade (gas imports for FRG and Italy; energy equipment exports for FRG, France, and Italy), levels of interdependence have remained relatively low. This situation will not necessarily persist. The "Yamburg" gas pipeline project has the potential for raising the level of Soviet-West European energy interdependence in both quantitative and qualitative terms.

One reason for the controversy surrounding this project is that it embodies the classic dilemmas of interdependence with the Soviet bloc. The U.S.S.R. has been anxious to enlist Western participation—particularly that of Japan, FRG, and Italy—because in terms of sheer construction and manufacturing capacity, the project may well be beyond the ability of the Soviet gas

industry alone to handle efficiently and expeditiously. The U.S.S.R. possesses the technical knowledge to construct such a pipeline itself, but could not without massive domestic economic adjustment produce pipe of adequate quantities and quality. Moreover, equipment could be paid for in exports of gas to the West. This is a clear case in which Western equipment and know-how could make a significant contribution to speed Soviet energy development.

In return, the pipeline will provide Western Europe with greatly expanded gas deliveries. **The gas may largely replace exports of Soviet oil. This may somewhat offset projected levels of dependence on Soviet energy as a whole, but that level is likely nevertheless to rise.** Likewise, the fact that the project offers opportunities for Western exporters of energy-related equipment raises the potential for Soviet manipulation of competing suppliers. Much of the equipment to

Photo credit: Oil and Gas Journal

Soviet gas pipelaying barge used in the construction of pipeline from Urengoy to the Ukraine

be used in this project—compressor stations, pipelaying, and telecommunications equipment, for instance, could be manufactured by firms in several Western countries. There is, therefore, strong competition among various potential Western suppliers. Moreover, discussions over the export pipeline are occurring in a period of heightened East-West tension. The difficulty of assessing these costs and benefits, together with the magnitude of the deal and its timing, make it a test case for East-West energy relations that may well set a precedent for future projects.

A great deal of confusion has surrounded the West Siberian gas export project.[45] One reason is that negotiations are still underway and many details have yet to be settled. A second problem is the lack of a single authoritative source of information; accounts must often be assembled from periodicals in a number of different countries, and these are sometimes inconsistent or contradictory. Third, and perhaps most im-

_____

[45]Material for this section is from Wharton Econometric Forecasting Associates, Inc., Centrally Planned Economies News Analysis, Aug. 20, 1981.

portant, incomplete information on changing Soviet plans, together with a general Western lack of familiarity with Soviet geography and geology (particularly the status and location of Soviet gas deposits), has led to the persistence of a number of misconceptions about the proposed route of the new pipeline and the source of the gas which it is to carry.

Yamburg is a supergiant gasfield located in West Siberia and extending north toward the Arctic Ocean. It is said to contain one of the world's largest untapped proven resources of gas—an estimated 2.6 trillion cubic meters. (This is equal to 2.1 billion tons or 15.6 billion barrels of oil equivalent.) But although this field has given its name to the new gas export pipeline which will supply Western Europe, it is itself not expected to produce gas until at least the end of the decade. Until that time gas destined for Western Europe will come from Urengoy, the world's largest gasfield, about 150 miles to the south. Initially, this gas will be transported through additions to the existing pipeline network. For instance, construction of a 56-inch pipeline which parallels the Northern Lights line to the Czech border is now well underway. (In April 1981, this new string had reached some 250 miles northeast of Moscow.) This line is scheduled to be put into service in 1984, and is only one of several new 56-inch pipelines planned for construction after 1982.

Another high-pressure 56-inch pipeline from Western Siberia is also planned. It is the latter which is commonly referred to as "Yamburg." The new export pipeline may eventually have two legs. It would seem that the U.S.S.R. now plans to build the first of these by the mid-1980's, reserving the latter for the end of the decade.

There are both economic and political advantages for the U.S.S.R. in exporting Urengoy gas. Unlike Yamburg, Urengoy is already a producing field. It will be significantly cheaper for the Soviets to further develop Urengoy than to initiate operations at Yamburg, and it thus makes good

economic sense to postpone development of the latter. Because key segments of the new line will follow the existing routes, construction and borrowing costs should be significantly reduced. This is an important consideration. Estimates of the cost of this project have ranged from $15 billion to $20 billion to as much as $40 billion. On the political side, this plan may well be designed, at least in part, to minimize political controversy over the entire export scheme. The Western equipment which the U.S.S.R. will need can now be purchased for expansion of the existing pipeline network, without necessarily specifying for which project the equipment will be used.

Regardless of where the gas is to come from, the new pipeline to Western Europe (hereafter referred to as Yamburg to conform to popular usage), together with the additional pipeline capacity already under construction, will allow a significant increase in Soviet natural gas exports to that region. Eventually this pipeline could bring Soviet gas to as many as 10 West European nations—West Germany, France, Italy, Belgium, Austria, Finland, the Netherlands, Switzerland, Sweden, and Greece. One Western estimate is that by 1985, Soviet natural gas exports to Western Europe will increase 120 percent over 1980 levels; by 1990, they could exceed 1980 levels by 270 to 330 percent.[46] These projections are shown in table 90. The 1988-90 projections shown here, 90 to 105 bcm, are the equivalent of 73 to 85 mmt/yr or 1.5-1.7 mbd of oil. In 1980, the U.S.S.R. exported about 66 mmt of oil (1.3

[46]Ibid.

Table 90.—Soviet Exports of Natural Gas to Western Europe (in billion cubic meters)

Year	Total	Existing transit pipeline and additions	Planned new West Siberian (Yamburg) pipeline
1980	24.5	24.5	—
1981-1983	25.0	25.0	—
1984	32-46	32-46	—
1985	53.0	53.0	—
1986	62-76	53-58	9-18
1987	71-91	53-63	18-28
1988-1990	90-105	53-68	37.0

SOURCE: Wharton Econometric Forecasting Associates, Inc.

mbd). The energy of value of Soviet gas exports to Western Europe could therefore exceed that of its oil exports by the end of the decade.

WEST EUROPEAN POLICY PERSPECTIVES

At the heart of controversies over Yamburg has been the question of whether or not Western nations might become unacceptably subject to Soviet economic, political, and energy leverage by virtue of their participation. If Western nations such as West Germany and France become more dependent on the U.S.S.R. for their gas, will they become in fact hostage to Soviet political pressure? This section reviews the relevant policy perspectives of groups in West Germany, France, Italy, and the United Kingdom. These reflect concern about increasing dependence on Soviet energy among West Europeans, but also fairly strong support for the project and considerable competition among the potential participants for contracts and sales. From the West European perspective, the Yamburg pipeline offers attractive export and import possibilities, and private firms likely to expand their sales if the project materializes have been at the center of negotiations.

West Germany

The FRG has taken a leading role in all aspects of the Yamburg discussions, with West German equipment suppliers—particularly the steel firms—playing key roles. West German banks and Ruhrgas, the country's largest gas importer and distributor, have also participated. Industrial and financial groups have evidently coordinated their efforts closely with the West German Government. But while the negotiations have primarily been carried on between Soviet officials and West German manufacturers and bankers, the scope of the project and its political sensitivity mean that government officials throughout Europe have been consulted.

In recent months, there has been a good deal of debate about Yamburg among informed policymakers in West Germany. The general consensus is that the pipeline is desirable for both economic and political reasons. This conclusion is based not only on the general orientations of West German industrialists and bankers toward trade with the East, but also on a fairly detailed assessment of security implications. Policymakers believe that trade acts as an incentive to restrained Soviet behavior in Europe, and many claim that the Yamburg negotiations may well have acted as a deterrent to a Soviet invasion of Poland.

Although the focus of much of the discussion between the United States and the FRG about Yamburg has been energy imports, the primary incentive for West Germany has been equipment exports. For West German firms such as Mannesman Steel Works, which last year exported to the U.S.S.R. 60 percent of the large-diameter pipe it produced, Yamburg represents the latest and largest in a long series of pipeline projects which help to to employ a few thousand workers. In fact, many West German government and business spokesmen generally view East-West trade as mutually beneficial, and the Yamburg project is particularly attractive because of the jobs it is likely to create.[47] It is not surprising that West German steel firms, including state-owned Salzgitter, have actively lobbied for the project.

A second important attraction of the pipeline is its potential for expanding energy supplies and diversifying energy sources. Several factors—levels of total imports from the U.S.S.R., projected West German domestic production, and imports from alternative suppliers—affect calculations of the significance of Yamburg for the FRG. West Germany already receives about 16 percent of the natural gas it consumes from the Soviet Union. The new pipeline would sig-

---

[47]See interview with Salzgitter Director Ernst Pieper, *Der Spiegel*, No. 8, Feb. 18, 1980. Salzgitter is a state-owned West German steel company.

nificantly increase Soviet gas supplies, but the exact levels of dependence likely to result are difficult to predict. This is partly due to the abandonment of the IGAT II project, which has now reduced anticipated gas supplies from the U.S.S.R. Yamburg could not only make up for IGAT II; it could also more than double the volume of Soviet gas imported by the FRG in 1979 (9.8 bcm).

Forecasts of FRG dependence likely to result from Yamburg vary not only according to assessments of IGAT II, but also according to forecasts of overall West German dependence on natural gas in the years ahead. If the FRG increases its consumption of natural gas from 60 to 80 bcm between 1980 and 1990, and if supplies from the U.S.S.R. rise to 24 bcm, then Soviet gas might represent more than 30 percent of West German gas consumption. The West German cabinet in 1980 announced that importing up to 30 percent of its natural gas from the Soviet Union would not constitute a security risk.[48] However, German gas firms do not need the permission of the government to import gas. Some observers estimate that by the year 2000 the FRG could be importing 40 percent of its gas from the U.S.S.R.,[49] but there is no concrete evidence to suggest that this is likely.

Evaluations of future dependence also hinge on assessments of the reliability and availability of alternative suppliers of gas. Some West German observers view the Soviet Union as a reliable supplier—at least in comparison to the Dutch, who have threatened to cut off gas because of a dispute over prices. Algeria is another alternative, but it has recently reneged on a contract for a natural gas liquefaction plant. Norway, which will be supplying small amounts of gas to the FRG beginning in 1985, could potentially provide more if the Norwegians decide to develop more of their gas,[50] but

Norway has proved extremely difficult to deal with in these matters. Nigeria will begin supplying West Germany after 1984; and a variety of other nations might also sell LNG to West Germany.

Another factor which will influence the impact of increased levels of Soviet gas on FRG energy import dependence is West German domestic gas production. While some observers anticipate that domestic output will continue to supply Germany with about 30 percent of its gas needs, others worry that the level of domestic production might fall, partly because important deposits may be technically difficult to exploit.

West German spokesmen are skeptical about the prospect of increased Soviet leverage over the FRG by virtue of its energy exports. First, at least until quite recently, the Soviet Union had the reputation of being a reliable exporter. It is true that in 1981 the U.S.S.R. announced a 30 percent cutback of gas deliveries to the FRG, but these reductions caused little difficulty because West Germany had alternative supply arrangements.[51] On the other hand, while West German observers point out that Algerian and Libyan suppliers have been less reliable than the Soviets in years past, the Soviet Union may be said to have a far greater political interest in West Germany. The incentives to use such threats for political purposes may therefore be stronger. Similarly, some argue that the Soviets would be unlikely to suddenly withdraw or threaten gas cutoffs to the FRG since the Soviets themselves need the equipment and the hard currency which gas exports will provide. While this argument makes sense in the short term, according to this logic, after the pipeline is in place, West Germany could be more vulnerable to a supply cutoff, or the threat of one.

A third line of reasoning holds that viable alternatives to Soviet gas provide a deterrent to the U.S.S.R.'s use of political pressure. The FRG could buy gas from a number

[48]*Der Spiegel,* No. 26, 1980.
[49]Wolfgang Mueller-Hassler, "Die Verantwortung der Gasmanner," *Frankfurter Allgemeine Zeitung,* Jan. 10, 1981, p. 11.
[50]"Norwegen will mehr Gas Liefern" *Süddeutsche Zeitung,* Dec. 23, 1980.

[51]Otto Graf Lambsdorff, "Plädoyer für sowjetishes Erdgas," *Rheinische Merkur, Christ und Welt,* No. 4, Jan. 17, 1981.

of nations, including Holland. Current plans are to reduce Dutch imports as more West Siberian gas begins to flow, but the Dutch have evidently agreed in principle to provide supplies in the event of shortfalls of Soviet gas. There is also the possibility of LNG from the Persian Gulf. Secondly, since the West German gas pipeline network is interconnected, shortfalls could be equalized through the country, and deficiencies in supply from the Soviet Union could be compensated for by increased supplies from the North Sea, to which a pipeline now extends.[52] Another protection lies in underground storage—about 2.5 bcm of gas are now stored underground and this is to be increased. (West Germany also has oil stockpiles for about 100 days.) Fourthly, many contracts with industrial users are "interruptible," meaning that industries are responsible for providing alternative energy for periods of up to 50 days if necessary. And finally, more dual-fired burners will be used in industry so that power stations can switch to oil or coal in connection with interruptible contracts. In short, according to spokesmen from the gas industry, the FRG could now survive a total cutoff in gas from the Soviet Union if it were forced to do so.[53]

One major, as yet unanswered, question surrounding the effect of a Soviet gas cutoff concerns which customers would be most affected. Most Soviet gas will go to Bavaria in southern Germany, site of the automobile and chemical industries. These may be less vulnerable to energy supply interruptions than the more energy-intensive steel industry, concentrated in the north. The proportion of gas going to households is also an important variable. Given present information, there is no way of evaluating the likely impact of cutoffs on these various consumers.

The tone of the West German debates about contingency planning and substitute supplies suggests that under current condi-tions, the West German Government believes that most consumers would not be greatly affected by a cutoff in supplies from the Soviet Union—all other things being equal. However, if such a shortfall were associated with a worldwide energy crisis, perhaps precipitated by an OPEC oil embargo, the ramifications could be much more serious. To the extent that West German dependence on Soviet gas increases, contingency planning becomes even more important. Timelags associated with substitution of alternative forms of energy for Soviet gas might cause hardship for certain consumers. While West Germany seeks to develop alternative supplies and contingency arrangements, these cannot completely eliminate the potential threat of a gas cutoff by the U.S.S.R.

At one time, financing was the most problematic aspect of the pipeline for the West Germans. The Soviet Union has been a reliable creditor, paying back loans for previous gas-pipe deals an average of 3 years early. The West German Government does not subsidize export credits. While the banks would probably prefer to use a system of floating credit rates for at least part of the loans, West German equipment suppliers object to this arrangement since floating rates would allegedly complicate supply contracts. The Soviet Union at first established a tentative arrangement for $5 billion in credits. This involved a consortium of more than 20 West German banks, led by the Deutsche Bank.[54] In early 1981, a deal was worked out whereby the banks would offer a fixed interest credit worth 9.75 percent (a nominal external rate of 7.75 percent plus a 2-percent increase in charges for West German equipment). The credit was to last for ten years. As interest rates rose in West Germany, however, the Deutsche Bank and others began to reconsider the Soviet credits.[55] The West Germans have now found a new, acceptable credit formula and negotiations

[52]Answer to State Secretary A. von Wuerzen to Bundestag member Lintner in Deutscher Bundestag, *Drucksache*, 9/35, pp. 9-10.

[53]Heinz Günther Kemmer, "Wenn Moskau an Gashahn dreht," *Die Zeit*, Jan. 23, 1981, p. 23.

[54]Kevin Dare, "Soviet Gas Project Loans Deadlock," *Financial Times*, Mar. 16, 1981.

[55]See John M. Geddes, "Germans Offer Loan to Soviets for Gas Pipeline," *Wall Street Journal*, Feb. 2, 1981.

have turned to the question of gas prices. Credits are to be provided in stages, an arrangement that not only allows for the contingency that interest rates may go down, but which also helps to reduce the political sensitivity of the deal. West German financing, instead of being concluded in dramatic, billion dollar segments, will take on a much lower profile, incremental aspect.

In sum, both the official position of the West German Government, and informal opinion in the FRG's business community, strongly favor the pipeline project. Although groups within opposition parties have expressed reservations and asked for risk assessments, no party has come out in open opposition.[56] Interest in contingency planning to minimize risks is growing. It appears that, failing a Soviet invasion of Poland or counterproductive pressure from Moscow over the terms of the deal, the pipeline will be built with strong support from West German firms.

France

French policymakers are interested in the pipeline project for the same reasons as the West Germans—expanded gas supplies and equipment export opportunities. Delivery of 8 to 10 bcm/yr of Soviet gas will add significantly to French dependence.[57] In 1979, France received 1.9 bcm—less than 10 percent of its imported gas and about 7 percent of its gas consumption—from the U.S.S.R. Some observers estimate that by 2000 the U.S.S.R. might be providing almost 30 percent of French gas.

The Soviets have been negotiating with French firms for credits, which could amount to $4 billion, and for equipment, but these discussions slowed in early 1981 in anticipation of the French election. Negotiations over French credits continue, with the Credit Lyonnais taking the lead for a consortium of French banks.

Like the West Germans, French policymakers have been generally positive toward the Yamburg project. They have tended to believe that the Soviet Union will have a long-term interest in assuring continued supplies of Western equipment, and that the U.S.S.R. will be a reliable supplier of energy. These beliefs were apparently unshaken by the fact that in the winter of 1979-80, gas shipments from the U.S.S.R. to France fell about 30 percent, due primarily to weather conditions. The shortfalls evidently did not cause any great hardships.

It would be technically possible for the Soviet Union to reduce or cut gas supplies to France without affecting West Germany, although the reverse is impossible—i.e., a cutoff of West Germany would also involve France. It would seem, however, that potential Soviet economic or political leverage over France is relatively small and there are at present no obvious incentives for such a cutoff. Additionally, the French have considered—perhaps more carefully than the West Germans—contingency arrangements in the event of gas supply shortfalls. France currently has 4 bcm of gas stored underground, and this stockpile will be doubled in the next 10 years. Moreover, French planners intend to place a ceiling on domestic gas use, allowing no more than 45 percent of the total to be used for household consumption. Gaz de France also hopes to increase interruptible supply contracts with industry during the next decade to about 30 percent of all industrial contracts. Company spokesmen say that all Soviet gas will be sold on the basis of interruptible supply contracts, and that none of it will be used for home heating.

Despite this stress on contingency planning and risk assessment, however, it would be a mistake to suggest that France would be unaffected by a Soviet gas cutoff. As is the case with West Germany, if Soviet gas supplies reach levels of 30 percent or more of total gas consumption, the effect of a cutoff

---

[56]Heinz Riesenhuber in CDU/CSU *Pressedienst*, Jan. 7, 1981.

[57]For estimates of Yamburg gas exports to various West European nations, see Robert W. Ball, "Europe Warms to Soviet Gas," *Fortune*, June 1, 1981, p. 78.

could be significant, particularly if it occurred in the context of a worldwide energy crisis.

On the whole, it appears that while the French Government has supported the export pipeline, it has been less enthusiastic than the FRG. The new Socialist government may in time develop a different attitude toward the project, but all the major French parties have in the past indicated their support. The French have left it to the West Germans to take the lead in negotiations, but they are nonetheless committed—both to the prospect of increased Soviet gas imports and to the idea of continued trade with the U.S.S.R.

Italy

Italian negotiations over Yamburg are at a more preliminary stage than those of either West Germany or France. Here, financing questions loom large, and only limited discussions have been held between Soviet officials and Italian equipment suppliers.[58] The pipeline was the main topic at the March 1981 meeting of the Italian-Soviet Economic Commission, but the results were inconclusive. The Italian Government has decided to set up an interministerial commission to study the project.[59] The question may eventually be taken to a vote in the Italian parliament—a move that would be unprecedented in Italian relations with the U.S.S.R.

Given these uncertainties, it is difficult to predict the significance of the project for Italy. Yamburg could yield between 5 and 10 bcm/yr of additional gas. Assuming a mid-level of 8 bcm, this would double Italian imports of Soviet gas (in 1979 these were 7 bcm).[60] If Italian gas consumption reaches the expected 40 bcm by 1985, Soviet supplies would account for 40 percent of Italy's gas consumption. Extensive Italian participation in Yamburg would thus ensure a con-

tinued high level of dependence on Soviet energy.

Italy does not now have elaborate plans to deal with potential reductions in Soviet energy supplies—a point worth noting, given the present level of Italian import dependence on Soviet gas. While the Italian gas corporation SNAM does not now have interruptible gas contracts, there are plans to introduce such arrangements if and when the pipeline project proceeds. There are also plans to increase stockpiles. At present, however, it would appear that the major contingency plan calls for use of the Algerian pipeline to pump additional supplies in the event of an emergency.

On the equipment export side, a number of Italian firms might participate, but the central issue has been financing. At a meeting of Italian and Soviet government officials in March 1981, the Soviets reportedly asked for credits to cover 85 percent of the financing of equipment exports worth $3 billion to $4 billion, at an interest rate of 7 percent.[61] The Italian Government is clearly doubtful that these conditions can be met. Firms such as Finsider, which sells large-diameter pipe, and Nuovo Pignone, which sells compressor stations, naturally favor the deal, but financing problems have precluded final agreement.

In sum, the Italians favor Yamburg, but with some reservations. If financing problems are solved—and this may be a major obstacle—then the predisposition of Italian leaders is to participate. But there is markedly less enthusiasm here than in West Germany or France. The Italians are both the most dependent of this group on Soviet energy, and the most cautious about continued and increased reliance on the Soviet Union as a gas supplier.

Britain

Of the nations included in this study, Britain has the least vested interest in the West Siberian pipeline project. But although the

---

[58]Il Fiorino, Mar. 8, 1981.

[59]Il Giornale, Mar. 17, 1981.

[60]See statement by Enrico Manca, Minister of Foreign Trade, in Il Messagero, Mar. 14, 1981.

[61]Il Fiorino, Mar. 10, 1981.

U.S.S.R. would not provide gas to the United Kingdom, British firms such as John Brown, Rolls Royce, and Cooper Industries could sell equipment for the project. British financial institutions have discussed the possibility of extending credits to this end, but talks remain at a preliminary stage.

The pipeline has received less attention in Britain than elsewhere in Europe or in the United States. Government spokesmen claim that the question of what constitutes a reasonable or dangerous level of dependence on Soviet energy is something that other governments must decide for themselves. There has been no official comment about the desirability of the project from the perspective of West European energy security. Indeed, the British case illustrates the clear connection between the energy and trade position of each Western nation and its interest in Soviet gas development. More nearly energy self-sufficient than any of the other nations, and traditionally less involved in East-West trade, the United Kingdom is understandably the least active participant in negotiations.

SUMMARY AND CONCLUSIONS

West Germany, France, and Italy all look to the Soviet Union not only as a way to increase energy supplies, but also as an attractive market for equipment exports. Barring unexpected political or economic developments in Europe, therefore, the West Siberian gas export pipeline will probably be constructed, and West Germany, France, and Italy will certainly become more dependent on Soviet gas. While it is very difficult to make precise determinations of the impact of West Siberian gas on the energy balance of each nation, reasonable estimates are that both West Germany and France could depend on the U.S.S.R. for about 30 percent of all gas imports, and that Italy could receive almost 50 percent of its imported gas from the Soviet Union. The corresponding percentages for total gas and total energy consumption would depend on the energy balance of each country at the time—a highly

uncertain matter. It must be remembered, however, that to some extent Soviet gas will *replace*, not supplement, Soviet oil in these countries, and that even with these import levels, energy dependence on OPEC is likely to remain much higher than dependence on the U.S.S.R. If the overall energy dependence of each nation examined here on the U.S.S.R. *doubled*, that dependence would range from about 3 percent in the case of Britain to nearly 20 percent in the case of Italy.

A sudden cutoff in gas supplies from the U.S.S.R. would impact each nation differently, but none would be immune from hardship—particularly in the context of a tightened world oil market or energy crisis. All would benefit by the development of more effective contingency plans to allow for substitution of alternative energy supplies in the event of a shortfall in Soviet gas. Such plans would diminish incentives for the Soviet Union to make use of its "gas weapon" to pressure Western Europe. Emergency planning would be most effective if it were undertaken by all the nations involved. Joint planning would reduce the ability of the Soviet Union to divide Western Europe by playing one country off against another. However, the prospects for coordination of West European policy toward trade and energy relations with the U.S.S.R. are not bright.

PROSPECTS FOR WEST EUROPEAN POLICY COORDINATION

One indication of Western Europe's willingness and ability to coordinate policy toward the U.S.S.R. was its response to the trade sanctions initiated by the United States against the Soviet Union following the invasion of Afghanistan. In January 1980, President Carter announced that U.S. exports of certain agricultural commodities and of high technology items would be restricted. In March 1980, these restrictions were intensified.

The extent to which America's allies were prepared to support these sanctions was unclear.[62] In the area of "high technology" (including computer systems, other advanced electronic equipment, and automated machine tools) a policy of "no exceptions" to CoCom controls was established. The expectation was that sales in these areas would be drastically reduced. But, except for the United Kingdom, West European nations, on the whole, were and remain unsympathetic to the political use of economic pressures against the U.S.S.R. One important exception was made to the CoCom "no exceptions" policy—exports of spare parts for oil and gas pipelines were not restricted.[63] CoCom's decision was based on the reasoning that it is not in Western Europe's interest to reduce the efficient functioning of the Soviet pipeline system which carries energy to the West.

In fact, the West European response to the economic sanctions initiated by the U.S. was largely one of "every man for himself." As table 91 shows, during 1980 all of the allied nations reviewed here increased their overall exports to the U.S.S.R. (Japan and Britain were the most supportive of U.S. sanctions, and the French and Italian governments did limit credits for the Soviet Union.)[64] West European businessmen openly criticized the sanctions as ineffective and contrary to the fundamental interest of the West. Both West Germany and France officially expressed their wariness of the sanctions; and discussions over the gas export

[62]U.S. Senate, Committee on Banking, Housing, and Urban Affairs, Subcommittee on International Finance, "U.S. Embargo of Food and Technology to the Soviet Union," Jan. 22 and Mar. 24, 1980, pp. 36 and 117.

[63]Richard N. Cooper, "Export Restriction on the U.S.S.R.," *Current Policy*, No. 211, Aug. 20, 1980.

[64]Federico Bungo, "Cossiga Infuria, Il Piano Manca," *L'Espresso*, June 15, 1980. See also *Il Fiorino*, June 15, 1980.

pipeline project and other deals continued throughout 1980. In a number of cases, West German and French firms won contracts which had been nearly completed by U.S., Japanese, or British firms. In Britain, the Thatcher government's support for the sanctions was openly questioned in a report for the House of Commons Foreign Affairs Committee. The report concluded that a Western embargo of technology needed by the Soviet Union for energy development might well prove counterproductive by stimulating Soviet aggression in the Persian Gulf. Thus, rather than responding with a set of joint policy initiatives, Western Europe remained divided.

This response not only illustrates the difficulty which the United States has had—and may well continue to have—in influencing West European trade relations with the U.S.S.R. It also reflects the limited ability of EEC nations to coordinate East-West trade policies. The EEC treaty calls for development of joint trade policies toward CMEA. But, despite repeated efforts on the part of EEC to persuade its members to negotiate one multilateral treaty with each CMEA nation, a joint approach has not emerged. EEC has legislation which prohibits the establishment of bilateral trade treaties, but its members have circumvented the substance of this position by concluding separate "cooperation" agreements with East European nations—for example, the FRG's 1978 25-year agreement with the U.S.S.R. for long-term cooperation in energy and trade.

This limited coordination also exists in energy policy, a fact largely the result of the differing resource endowments and differing

Table 91.—Exports of Western Nations to U.S.S.R. 1975-80 (millions of U.S. dollars)

	United States	Japan	France	West Germany	Italy	United Kingdom	Total 6
1975	1,625	1,147	2,824	1,020	964	8,914	
1979	3,604	2,461	2,007	3,619	1,220	694	13,605
1980	1,664	3,075	2,712	4,811	n/a	1,162	13,424 (without Italy)
	−54%	+24.9%	+35%	+32%	NA	+67%	

SOURCE: 1975-79, UN SITC; 1980, preliminary IMF data.

interests of the various EEC members.[65] Discussions of energy security within EEC have dealt with such issues as oil stockpiles, conservation, and substitution of coal and other energy sources for oil, but it is not clear that even the small degree of coordination achieved can be translated into the area of East-West energy relations.

Efforts to promote multinational energy cooperation have also resulted in discussions of an "all-European" energy conference. This idea originated in 1971 with a proposal by Soviet Prime Minister Kosygin for greater East-West cooperation in energy.[66] Proposals for such a conference have surfaced over the years at both the United Nations Economic Commission for Europe (ECE) (both the United States and the U.S.S.R. are members) and at the Conference on Security and Cooperation in Europe (CSCE), most recently at the 1980-81 Madrid CSCE meet-

ing. The United States strongly opposes such a conference and little progress has been made on the proposal, although a first step was taken in 1979 when the ECE established an ad hoc group, of "Senior Advisors to the ECE Governments on Energy." The major effort of the group so far has been to collect and study responses from member nations to an energy questionnaire. One problem has been that energy data from the Communist nations, particularly the U.S.S.R., has been either incomplete or unresponsive. In December 1980, ECE published a report on the energy problems of the ECE region,[67] but especially given the opposition of the United States, the future of this effort is uncertain. In fact, the all- European energy conference has become a political issue between the United States and its allies in Western Europe who favor its convening.

[65]See Georges Brondel, "Energy Policy and the EEC," *Energy Policy*, September 1978, p. 231.
[66]*Pravda*, Apr. 7, 1971.

[67]United Nations Economic and Social Council, Economic Commission for Europe, *Energy Problems and Co-operation in the ECE Region*, Dec. 15, 1980, p. 29.

CONCLUSIONS

Interaction between Western Europe and the U.S.S.R. over the course of the last decade—measured by both the level of exports of Western equipment to, and imports of energy and raw materials from, the Soviet Union—has increased. While trade with the U.S.S.R. does not represent a great portion of the total exports of any of the Western nations examined here, this trade does constitute a significant proportion of production for certain industrial sectors (e.g., steel). Similar patterns exist with respect to imports of Soviet energy. No nation included in this study receives more than about 10 percent of its total energy requirements from the U.S.S.R. However, in West Germany and Italy, dependence on Soviet gas now stands respectively at about 24 and 43.2 percent of gas imports, and 16 and 24 percent of total gas consumption. If the new West Siberian gas pipeline project is completed,

such dependence will rise significantly for West Germany, France, and probably Italy. The degree of overall dependence for each West European nation on Soviet energy will also rise, but given the fact that Soviet gas exports will largely replace Soviet oil exports, this increase may not be as large as would otherwise be expected. The existing situation is one of limited overall dependence of Western Europe on energy from and energy-related trade with the U.S.S.R.—but significantly greater dependence in certain industrial and energy sectors. If the West Siberian pipeline is built, Western Europe's interdependence in trade and energy with the Soviet Union will increase further.

To a large extent, this trend may be regarded as a fait accompli. There is strong commitment to energy and trade relations with the U.S.S.R. among all of the nations studied here, despite considerable variation in the salience of East-West energy and trade for public policy debate. This variation can be attributed to considerable differences in the energy and export situations of the countries. West Germany, by virtue of its historic ties, its geographic proximity, and its political concerns, e.g., with East Germany, is most active—both at the official and at the private diplomatic levels—in actively promoting interaction with the Eastern bloc. Italy, likewise, has developed a pattern of fairly strong energy and trade ties with the U.S.S.R. and its allies. France, and especially Great Britain, appear to perceive less need to promote energy imports (in the latter case because of North Sea oil, and in the former because of plans to rapidly develop nuclear power). France, unlike Great Britain, has played a strong role in exports of energy-related equipment. **The prospects for interaction between any of these nations and the U.S.S.R. cannot be understood outside the context of these differing national interests, experiences, and perspectives.**

This is not to suggest that there is unanimity in any West European nation over energy imports from the U.S.S.R., participation in the gas pipeline project or support of U.S. economic sanctions against the Soviet Union. Nevertheless, the general predisposition in each country examined here is to promote interdependence and to be unwilling to use trade as a lever in East-West political disputes. The latter point is well-illustrated by the fact that West Germany, France, and the United Kingdom all actually increased exports to the U.S.S.R. in 1980—despite U.S. calls for trade sanctions. Of course, a variety of domestic political forces could change West European official policies toward trade with the U.S.S.R., but interdependence with the U.S.S.R. is currently viewed as a fact of life. To a great extent, this view is based on a positive attitude (rather than simple resignation) toward the perceived potential benefits which East-West interaction generally, and cooperative

energy development with the U.S.S.R. specifically can confer. These benefits are perceived in both political and economic terms.

Unless the political situation changes dramatically, it is likely that the countries of Western Europe will participate in the development of the new West Siberian gas pipeline. If the project proceeds on the scale currently envisaged, it will lead to a significant growth in West European-Soviet energy independence. The value of the equipment needed for the pipeline is double the value of all the exports of the industrial West to the U.S.S.R. in the year 1979. Thus, completion of the project would be tantamount to a quantum increase in Western equipment exports to and credit financing for the U.S.S.R. In addition, while it is difficult to make precise estimates of the amounts of gas the pipeline will provide to any one country individually or to Western Europe as a region, dependence on Soviet energy will probably increase in the FRG, France, and Italy.

The gas pipeline project marks a significant new development in East-West relations. It will require a multinational effort on the part of Western Europe, and it is precisely this dimension which may be of greatest long-term significance to the United States. More than the increased trade and energy import opportunities which the project offers—and in both of these areas Western Europe's relationship with the U.S.S.R. would still be considerably weaker than its dependence on OPEC—a changed political climate would offer both potential risks and benefits. On the one hand, the project may provide an opportunity for the Soviet Union to lever individual West European nations—thereby challenging overall Western unity. On the other hand, if the project stimulates new types of Western policy coordination, it could change the overall context of East-West relations.

The critical question is whether Western nations, either individually or in concert, can act to limit the risks involved. (Any coordinated policy would necessarily involve Japan—the nation which supplied one-third of all energy-related equipment exports to the Soviet Union in 1979.) One area in which joint action could be useful is in assessing levels of energy dependence likely to result from new Soviet gas pipeline(s), and planning for contingency arrangements in the event of a supply shortfall. There is precedent for such an effort; IEA has already undertaken such discussions in the context of dependence on OPEC oil. A critical aspect of such a joint policy approach would be plans for gas and oil sharing in the event of a Soviet cutoff, including emergency provision of gas and other energy supplies from alternative sources.

A second area in which joint policy could profitably be developed is in further coordinating project negotiations at both official and private levels. This kind of joint action cannot be simply decreed; it must be built carefully and gradually. The absence of formal Western coordination may provide opportunities for the Soviet Union to play firms in one nation off against another. The U.S.S.R. could, for example, use an attractive offer of credit from one government to bargain with another. These tactics could produce a West Siberian project economically more advantageous to the U.S.S.R. than to the West. While firms generally tend to prefer open competition, in this case there are indications that, informally at least, some degree of cooperation has evolved among participating companies; informal exchange of information concerning prices, interest rates and credit terms has set an unspoken context within which each participant bargains with the U.S.S.R. Such communication, fostered by governments and by publically owned energy companies, could help to maintain Western unity. Recent moves in IEA to establish a system to monitor prices paid for oil are more formal but analogous mechanisms. Maintaining the flow of information about pipeline negotiations is thus an important component of a joint approach.

Some observers argue that a joint West European policy is not feasible—and that the past 10 years of limited progress confirms this view. The question is whether there is any viable alternative for limiting the risks of increasing energy dependence on the U.S.S.R. OTA's analysis suggests that under current conditions, predispositions in Western Europe would preclude the success of any attempt to "stop Yamburg." A more fruitful approach, therefore, would be to develop mechanisms for anticipating and ameliorating any negative consequences for the Western alliance which the project might engender.

Soviet Energy Availability and U.S. Policy

CONTENTS

Soviet Energy Availability and U.S. Policy

The Soviet energy situation was brought to the attention of the U.S. public in 1977 when the Central Intelligence Agency (CIA) forecast substantial and steep declines in Soviet oil output by 1985.[1] Although it has since modified its position, the CIA as late as April 1980 was predicting that the Council for Mutual Economic Assistance (CMEA) would be importing "at least" 1 million barrels of oil a day by 1985.[2] The possibility of impending Soviet energy shortages and of increased competition for oil on world markets thus raised a policy debate in the United States, a debate framed largely in terms of whether or not it is in the best interest of the United States to institute a policy of helping the Soviet Union increase its energy production.

Some favor a policy of promoting American exports of energy production technology to the Soviet Union in order to increase the world's total available supply of energy, to obviate extensive CMEA pressure on world energy markets, and/or to reduce the likelihood that the U.S.S.R. would intervene in the Middle East to acquire oil it could no longer produce in sufficient quantities at home. Adherents of the opposing view contend that assisting the development of Soviet energy resources would help to strengthen the economy of an adversary and/or that such assistance may convey direct or indirect military benefits. The concern here is with the transfer of dual-use technologies which have military application and/or the view that oil itself is a strategic commodity. Another dimension of this position is concerned with the prospects of increasing Soviet energy (i.e., gas) exports to Western Europe and the dangers of increased West European energy "dependence" on the U.S.S.R.

Whichever view one holds, the most direct means by which the United States might affect Soviet energy availability would be by deciding to export or withhold exports of energy (particularly petroleum) equipment and technology to the U.S.S.R. Alternatives for formulating a policy on U.S. energy-related exports to the Soviet Union can be broadly divided into four basic categories:

- policy options designed to bar the transfer of Western energy equipment and technology to the U.S.S.R.;
- policy options designed to use the inducement of increased exports or threat of curtailing production equipment and technology exports to exact political concessions from the Soviet Union, i.e., options designed to further a policy of linkage or leverage;
- policy options designed to facilitate Soviet energy resource development as quickly and efficiently as possible, in order to mitigate future energy shortages in the world as a whole; and
- policy options designed to reap whatever commercial advantages may be available from trade with the U.S.S.R. in all items except those of direct military relevance.

[1] Central Intelligence Agency, *Prospects for Soviet Oil Production*, April, 1977; *Prospects for Soviet Oil Production: A Supplemental Analysis*, July 1977.

[2] Testimony of Admiral Stansfield Turner, Director of the CIA, before the Committee on Energy and Natural Resources, U.S. Senate, Apr. 22, 1980.

CURRENT U.S. POLICY

THE EXPORT ADMINISTRATION ACT OF 1979[3]

U.S. exports of energy-related technology and equipment to the U.S.S.R. are regulated by the Export Administration Act of 1979 (Public Law 96-72). This act is the latest in a series of laws which for the past 30 years have sought to balance the dual objectives of promoting international commerce and safeguarding American national security. Controversy over the proper weight to be accorded each of these interests has been continuous, but over the years the thrust of U.S. trading policy has been gradually to expand opportunities for selling U.S. products and know-how to Communist nations.

Under the present legislation, U.S. firms seeking to do business with the Soviet Union must obtain validated export licenses if the goods or technology they plan to sell appear on the U.S. Commodities Control List (CCL). Most of the CCL consists of items which are also regulated by CoCom, the informal multilateral export control organization consisting of the United States and its NATO allies (minus Iceland, plus Japan). However, the United States does maintain unilateral controls over some 38 additional products and technologies. Many of these are energy related. The Secretary of Commerce, with the advice of the Secretaries of State and Defense, may delete such items from the CCL. Items may also be added if these are deemed to have significant military applications, to be in short supply, or to relate to specific foreign policy objectives. Inclusion in the CCL does not mean that the item is necessarily embargoed. Rather, it means that the potential exporter must file a license application with the Department of Commerce (DOC).

There are three circumstances under which a license application may be refused: the export will make a significant contribution to the military potential of another country; the item in question is in domestic short supply; or the restriction is necessary to significantly further the foreign policy of the United States.

Prior to passage of the 1979 Export Administration Act, the President's discretion to control exports for the latter reason was largely unlimited. Now, all foreign policy controls expire at the end of each calendar year. To renew them, the President must notify Congress and justify the reextention on the basis of criteria which include the probability that controls will achieve the intended foreign policy purpose in light of such factors as the availability of the goods or technology in question from other countries.

FOREIGN AVAILABILITY

This concept of "foreign availability" constitutes an important part of the Export Administration Act. Recognizing that the availability from other sources of items controlled by the United States undermines the impact of U.S. policies and places U.S. firms at a competitive disadvantage, section 5(f) stipulates that the Secretary of Commerce, in consultation with other Government agencies and technical advisory committees, should:

> ... review, on a continuing basis, the availability to countries to which exports are controlled ..., from sources outside the United States, including countries which participate with the United States in multilateral export controls, of any goods or technology the export of which requires a validated license ... (In the event) that any such goods and technology are available in fact to such destinations from such sources in sufficient quantity and of sufficient quality so that the requirement for a validated license for the export of such goods or technology is or would be ineffective in achieving the purpose set forth ... the Secretary

---

[3]For a legislative history of U.S. export control policy, as well as descriptions of the U.S. export licensing procedure and of CoCom regulations, see Office of Technology Assess-

ment, *Technology and East-West Trade* (Washington, D.C.: U.S. Government Printing Office, November 1979). This volume also contains the text of Public Law 96-72.

may not, after the determination is made, require a validated license for the export of such goods or technology during the period of such foreign availability, unless the President determines that the absence of such export controls under this section would prove detrimental to the national security of the United States.

The section goes on to require that the grounds for such a determination, together with an statement of the estimated economic impact of the decision, be published. The President is further enjoined to undertake negotiations with foreign governments to eliminate the availability. In the absence of such a Presidential determination, the Secretary of Commerce is directed to approve any validated license application which meets all other requirements and which is for export of goods or technology for which foreign availability has been established.

Determinations of foreign availability that are to be the basis for these licensing decisions must be supported by "reliable evidence, including scientific or physical examination, expert opinion based on adequate factual information, or intelligence information." The act specifically stipulates that "uncorroborated representations by applicants shall not be deemed sufficient evidence of foreign availability." Capability to monitor and gather information on foreign availability of all goods and technologies subject to U.S. export controls was to be established within the Office of Export Administration (OEA), the part of DOC responsible for export licensing, and each department or agency of the United States with export control responsibilities, including the intelligence services, were required to furnish OEA with appropriate foreign availability information.

However, it is clear by now that the entire concept of foreign availability is fraught with ambiguity and raises important practical difficulties. Nowhere, for instance, does the 1979 Export Administration Act define the terms "available without restriction," "available in significant quantity," or "comparable quality." Among the definitional questions pertaining to the meaning of

"availability" and "comparability" are the following:

- Must a foreign competitor have expressed a willingness to sell to the U.S.S.R. for its goods to be considered "available?" Must the U.S.S.R. have actually approached the competitor; or does the mere existence of goods and technologies outside the United States count as foreign availability?
- How do matters of price affect both availability and comparability: if a foreign item is cheaper or backed by foreign government export credits, how inferior need it be to the U.S. alternative before it is no longer counted as evidence of foreign availability?
- What are the parameters for assessing comparable quality; are these different for many pieces of equipment or technologies appearing on the CCL? Must items be identical to be considered comparable?
- Similarly, how are "significant quantities" to be determined? Are these relative to the amounts the Soviets wish to purchase in the immediate sale in question, to total world supply, or does their assessment involve comparison of the manufacturing capacities of U.S. industries and their foreign competitors?

Aside from these conceptual difficulties, there have been enormous practical problems involved in establishing a foreign availability assessment mechanism in DOC. Assembling sufficient information to answer the kinds of questions suggested above is a massive undertaking and as of this writing it does not appear that the executive branch has released the funds allocated by Congress to allot the staff and other resources necessary to complete this task in a systematic or comprehensive way.

Furthermore, even assuming that a clear conceptual framework for assessing foreign availability and the resources to handle the resulting data existed, it is not clear that present information-gathering mechanisms would be sufficient to satisfy the terms of the act. Indeed, most of the information required would have to be secured from private firms in foreign countries. Since a great part of this information might reasonably be expected to be company proprietary, serious practical—if not legal and ethical—problems might be encountered.

In short, satisfying the present legal criteria for ascertaining foreign availability will be expensive, time-consuming, and perhaps intrusive. The requirement in the act that this assessment be conducted "on a continuing basis" adds to these burdens. Given the fact that DOC's foreign availability capabilities have yet to be fully instituted, it is difficult to determine whether or not the provisions can be fulfilled in a cost-effective manner.

U.S. POLICY ON EXPORTS OF ENERGY-RELATED GOODS AND TECHNOLOGY

In July 1978, in response to the U.S.S.R.'s policies towards its dissidents, President Carter decided to invoke foreign policy controls and to place exports to the Soviet Union of technology and equipment for the exploration and production of oil and gas on the CCL. These items thereby became subject to U.S. unilateral control, i.e., U.S. exporters were required to obtain validated licenses for petroleum equipment and technology not included on the multilateral CoCom list. The absence of CoCom controls meant that firms in allied countries could continue to export such equipment and technology free of any restriction.[4] Two impor-

tant assumptions underlay Carter's decision: the Soviet Union had a critical need for the items in question, and it was largely dependent on the United States for their supply.[5]

Foreign policy controls on petroleum-related items were reaffirmed and reiterated in January 1980, following the Soviet invasion of Afghanistan. In his letter notifying Congress of the renewal of these controls, President Carter asserted that:

> The control on the export of petroleum equipment to the U.S.S.R. provides a flexible foreign policy tool. When necessary and appropriate it can be used to sensitize the Soviets regarding actions which are damaging to United States foreign policy interests . . . Discontinuation of this control would represent a change in policy not warranted by existing circumstances in our relationship with the U.S.S.R.[6]

At this writing, U.S. policy toward energy-related equipment and technology exports to the U.S.S.R. is under review. For the moment, applications for validated licenses for exports of oil and gas equipment and technology to the U.S.S.R. are decided on a case-by-case basis. Sales of end products alone have generally been approved, but those involving industrial manufacturing know-how are acted on with a presumption of denial.[7]

Any new policy direction, as noted above, would fit broadly into one of four basic categories. The following sections describe the four perspectives *from the point of view of their advocates,* and discuss the implications of implementing each.

---

[4]Under U.S. law, technology of U.S. origin requires a U.S. export license in order to be reexported from a third country.

[5]Samuel P. Huntington, "Trade, Technology, and Leverage: Economic Diplomacy," *Foreign Policy,* fall, 1978, p. 76.

[6]Letter of President Carter to Hon. Thomas P. O'Neill, Dec. 29, 1979, in *The Congressional Record,* Jan. 29, 1980, p. H381.

[7]*Business America,* Apr. 7, 1980, p. 12.

THE EMBARGO PERSPECTIVE

GOALS AND ASSUMPTIONS

In the past, legislation has been introduced in Congress which has been designed to severely curtail the ability of U.S. firms to sell energy-related equipment and technology to the U.S.S.R.[8] Those who favor this policy orientation usually hold one or more of the following views with respect to such sales:

1. Energy, and particularly petroleum, equipment and technology are dual-use items, i.e., they may have military applications.
2. Oil is itself a strategic commodity.
3. Helping the U.S.S.R. to maintain or improve its energy output bolsters the Soviet economy, contrary to U.S. national interest.

This perspective, like the linkage perspective discussed below, is often based on the premise that the denial of American equipment and technology will significantly inhibit the development of Soviet energy resources and Soviet energy output. In those cases where the United States is neither a sole nor a preferred supplier of equipment and technology, adherents of this position hold that the U.S. Government can and should undertake negotiations with its allies to enlist their cooperation in a technology embargo.

DISCUSSION

In a very few cases, energy-related technologies and equipment have had the potential for direct military use. The sophisticated computers and other seismic equipment, including large main-frame computers, array processors, and advanced automated data processing systems, sought by the U.S.S.R. certainly could convey military capabilities. Such computers and software are already under both U.S. national security and CoCom controls. It has been alleged that certain aspects of the technology required for the manufacture of oil drilling bits with tungsten carbide inserts are militarily relevant. These allegations have been the subject of considerable dispute and experts have disagreed over the military utility of this technology.[9] However, the final determination of U.S. export licensing authorities, including the Department of Defense, was that these technologies could be safely exported.

These instances are exceptions. The great majority of the energy equipment and technology which the U.S.S.R. purchases from the West consists of items which have raised few questions from the standpoint of their direct military relevance. Defining fuel itself as a strategic commodity, however, raises a different kind of problem—and invokes a rather different policy. A decision to restrict the export of items because of their economic or indirect, as opposed to direct, military significance would be tantamount to reversing the trend of the last 30 years of export control in the United States.

The Export Control Act of 1949, by allowing the control of items of "indirect" military utility, in fact was aimed at pursuing a policy of economic warfare against the Soviet Union.[10] This policy was abandoned, partly because it was recognized that it could be effective only if adhered to by America's allies. In other words, a wide array of items which the U.S.S.R. wished to purchase from the West had become available outside of the United States, in countries far more dependent than was the United States on foreign trade. The United States appeared to be unable to convince these alternative suppliers to impose the

---

[8]See, for example, The Technology Transfer Ban Act, H.R. 14085, introduced in the House of Representatives on Sept. 14, 1978.

[9]See "Transfer of Technology and the Dresser Industries Export Licensing Actions," Hearings before The Permanent Subcommittee on Investigations, Committee on Governmental Affairs, U.S. Senate, Oct. 3, 1978.

[10]OTA, op. cit., ch. VII.

same restrictive policies. Without such cooperation, American firms lost sales to European and Japanese competitors, and the U.S.S.R. was nonetheless able to obtain the nonmilitary goods and technologies it sought.

OTA has elsewhere explored the general East-West trade policies of those allied nations which are major Western trading partners of the U.S.S.R.[11] The basic conclusions of that analysis were that while America's allies do not deny the basic necessity of withholding items of direct military significance from the U.S.S.R., East-West trade has been economically more important to Western Europe and Japan than to the United States. These countries tend to view trading with the Soviet Union as primarily an economic issue, and to eschew the use of export controls for political purposes. The lukewarm response with which the post-Afghanistan technology embargo was greeted in Western Europe and, to a lesser extent, Japan—as well as the 1980 statistics reported in chapter 12 which show that trade between the Soviet Union and Japan, West Germany, France, Italy, and the United Kingdom actually grew during the "embargo period"— indicate that these basic orientations have not changed.

Chapters 11 and 12 discuss in detail the attitudes of Japan, West Germany, France, Italy, and the United Kingdom to specifically energy-related trade with the U.S.S.R. In general, it appears that for these nations, sales of energy-related technology and equipment to the U.S.S.R. pose no special foreign policy or national security concerns, nor have these transactions sparked intense de-

bate. Indeed, in some of these countries such sales are of significant economic importance. A U.S. policy of extending export controls to energy-related items with economic and political, but little or no direct military relevance, is therefore unlikely to encounter much sympathy or active cooperation.

The highly publicized gas pipeline deal, in which West European and Japanese export credits and equipment will be bartered for Soviet gas may change the context of this trade, however. There is little doubt that the magnitude of the proposed project and its importance to the Soviet economy make it a transaction of particular significance. OTA's research indicates that the potential Western participants are by no means insensitive to both the economic and security implications of embarking on this degree of cooperation and interdependence with the U.S.S.R. Nevertheless, these nations appear to have decided—both in principle, and now in practice—to proceed.

It is possible to posit circumstances under which the United States could persuade its allies to reverse these decisions. A major change in the international climate precipitated by a Soviet invasion of Poland, for instance, could certainly cause either a temporary or a permanent halt to the gas export pipeline project. In the absence of this kind of event, a U.S. policy initiative designed to discourage continued or increased allied energy- related trade with the U.S.S.R. might have its best chance of success if designed to offer allied governments positive alternatives, in the form of either realistic alternative energy supplies to replace Soviet gas or assistance in devising contingency plans for Soviet supply interruption.

---

[11]Ibid., ch. IX.

THE LINKAGE PERSPECTIVE[12]

GOALS AND ASSUMPTIONS

Linkage is a policy which seeks to use the prospect of expansion or curtailment of

trade as a "carrot or stick" to exact policy concessions from a trading partner. The perspective itself accommodates a number of different points of view. Those who favor pursuing a linkage strategy may disagree over the nature and scope of the goals which

---

[12]Ibid., ch. IV.

such a policy can further. These disagreements center on both the range of policies which linkage can or should hope to affect (i.e., should future trade be linked to the trading partner's domestic policies—treatment of dissidents in the case of the U.S.S.R.—or should it be restricted to attempting to affect only major foreign policies—such as the invasion of Afghanistan?) and on the kinds of trade which should be used as policy instruments (should the extension of credits and most-favored nation (MFN) become part of a linkage strategy; should all trade be affected—including grain—or should the policy apply only to technology trade?).

Adherents of adopting a linkage policy toward trade with the Soviet Union may also hold different basic perceptions of the nature of the U.S.-Soviet relationship and its potential. Some believe that trade can have a moderating effect on international politics by enmeshing trading partners in a "web of interdependence." Others see a fundamentally adversarial relationship between the United States and the Soviet Union. They may accept the fact that trade can be harnessed to political purposes, but are skeptical of the connection between trade and political moderation. Here trade is justified only if in return the trading partner makes policy concessions.

Regardless of these differences, however, the belief that a linkage policy can be effective entails acceptance of the basic proposition that the potential exports in question must be of sufficient value to the U.S.S.R., and the assumption that the United States either has a monopoly on these items, or failing that, is a strongly preferred supplier.

Although different Administrations have disagreed over the ways in which a linkage policy vis-a-vis the Soviet Union should be conducted, for some years America's trade with the Soviet Union has taken place within the context of linkage. U.S. efforts to use trade to moderate Soviet behavior have included linking the extension of MFN status and eligibility for official U.S. export credits

with the emigration of Soviet Jews (the Jackson-Vanik amendment); linking the export of a U.S. computer to the Soviet Union's treatment of its dissidents; and curtailing both shipments of U.S. grain and the export of technology after the Soviet invasion of Afghanistan.

There is little clear evidence so far that in any of these cases U.S. trade policy has had a measurable effect on Soviet foreign or domestic activities. Nevertheless, no overall determination of the success or failure of linkage as a basic strategy has yet been made, and the results of these policies have been subject to varying interpretations. The potential effectiveness of a policy specifically linking exports of U.S. petroleum equipment and technology is also the subject of some debate.

Opponents of such a policy may entirely reject the notion that trade can be an effective instrument to achieve political objectives. This view is held by a number of other Western governments and is often espoused by some American corporations. Others—often members of the petroleum equipment industry—contend that the United States has little or no leverage in this area because of the wide foreign availability of the equipment and technologies desired by the U.S.S.R.

On the other side, it has been contended that President Carter's inclusion of energy equipment and technology on the CCL was a major step in placing the United States in a position vis-a-vis the U.S.S.R. "in which the technological door can be more easily closed, or swung near to being closed, if that seems desirable or necessary."[13] This assertion is premised on the belief that for many items in the area of petroleum technology and equipment, including downhole pumps, gas-lift equipment, drill bits, well completion equipment, and offshore drilling technology, the United States has virtually been the Soviet Union's sole supplier, and that "this type of equipment is absolutely essential to the

---

[13]Huntington, op. cit.

Soviets if they are to stave off a significant decline in their oil production in the early or mid-1980's."[14]

Such statements are supported by CIA's 1977 report which identified items of technology and equipment particularly crucial to Soviet petroleum output. These included seismic exploration equipment; rock drill bits; oilfield pumps and gas-lift equipment; large diameter pipe; offshore technology; rotary rigs, drill pipe and casings; multiple completion equipment; and secondary and tertiary recovery equipment.

In December 1979, however, President Carter himself acknowledged that the list of items in which the United States was the sole or preferred supplier was somewhat narrower. His letter to Congress on December 29 stated that for most items of petroleum equipment, "adequate quantities of similar equipment are available from foreign sources." At the same time, "there is only limited foreign availability of some deep submersible pumps and seismic equipment."[15] The implication presumably remained that these items were critical enough to the U.S.S.R., and their supply controlled sufficiently by the United States, for the foreign policy controls to continue to be useful in furthering U.S. objectives.

DISCUSSION

As chapter 6 points out, the foreign availability assessment which was performed in the course of this study was inhibited by the same conceptual and practical difficulties described above, and its results should be considered suggestive rather than conclusive. With this caveat, OTA's findings tend to confirm President Carter's assertion that, with few exceptions, adequate quantities of the energy equipment sought by the U.S.S.R. are produced and available outside the United States, and that the quality of these foreign goods is general-

ly comparable to that of their U.S. counterparts. The most important exceptions to this general finding are electric submersible pumps and sophisticated seismic systems. But it does not necessarily follow that obtaining the latter items from U.S. firms is so critical to the U.S.S.R. at this time that the threat of their being withheld would result in significant Soviet policy concessions. Nor is it clear that the fate of Soviet petroleum production in this decade is entirely or even largely dependent on them.

The United States is the only producer of high capacity electric submersible pumps in the Free World. Several years ago, the U.S.S.R. purchased relatively large amounts of such equipment. It will be recalled, however, that although U.S. pumps are of substantially better quality than their Soviet counterparts, they never constituted more than a small portion of total Soviet stocks. Moreover, there is reason to believe that virtually all the American pumps in the U.S.S.R. are by now out of commission, and the Soviet Union has not replaced them. Indeed, the Soviet Union has bought no U.S. pumps for the past 3 years—nor has its oil output declined over this period. It seems hardly reasonable, therefore, to characterize the Soviet oil industry as dependent on this type of equipment. One or more of three things appears to have occurred: the Soviet Union has found at least a partial substitute for high-quality pumps (gas-lift equipment, purchased in France); in addition, it may have improved the quantity and/or the quality of its domestic pumps; or planners may have decided that less than state-of-the-art equipment is acceptable.

Similarly, it is generally recognized that the United States is a preferred supplier of seismic exploration equipment and that such equipment could significantly improve the quality and efficiency of Soviet seismic work. The United States also appears to be the Western nation best able to provide the U.S.S.R. with the full range of services and capabilities necessary for its exploratory efforts. Most of the oil hitherto discovered in

[14]Ibid.
[15]Letter of President Carter, op. cit.

the world, however, has been found in giant fields with exploration technology that significantly lagged the present state-of-the-art. In any case, long leadtimes are usually necessary before newly discovered deposits can be developed, and it is not clear that exploratory activities initiated now would produce significant results before the latter part of the decade. Moreover, even though systems with components assembled from a number of different suppliers may be less desirable than those purchased in their entirety, the U.S.S.R. might well be able to replace American equipment with a collection of items which, although not ideal, could function significantly better than Soviet domestic equipment.

There is a further issue. An important conclusion of this study is that the status of the Soviet gas industry may be more crucial than that of the oil industry to overall energy availability. Here, **the U.S.S.R. is quite dependent on the West—for the large diameter pipe and compressor stations it needs to construct gas pipelines—but the former item is not produced in the United States and there are multiple alternative suppliers for the latter.** It has been suggested that the United States may be the sole or preferred supplier of the heaviest pipelaying machinery used for installing gas pipeline, and that foreign manufacturing capabilities may be insufficient to fully supply the needs of the U.S.S.R. in this area. It is difficult to either establish or disprove the accuracy of this claim without access to detailed information about specific foreign corporations, but it must be recognized that the U.S.S.R. has in the past purchased pipelaying equipment from Japan.

The chances of the United States persuading its allies to join it in an energy-related policy of leverage against the U.S.S.R. are as small as those of obtaining agreement to an energy equipment and technology embargo. The point is not simply that the countries examined here—West Germany, France, Italy, Britain, and Japan —each have an economic stake in East-West trade greater than that of the United States, or that they have been traditionally reluctant to engage publicly in linkage practices. While the danger of energy dependence on the U.S.S.R. may seem to some to be the overriding political concern for the entire Western alliance, each nation approaches its trade and energy relations with the U.S.S.R. from its own political perspective. These differ among the allies themselves and from that of the United States. They range from West Germany's natural preoccupation with West Berlin in particular and European security in general to Japan's attempts to balance its policies towards both the U.S.S.R. and the People's Republic of China (see chs. 11 and 12 for a fuller discussion of these perspectives). It would seem that, regardless of U.S. judgments of the wisdom or accuracy of their views, these nations have determined that the risks of a certain degree of energy cooperation with the U.S.S.R. are outweighed by other political benefits.

In sum, the immediate leverage of the United States over the Soviet Union in the area of petroleum equipment and technology is probably limited by at least three factors. *First,* the United States is the sole supplier of very few petroleum-related items. *Second,* the U.S.S.R. has demonstrated some ability to do without these items, at least in the short term. *Third,* and perhaps most important, gas is the energy sector in which the U.S.S.R. is both most reliant on the West and most dependent for its energy future— and with the possible exception of construction equipment, the United States has little to offer in this area that is unique.

THE ENERGY COOPERATION PERSPECTIVE

GOALS AND ASSUMPTIONS

Adherents of this perspective may hold one or more of the following views:

1. Increased energy-related exports from the United States to the Soviet Union reduce the chances that the U.S.S.R. will experience the serious oil production problems predicted by the CIA, and therefore the chance that it will either have to import oil on world markets or have an incentive to intervene in the Middle East.
2. Such U.S. exports, by helping the U.S.S.R. to produce more oil, help to increase worldwide energy availability. This is a positive development no matter where such oil is located.
3. The trade ties established with the Soviet Union during the period of detente were a positive step toward drawing the U.S.S.R. into the world economy, a move which should increase that country's interest in maintaining world political and economic stability.

DISCUSSION

Here, the basic premise is the obverse of that of the embargo perspective, i.e., it is assumed that American technology and equipment could make a significant positive contribution toward increasing Soviet energy availability in the present decade. OTA's findings cast doubt on this assertion. It is certainly true that American and/or other Western petroleum equipment could assist the U.S.S.R. in overcoming many of the problems presently caused by equipment of inferior quality and insufficient quantity. It could also speed the development of offshore resources. But while it is undeniable that Western exports have made important, albeit unquantifiable, contributions to Soviet petroleum output in the past and could continue to do so in the future, policy changes in both the United States and the

Soviet Union would be required for such assistance to have maximum effect.

In its report, *Technology and East-West Trade*, OTA identified the lack of official U.S. export credits as the primary legal barrier to the expansion of trade between the United States and the Soviet Union. There is no reason to believe that this problem would not continue to hamper such expansion. But the willingness of the United States to sell on favorable terms is only half of what is needed for American exports to extensively aid the U.S.S.R. The Soviet Union must also be both able and willing to buy the items it needs in sufficient quantities, and to use them in an efficient and productive manner.

A frequent theme throughout this report has been the difficulties posed by the Soviet economic system in utilizing both domestic and foreign technology effectively. While it may be true that imported equipment is more productive than the closest Soviet equivalents, it is also usually the case that Western equipment and technology perform less well in the U.S.S.R. than in the country of origin or other Western nations. It cannot necessarily be assumed, therefore, that simple shipments of equipment or transfers of technology could easily solve Soviet energy problems.

This problem is exacerbated by the fact that the U.S.S.R. has traditionally been unwilling to allow the hands-on training by Western personnel which would make Western equipment and technology most productive. Nor has it appeared very willing to allow Western firms to participate extensively in Soviet energy development. Some overtures in this direction were made before the invasion of Afghanistan, but little has come of them. Not only would such active participation greatly expedite this development, but it would also give American and other Western companies the incentive, presently lacking, to become more extensively involved in the U.S.S.R.

Furthermore, hard currency shortages presently constrain the amounts of energy-related items which the U.S.S.R. can import. The Soviet Union has traditionally kept its trade with the West relatively small. Not only has it been unwilling to become dependent on the West, but it has been quite conservative in amassing a Western debt (especially compared to the nations of Eastern Europe). As chapter 8 points out, one consequence of this hard currency shortage is that different sectors of the Soviet economy compete for the ability to purchase from the West. Energy equipment and technology imports have thus been highly selective.

It is not entirely clear, moreover, that the U.S.S.R. will necessarily be propelled onto world markets for oil. Indeed, as chapter 2 has noted, if the CIA ever intended to foster this expectation, it no longer holds this view. OTA has identified worst case or "pessimistic" scenarios which show conditions under which the Soviet Union could have a net oil deficit, but a number of factors make this a highly uncertain basis on which to plan policy. First, more optimistic scenarios are probably more likely, i.e., the U.S.S.R. could continue to export oil for hard currency without extensive U.S. help. Second, the degree to which the U.S.S.R. is able to substitute gas for oil, both in domestic consumption and in exports, seems the more crucial variable. In other words, **the overall Soviet energy balance, not simply oil production, will be important in determining the ways in which the U.S.S.R. is able to handle its energy situation in the 1980's.** Third, hard

currency constraints would almost certainly minimize or even prevent such purchases.

It must also be pointed out that any Soviet decision to intervene in the Middle East—either militarily or through policy initiatives directed at OPEC governments—need not necessarily be driven by a domestic need for oil. The vital U.S. interest would seem more than sufficient to give the U.S.S.R. a reason for acting in this area should it wish to do so. The availability of additional oil, assuming that conditions allowed local cooperation or Soviet ability to operate the oilfields itself, might be an attractive bonus, but is is hardly a necessary condition.

Finally, institutions presently exist for fostering multilateral cooperation in energy supply issues. For instance, the Soviet Union has requested that the U.N. Economic Commission for Europe sponsor a high-level conference which would consider possibilities for multilateral energy cooperation. The United States has hitherto opposed the convening of such a conference. Presumably the reversal of this position would signal America's interest in participating in Soviet energy development. In addition, policymakers might wish to consider the formulation of a broader allied policy perspective on Soviet energy, arrived at either on a bilateral basis, through NATO, or through the International Energy Agency. It must be noted that the West European nations themselves have made little progress toward developing a unified East-West energy policy for their own region.

THE COMMERCIAL PERSPECTIVE

GOALS AND ASSUMPTIONS

This perspective rests on the assumption either that trade and politics should remain separate, i.e., that linkage is a misguided policy, and/or that regardless of the export control policy it adopts, the United States is unlikely to be able to significantly affect the

U.S.S.R.'s energy future in the present decade. The following reasoning applies in the latter case:

1. The United States retains control of very few of the energy technologies and little of the energy equipment which the U.S.S.R. purchases from the West, and

has little prospect of convincing America's allies to cease their own exports. An embargo of U.S. energy technology would, therefore, have little effect on the U.S.S.R., and the prospect of such an embargo confers very little leverage.

2. On the other hand, the ability of the United States to significantly enhance Soviet oil production, thereby relieving economic pressure on the U.S.S.R. and increasing the amount of oil in the world, is also constrained by the factors discussed in the previous section.

3. In any case, Soviet energy industries are enormous and the U.S.S.R. has a good record for being largely self-sufficient in areas where Western help is not easily forthcoming.

Given this line of reasoning, it becomes sensible to argue for the United States abandoning the area of energy as a promising context for its Soviet foreign policy, and reaping whatever economic benefits can be conferred by sales of energy and equipment to the U.S.S.R., so long as these have no direct military relevance. Such a policy would not necessarily have to be accompanied by the extension of export credits on favorable terms. The ability of American firms to compete with West European and Japanese companies for sales of energy-related items to the U.S.S.R. could be significantly enhanced simply by removing U.S. unilateral export controls on such items.

DISCUSSION

Given a desire to facilitate—or at least not to unduly impede—nonmilitarily sensitive exports to the U.S.S.R., there is room for significant improvement in the administration of export license applications. The export licensing system is complex, and given the volume of applications it handles, has worked with reasonable efficiency. Procedures could be instituted to streamline the system, however, without tampering with its basic structure or effectiveness. Such procedures might eliminate the present, seemingly unwarranted, occasional delays which

have subjected the entire export licensing system to criticism.

It must be recalled that Soviet trade with the West has never been large in absolute terms and that, except for grain sales, U.S. market shares in this trade have been relatively modest. The cost and difficulty of doing business with the U.S.S.R., American export license procedures, and the ineligibility of the U.S.S.R. for U.S. export credits have all been limiting factors. There is little or no reason to expect that this situation could change without dramatic changes in both U.S. export and Soviet import policies. Thus, while individual firms might well be able to conclude lucrative individual contracts for items of energy-related equipment or technology, it is highly unlikely that these sales would be large enough to affect the U.S. economy in general or even specific industries in any crucial fashion.

Aside from these economic considerations, there is a political dimension to the commercial perspective. Given the relatively limited opportunities for the United States alone to significantly influence Soviet energy availability in the present decade, and given the difficulties which would certainly arise in attempting to persuade America's allies to curtail their own energy relations with the U.S.S.R., U.S. policymakers might well choose not to expend political "chips"—either in negotiations with the USSR or with allied nations—by making Soviet energy development an area of contention. Removing energy-related export control issues from the political agenda, in other words, might possibly enhance the chances of obtaining allied cooperation in other aspects of East-West policy. If the commercial perspective is pursued for these motives, the extent of the trade it would engender becomes a secondary consideration.

A FINAL NOTE

The perspectives and policy options discussed in this chapter apply to the present state of the relationships between the Soviet Union and the West. But the judgments and decisions of both U.S. and Allied policymakers can and will tend to shift over time, in response to changing economic and political conditions. For instance, dramatic events involving the Soviet position in Eastern Europe could drastically alter the views of both Soviet and Western leaders on the options open to them, and on the national interests which would shape their choices. In contrast, the overall parameters of Soviet energy supply and demand are unlikely to change rapidly, because of the sheer size of the resources and infrastructure involved. Thus, even should their perceptions of national interests change, policymakers will still have to reckon with the limits imposed by the strengths and weaknesses to the Soviet energy industries.

Appendix

United States Senate

COMMITTEE ON BANKING, HOUSING, AND
URBAN AFFAIRS
WASHINGTON, D.C. 20510

May 19, 1980

The Honorable Morris K. Udall, Chairman
Technology Assessment Board
Office of Technology Assessment
U.S. Congress
Washington, D.C. 20510

Dear Mr. Chairman:

We understand that the Board is considering an OTA study of the relationship between U.S. technology exports and Russian energy development. The Banking Committee has jurisdiction over export control policy as well as export promotion. The Committee rewrote the Export Administration Act last year in order to establish a more efficient control policy. Additional legislation to improve export controls, especially those directed toward the Soviet Union, is under consideration by the Committee this year.

One of the most important unresolved issues is U.S. policy toward exports which could contribute significantly to Russian energy development, in particular, exploration, recovery and transportation of oil and gas from the large reserves located in the U.S.S.R. Among the unanswered questions are: (1) how much difference could U.S. technology make to the Russian situation over the next decade; and (2) how effective could U.S. export controls be in retarding Soviet energy development or, alternatively, facilitating energy development but avoiding transfer of technology which could enhance Soviet strategic capabilities?

We believe an unclassified assessment of the issue should be prepared and made available to the Congress and general public. Accordingly, we join the House Foreign Affairs

The Honorable Morris K. Udall
May 19, 1980
Page-2

affairs Committee in requesting OTA to conduct a study
of the contribution Amercian--as opposed to other Western--
technology might make over the next decade to Soviet energy
availability, with special attention to the potential impact
of alternative U.S. export control policies.

Jake Garn
Ranking Minority Member

Sincerely,

William Proxmire
Chairman

Congress of the United States
Committee on Foreign Affairs
House of Representatives
Washington, D.C. 20515

May 2, 1980

The Honorable Morris Udall
Chairman
Office of Technology Assessment
235 Cannon House Office Building
Washington, D.C. 20515

Dear Mr. Chairman:

Future relations between the United States and the Soviet Union will almost certainly be shaped in part by issues arising from energy supply and demand. The U.S. intelligence assessment is that the Soviet bloc will be forced to import petroleum by 1985, and the Soviets have acknowledged that they face impending problems in maintaining current levels of oil and gas production and delivery. The implications for the U.S. of Soviet entry onto world oil markets are enormous. Congress should, therefore, be aware of ways in which the U.S. could influence Soviet energy production.

One critical area of uncertainty is the manner and extent to which U.S. technologies could affect Soviet oil production and, thereby, Soviet energy policies.

Little is known about the potential contribution of American technology to Soviet energy development, or precisely how American equipment compares technically to that available from other Western countries. It is often difficult to identify sole or unique suppliers of equipment and technology, and to determine the costs to the importer of resorting to second best choices. Better analysis in this area could have important implications for U.S. policies on exporting energy technology.

As Chairman of the House Foreign Affairs Committee, I request that the Office of Technology Assessment conduct a full-scale assessment of the possible effects of American technology upon Soviet energy availability during this decade. Such a study could draw partly upon the material OTA has already assembled for its recent major study on Technology and East-West Trade.

The Honorable Morris Udall
May 2, 1980
Page Two

The study should address the following questions:

First, what equipment and technology are needed by the Soviet Union for its energy resources? In particular, what is the role of advanced computers and computer systems in expanded energy production? This may be partly illuminated by analysis of past energy technology purchases.

Second, what problems inhibit the applicability and the efficient use of imported energy technology in the U.S.S.R.? Such problems might range from geology, infrastructure, and lack of trained manpower to inappropriate institutional structures. What affect will particular foreign technologies have upon these problems?

Third, to what extent is the United States the sole or preferred supplier of energy technologies likely to be sought by the U.S.S.R., and what is the nature of and capabilities of those technologies?

Fourth, what would be the near or medium term (to 1990) impacts on Soviet oil (or other energy) production of either an expansion or contraction of American energy technology exports?

Your cooperation in this matter would be greatly appreciated.

With best wishes, I am

Sincerely yours,

Chairman

CJZ:giy

COMMITTEE ON SCIENCE AND TECHNOLOGY

U.S. HOUSE OF REPRESENTATIVES

SUITE 2321 RAYBURN HOUSE OFFICE BUILDING

WASHINGTON, D.C. 20515

(202) 225-6371

March 27, 1981

Honorable Ted Stevens, Chairman
Technology Assessment Board
Office of Technology Assessment
Washington, D. C. 20510

Dear Mr. Chairman:

We understand that the Office of Technology Assessment is currently engaged in a study of the relationship between Western technology exports and Soviet energy development. As you may know, the House Committee on Science and Technology has a continuing interest in the implications of technology trade between the United States and other nations.

One important unresolved issue in this area is U. S. policy towards exports which could contribute significantly to Soviet energy development, in particular, exploration, recovery and transportation of oil and gas from the large reserves located in the U.S.S.R. Among the unanswered questions are: (1) how much difference could U. S. technology make to the Russian situation over the next decade; and (2) how effective could U.S. export controls be in retarding Soviet energy development or, alternatively, facilitating energy development, but avoiding transfer of technology which could enhance Soviet strategic capabilities?

We believe an unclassified assessment of these issues will be valuable to the Congress and general public. Accordingly, we wish to join the House Foreign Affairs Committee and Senate Banking Committee in endorsing this study of the contribution American -- as opposed to other Western -- technology might make over the next decade to Soviet energy availability.

Sincerely,

DON FUQUA
Member of Congress

ALBERT GORE, JR.
Member of Congress

DF/AG:Jbe

About the Book

Technology and Soviet Energy Availability
Office of Technology Assessment

Endowed with abundant energy resources, the Soviet Union is the world's largest oil producer and a major exporter of both oil and gas. Energy exports provide over half of Soviet hard-currency receipts, and subsidized energy sales to Eastern Europe are vital tools of Soviet influence in that region. Despite this enviable position, there have been indications in the past few years that the U.S.S.R. may soon face an energy shortage.

In addition to examining the significance of U.S. petroleum equipment and technology for Soviet energy development, this book addresses the following questions: First, what opportunities and problems confront the U.S.S.R. in its five primary energy industries—oil, gas, coal, nuclear, and electric power—and what are plausible prospects for these industries in the present decade? Second, what equipment and technology are most needed by the U.S.S.R. in these areas, how much of each has been or is likely to be purchased from the West, and to what extent is the United States the sole or preferred supplier? Third, and perhaps most critical, how much difference could the West as a whole or the United States alone make to Soviet energy availability by 1990, and what are the implications of either providing or withholding such assistance for both the entire Soviet bloc and for the West?

The Office of Technology Assessment was created in 1972 as an advisory arm of the U.S. Congress. OTA's basic function is to help legislative policymakers anticipate and plan for the consequences of technological changes and to examine the many ways, expected and unexpected, in which technology affects people's lives. The assessment of technology calls for exploration of the physical, biological, economic, social, and political impacts that can result from applications of scientific knowledge. OTA provides Congress with independent and timely information about the potential effects—both beneficial and harmful—of technological applications.